T0306133

Biotic Evolution and Environmental Change in Southeast Asia

The flora and fauna of Southeast Asia are exceptionally diverse. The region includes several terrestrial biodiversity hotspots and is the principal global hotspot for marine diversity, but it also faces the most intense challenges of the current global biodiversity crisis.

Providing reviews, syntheses and results of the latest research into Southeast Asian earth and organismal history, this book investigates the history, present and future of the fauna and flora of this bio- and geodiverse region. Leading authorities in the field explore key topics including palaeogeography, palaeoclimatology, biogeography, population genetics and conservation biology, illustrating research approaches and themes with spatially, taxonomically and methodologically focused case studies. The volume also presents methodological advances in population genetics and historical biogeography. Exploring the fascinating environmental and biotic histories of Southeast Asia, this is an ideal resource for graduate students and researchers as well as for environmental NGOs.

David J. Gower is a researcher in the Department of Zoology at the Natural History Museum, London. An evolutionary and organismal herpetologist, his studies focus on caecilians, snakes and Triassic diapsid reptiles.

Kenneth G. Johnson is a researcher in the Department of Palaeontology at the Natural History Museum, London, studying the long-term biological and environmental history of coral reef ecosystems in Southeast Asia and the Caribbean.

James E. Richardson is a researcher at the Royal Botanic Garden Edinburgh and the University of the Andes in Bogotá. He studies the biogeographic history of tropical flowering plants.

Brian R. Rosen is a researcher in the Department of Zoology at the Natural History Museum, London, specialising in ecology, diversity and biogeography of reefs, and coral growth, form and taxonomy.

Lukas Rüber is a researcher in the Department of Vertebrates, Natural History Museum, Bern, Switzerland. He studies speciation, adaptive radiation, phylogeography, biogeography and systematics, focusing on fishes.

Suzanne T. Williams is a researcher in the Department of Zoology at the Natural History Museum, London. She studies global, regional and local factors important in shaping tropical marine biodiversity.

The Systematics Association Special Volume Series

The Systematics Association promotes all aspects of systematic biology by organising conferences and workshops on key themes in systematics, running annual lecture series, publishing books and a newsletter, and awarding grants in support of systematics research. Membership of the Association is open globally to professionals and amateurs with an interest in any branch of biology, including palaeobiology. Members are entitled to attend conferences at discounted rates, to apply for grants and to receive the newsletter and mailed information; they also receive a generous discount on the purchase of all volumes produced by the Association.

The first of the Systematics Association's publications *The New Systematics* (1940) was a classic work edited by its then-president Sir Julian Huxley. Since then, more than 70 volumes have been published, often in rapidly expanding areas of science where a modern synthesis is required.

The Association encourages researchers to organise symposia that result in multi-authored volumes. In 1997 the Association organised the first of its international Biennial Conferences. This and subsequent Biennial Conferences, which are designed to provide for systematists of all kinds, included themed symposia that resulted in further publications. The Association also publishes volumes that are not specifically linked to meetings, and encourages new publications (including textbooks) in a broad range of systematics topics.

More information about the Systematics Association and its publications can be found at our website: www.systass.org

Previous Systematics Association publications are listed after the index for this volume.

Systematics Association Special Volumes published by Cambridge University Press:

78. *Climate Change, Ecology and Systematics*
    Trevor R. Hodkinson, Michael B. Jones, Stephen Waldren and John A. N. Parnell
79. *Biogeography of Microscopic Organisms: Is Everything Small Everywhere?*
    Diego Fontaneto
80. *Flowers on the Tree of life*
    Livia Wanntorp and Louis Ronse De Craene
81. *Evolution of Plant–Pollinator Relationships*
    Sébastien Patiny

THE SYSTEMATICS ASSOCIATION SPECIAL VOLUME 82

Biotic Evolution and Environmental Change in Southeast Asia

EDITED BY

DAVID J. GOWER

Department of Zoology, The Natural History Museum, London

KENNETH G. JOHNSON

Department of Palaeontology, The Natural History Museum, London

JAMES E. RICHARDSON

The Royal Botanic Garden Edinburgh and the University of the Andes, Bogotá, Colombia

BRIAN R. ROSEN

Department of Zoology, The Natural History Museum, London

LUKAS RÜBER

Department of Vertebrates, Natural History Museum, Bern

SUZANNE T. WILLIAMS

Department of Zoology, The Natural History Museum, London

Shaftesbury Road, Cambridge CB2 8EA, United Kingdom

One Liberty Plaza, 20th Floor, New York, NY 10006, USA

477 Williamstown Road, Port Melbourne, VIC 3207, Australia

314–321, 3rd Floor, Plot 3, Splendor Forum, Jasola District Centre, New Delhi – 110025, India

103 Penang Road, #05–06/07, Visioncrest Commercial, Singapore 238467

Cambridge University Press is part of Cambridge University Press & Assessment,
a department of the University of Cambridge.

We share the University's mission to contribute to society through the pursuit of
education, learning and research at the highest international levels of excellence.

www.cambridge.org
Information on this title: www.cambridge.org/9781107001305

First published 2012

A catalogue record for this publication is available from the British Library

ISBN 978-1-107-00130-5 Hardback

Additional resources for this publication at www.cambridge.org/9781107001305

Contents

Colour plate section appears between pages 176 and 177

Contributors

CHITTIMA ARYUTHAKA Department of Marine Science, Kasetsart University, Thailand

WILLIAM J. BAKER Royal Botanic Gardens, Kew, UK

CHRIS BANKS Department of Geography, University of Leicester, UK (Current address: National Oceanography Centre, Southampton, UK)

DAVID R. BELLWOOD School of Marine and Tropical Biology, and the Australian Research Council Centre of Excellence for Coral Reef Studies, James Cook University, Australia

DAVID BICKFORD Department of Biological Sciences, Faculty of Science, National University of Singapore, Singapore

RAFE M. BROWN University of Kansas, Biodiversity Institute, USA

MARK DE BRUYN Molecular Ecology and Fisheries Genetics Laboratory, Environment Centre for Wales, Biological Sciences, Bangor University, UK

PATRICK CAMPBELL Department of Zoology, The Natural History Museum, London, UK

CHARLES H. CANNON Xishuangbanna Tropical Botanical Garden, Chinese Academy of Sciences, China, *and* Department of Biology, Texas Tech University, USA

GARY R. CARVALHO Molecular Ecology and Fisheries Genetics Laboratory, Environment Centre for Wales, Biological Sciences, Bangor University, UK

CRAIG M. COSTION Australia Tropical Herbarium, E2 Building, James Cook University Cairns Campus, Australia, *and* University of Adelaide, Australian Centre for Evolutionary Biology and Biodiversity, School of Earth and Environmental Sciences, Australia

THOMAS L. P. COUVREUR The New York Botanical Garden, USA (Current address: Institut de Recherche pour le Développement (IRD), UMR-DIADE, France

BEN J. EVANS Department of Biology, McMaster University, Canada

NICHOLAS J. EVANS Department of Zoology, The Natural History Museum, London, UK

MATTHIAS GLAUBRECHT Museum für Naturkunde, Leibniz-Institut für Evolutions- und Biodiversitätsforschung an der Humboldt-Universität zu Berlin, Germany

DAVID J. GOWER Department of Zoology, The Natural History Museum, London, UK

ROBERT HALL Southeast Asia Research Group, Department of Earth Sciences, Royal Holloway University of London, UK

FABIAN HERDER Zoologisches Forschungsmuseum Alexander Koenig, Sektion Ichthyologie, Germany

ALJOSJA HOOIJER Deltares, The Netherlands

AGATA HOSCILO Department of Geography, University of Leicester, UK

CHAWAPORN JITTANOON Department of Marine Science, Kasetsart University, Thailand

KENNETH G. JOHNSON Department of Palaeontology, The Natural History Museum, London, UK

MICHAEL A. KENDALL Plymouth Marine Laboratory, Prospect Place, UK

PETER B. MATHER Biogeosciences, Faculty of Science and Technology (FaST), Queensland University of Technology, Australia

YAOWALUK MONTHUM, Department of Marine Science, Kasetsart University, Thailand

ROBERT J. MORLEY Southeast Asia Research Group, Department of Earth Sciences, Royal Holloway, University of London, UK, and Palynova Ltd., Littleport, UK

ALEXANDRA N. MUELLNER Department of Molecular Evolution and Plant Systematics, Leipzig University, Germany

VINCENT NIJMAN Oxford Wildlife Trade Research Group, School of Social Sciences and Law, Oxford Brookes University, UK

LES R. NOBLE, Department of Zoology, The University, Aberdeen, UK

KEVIN M. O'NEILL, RAF, High Wycombe, UK

SUSAN PAGE Department of Geography, University of Leicester, UK

GORDON L. J. PATERSON, Department of Zoology, The Natural History Museum, London, UK

SINLAN POO Evolutionary Ecology and Conservation Lab, Department of Biological Sciences, Faculty of Science, National University of Singapore, Singapore

MARY ROSE C. POSA Department of Biological Sciences, Faculty of Science, National University of Singapore, Singapore

RICHARD REE Field Museum of Natural History, Department of Botany, Chicago, USA

WILLEM RENEMA Nederlands Centrum voor Biodiversiteit Naturalis, the Netherlands

JAMES E. RICHARDSON Royal Botanic Garden Edinburgh, UK, *and* Universidad de los Andes, Colombia

JACK RIELEY School of Geography, University of Nottingham, UK

KRISTINA VON RINTELEN Museum für Naturkunde, Leibniz-Institut für Evolutions- und Biodiversitätsforschung an der Humboldt-Universität zu Berlin, Germany

THOMAS VON RINTELEN Museum für Naturkunde, Leibniz-Institut für Evolutions- und Biodiversitätsforschung an der Humboldt-Universität zu Berlin, Germany

BRIAN R. ROSEN Department of Zoology, The Natural History Museum, London, UK

LUKAS RÜBER Department of Vertebrates, Natural History Museum, Bern, Switzerland

CHRISTOPH D. SCHUBART Institut für Biologie 1, Universität Regensburg, Germany

CHRIS R. SHEPHERD TRAFFIC Southeast Asia, Malaysia, *and* Oxford Wildlife Trade Research Group, School of Social Sciences and Law, Oxford Brookes University, UK

BRYAN L. STUART North Carolina Museum of Natural Sciences, USA

MATTHEW TODD Oxford Wildlife Trade Research Group, School of Social Sciences and Law, Oxford Brookes University, UK

CAMPBELL O. WEBB Arnold Arboretum of Harvard University, USA

SUZANNE T. WILLIAMS Department of Zoology, The Natural History Museum, London, UK

JOHN VAN WYHE Department of Biological Sciences, National University of Singapore, Singapore, and Department of History, National University of Singapore, Singapore

Foreword

Surely there is nowhere more exciting – or more daunting – than Southeast Asia to demonstrate that the past is the key to understanding the nature of our world. There is a vast complexity in the multiple physical and biological processes that have formed what we see today. The region comprises islands large and small, islands new and old, islands that once were mainland, and highlands that used to be islands. There are both young and ancient rocks, forming staggeringly beautiful limestone towers and picture-book volcanoes. Over eons the parcels of land have moved vertically and horizontally, their climates have changed, their coastlines have been stranded and inundated, and their coral reefs have bloomed and died. There are places that experience deluges of rain, and places that know the harshness of long droughts. The region is bathed in a warmth that can support year-round, vigorous growth. There are caves whose speleothems capture within them the details of former climates, and the gentle showering of pollen from the changing constituents of the vegetation is preserved in swamps. The region has seen successive waves of people introducing diverse cultures, languages and behaviours, modifying the environment to their needs as they travelled. And it has an abundance and diversity of animals and plant life, which has long attracted scientists to collect and to observe them in their natural state. Doing this, scientists have for many years noticed and endeavoured to explain the patterns in the distribution of those species that most captured their interest.

It is the revelations from the latest research on these patterns that this book discusses. I find it very exciting to read how much our understanding has improved and deepened over the last few decades and to see the increasing integration of the geological and biological sciences in pursuit of a common desire. It is not just the sheer intellectual endeavour undertaken to produce this book which impresses, but also the application of novel molecular approaches to phylogenies in a variety of groups, the use of powerful, new statistical techniques, and the comparisons among diverse datasets which, but for this book, might remain on separate bookshelves.

The region has been dubbed by some a 'natural laboratory' for biogeography, but this is hardly appropriate given the enormous, relentless, landscape-level changes that are happening as a consequence of political expediency, perverse incentives and poorly-enforced laws. I know the anguish of trying to find even a remnant of biologically intriguing forest battered, cut and burned out of existence to establish the serried matrices of oil palm plantations. Such vast swathes of crops now cover the region, that the detail of distribution patterns for many species is no more than conjecture. Of course, as if this were not enough, we are having to contend with the changes in climate, aggravated by land-use change, which are adding new pressures to the remaining natural ecosystems. While the world continues to argue about mechanisms for mitigation to pull us back from the brink, strategies for adaptation are needed, and understanding the origins of the natural features of the region and how they have changed could help us to fashion plans for the future.

One can easily be lulled into thinking that the increasingly sophisticated mapping associated with the various global species datasets reflects a close-to-perfect knowledge. In fact, there is a great deal of assumption and educated guesswork involved, often using out-of-date data. There appears to be a lot of biological research going on, but – even 160 years after Wallace arrived in the region – there is still a real need for careful collecting, inventory work and careful molecular and more traditional taxonomy. In my experience, when collections are made almost anywhere in the region in fairly natural habitats, new species are found, as are new records of known species which give fresh insights into their ecological preferences and add weight (or otherwise) to explanatory hypotheses on their distributions.

Despite the intense scholarship exhibited in this book, there is little doubt that there will eventually be a need for another similar volume that will recount yet more advances in the geology and biology of Southeast Asia, and in increasingly comprehensive ways to interpret the findings. I hope that this wide-ranging and detailed book will inspire a new cohort of scientists to further, deepen and reassess the corpus of knowledge laid out so clearly within its pages.

Tony Whitten
Regional Director, Asia-Pacific
Fauna and Flora International
Cambridge
UK

Preface

This Systematics Association Special Volume was prompted by the organisation and staging of a conference held at Royal Holloway, University of London, in September 2009: Southeast Asian Gateway Evolution (SAGE), 2009. The meeting was organised by geologists at Royal Holloway, University of London, and biologists (palaeontologists and neontologists) at the Natural History Museum, London, who shared an interest in Southeast Asia. The region is megadiverse biologically, and of unparalleled tectonic activity and complexity. The region is the focus of the highest levels of marine biodiversity globally, and in terms of terrestrial biota it is an almost continuous patchwork of 5 of the 35 or so global hotspots. The geological history of the region has undoubtedly had substantial impact on the biodiversity, and the distribution and evolutionary relationships of organisms in space and time has informed reconstructions of the geological history.

The meeting attracted 150 scientists, who together presented 85 talks and 50 posters, and it was a great success. Scientists today spend much of their time working in fairly narrowly focused topics or, even if conducting relatively broad studies, working within narrow paradigms and ways of communicating. The organisers hoped that delegates would benefit from scientific exchange beyond their areas of expertise, but scope for detailed interaction between biologists and geologists was patchy. The majority of the biologists attending were working in biogeography, systematics and evolution; the majority of the geologists in structural geology. The lexicons of each tribe were largely non-overlapping. The interest in geology for most of the biologists was restricted to a desire to see accurate reconstructions of the disposition of land masses through time in order to direct and/or test hypotheses of biotic evolution, and if the geologists had a professional interest in the biology, then this was largely in key fossils or historical biogeographic data that could similarly assist palaeogeographical reconstructions. Many delegates still benefited from the mixed attendance – if only to gain some insights into the language, working methods and outlook of relevant but distant disciplines. There were also several more obviously integrative or cross-disciplinary presentations.

This volume is not a conference proceedings in the usual sense, but instead a collection of chapters inspired by topics covered by and themes running through many of the more biological aspects of the meeting. Some of the authors of contributions included here were unable to attend the SAGE 2009 meeting, but were recruited because the editors were looking to assemble a collection of works that covered a range of topics that seemed to us to be particularly relevant. The contributions vary from substantial syntheses or reviews, to more tightly focused case studies that illustrate particular advances and/or approaches that will have a broader relevance. We draw attention to a parallel volume on the more geological aspects, published in 2011 as Special Publication 355 of the Geological Society of London and edited by Robert Hall, Mike Cottam and Moyra Wilson: *The SE Asian Gateway: History and Tectonics of Australia–Asia Collision*. Readers might also note that a SAGE 2013 meeting is planned for Berlin, Germany.

We are very grateful to our colleagues in the SE Asia Research Group at Royal Holloway for undertaking the majority of the SAGE 2009 meeting organisation, especially Mike Cottam, Robert Hall and Simon Suggate. Funding for the SAGE meeting was provided by Royal Holloway's Southeast Asia Research Group, The Geological Society of London, The Systematics Association, The Linnean Society and the Malacological Society of London. Help and advice in producing the volume was provided by several people, including Dominic Lewis, Zewdi Tsegai, Megan Waddington (all CUP) and Alan Warren (former Editor in Chief, the Systematics Association). The editors are extremely grateful to the expert reviewers of each chapter, those anonymous as well as Alexandre Antonelli, William Baker, Celine Becquet, Elizabeth Bennett, Ben Cook, Gathorne Cranbrook, Michael Crisp, Martin Haase, Bert Hoeksema, Hugh Kirkman, Robert Morley, Arne Nolte, Timothy Page, Toby Pennington, Dinarzarde Raheem, David Reid, Isabel Sanmartín, Menno Schilthuizen, Matt Struebig, Richard Thomas, Harold Voris, Peter van Welzen, Clive Wilkinson and Henk Wösten. We thank Burt Jones and Maurine Sherlock for use of the cover image and Elena Perez for producing the index. Finally, we thank all 49 contributing authors to the book, several of whom also peer-reviewed other chapters. We encourage all readers to support the activities of the Systematics Association (www.systass.org).

Please note that the contributors have made abstracts (in English and in Bahasa Indonesia) of chapters freely available on the internet at:

www.cambridge.org/9781107001305

Translations of abstracts were produced by Evy Arida, Museum Zoologicum Bogoriense, Java, Indonesia, with help from Fauzie Hasibuan, Geological Agency of Indonesia, Bandung, Java, Indonesia.

Cover image

The cover image encapsulates some of the diverse topics of our original meeting and this volume. The photograph was taken off Wagmab Island, Southeast Misool, Raja Ampat, Papua, Indonesia. The bommie of the fringing reef flat in the foreground comprises several corals, including species of *Pocillopora*, *Acropora*, *Favia*, *Goniastrea* and massive and finger forms of *Porites*. Other animals here include the damselfish *Dascyllus melanurus* and *Chrysiptera cyanea*, the starfish *Linckia laevigata* and giant clam *Hippopus hippopus*. In the background is typical coastal limestone vegetation of northern New Guinea, likely including species of common genera such as *Diospyros*, *Syzygium* and *Ficus*. The exposed rocks are Eocene–Oligocene limestones of the Zaag Limestone Formation, part of the largely undeformed Australian passive margin in the Bird's Head of western New Guinea that borders the Banda region of eastern Indonesia.

1

Introduction

David J. Gower, Kenneth G. Johnson,
James E. Richardson, Brian R. Rosen,
Lukas Rüber and Suzanne T. Williams

1.1 Introduction

Southeast Asia has long been of major interest for many scientific disciplines, both separately and in the way that different disciplines have interacted with each other in trying to address the numerous fascinating questions that the region poses. Two particular features stand out. Biotically, it is one of the most organically diverse regions in the world for numerous different groups of organisms. Moreover, this high diversity is true through a large range of environments, aquatic to terrestrial, and freshwater to marine. Second, Southeast Asia's geology, scenery, geomorphology and range of physical habitats are complex and also very diverse. Given the size of the region (defined loosely below) some of this great biotic variety can obviously be attributed to scale as well as its tropical climate (both of these being traditional explanations for biotic diversity), but because its latitudinal span (approximately 20°N–10°S) is entirely intra-tropical, latitudinal effects on their own would seem to be insufficient to account for its richness. By comparison, the region's longitudinal spread is almost twice as great, and in elevation it ranges from marine deeps and troughs (*c.* -6000 m) to land altitudes of *c.* 5000 m with glaciers (albeit in rapid retreat). This invites the widely held working hypothesis that its physical complexity must in some way be 'driving' many of its key biological features including its diversity. Southeast Asia also has some of the densest human populations in the world, and some of its member countries now have

Biotic Evolution and Environmental Change in Southeast Asia, eds D. J. Gower et al. Published by Cambridge University Press.

rapidly expanding economies. These factors, combined with global environmental change, have been having an increasing impact on the biology and physical habitats of the region, and have generated substantial conservation concerns and initiatives.

There are therefore three scientific focal interests in the region, and multidisciplinary feedbacks between them: (1) understanding the origin of its main biotic features, (2) understanding its geological origins, and (3) trying to manage and mitigate the human social and economic impacts on these other two features. These were all addressed during the original SAGE meeting in September 2009 (see *Preface*), but here we concentrate on (1) and (3) because the earth science proceedings were made the subject of a separate volume (Hall et al. 2011). As already explained in the *Preface*, the present volume is not a proceedings as such, but a collection of invited overview and case-study articles that treat these other two main themes.

In this introduction we aim to highlight major themes that we have identified in the various chapters in this volume, and to link aspects of these contributions, before attempting to summarise what the volume contributes as a whole. Multiauthored volumes stemming from conferences inevitably reflect what people are working on and wish to communicate, though accidents of history dictate that an assembled collection of papers will not precisely match what took place in a conference or what is taking place in the wider research community. In particular, Southeast Asia is a very large region and tackling its biotic history is a major undertaking, so we are well aware that our 16 main chapters give only a limited coverage. Some of the more obvious geographical (e.g. New Guinea, the Philippines) and taxonomic (e.g. birds, micro-organisms, many groups of invertebrates) gaps do reflect the coverage of our 2009 meeting, but some of the methodological gaps (e.g. spatial ecology, ecological niche and climate envelope modelling) have appeared by accident rather than intention during the production of this volume.

In organising the SAGE 2009 meeting, we had to contend with finding a working definition of the spatial limits of 'Southeast Asia', and this was largely true for this volume, too. Our arbitrary decision was a pragmatic mixture of the physical and political. We set the northern and western limits to include mainland Indochina south of China, from Burma/Myanmar in the west to Vietnam in the east. Travelling south and east we included the Thai–Malay peninsula, the Sunda shelf islands, the islands of Wallacea (between Sunda and New Guinea), the Philippines and New Guinea. Thus, the region can be summarised as tropical Indochina + Malesia (the latter comprising Sunda, Wallacea, the Philippines and New Guinea). Given the somewhat arbitrary nature of this boundary and its porosity in terms of biological and geological history, it is unsurprising that individual chapters in the volume include other terminologies/definitions, and also occasionally extend to neighbouring regions, such as northern Australia and the West Pacific. In any case, it is not really useful to be rigid about the boundaries: the biota of Southeast Asia will not be fully understood by studying this region in isolation or according to boundaries which may have little biotic significance.

1.2 Overview

The scope of this volume is broad and varied in terms of time, space, biotic diversity and topic. The running order could have been structured in alternative ways but we felt it appropriate to begin the main contributions with a science historian's human perspective on our region. It is a reflection of the long-recognised biotic importance of Southeast Asia that John van Wyhe (Chapter 2) debates the region's significance for the origins of the theory of evolution by natural selection, by delving into the rich and fascinating, but still contentious, history of the respective roles of Charles Darwin and Alfred Russel Wallace. Southeast Asia is generally accepted as crucial to prompting Wallace's establishment of the field of historical biogeography, but van Wyhe also makes a strong case for Southeast Asia supplanting the Galápagos as the real field site of the discovery of evolution by natural selection. Despite this, van Wyhe rejects conspiracy or underhand behaviour on Darwin's part in denying Wallace fair credit for the uncovering of evolution by natural selection. Van Wyhe bases his assessment in part on intriguing detective work, to give a fresh interpretation of the evidence for the much-debated date upon which Darwin received Wallace's famous Ternate essay. Van Wyhe points out that there is no concrete evidence for the idea that when Darwin received this essay he delayed forwarding it to Lyell in order to incorporate first some of Wallace's manuscript ideas into his own work. Van Wyhe shows that the relevant movements of mail, ships and of Wallace himself in Southeast Asia all allow for, and indeed point to, a much later posting date by Wallace for his essay and hence later receipt by Darwin.

The presence of Wallace and Darwin (among others) in this volume also lies behind the old and broadly accepted idea that geological history and biotic history must be related. In this context, further essential background is provided in Chapter 3, in which Robert Hall gives an excellent updated outline of the complex geological history of Southeast Asia. Hall concentrates on the kind of information that biogeographers and other biologists will find most directly useful, especially in setting out Southeast Asia's dramatically changing palaeogeography during the last 70 million years or so. Further information about the geology of the region is also to be found in Hall's introductory paper in the geological SAGE volume (Hall et al. 2011), and the other papers in the rest of that volume. Previous Hall articles on Southeast Asian earth history are cited frequently in this volume and in many articles published elsewhere by (especially terrestrial) biologists trying to interpret the biotic history of the region. We therefore anticipate that Hall's chapter here will prove similarly influential, with its summaries of new evidence and interpretations, including some important new information for biologists (e.g. the nature of the first Australian continent contact with Eurasia, and possible emergent terrestrial dispersal routes during the Neogene).

Although the idea that earth and life history are interrelated is widely if rather vaguely assumed, concrete evidence for direct cause and effect at the case-history level (e.g. when addressing the biogeography of particular organisms of Southeast Asia) can often be difficult to establish. This problem is central to much of what life scientists are trying to understand about the region, so we feel it would be useful to say a little more about this here. It has been common in the past, at least for biogeographers, to use an integrated narrative approach in constructing plausible scenarios that deferred to existing geological interpretations. But problems have then surfaced when the biotic patterns do not seem to fit, or more embarrassingly, when geologists change their own former interpretations which previously seemed to be compatible with the biotic ones, thereby 'wrong-footing' the biologists – notably as happened when plate tectonics replaced the former fixed-continent paradigm of earth history. It is worth remembering that historical interpretations in both disciplines are open to change.

It is against this background that, following our invitation, Hall has outlined the geology without giving any more than the broadest suggestions of biological implications such as habitable environments, dispersal routes and 'Noah's Arks'. It is up to biologists to use his chapter and other works to go further with more specific biotic interpretations of the geology for themselves. Moreover, apparent anomalies between geology and biology do not automatically signify that biologists have got it wrong and must bend their interpretations to fit the geology. Biologists in general might be too willing to defer to the veracity and robustness of historical geological hypotheses (a 'hard science'). Indeed, Hall clearly demonstrates that some of the geological data are open to individual interpretation and also draws attention to at least one instance where geologists would welcome data from historical biogeography. We know from Hall personally that he is always interested to learn of such anomalies, because they can point to further lines of geological investigation.

We therefore also suggest that readers should not limit their geological curiosity to Hall's chapter, but should also use it as a gateway into further relevant geological reading, and especially into reading critically the geological literature that presents the primary evidence for events of particular interest. It should be borne in mind that although Hall's maps provide current land/sea reconstructions, the inference of the presence and position of a particular terrane in time and space does not automatically imply the presence of land or habitable terrestrial environments, or of a definite new dispersal route. The degree of constraint on geological reconstructions varies from area to area, and in any case (as above) they should be viewed as hypotheses in progress and of varying robustness. Careful comparison with evolutionary biological narratives might be informative. Good historical biologists recognise the power of making predictions in order to discriminate between competing hypotheses, and attempt to explore the robustness of their results/interpretations to different analytical regimes and starting parameters. The same approach could usefully be extended to the basis for, and robustness of, palaeogeographic reconstructions. A greater dialogue between life and earth scientists would help both, where these subjects overlap.

An additional important abiotic element for which biologists like to have data is climatic history. In Chapter 4, Robert Morley addresses Southeast Asian palaeoclimate in the Cenozoic (the last 65.5 Myr). Palaeoclimate in the region is complex, linked with the complex geological (and oceanographic) history, and what we know clearly indicates that palaeoclimate has had a strong influence on vegetation (and thus fauna). Morley provides an updated overview of Cenozoic palaeoclimate as interpreted largely from the palynological record but also from lithological (rock) and global temperature data. Morley has great experience of working in Southeast Asia in the commercial hydrocarbon industry, so the reader is able to benefit from his access to, and deep understanding of, extensive datasets not immediately or yet available/accessible to academics. Morley's chapter contains too much information for a succinct synopsis here, but we draw particular attention to important and substantial new insights into the palaeoclimate record of Borneo. This chapter is a major new review, the first for more than a decade, and we anticipate that it will prove highly influential in studies of the abiotic and biotic history of Southeast Asia for years to come. As with Hall's geological overview (Chapter 3), the biologist should gain insights into the nature of the underlying datasets and interpretations as well as an informative summary of what is currently known about Southeast Asian palaeoclimate.

In Chapter 5, Charles Cannon offers a more spatiotemporally focused review to contribute insights into the historical biogeography of forests of Sundaland through the Quaternary (approximately the last two million years). Key themes include a focus on the utility of spatial modelling in generating testable hypotheses, particularly of comparative analyses of different major forest types; and the need for novel approaches to incorporate the substantially atypical life history of rainforest trees into interpretations of forest historical biogeography. In synthesising and interpreting available information, Cannon also highlights ways in which Sundaland represents a superb natural laboratory for testing far-reaching assumptions regarding community assembly processes, historical population size and the formation of refugia. Finally, this chapter draws attention also to outstanding major gaps in knowledge of the geomorphology of the Sunda Shelf and the interactions between forests and soils.

Wallace's Southeast Asian presence surfaces again with consideration of his eponymous Line, this being probably the most famous of all biogeographic boundaries. In Chapter 6, James Richardson and colleagues provide a modern, detailed assessment of its role in the distribution of plants across Southeast Asia. Drawing especially from (molecular) phylogenies, Richardson et al. assess the likely geographical origins of plant groups and demonstrate a clear bias in the direction of trans-Wallace's Line dispersal in favour of west to east rather than vice versa. They propose an explanation in terms of a combination of an area effect (area available for colonisation) and phylogenetic niche conservatism, and make testable predictions based on this. The authors highlight the need for further progress in plant systematics (taxonomy and phylogeny) to advance understanding of historical biogeography. Although dated molecular trees offer great potential for understanding Southeast Asian biogeography, we are in the early stages of utilising these approaches and interesting times lie ahead.

William Baker and Thomas Couvreur (Chapter 7) take an organismal approach to historical biogeography in considering a single group of plants, the palms, and focusing especially on Malesia – a hotspot of palm diversity (nearly 1000 species) within Southeast Asia. This study draws on a recently established, highly resolved and well-supported framework phylogeny, a complete species checklist (including distributional data), and a fossil-calibrated timescale for palm diversification. Consideration of these data allows Baker and Couvreur to summarise five main distributional patterns for different groups of palms, and to establish that similar distributional patterns in various lineages are not necessarily caused by common biogeographic events, such that the biogeographic history of the group as a whole within the region is complex. The data indicate a strong role for dispersal as well as geological history. Clearly, an organismal approach still holds value in documenting and explaining Southeast Asian biotic history. Baker and Couvreur suggest that more densely sampled, dated phylogenies and continued integration with fossil and abiotic data will be required to make substantial further progress.

Although historical biogeographers have long been drawn to and inspired by Southeast Asia, it is our opinion that studies based on the phylogeny and distribution of extant organisms have still not got to grips with the great diversity of the system and its highly complex, dynamic abiotic history. One issue is that there are so many geological, sea-level and climatic events in Southeast Asian history that it is possible to create post hoc explanations (narratives) for almost any (especially undated) phylogeny. In addition to thinking more clearly about hypotheses a priori, further methodological and theoretical advances also need to be made. In Chapter 8, Campbell Webb and Richard Ree argue that the features of the Southeast Asian (especially Malesian) system make it unlikely that model-based biogeography methods will accurately reconstruct lineage history unless they manage to incorporate information about temporal changes in area size and connectivity. The authors address the example of the angiosperm *Rhododendron* sect. *Vireya* using a recently developed maximum likelihood method, and a new range simulation-based method for ancestral area reconstruction. Incorporating landscape history into historical biogeography inference is non-trivial, and the new method outlined here should be seen as an important step that might offer advantages over currently dominant alternative methods, not only in achieving more accurate results, but also in gaining insights into the links between initial assumptions and resulting inferences. The authors argue that the new methodological developments open the way to using well-founded dated phylogenies to test competing palaeogeographic models. Webb and Ree conclude by echoing Richardson et al.'s (Chapter 6) and Baker's (Chapter 7) requests for more and better taxonomically and geographically sampled molecular phylogenies for Southeast Asian organisms. We also emphasise the need to incorporate data from outside the region in order to place the Southeast Asian biota in a broader context.

Southeast Asia is currently the most biodiverse region on Earth for numerous groups of marine organisms. In Chapter 9, David Bellwood and co-authors first characterise the nature of the current hotspot by reviewing data for patterns of diversity of corals and fishes. They then go on to review the origins, assumptions, predictions, support for and potentially useful tests of five main hypotheses that have been proposed to explain the famously concentric ('bull's-eye') diversity pattern of the hotspot. An important part of Bellwood et al.'s synthetic and critical review comes from the broader temporal perspective provided by the Cenozoic fossil record. Much of shallow marine biodiversity can be related to reef development, or more generally to areas of carbonate deposition, (though the reciprocal of that, that development of reefs requires high diversity, is less true). In fact, the global hotspot has not always been in Southeast Asia, as many assume. Interestingly, during the Cenozoic when the Europe and Mediterranean region tectonically resembled modern Southeastern Asia, and was warmer than today with widespread carbonate environments, this instead was the hotspot of the times for many shallow marine groups. Dating the switchover to Southeast Asia is the object of major current research but is provisionally thought to have been during the Oligocene, and is seemingly related in part to the eventual proximity of Australia to Southeast Asia prior to their eventual collision.

Bellwood et al. therefore note that marine global hotspots originate, proliferate, senesce and die, and that the origination of new global hotspots has been associated with major tectonic collisions that have resulted in the formation of shallow, enclosed continental seas and island arcs. They propose that these circumstances, combined with increased effects of sea-level changes during the later Cenozoic, created an unstable 'dynamic mosaic' of metapopulations which led to frequent reorganisation of biogeographical ranges, and increased speciation rates, and so acted as a 'diversity pump'. Indeed, because Southeast Asia is characterised by high diversity in numerous other (e.g. terrestrial) groups of organisms too, a dynamic mosaic model may provide a unifying theory to explain its high diversity, and possibly high diversity patterns, generally. Attractive though this idea may be however, much detailed work is needed to test it. In any case, Bellwood et al. also point out a difference between marine and terrestrial biodiversity patterns in Southeast Asia, with the marine biota forming a single massive hotspot (covering two thirds of the world's tropical equatorial oceans) focused on the Indo-Australian Archipelago, and the terrestrial biota characterised by greater endemism and multiple localised hotspots. We note that analyses of terrestrial biodiversity patterns in Southeast Asia lag behind those of marine biota in that the accuracy and precision of terrestrial species inventories across the region are extremely uneven (see also comments below on Chapter 14), perhaps because of the greater spatial disconnectivity.

A much more spatiotemporally restricted focus is presented in Chapter 10, in which Gordon Paterson and colleagues assess the impacts of major tsunami events on marine biodiversity. These are rare events relative to human lifespans, with only 12 large, transoceanic tsunamis recorded globally since 1755, and so they are not easy to study, especially given that biodiversity impacts have not been uppermost in people's minds in the post-tsunami human response, and that terrestrial and human consequences have been of greater immediate import. Following a review of the brief previous literature of recorded and theoretical impacts, Paterson et al.'s contribution focuses on a detailed assessment of the impact of the 2004 Asian Tsunami. Two main approaches are considered: first, an overview of reported impacts across the full range of major marine habitats/ecosystems in Southeast Asia and, second, consideration of detailed pre- and post-tsunami monitoring datasets in Thailand. The authors conclude that, based on surveyed communities and localities, the marine biota did not experience a major large-scale disaster. Although detrimental effects occurred, these were often local and extremely patchy, and the overriding impression is one of ecosystem resilience in the face of this rare impact.

A feature of the SAGE 2009 conference that prompted this volume (see *Preface*) was the disproportionate (in terms of land area and total species diversity) number of historical biogeography contributions on the terrestrial biota of the island of Sulawesi. Given that this bias in research focus reflects reality (certainly we believe that there are far fewer studies of historical biogeography of the much larger and biodiverse islands of Borneo and New Guinea), why might this be? The unusual shape and high topographic diversity of Sulawesi immediately suggest that its biota might be interestingly structured spatially, and this seems to have been greatly enhanced by its complex and multipartite geological history (see this volume, Chapter 3), and also its more distant position from the two major (Sunda; Australasia) continental shelves in the region. As reviewed by Ben Evans in Chapter 11, an interesting finding from previous research is that multiple distantly related lineages such as toads, monkeys, tarsiers, bats, lizards and frogs, have similar distributions of substantially differentiated populations in areas largely corresponding to Sulawesi's constituent palaeoislands. As well as greatly simplifying the main biogeographical hypotheses (a confounding factor in studies of Southeast Asian biogeography more generally, see comments on Chapter 8 above), the special situation encountered in Sulawesi provides a fertile case for the development of some cutting edge methodological research to refine models used in divergence population genetic analysis. Evans illustrates this in an innovative approach to the potential violation of the standard model of divergence population genetics by asymmetric population structure. This is especially pertinent to studies of dispersal to Sulawesi because many groups here appear to have more genetically structured populations than on similarly sized portions of Borneo (from where several lineages are believed to have dispersed). Evans tackles this issue using a dataset for Sulawesi and Borneo macaque monkeys, and simulating outcomes using a recently developed coalescent approach. Evans' results (extreme asymmetry does not lead to strong biases) illustrate how even relatively simple demographic models can provide a useful framework for understanding complex biological systems, and they will have implications for similar studies in Southeast Asia and beyond.

Staying on the island of Sulawesi, Thomas von Rintelen and colleagues (Chapter 12) provide an insightful overview of the radiations of animals in the ancient lakes on the island. These are relatively very well studied (largely by the authors of the chapter) and taxonomically diverse (crabs, shrimps, gastropods, fish) multi-lineage radiations that allow powerful comparative studies of biotic diversification. Von Rintelen et al. summarise the abiotic history and characteristics of the lakes before reviewing the diversity, natural history and phylogeny of each of the animal lineages. The result is a synthesis in which a wide range of evolutionary aspects are addressed, including lake colonisation, adaptive radiation, co-evolution, sympatric speciation, hybridisation and possible sexual selection. In our view, von Rintelen and colleagues establish these lakes as a stunning and under-appreciated island-like system (within an island) that deserves greater general attention.

A well-developed understanding of the biotic history of Southeast Asia will require investigations of a wide range of organisms with varying natural histories and thus dispersal abilities. However, if a biogeographer were aiming to understand how historical events impacted the evolution of the terrestrial biota, then they might choose to study lineages with high fidelity to terrestrial environments and poor ability to disperse across sea barriers – such as obligate freshwater animals. This is the premise of Mark de Bruyn and colleagues' (Chapter 13) review of the spatiotemporal history of the freshwater fauna of the Indo-Australian Archipelago. De Bruyn et al. set the scene by reviewing research on the spatial connectivity of Southeast Asian freshwater environments through time, as influenced by tectonics and (especially more recently, in the Pleistocene) changing sea level. This is followed by a review of historical biogeography data by taxon, paying particular attention to insights from molecular biogeography and phylogeography, the latter focusing on spatial genetic structure below the species level. The overall picture gained from molecular-based analysis combined with insights into the natural history of each lineage is one of promising potential. Available studies are relatively few thus far, but the results often include robust patterns of genetic (dis)connectivity among areas. De Bruyn et al. consider this to suggest that, if more studies on more organisms are conducted, then the ability to dissect the causes (chiefly centre of species accumulation versus eustatically driven vicariance/refugium hypotheses) of the very high freshwater biodiversity in Southeast Asia will improve.

One theme in several of the organismal biogeography chapters in this volume is that interpretations rely to a great extent on input data from species inventories and phylogenies, and that much more 'basic' biodiversity exploration is required (urgently, given the conservation crisis – see below). As for many groups of organisms, Southeast Asia harbours an exceptional diversity of amphibians. In Chapter 14, Rafe Brown and Bryan Stuart show how relatively traditional investigations of historical patterns of discovery of biodiversity can still provide useful insights into how (in)complete inventories of particular organisms and regions might be. They document patterns in the last 200 years and demonstrate that the more than 600 species have been described in fits and starts, and unevenly across the region with, for example, no discoveries from Sulawesi since 1930 and extremely high rates of new discoveries from northern Indochina that still show no sign of abating. The Sulawesi situation is considered not to reflect a completion of that island's inventory, and the stasis in new species descriptions is interesting when contrasted with the relatively great attention currently being paid to Sulawesi's biogeography (see comments on Chapter 11 above). The rate of recent descriptions is high, in association with incorporation of increasingly diverse types of evidence (morphology, acoustics, DNA), and Brown and Stuart conclude that currently we are in a modern 'Age of Discovery' for Southeast Asian amphibians (at the same time as the region faces a conservation crisis). Most of the widespread species were described by the end of the 1800s, with most recent discoveries being of taxa in small areas of single biogeographical zones. Despite evidence for some long-distance dispersal over seawater, amphibians are generally thought to be 'good' subjects of terrestrial biogeography studies. Amphibians could play a major role in future understanding of Southeast Asian biogeography, but clearly systematics will remain for some time an important component of the work that needs to be done.

As well as its extraordinary biological and geological diversity, Southeast Asia sadly is known as the theatre of probably the world's worst current biodiversity crisis, and it would have been negligent to avoid this topic here. It is a sign of the times that there are comments about conservation biology (without any editorial prompting whatsoever) in most chapters throughout this volume, including those chapters without any special conservation focus. In addition, the Southeast Asian biodiversity crisis is addressed more directly here in three contributions. In Chapter 15, Vincent Nijman and colleagues present data from a series of case studies that illustrate the shocking scale of the wildlife trade and its negative impact on biodiversity and conservation in the region. Nijman et al. focus on the skin, meat and pet trade in four groups of reptiles in four Southeast Asian countries and present data on numbers of animals and animal parts found in surveys of markets, official data from seizures of illegal holdings and the feasibility of the supposed captive breeding of legally protected species. The data indicate that a large amount of illegal trade is occurring, relatively openly in some cases. Although this is a substantial additional conservation threat for many already endangered species in Southeast Asia, the authors point out that the trade (and therefore its causes and solutions) is international and stretches far beyond the bounds of the region.

In Chapter 16, Susan Page and colleagues focus on an exceptionally threatened Southeast Asian habitat, peat swamp forests. These are a Southeast Asian speciality, with the region holding more than half of all the world's tropical peat swamp forests. In Southeast Asia, these habitats are massively threatened and rapidly diminishing. Page et al. point out that these forests occur in unique environmental conditions and support a high biodiversity with many specialists. Peat swamp forests also contain a globally significant carbon store. The chapter reviews data on the key biotic and abiotic features of Southeast Asian peat swamp forests before focusing on the causes and consequences of human impacts. Land-use change, drainage and fire have contributed to a substantial loss and degradation of this habitat in the region in recent years, with remaining patches being small, disturbed and fragmented. Southeast Asian peat swamp forests and their biotas are critically endangered, with the degradation having major environmental impact, including lowering surface water levels, carbon loss and increased CO_2 emissions. The prospects are not good, and urgent and effective action is required to save the remaining irreplaceable habitat patches.

The Southeast Asian biodiversity crisis has many aspects to it, and various articles have been written elsewhere, but in Chapter 17, David Bickford and co-authors undertake an ambitious synthesis of what is known. Their review reiterates the exceptional biodiversity of the region and underlines the severity of the current crisis. In so doing, they provide some new insights into the main causes, synergies and effects of the crisis. In addition, Bickford et al. attempt to move forwards by outlining the main challenges and suggesting directions that might be taken to avert further disaster. This is a daunting and complex task, but a combination of voicing concerns strongly and clearly (on the basis of sound scientific data and interpretation) plus making positive suggestions for amelioration would seem beneficial. Amid the doom, Bickford et al. point out that large steps have already been made in societal awareness, capacity building and science-based conservation management.

1.3 Overall contribution

What is the overall contribution of this volume and what might it tell us about changing knowledge and approaches in the field? One way to assess this is to compare it with other multi-author volumes on Southeast Asian biodiversity. Probably the most similar previous volume is *Biogeography and Geological Evolution of SE Asia* (Hall and Holloway 1998), but also worth considering here are *Wallace's Line and Plate Tectonics* (Whitmore 1981) and *Biogeographical Evolution of the Malay Archipelago* (Whitmore 1987). Two other books can also be mentioned, although their overlap with the scope of ours is much less complete. *Gondwana Dispersion and Asian Accretion* (Metcalfe 1999) contains contributions on geology, with considerations of biogeography based on palaeontological data (under-represented in our volume). Kershaw et al.'s (2002) *Bridging Wallace's Lines* is much more spatially focused than our volume, with the geological and palaeoenvironmental contributions designed to provide a backdrop to assessments of human evolution and cultural history in the region.

The first thing to note in comparing our volume with the first three mentioned above is the substantial change in background scientific knowledge and techniques over the period encompassed by these publications. For example, plate tectonics had been established for less than 20 years when the first of Whitmore's (1981) volumes was published, and generating DNA sequence data using the polymerase chain reaction (PCR) had yet to be invented. Many of the 16 main contributions in Whitmore's (1981, 1987) volumes are on topics that are substantially updated in our volume, including geological history, palaeoclimatology, plant distributions across Wallace's Line, biotic history of Sulawesi and palm biogeography. Only a single chapter in Whitmore's two volumes includes any phylogenetic trees (Holloway 1987), and one of the most notable developments in Hall and Holloway's (1998) volume was a much greater number of cladograms. Hall and Holloway's (1998) volume included only brief consideration of molecular genetic data, one chapter looking at genetic structure and gene flow (without phylogenetic trees; Benzie 1998) and another assessing biological history with a molecular phylogeny that was dated using a strict molecular clock based on average substitution rates (Butlin et al. 1998). Molecular approaches were much more prevalent in the SAGE 2009 meeting and in this volume, where they are applied in relation to phylogeography, biogeography, adaptive radiation and in the interpretation of branch length data as well as area relationships.

Concern about the philosophy behind methodological issues in historical biogeography is a repeated theme in Hall and Holloway (1998). Of the 13 or so biogeographical contributions, an approximately equal number favoured either cladistic biogeography or more traditional interpretations of distribution ranges (at least partly in the absence of phylogenies). A substantial change has occurred in this respect over the past decade or so, a change that was reflected in the contributions at the SAGE 2009 meeting and in this volume. We identify a move away from region-wide biogeography using either of the two dominant methods seen in Hall and Holloway (1998), and much less concern with philosophical debates about methods. The interest in biotic history remains, but with a broader range of perspectives and a more pluralist approach. A majority of the neontological contributions at the SAGE 2009 meeting were based on genetic data, but these were used in DNA taxonomy, population genetics, phylogeography, speciation/diversification dynamics and dating as well as phylogenetics above the species level. Distributions were still being interpreted, while recognising that they have limitations without associated phylogenetic data. Investigators were less readily labelled as proponents of one particular methodological faction, and there were a minority of workers carrying out what might be termed 'formal' biogeographical analyses such as computed dispersal–vicariance analysis. Many workers were addressing finer spatial and/or temporal scales with their selected taxon sampling rather than attempting to solve Southeast Asian biogeography in one go. Those researchers in 2009 addressing the region as a whole were more likely to comment on the need to generate and compare multi-taxon dated phylogenies than to attempt this in full at this stage. It is not entirely clear whether this stepping back from tackling the very big picture is explained by temporary scientific fashion, or a judicious pause while waiting for one or more of: more comparative timetree/chronogram data; more suitable/implemented methods; more precise palaeogeographical reconstructions; more complete inventories of alpha diversity. In the future, we might expect or hope for a return to some of these larger-scale regional studies aiming at a more synthetic history based on improved primary data and carefully considered biogeographical analyses. In addition to more phylogenetic and temporal data, we recognise the clear ongoing need to improve inventories of alpha diversity. We also identify a need to increase the generation and integration of appropriate palaeontological data into more complete analyses of Southeast Asian biotic history.

In reflecting upon the SAGE 2009 meeting, and considering what topics this volume might cover, we perceived something of a disconnection between studies of marine and terrestrial organisms. Marine biologists have made considerable progress in establishing the main biodiversity patterns in the region, and in testing hypotheses that explain when and how these were assembled (see Chapter 9).

Terrestrial biologists seem to have been much less concerned with this endeavour, to the extent that there remains a notable gap in available knowledge on levels of biodiversity for most groups across the region that would allow us to understand more precisely how (un)evenly terrestrial diversity is distributed in Southeast Asia and how this might be explained.

There is considerable evidence that temperate biotas were greatly impacted by late Neogene environmental fluctuations (e.g. Hewitt 2004). Although the idea that cyclical sea-level change and eustatically driven habitat fragmentation/reconnection might have elevated diversification rates in Southeast Asia in the late Neogene (especially Pliocene and Pleistocene) is intuitively attractive, existing timetree/chronogram data have indicated much older origins of many extant lineages in the region. Perhaps more phylogeographic studies of Southeast Asian organisms are needed before a firm conclusion can be reached, but thus far a substantial signal of abiotically driven late Neogene diversification has yet to be established. There is clearly some way to go before it can be determined whether possible large-scale differences between marine/terrestrial and/or tropical/temperate diversification are real, but we suggest that Southeast Asia is a good place to carry out the studies that could lead to a more unified understanding.

The Southeast Asian biodiversity crisis and conservation aspect of many of the contributions in this volume have already been noted. This is a substantial difference to the earlier volumes outlined above. It is clear that the change is real in terms of the perspective of many biologists interested in the area. This is underpinned by a similarly prominent showing of conservation issues at the 2009 SAGE meeting, and also in the 39 articles published in 2010 in numbers 2 (pp. 313–600) and 4 (pp. 913–1204) of the journal *Biodiversity and Conservation* under the titles 'Tropical islands biodiversity crisis: the Indo-West Pacific' and 'Conserving Southeast Asia's imperilled biodiversity: scientific, management, and policy challenges', respectively. Many of the desiderata for improving our understanding of the biotic history of Southeast Asia (e.g. improved inventories of alpha diversity; understanding of biotic responses to environmental change; understanding genetic structuring of populations/species) are also paramount to good conservation biology, and we anticipate a shared common interest for the foreseeable future.

References

Benzie, J. A. H. (1998). Genetic structure of marine organisms and SE Asian biogeography. In *Biogeography and Geological Evolution of SE Asia*, ed. R. Hall and J. D. Holloway. Leiden, The Netherlands: Backhuys, pp. 197–209.

Butlin, R. K., Walton, C., Monk, K. A. and Bridle, J. R. (1998). Biogeography of Sulawesi grasshoppers, genus *Chitaura*, using DNA sequence data. In *Biogeography and Geological Evolution of SE Asia*, ed. R. Hall and

J. D. Holloway. Leiden, The Netherlands: Backhuys, pp. 355–59.

Hall, R. and Holloway, J. D. (eds.) (1998). *Biogeography and Geological Evolution of SE Asia*. Leiden, The Netherlands: Backhuys.

Hall, R., Cottam, M. and Wilson, M. E. J. (2011) *The SE Asian Gateway: History and Tectonics of the Australia–Asia Collision*. London: Geological Society of London.

Hewitt, G. (2004) Genetic consequences of climatic oscillation in the Quaternary. *Philosophical Transactions of the Royal Society B*, **359**, 183–95.

Holloway, J. D. (1987) Lepidoptera patterns involving Sulawesi: What do they indicate of past geography? In *Biogeographical Evolution of the Malay Archipelago*, ed. T. C. Whitmore. Oxford: Oxford University Press, pp. 103–118.

Kershaw, P., David, B., Tapper, N., Penny, D. and Brown, J. (eds.) (2002). *Bridging Wallace's Lines. The Environmental and Cultural History and Dynamics of the SE-Asian–Australian Region*. Reiskirchen, Germany: Catena Verlag.

Metcalfe, I. (ed.) (1999). *Gondwana Dispersion and Asian Accretion*. Rotterdam, The Netherlands: A. A. Balekma.

Whitmore, T. C. (ed.) (1981). *Wallace's Line and Plate Tectonics*. Oxford: Oxford University Press.

Whitmore, T. C. (ed.) (1987). *Biogeographical Evolution of the Malay Archipelago*. Oxford: Oxford University Press.

2

Wallace, Darwin and Southeast Asia: the real field site of evolution

JOHN VAN WYHE

2.1 Darwin in the Galápagos

The Galápagos Islands are famous the world over as the supposed site of the discovery of the theory of evolution by natural selection. According to legend it was on these rocky islands 600 miles off the west coast of South America where in 1835 Charles Darwin, during the voyage of the *Beagle*, observed the gradually differing beaks of the Islands' unique finches and experienced a 'eureka moment'. Although this is a charming and widely believed story, historians of science have known since the work of Frank Sulloway (1982a) that there was no such eureka moment on the Galápagos islands. Indeed not only did Darwin not discover evolution or natural selection on the Galápagos, but his later *Beagle* notes reveal that he still believed in some form of divine creation of species after his visit there.

Even more importantly Darwin could not have experienced the Galápagos species in the way legend suggests. First of all, while in the Galápagos Darwin could not be sure that the Islands' birds were endemic because he had not visited the South American mainland at the same latitude. Furthermore, he did not even know that the famous finches which, since 1935 have been called 'Darwin's finches', were not true finches (family Fringillidae) at all (Lowe 1936). In fact, Darwin believed he had collected a number of different kinds of birds: finches, grosbeaks, American warblers, wrens and warblers (Barlow

Biotic Evolution and Environmental Change in Southeast Asia, eds D. J. Gower et al. Published by Cambridge University Press.

1963, Steinheimer 2004). It was only after his return to London in 1836 that the expert ornithologist John Gould, with access to a worldwide collection, was able to identify the birds as both endemic to the Galápagos, some apparently specific to particular islands, and indeed as mostly belonging to a single genus, *Geospiza*. This is distinct from true finches, and this genus is now placed in the family Emberizidae (Steinheimer 2004). Only then, in London, did the 'finches' begin to strike Darwin as having evolutionary significance. Darwin was similarly struck with new identifications in London of his South American mammal fossils. Hence the theory of evolution by natural selection was first formulated in London between 1837–39. If the public is to be encouraged to associate an exotic field location with the discovery of evolution by natural selection, then the Galápagos are inadmissible.

Yet it is also true that while in the islands Darwin did note that some of the mockingbirds were distinct on different islands (Chancellor and van Wyhe 2009). It was the mockingbirds rather than Darwin's finches that were the basis of the famous passage in his ornithological notes written when the *Beagle* was almost home in 1836:

> If there is the slightest foundation for these remarks, the Zoology of Archipelagoes will be well worth examining; for such facts undermine the stability of species. (Barlow 1963: 262)

In later life Darwin was occasionally asked about the origins of his theory. He never stated that he became an evolutionist while actually in the Galápagos. Instead he always listed three kinds of evidence found in the field as in this 1864 letter to German zoologist Ernst Haeckel:

> In South America three classes of facts were brought strongly before my mind: 1stly the manner in which closely allied species replace species in going Southward.
>
> 2ndly the close affinity of the species inhabiting the Islands near to S. America to those proper to the Continent. This struck me profoundly, especially the difference of the species in the adjoining islets in the Galápagos Archipelago.
>
> 3rdly the relation of the living Edentata and Rodentia to the extinct species. I shall never forget my astonishment when I dug out a gigantic piece of armour like that of the living Armadillo. (Darwin in Burkhardt et al. 2001: 302)

Darwin's recollections tended, however, to leave the location where he made his conclusions ambiguous. Hence in later years it was possible for readers to conclude that he meant that the Galápagos facts had influenced him at the time, just as we can now read them as referring to retrospective importance. The point here is not to downplay the importance of the Galápagos fauna and flora for turning Darwin towards an evolutionary explanation, but to stress that these Islands were not the locus of discovery or formulation of the theory – though it is precisely this notion that remains so widely believed.

During the latter half of the nineteenth century the Galápagos were associated with Darwin by many readers because it was his stirring account in *Journal of Researches* which had first made the wildlife of the Islands widely known (Darwin 1839, 1845). Yet it remains a little known fact that in the decades following the publication of *The Origin of Species* (Darwin 1859), and even after Darwin's death, the Galápagos were not closely associated with the origins of his theory. In fact, they were seldom even mentioned at all in accounts of Darwin's life, such as his obituaries and early biographies.

An exception is found in the monograph of ornithologist and entomologist Osbert Salvin (1835–98) on the birds of the Galápagos in 1876. He noted (Salvin 1876: 461) that the ornithological study of the Galápagos began with Darwin's visit:

> Unfortunately, at the time of his visit, Mr. Darwin did not fully appreciate the peculiar distribution of the bird-fauna throughout the different islands, and the particular island where each specimen was obtained was not always noted at the time. Enough, however, was recorded to form a basis for deductions, the importance of which in their bearing upon the study of natural science has never been equaled.

Salvin's remarks were highly unusual, at that time, for attributing such importance to the birds of the Galápagos. Presumably, he had closely read Darwin's second edition of the *Journal of Researches* (Darwin 1845) with its famously suggestive passage about the finches:

> Seeing this gradation and diversity of structure in one small, intimately related group of birds, one might really fancy that from an original paucity of birds in this archipelago, one species had been taken and modified for different ends. In a like manner it might be fancied that a bird originally a buzzard, had been induced here to undertake the office of the carrion-feeding Polybori of the American continent. (Darwin 1845: 380)

Perhaps Salvin concluded, as many subsequent readers have, that the passage added to the 1845 edition (by far the most widely available version of Darwin's *Journal*) was a record of impressions Darwin had had at the time (something Darwin never suggested). Or Salvin may have considered Darwin was spurred to his evolutionary views by the birds of the Galápagos.

Incidentally, the famous passage quoted above was not Darwin's first published remark showing his interest in the origin of species. This actually appeared in the conclusion of his coral reef paper (van Wyhe 2009: 37–9), where he says,

> some degree of light might thus be thrown on the question, whether certain groups of living beings peculiar to small spots are the remnants of a former large population, or a new one springing into existence. (Darwin 1837: 554)

Salvin did not explicitly suggest that Darwin appreciated the evolutionary significance of the birds of the Galápagos while Darwin was actually in the Islands, though he had *sufficient records* to deduce important conclusions later. Salvin also did not lend any particular weight to Darwin's finches as opposed to the other kinds of birds collected. Although Salvin called the Galápagos 'classic ground', it was not because he believed it to be the site of Darwin's conversion or discovery. It was classic because it was so peculiar and made famous by the great naturalist. Similarly, Alfred Russel Wallace referred to Amboyna as 'classic ground' in his *Malay Archipelago*:

> since it was there that I first made the acquaintance of those glorious birds and insects which render the Moluccas classic ground in the eyes of the naturalist, and characterise its fauna as one of the most remarkable and beautiful upon the globe. (Wallace 1869, volume 1: 477–8)

In the second and third decades of the twentieth century, with renewed scientific studies and expeditions to the Galápagos, the celebrated Darwin came to be more often associated with the Islands. Only after this stage-setting was a dramatic change to occur in 1935. This was the centenary of Darwin's 1835 visit to the Galápagos. The event was celebrated because the famous author of *The Origin of Species* had visited there during the voyage of the *Beagle*, not because it had long been believed that the Galápagos were the site of discovery of evolution. Indeed almost no writers had ever claimed this for the Galápagos before this centenary. A few celebrations and commemorations of various kinds were independently organised around the world. Proposals for founding a research station on the Galápagos were made, and the Ecuadorian government passed laws to protect wildlife. A monument to Darwin was erected on the islands by a specially formed Darwin Memorial Expedition led by the American travel writer Victor von Hagen. The monument read,

> Charles Darwin landed on the Galapagos Islands in 1835 and his studies of the distribution of animals and plants thereon led him for the first time to consider the problem of organic evolution. Thus was started that revolution in thought on this subject which has since taken place.

The inscription itself is one of the earliest claims that something theoretically new had occurred to Darwin there in the field.

At the meeting of the British Association for the Advancement of Science in Norwich on 4–11 September 1935, the Zoological section discussed and heard papers under the heading 'Centenary of the landing of Darwin on the Galapagos Islands, and of the birth of the hypothesis of the "Origin of Species"'(BAAS 1935). Now, for the first time, Darwin was widely represented as discovering evolution *on* the Galápagos. The title of this section alone was reproduced in newspapers and journals throughout the world. What may have begun as a 'sexed up' section heading in fact carried a version of the story: Darwin's belief in evolution, or discovery of natural selection (renditions naturally varied) occurred in the field – and specifically in the Galápagos Islands.

Nora Barlow, Darwin's granddaughter and historian, wrote an important letter to *Nature* published 7 September 1935 (i.e. during the BAAS meeting). She asked,

> At what period during the *Beagle* voyage did [Darwin's] views crystallise? [...] I have [...] been fortunate in finding among the contemporary ornithological notes a passage bearing directly on the subject, where the significant phrase 'for such facts would undermine the stability of species' occurs. Here we have the earliest date yet obtained, I think, for an admitted upheaval of his thoughts along evolutionary lines. The ferment had already begun to work in September 1835. (Barlow 1935)

We now know that this passage was written on the *Beagle* around mid-June to August 1836 between South Africa and Britain (Sulloway 1982b). It is hardly surprising that Barlow took this passage about Galápagos birds to mean that Darwin had written it during his visit, especially as Barlow was writing in the year of commemoration of the Galápagos visit. It would take until 1982 for Sulloway (1982b) to dispel what became a widespread legend.

Thus the Galápagos are not the physical location where Darwin discovered evolution or natural selection. Hence it is a pity that they should be so often misrepresented as a pilgrimage site for nature lovers. South America, with its fossil mammals and geographical distribution puzzles, was probably just as important if not more so for Darwin's evolutionary conversion, although, like the Galápagos, retrospectively. Of course, the Galápagos present wonderful and rare examples of evolution in isolation and of adaptive radiation, but evolution was not discovered there, and it was not the Galápagos evidence alone that brought evolution to light.

2.2 Wallace in Southeast Asia

If not Galápagos, what other region might have a bona fide claim as a field site of the discovery of the theory of evolution by natural selection? Whereas Darwin's evolutionary conclusions were formalised and written down on reflection after his travels in London, Alfred Russel Wallace, who lived and worked in Southeast Asia between 1854 and 1862, had made his own brilliant independent discovery of natural selection in the field there. This is not entirely surprising considering the rich biodiversity and striking patterns of zoological distribution in Southeast Asia compared to the barren and comparatively scantly populated Galápagos.

Wallace was born in 1823 near the village of Usk on the Welsh borders. He is sometimes described as Welsh, though his parents were English and he regarded himself as English (Smith 1998) Even more commonly, Wallace is described by some modern commentators as working class. This too is incorrect. Wallace was the son of a gentleman who gradually lost the family's wealth. Thus although Wallace attended a classical grammar school somewhat similar to Darwin's, the family's decline in fortune meant that Wallace received no further formal education after the age of 14. Although in his adolescence he moved among circles of working men, Wallace always observed them as an outsider. This is clear from the way he carefully noted their different language and habits in his autobiography. There is a vast scholarly literature on the meanings and definitions of social class in Victorian Britain which shows that class was by no means simply a product of financial wealth (see for example Cannadine 1999).

But for most of his life Wallace had to work for a living. Thus the hard-working, self-educated Wallace is often contrasted with the extremely affluent and Cambridge-educated Darwin who never had to work a day in his life for pay. Yet the important thing about Wallace is not that he was disadvantaged. While it is widely recognised that Wallace was mostly self-taught in science, the same can be argued for Darwin too, because he derived only a small proportion of his scientific knowledge and skills from his formal education. Wallace was remarkably intelligent and talented - and he put these talents to good use in his committed pursuit of natural history. After a series of jobs, mostly land surveying in England and Wales, Wallace began his real career in natural history when he set out for the Amazon Basin in 1848 along with his friend Henry Walter Bates. After 4 years of collecting and observing, Wallace set sail for Britain in 1852. Tragically his ship caught fire and almost all of his collections and notes were destroyed. Fortunately the collection was insured by Wallace's agent, Samuel Stevens, for £200. If Wallace collected any notes or material for his interest in the origin of species, none has survived and he never referred to any in his later writings.

Wallace's subsequent publications therefore suffered from the dearth of data he was able to bring home. His first book, *Palm Trees of the Amazon* (Wallace 1853a), described the distribution and uses of the palms he had observed and was illustrated from his own sketches. The book was criticised by some contemporaries because of its scanty detail, inaccuracies in some of the drawings and sometimes amateurish descriptions, all resulting from his lack of formal training as a botanist. His other book fared better. *A Narrative of Travels on the Amazon and Rio Negro* (Wallace 1853b), although also criticised for its paucity of particular details, was better received and sold better. Wallace also read papers before scientific societies and made important connections in the London scientific community.

After only 18 months in England, Wallace again set off for the tropics to work as a specimen collector. As he recalled in his autobiography,

> During my constant attendance at the meetings of the Zoological and Entomological Societies, and visits to the insect and bird departments of the British Museum, I had obtained sufficient information to satisfy me that the

> very finest field for an exploring and collecting naturalist was to be found in the great Malayan Archipelago, of which just sufficient was known to prove its wonderful richness, while no part of it, with the one exception of the island of Java, had been well explored as regards its natural history. Sir James Brooke had recently become Rajah of Sarawak, while the numerous Dutch settlements in Celebes and the Moluccas offered great facilities for a traveller. So far as known also, the country was generally healthy, and I determined that it would be much better for me to go to such a new country than to return to the Amazon, where Bates had already been successfully collecting for five years, and where I knew there was a good bird-collector who had been long at work in the upper part of the river towards the Andes. (Wallace 1905, volume 1: 326)

Wallace used Singapore as a base for his early expeditions into the surrounding region. Over the next 8 years he visited all of the major islands of the region, travelling as far east as New Guinea. He collected 125 660 natural history specimens, of which there were over 80 000 beetles. He proposed a line of demarcation between the fauna of Australia and Asia, now known as the Wallace's Line. As he described it,

> We have here a clue to the most radical contrast in the Archipelago, and by following it out in detail I have arrived at the conclusion that we can draw a line among the islands, which shall so divide them that one-half shall truly belong to Asia, while the other shall no less certainly be allied to Australia. (Wallace 1869, volume 1: 13)

It was some of the striking puzzles of geographical distribution like these, in addition to the scientific literature of the day, that prompted Wallace to revisit the questions of species origins.

In 1855, while living in Sarawak on the island of Borneo, Wallace wrote his first theoretical paper on species: 'On the law which has regulated the introduction of new species' which appeared in the *Annals and Magazine of Natural History* (Wallace 1855). In this essay Wallace argued that, 'Every species has come into existence coincident both in time and space with a pre-existing closely allied species.' The paper, however, did not explicitly state that species transmuted one into another. Therefore it was possible for some readers, such as Darwin, to conclude that Wallace referred to a series of creations. Hence only much later in *The Origin of Species*, Darwin (1859: 355) wrote, 'I now know from correspondence, that this coincidence [Wallace] attributes to generation with modification.' Others, less accustomed to accepting the evidence for transmutation, such as the geologist Charles Lyell, found the implications of the Sarawak paper hard to avoid. Lyell opened his own species notebooks (Wilson 1970). Lyell also urged Darwin to publish his views in outline first rather than continuing to complete his studies and publish on a very large scale (van Wyhe 2007). Hence on 14 May 1856 Darwin 'Began by Lyells advice' a more condensed version of his original plan (Darwin 1809–1881 in van Wyhe 2006). This condensed version is still known as the 'big book' and would have extended to three volumes (Stauffer 1975: 11). By the spring of 1858 Darwin had completed more than 10 chapters, covering two thirds of the topics later discussed in *The Origin of Species*.

In 1858 Wallace was living in the Moluccas, then part of the Dutch East Indies. It was here that he conceived of an explanation for the origin of new varieties and species that was strikingly similar to Darwin's natural selection. Wallace was particularly preoccupied with the origins and relationships between local human races. According to his own much later recollections he was suffering from a recurrent bout of fever when the idea came to him. Years before he had read Thomas Malthus' arguments that inevitable geometrical population growth was prevented only by severe checks. Hence, remembering Malthus, Wallace conceived of 'a general principle in nature' that permitted only a 'superior' minority to survive 'a struggle for existence' (Wallace in Darwin and Wallace 1858). Wallace elaborated this theory in his so-called Ternate essay (February 1858) 'On the tendency of varieties to depart indefinitely from the original type'. As he wrote in the essay itself,

> The numbers that die annually must be immense; and as the individual existence of each animal depends upon itself, those that die must be the weakest – the very young, the aged, and the diseased, – while those that prolong their existence can only be the most perfect in health and vigour – those who are best able to obtain food regularly, and avoid their numerous enemies. It is, as we commenced by remarking, 'a struggle for existence,' in which the weakest and least perfectly organised must always succumb. (Wallace in Darwin and Wallace 1858: 56–7)

Clearly Wallace's discovery was one made in the field itself, not in an urban study after returning to England.

2.3 Facts, fairness and conspiracies

Much of the subsequent part of this history has long been shrouded in controversy, wild speculations and even accusations of dishonesty and plagiarism. (Notably such views have come into existence only after all living participants died. Nothing so inconsistent with the evidence could have been put forward were any living protagonists still available.) It is sometimes claimed that Wallace wrote a publishable article on the theory of natural selection before Darwin did. Yet Wallace did not intend his essay for publication. At any rate, many commentators have claimed that Darwin's (1844 [1909]) essay was publishable and wondered why he did not do so (see van Wyhe 2007). So both men prepared theoretical draft essays not intended for publication, but of a standard that were considered publishable by others. Wallace sent the Ternate essay to Darwin whom he knew from correspondence was close to completing a large work on evolution. Wallace, however, did not plan to publish on the subject until his return to England. In an 1857 letter to Darwin, Wallace indicated his intention to prepare a work on species after returning, when he would have access to essential English libraries and collections (Burkhardt and Smith 1990: 457). Wallace later wrote to Alfred Newton in 1887, 'I *had* the idea of working it out, so far as I was able, when I returned home' (F. Darwin 1892: 190). Wallace returned home in 1862 – an estimated 2 years after Darwin would have completed and published his big book on species, if not interrupted by Wallace's unexpected revelation in his letter with the enclosed Ternate essay in June 1858 (van Wyhe 2007).

The repeated attempts to impugn Darwin and elevate Wallace seem to be inspired by a sympathy for the apparent underdog, Wallace. Although thus nobly motivated, the archival documentation for the development of Darwin's theoretical work from 1837 is extensive and irrefutable. There can be no question about which man had the idea of common descent, struggle for existence, natural selection or the principle of divergence first.

The biggest mystery over the past half century is the date that Wallace sent the essay to Darwin. Here the evidence has appeared confusingly ambiguous. The Ternate essay is dated February 1858. The original manuscript and its covering letter do not survive. If the essay was sent to Darwin on the next monthly mail steamer after February, as Wallace recollected more than a decade later, this would have been 9 March 1858. A letter from Wallace to F. Bates sent on this steamer still survives and bears postmarks showing that it arrived in London on 3 June 1858 (see McKinney 1972). Davies (2008) has shown that all the intermediate mail steamer connections fit for these dates. Darwin's letter forwarding Wallace's essay to Lyell, which claimed receipt 'to-day', was dated '18' June 1858 (Burkhardt and Smith 1991: 107).

Therefore, several writers have asked, if both the Bates and Darwin letters left Ternate on the same steamer, how could Darwin receive his on 18 June and not 3 June? This supposed discrepancy has been the source of much confusion. These two weeks are of consequence because some commentators believe that Darwin delayed forwarding Wallace's essay to Lyell in order to appropriate, unacknowledged, ideas from Wallace's manuscript into his own (Brackman 1980, Brooks 1984, Davies 2008).

However, the conspiracy theorists have simply assumed that Wallace sent the essay to Darwin in March 1858 – but without any direct evidence. They have therefore failed to realise that Wallace wrote his lost letter *in reply* to one he received from Darwin on that very same 9 March steamer. There is no evidence from Wallace's surviving correspondence that he could reply by the same steamer while in the Moluccas and I suspect it was not possible for Wallace to respond before the following steamer. This letter from Darwin was probably the most flattering Wallace had ever received. Darwin reassured Wallace that he and even the great

Lyell were greatly impressed with the Sarawak paper. Wallace had heard nothing positive about his most ambitious paper (Wallace 1855) for 3 years. Suddenly the most eminent man of science he had ever corresponded with praised it very highly. Hence I believe that it was only after receiving Darwin's letter on 9 March 1858 that Wallace decided to send to Darwin, and through him, to the more famous and influential Lyell, the ambitious Ternate essay. Wallace must have left his letter and essay for Darwin at the Ternate post office before departing on his next expedition. We can be reasonably sure of this because the receipt date claimed by Darwin, 18 June 1858, is exactly the right day for receipt of a letter from the Dutch East Indies aboard a P&O mail steamer that docked at Southampton on 16 June. Her letters arrived in London on 17 June. This mail had an unbroken series of mail connections all the way back to Ternate, having departed in early April 1858. Hence the mystery is solved. It was always just an assumption that the letters to Darwin and Bates were sent on the same day, but in recent years some writers have begun to treat it as if it were a historical fact.

Apart from the February date on the Ternate essay, the only reason to conclude that Wallace had sent the essay earlier was Wallace's own recollections. Historians do not attribute the same kind of accuracy or reliability to recollections as to contemporary records. Wallace must have been mistaken to recall that he sent the essay to Darwin days after composing it. It seems clear from his correspondence, however, that Wallace did compose the essay in February 1858. Almost all writers on Wallace since McKinney (1972) have accepted that the Ternate essay was not actually written on Ternate but in fact on the nearby island of Gilolo, because Wallace's field journal appears to indicate that he was absent from Ternate during the entirety of February. In fact, the field journal was written retrospectively and conflicts with contemporary correspondence, so Wallace may have been on Ternate during part of February 1858. At any rate it is unfortunate how unanimous commentators have become in discounting the accuracy of Wallace's recollections of composing the essay on Ternate, but no similar scepticism has been applied to the detail that the essay was sent a few days later. There is no contemporary evidence for this point, and the earliest recollection by Wallace was written 11 years later.

The recurring accusations that Darwin did or could have borrowed ideas, namely the principle of divergence, from Wallace's writings were conclusively refuted by David Kohn (1981). Kohn showed that what many writers mistakenly refer to as 'divergence' in the writings of Darwin and Wallace is in fact two different concepts. Kohn called these 'taxonomic divergence' and 'a principle of divergence'. Taxonomic divergence is merely the observation that 'taxa can be arranged in a branched – hence diverging-scheme' (Kohn 1981: 1105). Darwin made this observation as early as1837, and this is reflected in his famous family tree sketch ('I think') on p. 36 of his *Notebook B*, which depicts daughter varieties diverging off a central ancestral trunk (CUL-DAR121 in van Wyhe 2002). Wallace also mentioned taxonomic divergence in one line of his Sarawak paper, but gave no explanatory principle for it.

A 'principle of divergence', according to Kohn, explains 'how divergence occurs'. Darwin developed this by the mid-1850s and explicitly described it in a letter to Asa Gray in September 1857 (Darwin and Wallace 1858). The same treatment of divergence appeared in Darwin's (1856–58) draft chapters for *Natural Selection* (Stauffer 1975). After these documents were written, Darwin received Wallace's Ternate essay. The essay contained only one statement on how divergence occurs:

> But this new, improved, and populous race might itself, in course of time, give rise to new varieties, exhibiting several diverging modifications [...]. Here, then, we have *progression and continued divergence.* (Wallace in Darwin and Wallace 1858: 59)

As Kohn has demonstrated, there were differences between Wallace's 1858 principle of divergence and Darwin's 'principle of divergence' in his longer 1857 account in the letter he sent to Asa Gray, an abstract of which was published in the Darwin and Wallace (1858) paper. As Kohn explained,

> [Wallace] offered an explanation that is ecologically static, where a new species forms only by the extinction of its parent. There is none of the creation of new evolutionary opportunities by the subdivision of the environment that characterised Darwin's principle of divergence. (Kohn 1981: 1106)

Kohn also addressed another area of contention amongst modern commentators, namely that there was something untoward about the arrangement of the joint publication by Darwin and Wallace (1858):

> Darwin's friends acted to protect his interests by arranging simultaneous publication. [...] Darwin was sufficiently self-interested to encourage joint publication and produce both an extract of his 1844 Essay to prove the longevity of his claim to natural selection and the 1857 abstract prepared for Gray to prove the priority of his claim to the principle of divergence. But Darwin's claims were valid and the mere fact that his friends acted to defend them is not a conspiracy. Hooker and Lyell, however, did go one step further. [...] they manipulated the order of submission (without Darwin's knowledge) by putting Darwin's pieces before Wallace's paper. By placing the documents in the chronological order of their composition they favored Darwin's priority over Wallace's. No doubt they colored the judgment of history. (Kohn 1981: 1108)

Some modern commentators voice strong opinions about fairness and credit in the arrangement, opinions which are greatly misleading to readers without historical training. Many of the accusations and opinions are not only anachronistic judgements of the actions of Victorian men of science by current (or rather, a writer's own) standards, but also quite uninformed about acceptable practice at the time. For example, the young Darwin too had early scientific writings published by a correspondent at home without his knowledge (Darwin 1835). When he found out, Darwin reacted similarly to Wallace, pleased that his writing was considered worthy of publication and discussion by his seniors, and a bit embarrassed he had not been able to correct the proofs. But it was perfectly normal practice and there were no grounds for complaint.

And at any rate, modern commentators greatly exaggerate the significance and import of the reading of the Darwin and Wallace papers at the Linnean Society in July 1858. The contemporary accounts do not suggest anything like the significance that has been retrospectively applied to the occasion, which indeed could only have been appreciated retrospectively. A great scientific theory does not shake the community because it has been outlined or proposed, but because it gradually convinces the majority of practitioners and successfully withstands criticism and questioning. All of this of course the theory did undergo – but not in the abbreviated and outlined form of the Darwin and Wallace publication which was largely overlooked, a few exceptions notwithstanding. It was instead Darwin's condensed 'abstract' of his big book, *The Origin of Species* (1859), which did the work, with an ever-growing group of prestigious converts, of convincing the international scientific community within the space of 20 years, that evolution was a fact. Indeed Wallace lamented the fact in a letter to Darwin after a meeting of the British Association for the Advancement of Science at Norwich in 1868:

> The worst of it is, that there are no opponents [of evolution] left who know any thing of Nat. Hist, so that there are none of the good discussions we used to have. (Wallace in Burkhardt et al. 2008: 705)

Hence Darwin and Wallace deserve equal credit for the first publication of the theory of evolution by natural selection. So why does Darwin get so much more attention? He is not remembered for his half of the Darwin–Wallace paper at the expense of Wallace. Wallace himself addressed the question after Darwin's death:

> Darwin's name and fame are more widely known than in the case of any other modern man of science [...] The best scientific authorities rank him far above the greatest names in natural science – above Linnaeus and Cuvier, the great teachers of a past generation – above De Candolle and Agassiz, Owen and Huxley, in our own times. Many must feel inclined to ask, – What is the secret of this lofty pre-eminence so freely accorded to a contemporary by his fellow-workers? What has Darwin done, that even those who most strongly oppose his theories rarely suggest that he is overrated? Why is it universally felt that the only name with which his can be compared in the whole domain of science is that of the illustrious Newton? [Wallace answered:] [...] [Darwin] has given us new conceptions of the world of life, and a theory which is itself a powerful instrument of research; has shown us how to combine into one consistent

> whole the facts accumulated by all the separate classes of workers, and has thereby revolutionised the whole study of nature. (Wallace 1883: 420, 423–4)

Darwin is so celebrated because he convinced the international scientific community that evolution is a fact not only with *The Origin of Species* (Darwin 1859) and its five subsequent editions, 1860–76 (van Wyhe http://darwin-online.org.uk), but also with his detailed account of how natural selection could explain the most minute details of the reproductive structures of orchids (Darwin 1862), the massive compilation of evidence in the *Variation Of Animals And Plants* (1868), the *Descent of Man* (1871) and *The Expression Of The Emotions* (1872), his other botanical volumes, and many scientific papers (van Wyhe 2009). Wallace himself, on numerous occasions, insisted on the same point. He titled his own volume on the theory, published after Darwin's death in 1882, *Darwinism* (Wallace 1889). It was one of the best overviews of the evidence for the theory published in the nineteenth century.

It is sometimes claimed by Wallace's admirers that Darwin's reputation has somehow usurped that of Wallace with the passage of time. This is incorrect as the above quotation from Wallace attests. Darwin was vastly more widely known and overshadowing in his reputation during his and Wallace's lifetime. This has remained essentially unchanged ever since, as seen in the unprecedented succession of anniversary celebrations starting in 1909, 1959, 1982 and most notably of all, 2009.

True, Wallace was in his own lifetime a leading figure in science. Yet he was never considered by his contemporaries as of the same rank as Lyell, de Candolle, Hooker, Agassiz or Darwin. And as such, Wallace's reputation today is exactly what would be expected of a well-known figure from his time, becoming less and less remembered because there are so many other major figures from so many eras. And yet Wallace enjoys today a special following, not least in Southeast Asia. Few figures receive almost yearly book-length biographies a century after their death as Wallace does. And long may it continue. There is much yet to be uncovered and explained about the life and work of Wallace. With the welcome founding of a new Wallace correspondence project at the Natural History Museum, London, as well as my own Wallace Online project at the National University of Singapore, we can expect a new era in Wallace research in the coming years.

At any rate, Wallace clearly uncovered evolution by natural selection *in the field* in Southeast Asia. Hence it has a far better claim - indeed an undeniable one - to be the field site for the discovery of the theory of evolution by natural selection - something long mistakenly attributed to the Galápagos Islands. Perhaps in time the myth of the Galápagos discovery will recede, and hopefully the ever-increasing interest in Wallace will help restore the Southeast Asian region to its proper place in the history of scientific discovery, in addition to its unending importance for biogeography, biodiversity and much more besides.

References

Note: many of the cited references have also been reproduced and transcribed on the *Darwin Online* website edited by the author (van Wyhe 2002–). This is indicated by the abbreviation '[DO]' after the references, as applicable.

BAAS (British Association for the Advancement of Science) (1935). Sectional Transactions. D. Zoology. Afternoon. Centenary of the landing of Darwin on the Galapagos Islands, and of the birth of the hypothesis of the 'Origin of Species'. In *The British Association For The Advancement Of Science. Report Of The Annual Meeting, 1935 (105th Year). Norwich September 4–11*. London: Office of the British Association, pp. 387–90.

Barlow, N. (1935). Charles Darwin and the Galapagos Islands. *Nature*, **136**, 391.

Barlow, N., ed. (1963). Darwin's ornithological notes. *Bulletin of the British Museum (Natural History). Historical Series*, **2** (7), 201–78. [DO]

Brackman, A. C. (1980). *A Delicate Arrangement*. New York: Times Books.

Brooks, J. L. (1984). *Just Before the Origin*. New York: Columbia University Press.

Burkhardt, F. M. and Smith, S., eds. (1990). *The Correspondence of Charles Darwin*, Volume 6: 1856–1857. Cambridge: Cambridge University Press.

Burkhardt, F. M. and Smith, S., eds. (1991). *The Correspondence of Charles Darwin*, Volume 7: 1858–1859; Supplement To The Correspondence, 1821–1857. Cambridge: Cambridge University Press.

Burkhardt, F. M., Porter, D. M., Dean, S. A., White, P. S. and Wilmot, S., eds. (2001). *The Correspondence of Charles Darwin*, Volume. 12: 1864. Cambridge: Cambridge University Press.

Burkhardt, F. M., Secord, J. A., Dean, S. A., et al., eds. (2008). *The Correspondence of Charles Darwin*, Volume 16, 1868. Cambridge: Cambridge University Press.

Cannadine, D. (1999). *The Rise and Fall of Class in Britain*. New York: Columbia University Press.

Chancellor, G. and van Wyhe, J., eds. (2009). *Charles Darwin's Notebooks from the Voyage of the Beagle*. Cambridge: Cambridge University Press. [DO]

Darwin, C. R. ([1835]). Extracts from letters addressed to Professor Henslow. Cambridge: Privately printed. [DO]

Darwin, C. R. (1837). On certain areas of elevation and subsidence in the Pacific and Indian oceans, as deduced from the study of coral formations. *Proceedings of the Geological Society of London*, **2**, 552–4. [DO]

Darwin, C. R. (1839). *Narrative of the Surveying Voyages of His Majesty's Ships Adventure and Beagle between the Years 1826 and 1836, Describing their Examination of the Southern Shores of South America, and the Beagle's Circumnavigation of the Globe, Volume III. Journal And Remarks. 1832–1836*. London: Henry Colburn. [DO]

Darwin, C. R. (1844 [1909]). Two essays written in 1842 and 1844, In *The Foundations of the Origin of Species*. ed. F. Darwin. Cambridge: Cambridge University Press.

Darwin, C. R. (1845). *Journal of Researches into the Natural History and Geology of the Countries Visited During the Voyage of H.M.S. Beagle Round the World, Under the Command Of Capt. Fitz Roy, R.N.*, 2nd edn. London: John Murray. [DO]

Darwin, C. R. (1859). *On the Origin of Species by Means of Natural Selection, or the Preservation of Favoured Races in the Struggle for Life*. London: John Murray. [DO]

Darwin, C. R. (1862). *On the Various Contrivances by which British and Foreign Orchids are Fertilised by Insects, and on the Good Effects of Intercrossing*. London: John Murray. [DO]

Darwin, C. R. (1868). *The Variation of Animals and Plants Under Domestication*, 2 Volumes. London: John Murray. [DO]

Darwin, C. R. (1871). *The Descent of Man, and Selection in Relation to Sex*, 2 Volumes. London: John Murray. [DO]

Darwin, C. R. (1872). *The Expression of the Emotions in Man and Animals*. London: John Murray. [DO]

Darwin, C. R. and Wallace, A. R. (1858). On the tendency of species to form varieties; and on the perpetuation of varieties and species by natural means of selection. *Journal of the Proceedings of the Linnean Society of London. Zoology*, **3** (20 August), 46–50. [DO]

Darwin, F., ed. (1892). *Charles Darwin. His Life Told in an Autobiographical Chapter, and in a Selected Series of his Published Letters*. London: John Murray. [DO]

Davies, R. (2008). *The Darwin Conspiracy*. London: Golden Square Books.

Kohn, D. (1981). On the origin of the principle of diversity. *Science*, (New Ser.) **213**(4512), 1105–108.

Lowe, P. R. (1936). The finches of the Galapagos in relation to Darwin's conception of species. *Ibis*, (13th ser.) **6**, 310–21.

McKinney, H. L. (1972). *Wallace and Natural Selection*. New Haven and London: Yale University Press.

Salvin, O. (1876). On the Avifauna of the Galapagos Archipelago. *Transactions of the Zoological Society of London*, **9**, 447–510.

Smith, C. S. (1998). The Alfred Russel Wallace page. http://people.wku.edu/charles.smith/index1.htm

Stauffer, R. C., ed. (1975). *Charles Darwin's Natural Selection; Being the Second Part of his Big Species Book Written from 1856 to 1858*. Cambridge: Cambridge University Press. [DO]

Steinheimer, F. D. (2004). Charles Darwin's bird collection and ornithological knowledge during the voyage of *H.M.S. Beagle*, 1831–1836. *Journal of Ornithology* 145, 300–320 [and online appendix, pp. 1–40 in *Journals: Supplementary Information to Published Papers*, at http://www.do-g.de]. [DO]

Sulloway, F. (1982a). Darwin and his finches: the evolution of a legend. *Journal of the History of Biology*, **15**, 1–53.

Sulloway, F. (1982b). Darwin's conversion: The *Beagle* voyage and its aftermath. *Journal of the History of Biology*, **15**, 325–96.

Wallace, A. R. (1853a). *Palm Trees of the Amazon and their Uses*. London: J. Van Voorst.

Wallace, A. R. (1853b). *A Narrative of Travels on the Amazon and Rio Negro, with an Account of the Native Tribes, and Observations on the Climate, Geology, and Natural History of the Amazon Valley*. London: Reeve and Co.

Wallace, A. R. (1855). On the law which has regulated the introduction of new species. *Annals and Magazine of Natural History*, (2nd ser.) **16**,184–96. [DO]

Wallace, A. R. (1869). *The Malay Archipelago: The Land of the Orang-Utan and the Bird of Paradise. A Narrative of Travel, with*

Studies of Man and Nature, 2 Volumes. London: Macmillan and Co.

Wallace, A. R. (1883). The debt of science to Darwin. *Century Magazine*, **25**, (3 January), 420–32. [DO]

Wallace, A. R. (1889). *Darwinism: An Exposition of the Theory of Natural Selection with some of its Applications*. London: Macmillan.

Wallace, A. R. (1905). *My Life. A Record of Events and Opinions: With Facsimile Letters, Illustrations and Portraits*, 2 Volumes. London: Chapman and Hall. [DO]

Wilson, L. G., ed. (1970). *Sir Charles Lyell's Scientific Journals On The Species Question*. New Haven, CT, and London: Yale University Press.

Wyhe, J. van, ed. (2002). *The Complete Work of Charles Darwin Online*. http://darwin-online.org.uk [DO]

Wyhe, J. van (2006). *Darwin, C. R. 'Journal' (1809–1881). CUL-DAR158.1–76*.http://darwin-online.org.uk/content/frameset?viewtype=sideanditemID=CUL-DAR158.1-76andpageseq=1

Wyhe, J. van (2007). Mind the gap: did Darwin avoid publishing his theory for many years? *Notes and Records of the Royal Society*, **61**, 177–205. [DO]

Wyhe, J. van (2009). *Charles Darwin's Shorter Publications 1829–1883*. Cambridge: Cambridge University Press. [DO]

3

Sundaland and Wallacea: geology, plate tectonics and palaeogeography

Robert Hall

3.1 Introduction

The Southeast Asian gateway is the connection from the Pacific to the Indian Ocean, which during the Cenozoic diminished from a wide ocean to a complex narrow passage with deep topographic barriers (Gordon et al. 2003) as plate tectonic movements caused Australia to collide with Southeast Asia (Hamilton 1979, Hall 1996, 2002). It is one of several major Cenozoic ocean passages but has received much less attention than others, such as the Drake Passage, Tasman, Panama or Tethyan Gateways (e.g. von der Heydt and Dijkstra 2006, Lyle et al. 2007, 2008). Unlike others that have closed, the Southeast Asian gateway is still partly open and ocean currents that flow between the Pacific and Indian Oceans have been the subject of much recent work by oceanographers (e.g. Gordon 2005). We now know that the Indonesian Throughflow (Godfrey 1996), the name given to the waters that pass through the Earth's only low latitude oceanic passage, plays an important role in Indo-Pacific and global thermohaline flow, and it is therefore probable that the gateway is important for global climate (Schneider 1998). It is also known that today the region around the Southeast Asian gateway contains the maximum global diversity for many marine (e.g. Tomascik et al. 1997, Bellwood et al. this volume, Chapter 9) and terrestrial organisms (e.g. Whitten et al. 1999a, b). It is not known precisely when and why this diversity originated, if there is a connection

Biotic Evolution and Environmental Change in Southeast Asia, eds D. J. Gower et al. Published by Cambridge University Press.

between biotic diversity and oceanography, what the role is of the throughflow in the modern climate system, and how the restriction and almost complete closure of the passage between the Pacific and Indian Oceans may be linked to the history of climate change. However, all of these are likely consequences of geological changes related to the closure of the ocean that separated Australia and Southeast Asia at the beginning of the Cenozoic.

This chapter aims to give an overview of the geological history of the region around the gateway for life scientists; more detailed studies are included in a companion *Special Publication of the Geological Society of London* (Hall et al. 2011), which includes papers by predominantly physical science contributors to the meeting. The main intention of this chapter is to explain why the palaeogeography of the region has changed and what were the major causes of the changes. It is hoped that the palaeogeographical maps that accompany the chapter will be useful to life scientists, the explanations given here will suffice for most readers, and that those who seek more complete explanations will find them in the references cited here, Hall (2011), Hall et al. (2011) and other papers in that volume.

There are many locations referred to in the text that cannot be shown on the limited number of figures possible in this chapter. In order to aid the reader, a number of supplementary maps have been placed at http://searg.rhul.ac.uk/FTP/sage_biomaps/, which can be downloaded using any web browser. Also, readers requiring an explanation of some of the geological terminology can refer to the geological section of the glossary in Hall and Holloway (1998: 405–407), available at http://searg.rhul.ac.uk/publications/books/biogeography/biogeog_pdfs/glossary.pdf.

In geological terms, eastern Indonesia separates Asia from Australia. It is a tectonically complex region at the centre of the convergence between the Eurasian, Australian, Philippine Sea and Pacific plates (Fig 3.1). As Australia collided with the Southeast Asian margin in eastern Indonesia the wide and deep oceanic passage between the Pacific and Indian Oceans closed, although a physical oceanic connection remained (Fig 3.2). This is now the passage for water, which moves from the Pacific to the Indian Ocean by complex routes reflecting the development of the collision zone since the Early Miocene. Eastern Indonesia has a biota and diversity as fascinating as the geology and has become known to biologists as Wallacea (Fig 3.3) after Alfred Russel Wallace who contributed so much to our understanding of evolution and our knowledge of the region (van Wyhe this volume, Chapter 2). Wallace (e.g. 1869) recognised very early that distributions of plants and animals reflected changing distributions of land and sea but it was almost another century, following the plate tectonic revolution, before the amount and significance of geological change was fully appreciated.

The plate tectonic development of this region provides the most important framework for interpreting its biogeography, but the explanatory power of this

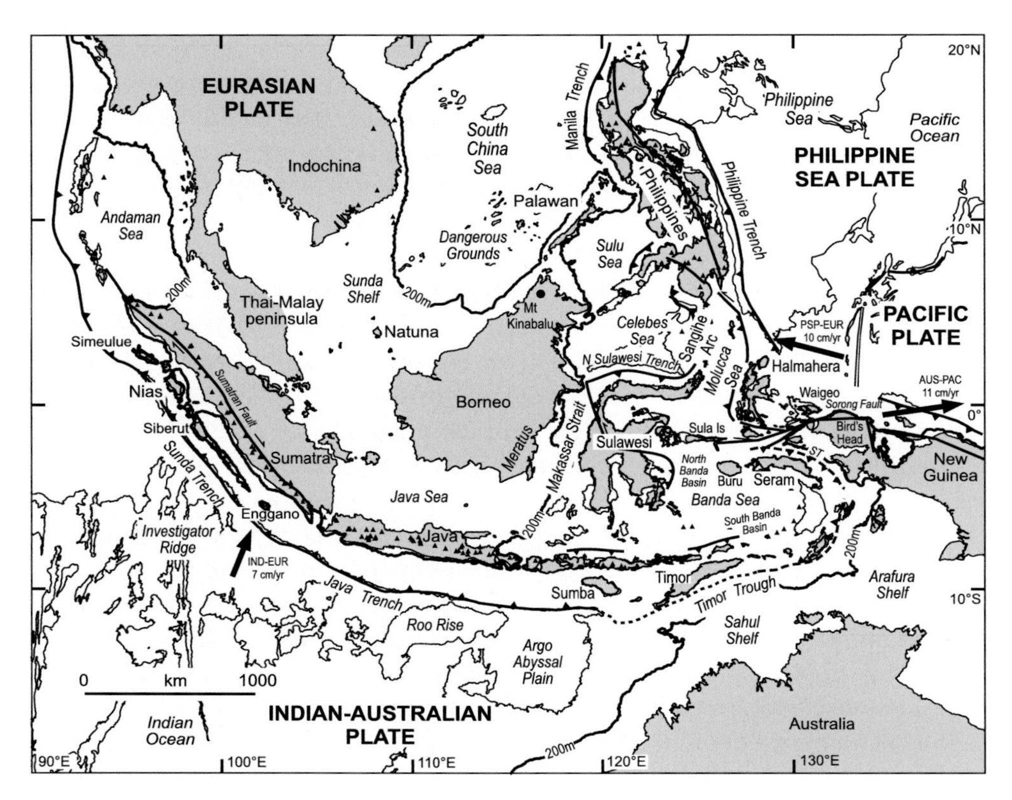

Figure 3.1 Simplified geography of Wallacea and surrounding regions. Small black filled triangles are volcanoes following the Smithsonian Institution, Global Volcanism Program (Siebert and Simkin 2002), and bathymetry is simplified from the GEBCO (2003) digital atlas. Bathymetric contours are at 200 m and 5000 m. The 200 m bathymetric contour at the edge of the Asian margin, Sunda shelf and the Sahul–Arafura Shelf is shown by a heavy line. The arrows indicate the relative motions between the major plates. The lines with small triangles on them represent subduction zones and the triangles are on the upper plate side and indicate direction of movement of the lower plate.

hypothesis is sometimes exaggerated. Plate tectonics has been the engine, but the consequence has been more than merely movements of pieces across the Earth's surface. Plates, and possibly microplates, have carried their own biotas, but the land mass of Australia has crossed climatic zones, mountains have risen and disappeared, deep waterways have closed and opened, and islands have provided connections. We are still far from a complete understanding of the geology of this region, still less the links between geology, palaeogeography, ocean–atmosphere circulation and climate which have influenced biological change, biogeographical patterns and biodiversity. It seems more and more likely that the geological history of Wallacea is important not only for understanding the impact on life in this

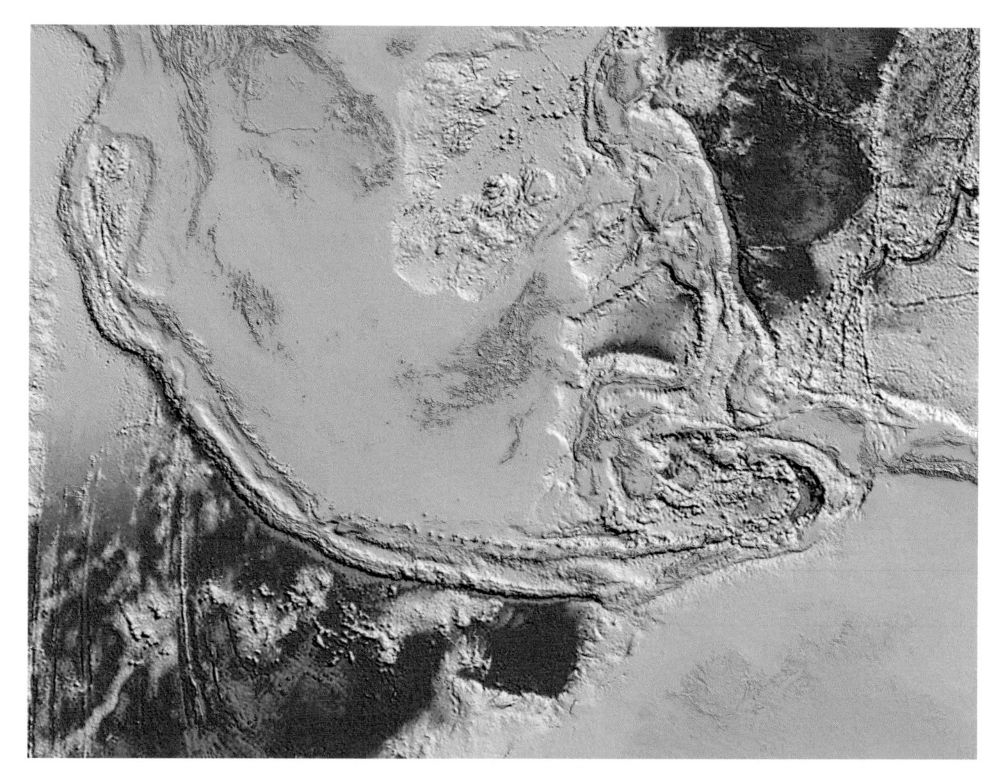

Figure 3.2 Digital elevation model showing satellite gravity-derived bathymetry combined with SRTM (Shuttle Radar Topography Mission) topography (Sandwell and Smith 2009) of the region shown in Fig 3.1. The image highlights the shallow shelves surrounding Southeast Asia and Australia with the complex geology of eastern Indonesia and the Philippines. See plate section for colour version.

equatorial region, but also for the planet because of the probable consequences for global climate.

The core of Southeast Asia, Sundaland (Fig 3.3), was initially assembled in the Late Palaeozoic and Early Mesozoic by a process that continued during the Mesozoic – addition of continental fragments carried from Gondwana to Southeast Asia (e.g. Metcalfe 2011). Rifting of fragments, now in Indonesia, from western Australia determined the shape of the Australian margin and influenced its later collision history. Their arrival affected the character of Sundaland, in terms of its strength and elevation, and terminated subduction for a period during the Late Cretaceous and Early Cenozoic. Subduction resumed in the Eocene and continues to the present day. It has been the most important geological process in the development of the region – but unfortunately it has destroyed a good deal of the evidence required to reconstruct the

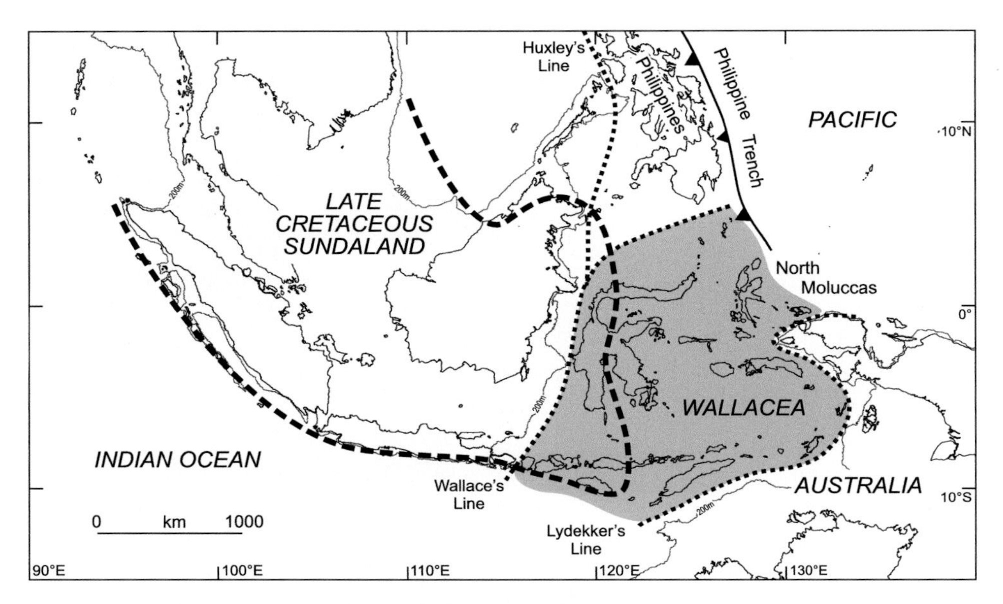

Figure 3.3 Wallacea and Sundaland. The 200 m bathymetric contour is drawn for Southeast Asia and Australia and shown as a light dotted line. The heavy dashed line is the limit of the Sundaland continent after the early and mid-Cretaceous collisions. Biogeographical lines in and around Wallacea have been drawn in many different positions and have stimulated much debate (e.g. Mayr 1944, Simpson 1977). Part of Wallacea is geologically part of the Cretaceous eastern margin of Sundaland. The western boundary of Wallacea is most commonly drawn through the Makassar Straits and across the Celebes Sea at the original Wallace's Line (Wallace 1869). In geological terms the Philippines has many similarities to the North Moluccas, and Wallacea could be continued north and bounded by the Philippine Trench to the east and to the west by the line that Huxley (1868) suggested as an alternative position for the biogeographic boundary at the eastern edge of Asia. Dickerson (1928) proposed the term Wallacea and did include the Philippines in it although this area is now usually excluded. The southeastern boundary of Wallacea mainly follows Lydekker's Line (Lydekker 1896, Simpson 1977) except close to the Bird's Head where geologically the line would be drawn slightly further south as shown here.

region. The Cenozoic development was strongly influenced by what was present in Southeast Asia before collision, as well as the form of the Australian margin. The heterogeneous nature of the Sundaland basement, combined with a high regional heatflow due to magmatism and subduction, and subduction-related forces at plate margins, determined the way in which the Australia–Sundaland collision proceeded and the response of the upper crust to the movements of major plates. In Wallacea topography and bathymetry changed very rapidly during the late Neogene.

3.2 Background

The palaeogeographical maps in this chapter (Figs 3.4 to 3.12) are an attempt to display the surface of this region over time and have developed from earlier tectonic reconstructions (Hall 1996, 2002, Hall et al. 2009a) and maps (Hall 1998, 2001, 2009a). The maps cover a large area, have substantial intervals of time between them (10 to 5 Ma) and are inevitably generalised. They differ from earlier maps (e.g. Hall 2001, 2009a) slightly, but possibly significantly for some biogeographers, in showing more small areas of land within Wallacea in the Neogene, as a result of changing ideas discussed in the text below. Drawing them emphasises the difficulties for one person in acquiring and interpreting a vast amount of information and it should not be surprising that they differ considerably from maps that cover larger areas over even greater periods of time with greater intervals between them (e.g. Stampfli and Borel 2002, Scotese 2010). It is impossible in a short space to identify and explain the differences between the maps here and those of other authors but notable differences are in the orientation and greater southward extent of the Sundaland margin in the Early Cenozoic, the importance of the Banda embayment in the Australian margin and the effective elimination of the wide deep marine gap between Australia and Southeast Asia in the Early Miocene.

Because this chapter is aimed at a reader from the life sciences I have omitted detailed arguments for the geological interpretations and restricted the number of references cited. In many places I have cited recently published papers that include references to primary sources rather than the primary sources themselves, simply to limit the number of references cited. The companion volume (Hall et al. 2011) to this book is concerned with the geology of the region and some of the physical consequences of geological change. The reader is referred to Hall (2011) and several other papers cited below in that volume for geological detail particularly relevant to this chapter. Here, I summarise briefly the region's geological history, and draw attention to changed and new ideas of its development, especially for Wallacea. In particular, I highlight young and rapid vertical movements of the crust in Wallacea that may have caused greater changes in palaeogeography than previously recognised.

3.3 Triassic to Cretaceous: assembly of Sundaland

It is now generally accepted that Sundaland (Fig 3.3) was assembled from continental blocks that separated from Gondwana in the Palaeozoic and amalgamated with Asian blocks in the Triassic (Metcalfe 1996, 1998, 2011). The Indochina–East Malaya block separated from Gondwana in the Devonian, and by the Carboniferous was in warm tropical low latitudes where a distinctive Cathaysian flora developed. In contrast, Carboniferous rocks, including glacio-marine diamictites, indicate the Sibumasu block was at high southern latitudes during the Carboniferous. It collided with Indochina–East Malaya, already amalgamated with the South and North China blocks, in the Triassic. Discussion continues about details of different tectonic models, timing of events and some of the consequences. The major result of subduction and collision was a promontory with a Proterozoic continental basement, intruded by widespread Permian and Triassic granites, which formed an elevated land mass for most of the Mesozoic (Abdullah 2009). This was the area that now includes Sumatra, the Malay peninsula, and the Sunda Shelf east of the peninsula.

Other parts that are commonly included in Sundaland have a less certain origin. Metcalfe (1996, 1998) interpreted a number of smaller continental blocks (or terranes) around the Indochina–East Malaya–Sibumasu core within an area shown as accreted crust. The time when they were accreted is unclear, as is their origin, and some are suggested to have come from Asia and others from Australia. The uncertainties are a result of limited exposure in areas now submerged or covered with younger rocks, or which are under studied. Borneo includes the largest of these blocks with rocks exposed that are older than Mesozoic, and it is often assumed that the Southwest Borneo continental core was attached to Sundaland before the Cretaceous. The surrounding region, from Sarawak, Sabah, East Kalimantan and East Java to South Sumatra has been interpreted as Cretaceous and Tertiary subduction complexes (e.g. Hamilton 1979) including some microcontinental fragments. Different interpretations of the many fragments are reviewed in Hall (2011). Here I summarise my own views, illustrated in Hall et al. (2009a) and Hall (2011), that show where the microcontinental fragments originated and how they moved into Southeast Asia.

Since Metcalfe (1990, 1996) suggested it, on the basis of Triassic (quartz-rich) turbidites above a pre-Mesozoic basement, it has become conventional wisdom that microcontinental fragments rifted from Northwest Australia are now in West Burma, although this was considered by Metcalfe himself as 'speculative', since he observed that there was 'as yet no convincing evidence for the origin of this [West Burma] block'. I follow Mitchell (1984, 1992) in interpreting West Burma as part of Asia and Barber and Crow (2009) who also interpreted it as part of Asia and as a continuation of the West Sumatra block. These fragments were part of Sundaland from the Late Palaeozoic and were separated by opening of the Andaman Sea. The Sikuleh and Natal blocks in Sumatra are not continental fragments derived from Australia, but were part of the Woyla intra-oceanic arc (Mitchell 1993, Barber 2000, Barber and Crow 2009) thrust onto the Sumatran Sundaland margin in the mid Cretaceous.

I consider that blocks rifted from Northwest Australia in the Jurassic are now part of Sundaland. The Southeast Asian promontory east of the Indochina–East Malaya block grew by the addition of continental crust during the early to mid-Cretaceous, to form the Early Cenozoic Sundaland continent (Hall 2009c). Some continental fragments have an Asian origin, but most are Australian. An Asian fragment or fragments collided with east Sundaland in the mid Cretaceous, including the area of offshore Vietnam and Sarawak and onshore northern Borneo, within which are the Semitau and Luconia terranes of Metcalfe (1996, 1998). The Dangerous Grounds is in many ways a continuation of this region because it is underlain by stretched continental crust that in the Late Cretaceous was part of the South China continental margin. An Asian origin is supported by obvious Cathaysian characteristics of faunas and floras from the Dangerous Grounds (Kudrass et al. 1986), Northwest Kalimantan (Williams et al. 1988) and Sarawak (Hutchison 2005). The Dangerous Grounds was partly separated from Asia and Sundaland by opening of the South China Sea, subduction of the proto-South China Sea beneath North Borneo, and collision with Borneo.

Fragments rifted from Northwest Australia in the Late Jurassic docked against the East Malaya block in the early to mid Cretaceous (Hall et al. 2009a). Southwest Borneo is the largest and was the first of these to arrive. It separated at about 160 Ma to form the Banda embayment and was added to Sundaland in the Early Cretaceous. A small Inner Banda block is interpreted to have followed the Banda block, but moved relative to it during a later collision event, and may now underlie part of Sabah and northern West Sulawesi. The East Java–West Sulawesi block is interpreted as the Argo block (Mt Victoria Land of Veevers 1988, or Argoland of Powell et al. 1988), which was the offshore continuation of the Canning Basin, whose detrital sediments are the source of Palaeozoic to Archaean zircons now found in East Java (Smyth et al. 2007). The East Java–West Sulawesi block separated from NW Australia at about 155 Ma as rifting propagated west and south (Pigram and Panggabean 1984, Powell et al. 1988, Fullerton et al. 1989, Robb et al. 2005). East Java and West Sulawesi may be a number of separate fragments, rather than a single block, added to Sundaland at about 90 Ma at a suture running from West Java towards the Meratus Mountains and then northward (Hamilton 1979, Parkinson et al. 1998). Collision of the Woyla intra-oceanic arc with the Sumatran Sundaland margin occurred at the same time as East Java–West Sulawesi docked.

3.4 Mid Cretaceous: collision and termination of subduction

The rifting of fragments from Australia determined the shape and character of the Australian margin, which were to be a major influence on the Neogene development of Australia–Sundaland collision. The multiple collisions of the continental blocks from Asia and Australia also had a profound effect because they terminated

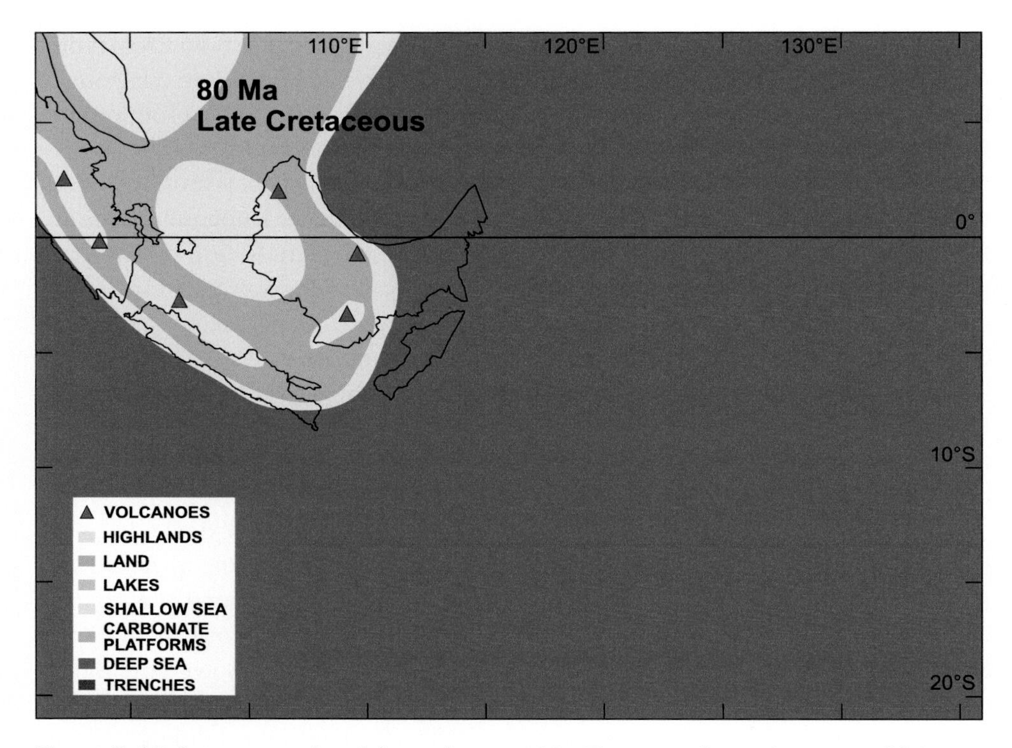

Figure 3.4 Palaeogeography of the region at 80 Ma. The very schematic aspect of this map results from the almost complete absence of rocks of this age from Southeast Asia, because much of the region was emergent. It was probably also significantly elevated judging from the character of sedimentary rocks in Sarawak where there are some of the few rocks of this age that are preserved. They are poorly sorted clastic sediments derived from a granitic source (e.g. Wolfenden 1960) implying deep erosion to expose continental basement rocks from several kilometres depth. See plate section for colour version.

subduction (Smyth et al. 2007, Hall et al. 2009a) around Sundaland in the mid-Cretaceous for 45 Myr. For the period from about 90 Ma to 45 Ma around most of Sundaland there was no subduction. Australia was not moving north and there was an inactive margin south of Sumatra and Java until 70 Ma, there was also no subduction beneath North Borneo. No significant volcanic activity is recorded during the period 90 to 45 Ma and most granite magmatism also ceased (Hall 2009b, 2009c).

By the mid Cretaceous (*c.* 90 Ma) there was a large promontory of continental crust that extended from Thailand and Indochina southwards to Sumatra, Java and Borneo, which included West Sulawesi. The palaeogeography of this region is difficult to reconstruct, but it is likely that much of Sundaland was emergent (Fig 3.4) and there were large rivers draining it. Upper Cretaceous and Paleocene rocks

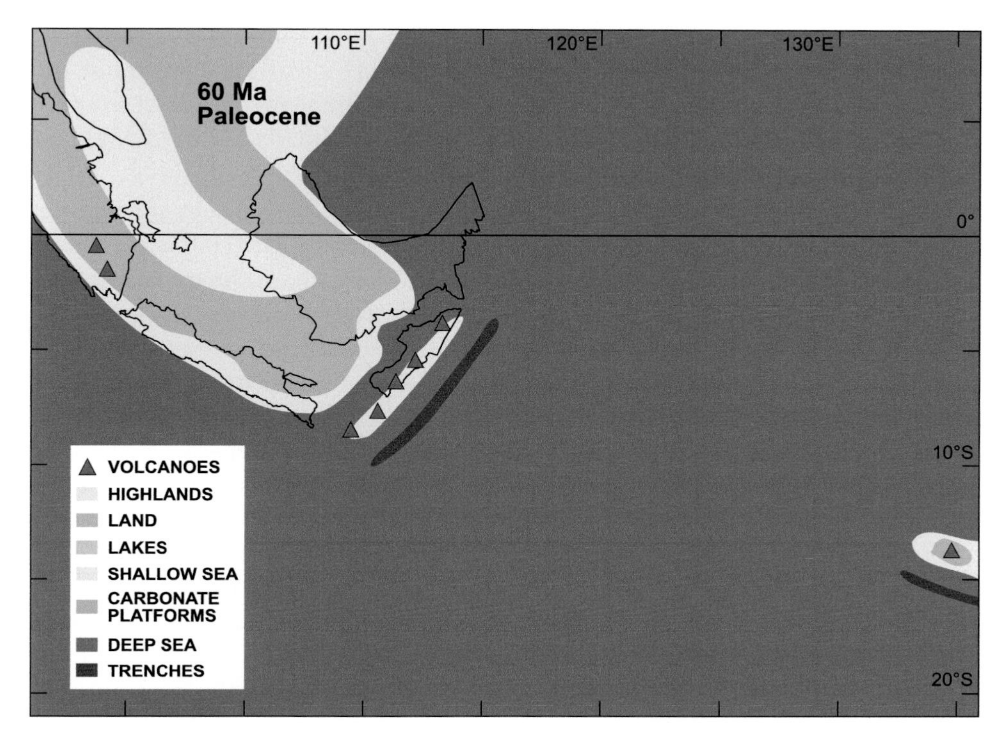

Figure 3.5 Palaeogeography of the region at 60 Ma. Again, there are few rocks of this age in Southeast Asia, because much of the region was emergent. There was a short-lived volcanic arc at the eastern edge of Sundaland in present-day West Sulawesi and Sumba. Other parts of the Sundaland margins were inactive, although there is some uncertainty about the margin in Sarawak and northwards at this time, because much of the region is offshore and covered by younger sediments. See plate section for colour version.

are almost unknown within Sundaland, although there may be terrestrial conglomerates and sandstones in Sarawak (Muller 1968, Hutchison 2005). Terrestrial deposits from the Malay peninsula, which are Jurassic to Lower Cretaceous, became inverted and were eroded after the Early Cretaceous (Abdullah 2009). Clements et al. (2011) have suggested that a major regional unconformity is a response to the mid-Cretaceous termination of subduction. A prolonged period of emergence and widespread reworking of older rocks is suggested by the character of the oldest sediments in the many Cenozoic sedimentary basins (Hall and Morley 2004, Hall 2009b) throughout Sundaland that formed when sedimentation resumed. Typically sediments are quartz-rich clastics, very mature in terms of grain shape and composition, suggesting multiple episodes of recycling. Although most of Sundaland was emergent in the Paleocene (Fig 3.5) it is likely that at the margins, for example in parts of Borneo, West Sulawesi, Sumba, Central Java and

South Sumatra, deep water sedimentation continued (e.g. Burollet and Salle 1981, Hasan 1990, Wakita et al. 1994, Moss 1998, Wakita and Metcalfe 2005) through the Late Cretaceous into the Early Cenozoic. There was northwest-directed subduction beneath Sumba and West Sulawesi in the Paleocene (63 Ma to 50 Ma) and calc-alkaline volcanism (Hall 2009b, Hall et al. 2009a).

North of India between 90 and 45 Ma there was a different tectonic setting. Subduction continued beneath the intra-oceanic Incertus Arc (Hall et al. 2009a) that had been the westward continuation of the Woyla Arc. India moved rapidly north, especially from about 80 Ma, with north-directed subduction within Tethys at this intra-oceanic arc and at the Asian margin. The different movements of the Australian and Indian Plates were accommodated by a transform boundary. There is no evidence from Sundaland that India made contact with Sumatra as it moved northwards, as recently suggested (Ali and Aitchison 2008), but there is increasing evidence for a collision between the Incertus intra-oceanic arc and the northern margin of Greater India (Aitchison et al. 2007a, Hall et al. 2008, 2009a, Khan et al. 2009) at about 55 Ma. The arc was later carried north with India and may be represented in Tibet as the Zedong terrane (Aitchison et al. 2007b) or further west as the Kohistan–Ladakh Arc (Khan et al. 2009). This collision may have biogeographic significance because it could have provided an opportunity for dispersal from India into Sundaland in the Early Eocene; a complete land bridge is improbable but a connection via islands is possible. (In this chapter I use 'dispersal' to mean a change in the gross distributional pattern of an organism without implying any particular biological processes which might have caused that change.)

3.5 Eocene to Miocene: resumption of subduction

Until recently, most reconstructions have assumed subduction continued all round Sundaland uninterrupted from the Cretaceous into the Cenozoic (e.g. Haile 1974, Hamilton 1979, Hall 1996, 2002, Hutchison 1996, Metcalfe 1996, Barber et al. 2005, Whittaker et al. 2007). However, as outlined above, there is little direct evidence to support this and almost no subduction-related volcanic record, in contrast to the period before 90 Ma and the period after 45 Ma (Hall 2009b, 2009c). I interpret subduction to have ceased during the Late Cretaceous and Paleocene and to have resumed in the Eocene when global plate models indicate that Australia began to move northwards at a significant rate. The one part of the Sundaland margin in which there is evidence for older subduction-related volcanism is Sumba and parts of west Sulawesi. Here there is evidence for Paleocene and Eocene volcanic activity. Reconstructions suggest northwest-directed subduction of the Australian Plate beneath Sundaland between about 63 Ma and 50 Ma accompanied by slight extension and significant dextral strike-slip motion at the Sumatra and Java margin.

3.5.1 Sundaland palaeogeography

At present the interior of Sundaland, particularly the Sunda Shelf, Java Sea and surrounding emergent, but topographically low, areas of Sumatra and Borneo is largely free of seismicity and volcanism (Hamilton 1979, Engdahl et al. 1998, Hall and Morley 2004, Simons et al. 2007). This region formed an exposed land mass during the Pleistocene and most of the Sunda Shelf is shallow, with water depths less than 200 m and little relief. This has led to a misconception that it is a stable area and it is often described as a shield or craton. Cratons are old continental regions underlain by a thick cold lithosphere stabilised early in the Precambrian and have behaved as strong areas since then. They are typically flat, deformation is largely restricted to their margins, and they may flex on a very long wavelength scale with low amplitude vertical movements. Sundaland is a continent but not a craton (Hall and Morley 2004, Hyndman et al. 2005, Currie and Hyndman 2006). During the Cenozoic, different parts of the continent became substantially elevated and shed sediment to the many sedimentary basins (Hall and Morley 2004) within Sundaland that have subsided substantially and rapidly.

The resumption of subduction was accompanied by widespread extension and basin formation within the Sundaland continent from the Middle Eocene (Fig 3.6). Despite rapid subsidence, most basins were not bathymetrically deep features and contain fluviatile and marginal marine deposits. A major exception is to the east of Borneo, where Middle Eocene rifting led to separation of West Sulawesi from East Borneo and formation of the Makassar Straits. It is uncertain whether the straits are underlain by oceanic or continental crust because there is a very thick sequence of sediments, up to 14 km, above the basement in the central parts of the northern straits. From a biogeographic point of view this uncertainty should not be an important issue because extension had formed a significant marine gap (topographic barrier), wider than today, from the Eocene onwards. The South Makassar Straits are probably underlain by thinned continental crust (Hall et al. 2009b), and in the centre may have subsided to depths of up to a kilometre below sea level with shallow marine carbonate platforms to the west and east. At times of low global sea level the marine gap may have been a few tens of kilometres wide. The North Makassar Straits are currently about 2500 m deep and much wider. I consider it likely that they are underlain by highly thinned continental crust as discussed in Hall et al. (2009b). Since the Eocene, the marine gap between land on Borneo and Sulawesi has been at least as great as today.

The palaeogeography changed over time with eustatic sea-level fluctuations, but from the Eocene to Early Miocene (Figs 3.6 to 3.9) most of western Sundaland was terrestrial with deposition in sedimentary basins dominated by fluviatile input, with areas of shallow marine seas increasing with time. It is likely that the Malay peninsula was one important elevated region within Sundaland supplying

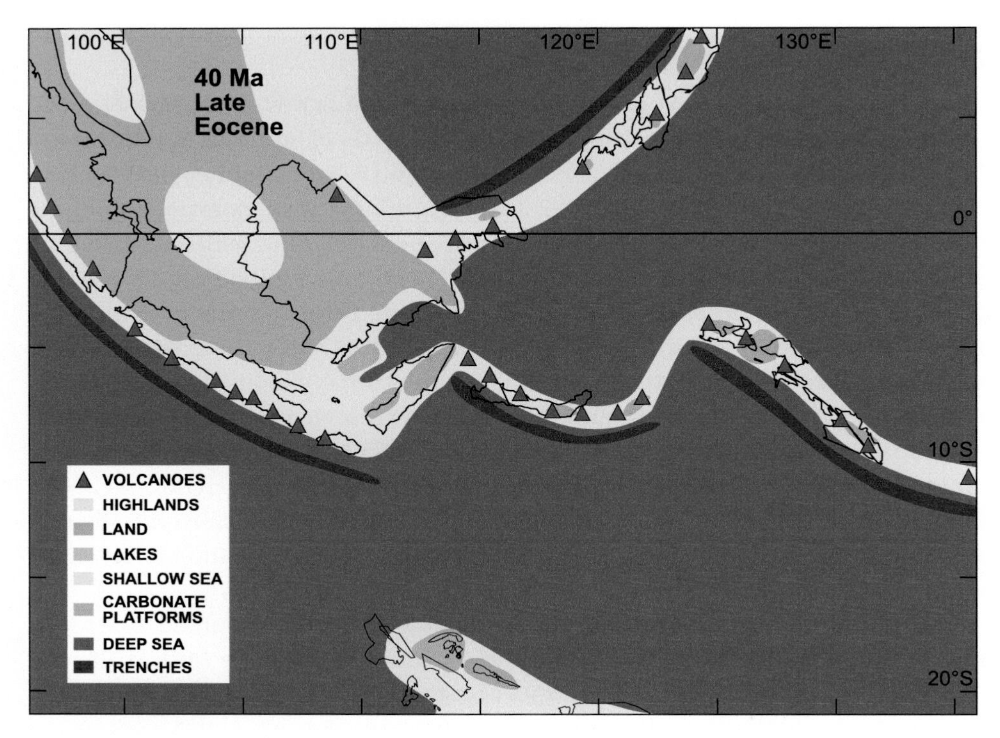

Figure 3.6 Palaeogeography of the region at 40 Ma. Subduction resumed around much of Sundaland in the Middle Eocene and many new sedimentary basins began to form, in which terrestrial sediments were deposited, derived from local sources. The Makassar Straits were already a significant marine gap, especially in the northern straits, at the eastern edge of Sundaland. See plate section for colour version.

sediment to Sunda Shelf basins, and the Schwaner Mountains of West Borneo was another. Parts of the present Java Sea may also have been elevated and provided sediment. In the area west of the present Meratus Mountains, known as the Barito Basin, there was a wide river system, where coals and fluvial and estuarine sediments were deposited. This appears to have flowed north during the Eocene from the present Java Sea (D. Witts pers. comm. 2010) towards the Makassar Straits, with limestones deposited during periods of higher sea level. Clastic sediment was also transported into the North Makassar Straits from central or West Borneo.

In the northern Borneo part of the Sundaland land mass there was a narrow shelf and slope at the margin of the deep-water area of the proto-South China Sea (van Hattum 2005). Sediment was carried offshore from rivers flowing from the Malay peninsula and the Schwaner Mountains and deposited on a shelf, in the area of present offshore Sarawak, and carried into a deep-water sediment accumulation, the 'Crocker Fan', in Sabah on the south side of the proto-South China Sea.

3.5.2 Sundaland margins

At the southern margin of Sundaland between the Eocene and Early Miocene (Figs 3.6 to 3.8) were volcanic arcs, and within Borneo there was volcanic activity, mainly related to southward subduction of the proto-South China Sea. The volcanoes mainly formed islands rather than continuous and extensive areas of land. It is probable that in West Sulawesi there were no volcanoes.

In Sumatra, volcanic activity became widespread from the Middle Eocene (Crow 2005). The Eocene arc was in a similar position to the Mesozoic arcs (McCourt et al. 1996) and was initially constructed on the edge of the land mass with some volcanic centres in a terrestrial setting and others forming islands on a coastal plain (M. J. Crow pers. comm. 2010). A possible Toba-scale caldera may have spread ash over a major part of Central Sumatra in the Late Eocene (Crow 2005). Later regional subsidence is suggested by Barber et al. (2005) to have led to marine transgression, with deepening and widening of marine basins in both the forearc and backarc leaving the volcanic Barisan Mountains as a chain of large islands south of the elevated Malay peninsula by the Early Miocene. The Sumatran arc remained in essentially the same position during the whole of the Cenozoic.

In Java a volcanic arc ran the length of the island from the Middle Eocene (Smyth et al. 2007, Hall and Smyth 2008) well to the south of the Cretaceous active margin and close to the present south coast of Java. It formed a series of small volcanic islands rather than a large single island and was south of the Sundaland coast and separated from it by a marine gap. The shelf edge ran roughly east–west through northern Java and was locally quite steep. The extensive emergent area of Sundaland to the north of West Java was crossed by large rivers that drained the Malay–Thai peninsula area and West Borneo feeding sediment to the coast, and further east were shallow marine carbonates on a broad flat shelf covering much of the area as far east as Sulawesi.

The Eocene and Oligocene volcanic activity records north by northeast-directed convergence of the Indian–Australian Plate with respect to Sundaland, which meant that in Sulawesi the subduction direction was almost parallel to the continental margin. The volcanic activity that began in the Paleocene ceased in Sumba and West Sulawesi in the Late Eocene (van Leeuwen et al. 2010) or Early Oligocene (Abdullah et al. 2000). There was a transform margin (van Leeuwen et al. 2010) at about the position of the Walanae fault zone in South Sulawesi and an offset in the arc, which continued eastwards into the Pacific via the North Arm of Sulawesi, the East Philippines and Halmahera. South Sulawesi was the site of an extensive carbonate platform for most of the Cenozoic although there may have been some arc activity and possibly a marginal basin along the east side of the transform margin. To the north there must have been some islands in West Sulawesi

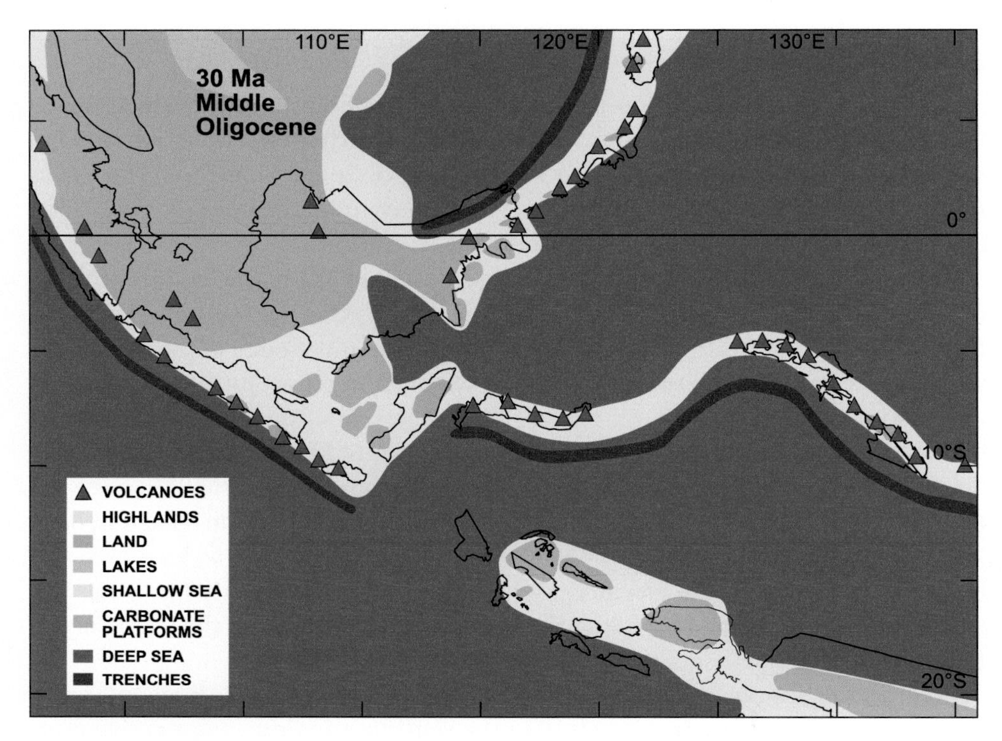

Figure 3.7 Palaeogeography of the region at 30 Ma. Although much of Sundaland was emergent it is likely that topography was significantly lower than earlier in the Cenozoic. Rivers carried recycled clastic sediments to internal basins and the continental margins. On the Sunda Shelf there were large freshwater lakes, not linked to the ocean, which are shown in a different shade from normal salinity seas. See plate section for colour version.

supplying detritus to the Makassar Straits. In the Eocene there appears to have been more land with coal swamps and small marine limestone-capped tilted fault blocks, which became submerged as the Makassar Straits subsided. Small areas of land must have remained until the Pliocene, since clastic shelf sediments are preserved in West Sulawesi and indicate a source east of the Makassar Straits, although where is not known.

From the Eocene to Early Miocene, the Proto-South China Sea was subducted southwards beneath northern Borneo. As it was subducted, the present South China Sea formed to the north of it between the Oligocene and Middle Miocene (Taylor and Hayes 1983, Briais et al. 1993). The Crocker Fan was deposited in deep water at the active subduction margin (Tan and Lamy 1990, Tongkul 1991, Hazebroek and Tan 1993, Hutchison et al. 2000). Much of the sediment came from the Malay–Thai peninsula and Schwaner Mountains but some has an ophiolitic

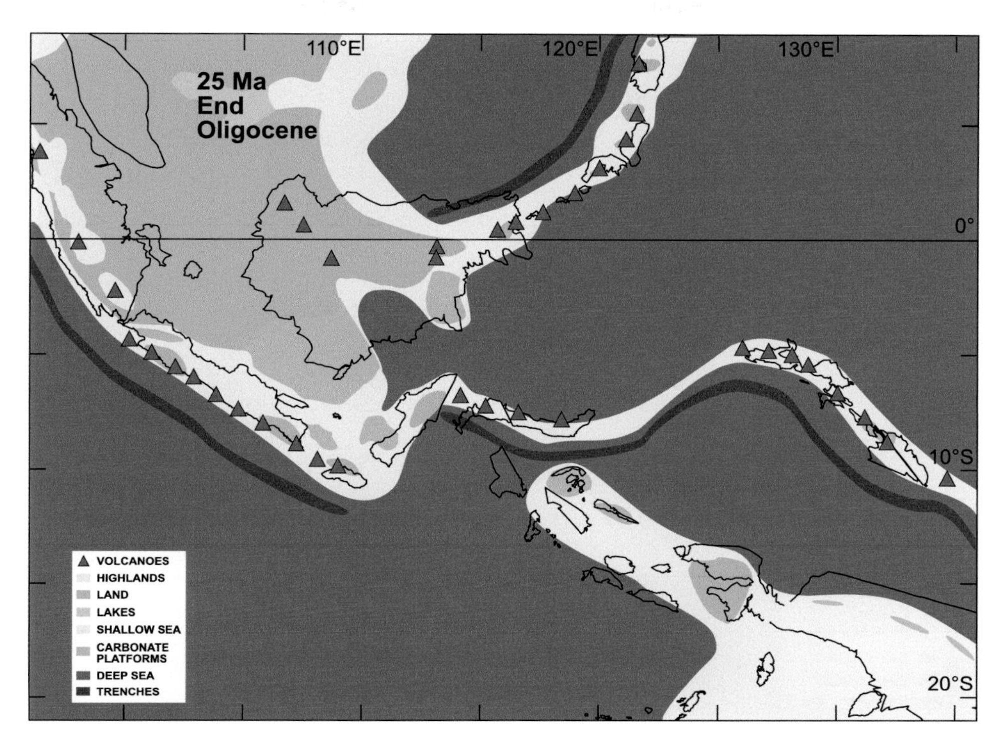

Figure 3.8 Palaeogeography of the region at 25 Ma. On the Sunda Shelf lakes (shown in a different shade) were intermittently connected to the sea and were brackish. The Sula Spur was about to collide with the volcanic arc of Sulawesi's North Arm resulting in ophiolite emplacement and uplift, and becoming the first part of the Australian continent to connect Australia and Southeast Asia, although there was no continuous land connection. See plate section for colour version.

provenance indicating some land areas in Sabah (van Hattum 2005, van Hattum et al. 2006).

3.6 Miocene to Recent: Australia collision in Wallacea

At the beginning of the Miocene there was collision between Sundaland and Australia (Figs 3.8, 3.9), and later in the Early Miocene there was collision in north Borneo with the extended passive continental margin of South China (Hutchison et al. 2000, Hall and Wilson 2000). Continental fragments have since been accreted to, or rearranged in, East Indonesia. These collisions led to mountain building in Sulawesi, the Banda Arc, and Borneo. In addition, the arrival of arcs from the Pacific in East Indonesia led to the emergence of islands in east Indonesia.

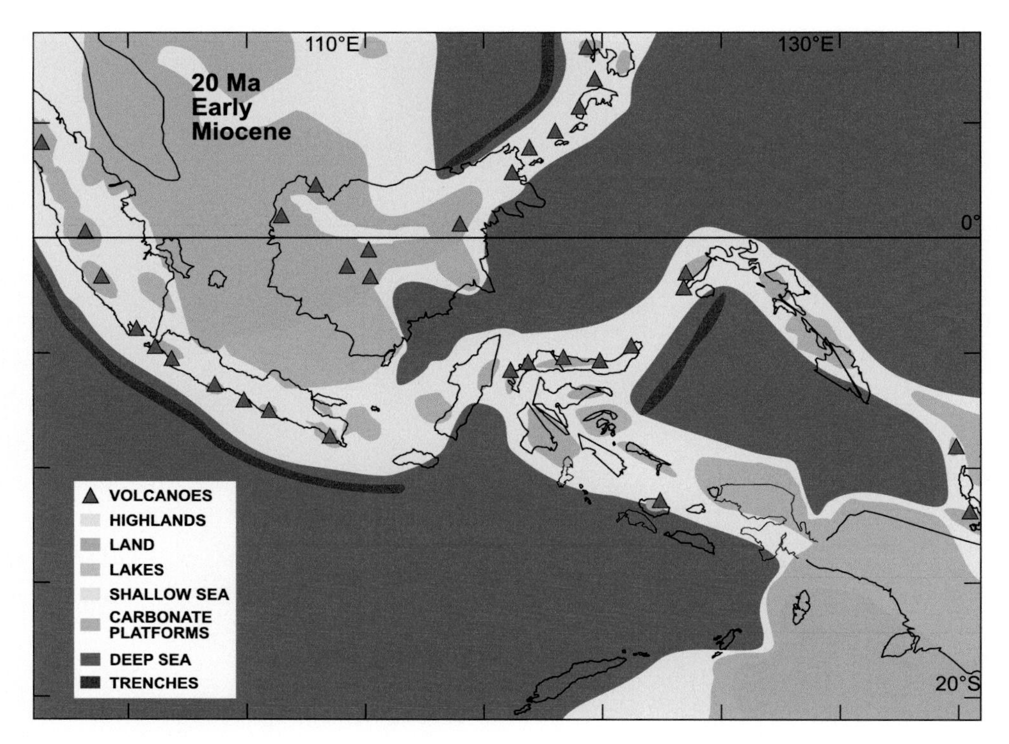

Figure 3.9 Palaeogeography of the region at 20 Ma. There was significant marine incursion onto the Sunda Shelf and extensive areas of carbonate build-ups throughout the region on the wide shallow shelves. Borneo became an important source of clastic sediments which began to pour into the deep offshore basins to the north, east and southeast. It is likely that the flow of water from the Pacific to the Indian Ocean was significantly reduced from 20 Ma until after 10 Ma. Much of Wallacea between Sulawesi and the Bird's Head was the site of shallow marine carbonate deposition and may have been emergent in some areas. See plate section for colour version.

3.6.1 Banda: Early Miocene collision

One important influence on the way in which the Australia–Southeast Asia collision developed was the nature of the Australian margin. The Jurassic rifting had led to formation of a continental promontory, the Sula Spur (Klompé 1954), that extended west from New Guinea on the north side of the Banda embayment. Parts of it are now present in East, Central and Southeast Sulawesi, the Banggai-Sula Islands, Buru, Seram but also in submerged ridges of the Banda Sea, small islands of the outer arc, and possibly in Timor. Because of collision-related deformation and subsequent fragmentation it is impossible to know the detailed palaeogeography of the spur, and in particular if there was any land. The limited evidence records a change from terrestrial to marginal marine sedimentation in the Early to Middle Jurassic and marine transgression later in the Jurassic. Cretaceous rocks from different parts of the spur are marine and suggest quiet deep waters on a continental margin (Cornée et al. 1995) west of the Bird's Head. The evidence suggests that the Sula Spur was submerged during the late Jurassic and Cretaceous. However, in the Banggai-Sula Islands, Cretaceous–Paleocene rocks are overlain in the west by flat-lying Eocene to Recent shallow marine carbonates with a slight angular unconformity, and further east there are no carbonates but the Mesozoic sequence has been eroded to expose a basement window. Seismic lines offshore show no significant erosional products to indicate that the elevation of the Sula Islands (over 1000 m) occurred during the Late Cenozoic. These observations suggest that parts of the islands, and possibly other parts of the spur, could have become emergent during the Eocene and remained as land until the present (Figs 3.6 to 3.12).

A widely held view of the region is that small fragments of continental crust were sliced from the Bird's Head during the Neogene, moved along a left-lateral strike-slip zone and have travelled west to collide with Sulawesi. This concept originated with Hamilton (1979) and has become famous as the 'bacon-slicer' and is incorporated in many tectonic models and reconstructions. For geologists, the supposed collisions (often vaguely dated) are used to explain all sorts of tectonic events throughout Wallacea and Sundaland, despite the very small size of the blocks. The model has been popular with biogeographers as it may explain dispersal of organisms from Australia as passengers on small 'arks' (e.g. McKenna 1973). I now doubt it for many reasons. For example, at the front of the blocks now in Sulawesi there should be a volcanic arc formed above the subducting ocean as the blocks moved west, typically identified with the West Sulawesi plutonic–volcanic arc, or a volcanic arc in the Togian Islands of Tomini Bay for the Banggai-Sula block. It is now known that there was little or no Mio-Pliocene volcanic activity in West Sulawesi, and that which there was in West Sulawesi and the Togian Islands does not have a subduction character (Cottam et al. 2011). There is much other evidence inconsistent with this simple tectonic model, such as the absence of through-going strike-slip faults (Watkinson et al. 2011), timing of slicing and collisions, and the presence of young ocean basins in the collision zone.

Instead, I suggest that soon after 25 Ma the Sula Spur began to collide with the North Sulawesi volcanic arc and this was the first part of the Australian continent to make contact with the Sundaland margin. An animated reconstruction of this collision and its consequences is given by Spakman and Hall (2010). Ophiolites in East Sulawesi (Kündig 1956, Silver et al. 1983) derived from the ocean north of the Sula Spur and probably from the North Sulawesi forearc, were thrust onto the continental crust of the Sula Spur. Ophiolite debris is found in probable Early Miocene terrestrial and marginal marine sediments (Surono 1995) in Southeast Sulawesi. By the Early Miocene there was Australian crust in East and Southeast Sulawesi, which was connected to Australia via the Bird's Head, although this does not mean there was an emergent or continuous terrestrial connection (Figs 3.8, 3.9).

Between 25 and 15 Ma the continued northward movement of the Indian–Australian plate was absorbed in several ways: subduction of Indian Ocean crust at the Java Trench; subduction of the Proto-South China Sea; widespread non-rigid counter-clockwise rotation of Sundaland (Borneo, West Sulawesi, Java); internal deformation of Sundaland; and contraction, uplift and erosion in East and Southeast Sulawesi. Sulawesi remains something of an enigma during this period. In West Sulawesi there is an incomplete stratigraphic record but there is little indication of mountain building, no sign of a significant sediment input, and no major deformation. In South Sulawesi carbonate deposition continued, although it changed character from an extensive platform to dispersed build-ups. The nature of the boundary between West and East Sulawesi is obscure, and in East and Southeast Sulawesi there is little stratigraphic record.

One key to dating tectonic events in Sulawesi lies in the clastic sediments deposited after collision (Hall and Wilson 2000). Sediments first described by Sarasin and Sarasin (1901) in the Southeast Arm of Sulawesi were named the Celebes Molasse because of their supposed similarity to the molasse of the Swiss Alps. Wanner (1910) correlated similar rocks of the East Arm with the Southeast Arm and van Bemmelen (1949) used the term for clastic sediments of young Neogene age that are found all over Sulawesi. The Celebes Molasse is usually considered to have been deposited in the Early to Late Miocene interval, implying mountains, uplift and erosion, but most of the rocks assigned to it are undated, and the most reliably dated are Late Miocene and Pliocene. The inter-arm bays of Gorontalo and Bone contain thick sediment sequences that could have been eroded during the Early to Late Miocene from a mountainous region in East, Central and Southeast Sulawesi, and in many parts of Central and Southeast Sulawesi there are low grade metamorphic rocks and ophiolites suggesting substantial erosional removal of sedimentary cover rocks, but our present knowledge allows little more than the statement that there was an elevated land mass in Sulawesi during much of the Miocene.

3.6.2 Banda: Miocene extension

There was an important change in the Middle Miocene at about 15 Ma to widespread extension and major subsidence. This change was caused by the second major influence on the Australia–Sundaland collision: the Banda embayment. This was the oceanic area south of the Sula Spur left after the continental fragments now beneath Borneo, West Sulawesi, Sumba and East Java rifted from Northwest Australia in the Late Jurassic. It was part of the Australian plate and contained oceanic crust of Late Jurassic age. Its last remnant is the Argo Abyssal Plain southwest of Timor. From the Late Jurassic to the Neogene the embayment was surrounded by a passive continental margin that can be traced from the Exmouth plateau, via Timor, Tanimbar and Seram, to Southeast Sulawesi. Subduction of this embayment began, the Sula Spur became fragmented, new ocean basins were created, and continued rollback ultimately led to formation of the Banda Arc as we know it today (Spakman and Hall 2010).

Subduction is often thought of as a consequence of two plates converging, but can be viewed equally well as one plate sinking under the influence of gravity (e.g. Elsassar 1971, Molnar and Atwater 1978, Carlson and Melia 1984). As the lower plate sinks, the subduction hinge moves oceanward, away from the arc, in a process that has been described as subduction rollback, hinge retreat or rollback, trench retreat or rollback, or slab rollback (Hamilton 2007). The rollback induces extension in the upper plate to fill the space created.

As Australia moved north, the Java Trench subduction zone became aligned with the northern margin of the embayment, a tear fault developed from the western edge of the Sula Spur and propagated eastward along the continent–ocean boundary from about 15 Ma. As the tear moved east, the oceanic embayment began to sink rapidly and the subduction hinge began to roll back into the Banda embayment. The effect of rollback was dramatic extension of the region above the Banda slab, which included parts of the pre-collision Sundaland margin in West Sulawesi and the collided Australian crust of the Sula Spur. There was subsidence in the Banda forearc near Sumba (Fortuin et al. 1997, Rigg and Hall 2011), oceanic spreading began in the North Banda Sea (Hinschberger et al. 2000), Banda arc volcanism started, there was extension-related volcanic activity in West Sulawesi (Polvé et al. 1997), and extension formed a core complex in the Sulawesi North Arm (van Leeuwen et al. 2007). Core complexes are metamorphic rocks exhumed rapidly from the middle and lower continental crust by low angle faults that cut through the entire crust. As extension occurs on the faults the lower (footwall) block, with ductile deformed high-grade metamorphic rocks or granites, is exposed at the surface and may be overlain by remnants of much lower grade metamorphic or unmetamorphosed rocks of the upper (hangingwall) block, typically characterised by brittle deformation.

Extension is interpreted to have occurred in three important phases. The earliest phase led to formation of the North Banda Sea between 12.5 and 7 Ma (Hinschberger et al. 2000). Continental crust from the Sula Spur was extended above the subduction hinge and separated from what remains in East and Southeast Sulawesi. Some of this crust remains in the Banda Ridges, and some formed part of the basement of the Banda volcanic arc and its forearc east of Flores. The eastern part of this volcanic arc, from east of Wetar to Seram, was active for a short period (*c.* 8–5 Ma) before a second major phase of extension led to formation of the South Banda Sea (Hinschberger et al. 2001). During opening of the South Banda Sea, the arc and continental crust was further extended to form the Banda forearc and is now found in Timor and several of the small outer arc islands from Leti to Babar (e.g. Bowin et al. 1980).

3.6.3 Banda: Pliocene collision

Volcanic arc activity continued in the Inner Banda Arc from Flores at least as far east as Wetar, but continued rollback of the subduction hinge led to collision between the southern passive margin of the Banda embayment and the volcanic arc which began in East Timor at about 4 Ma (Audley-Charles 1986, 2004) and led to termination of volcanic activity from Alor to Wetar. A remnant of the oceanic embayment remained to the east of Timor after collision and was subducted as rollback continued. This final phase of extension of the upper plate above the retreating hinge formed the Weber Deep, which subsided from forearc depths of about 3 km to its present-day depth of more than 7 km in the last 2 Myr. The very young volcanoes in the eastern part of arc from Damar to Banda (Abbott and Chamalaun 1981, Honthaas et al. 1998, 1999) record the final stage of rollback that accompanied formation of the Weber Deep.

In Timor and Sumba, the arc-continent collision was marked by a cessation of volcanic activity in the inner Banda arc in Wetar and Alor at 3 Ma (Abbott and Chamalaun 1981, Scotney et al. 2005, Herrington et al. 2010) and by the rapid uplift that followed collision which moved sedimentary rocks deposited at depths of several kilometres below sea level to their present positions of several kilometres above sea level (e.g. Fortuin et al. 1997, Audley-Charles 2011). Islands such as Savu and Roti have emerged even more recently and continue to rise.

A number of points are worth reiterating for the Banda region. The term collision is often used in different ways. The first contact between the Australian continent (Sula Spur) and the Sundaland margin was in the Early Miocene in Sulawesi, but collision between the continental margin (Northwest Shelf) and the Banda volcanic arc in Timor did not occur until the Pliocene. This can now be understood as a consequence of the shape of the Australian continental margin and the rollback of the subduction zone into the Banda embayment (Spakman and Hall 2010). Land has emerged at different times since the early Miocene, and many parts of Wallacea have emerged at very high rates. For example, in Timor Audley-Charles (1986) estimated average rates of uplift of 1.5 km/Myr and Quaternary limestones are mapped in West Timor at elevations above 1 km (Suwitodirdjo and Tjokrosapoetro 1975). Other large islands such as Seram, Sumba and parts of Sulawesi have emerged at similar rates. However, at the same time, basins such as the South Banda Sea and Weber Deep have subsided at similar rates. The geography of this critical region has changed dramatically in the last 5 Myr.

3.6.4 Sulawesi

During the Miocene and Pliocene there were significant vertical movements and the palaeogeography of Sulawesi changed significantly (Figs 3.10 to 3.12). This is

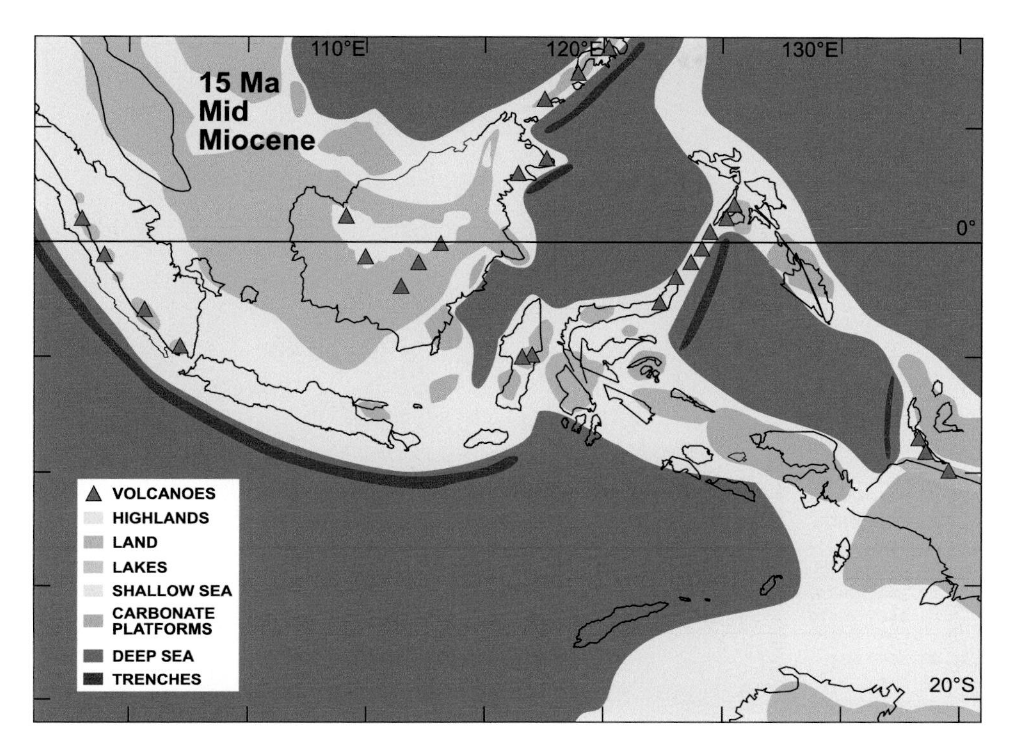

Figure 3.10 Palaeogeography of the region at 15 Ma. Soon after 15 Ma the Java Trench subduction zone began to propagate east as it tore along to the continent ocean boundary south of the Sula Spur and Bird's Head. Rollback of the subduction hinge caused major extension in Sulawesi and fragmented the Sula Spur. See plate section for colour version.

true for most of North and East Sulawesi, possibly for Southeast Sulawesi, and certainly for the major enigmatic inter-arm basins of Gorontalo Bay and Bone Bay. West Sulawesi and the North and East Arms are striking in their exceptional elevations (up to 3 km) within short distances of the coast, and the narrow width of these elevated areas. The timing of subsidence, uplift and exhumation is uncertain. In North Sulawesi metamorphic ages may record Early Miocene collision of the Sula Spur. Throughout North and West Sulawesi there is evidence for extension from the Middle Miocene, beginning at about 15 Ma, which I interpret to be driven by subduction rollback in the Banda embayment. The relief and distribution of land are uncertain.

The Makassar Straits was a pre-existing deep-water area into which sediment was transported, but seismic lines across the northern margin of the Paternoster platform indicate at least 1 km of additional subsidence of the North Makassar basin at the end of the Miocene. The subsidence is the same age as the rapid exhumation on land and there is probably a causal link between them. Throughout West,

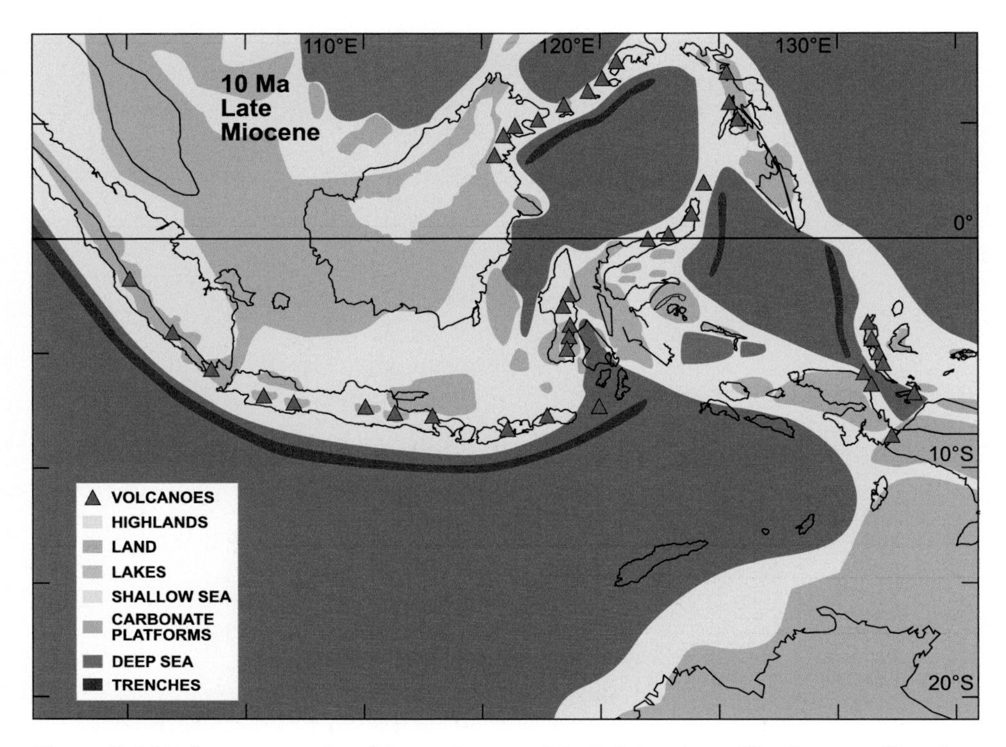

Figure 3.11 Palaeogeography of the region at 10 Ma. Subduction rollback was well under way and extension in Sulawesi led to volcanic activity, subsidence in Bone Gulf and oceanic crust formation in the North Banda Sea. Borneo was now a significantly emergent and elevated area. Wallacea probably had a much more complex palaeogeography than shown as the Sulu Spur was fragmented. Note that this fragmentation was not the result of slicing of fragments from the Bird's Head but driven by rollback of the subduction hinge into the Banda embayment. See plate section for colour version.

North and East Sulawesi there is evidence for a significant vertical motion on land at about 5 Ma. There was clearly a major increase in output of clastic sediment at this time (e.g. van Bemmelen 1949, Garrard et al. 1988, Davies 1990, Calvert 2000, Calvert and Hall 2007).

In West Sulawesi shallow marine Miocene sedimentary rocks deposited on a shelf on the east side of the Makassar Straits are overlain by Pliocene coarse terrestrial clastics derived from the east (Calvert and Hall 2007). Since 5 Ma there has been a major increase in land area and a significant change in elevation. The deep valleys incised into steep mountains expose deep crustal rocks such as garnet granulites and eclogites, intruded by young granites, in the Palu area (Watkinson 2011), and probably throughout West Sulawesi. Rapid uplift and exhumation, which began about 5 Ma recorded by K-Ar and apatite fission track

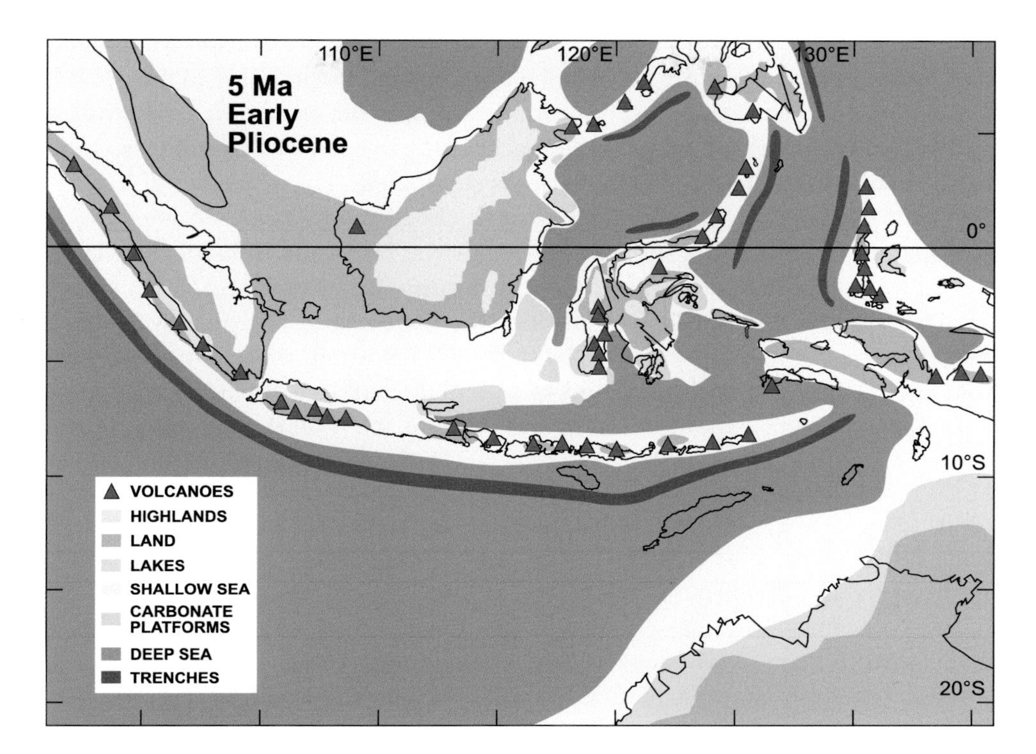

Figure 3.12 Palaeogeography of the region at 5 Ma. New deep oceanic basins had formed throughout Wallacea. Oceanic spreading in the North Banda Sea had ceased but had begun in the South Banda Sea. Deepening of the Flores Sea is inferred. At the same time substantial and rapid elevation began in Sulawesi, as rollback commenced at the North Sulawesi Trench driving core complex formation in the neck and East Arm, but Goontalo Bay was about to subside to present depths of up to 2 km below sea level. Uplift was about to begin in Timor as the Banda volcanic arc collided with the Australian continental margin in Timor. The last 5 Myr has seen significant changes in palaeogeography and dramatic uplift and subsidence. Seram and Timor are the two largest islands in the Banda region to have emerged in this interval. See plate section for colour version.

ages (Bellier et al. 2006), provided significant sediment to a young west-vergent offshore fold and thrust belt in the Makassar Straits.

In Gorontalo Bay, spectacular very young and rapid subsidence is recorded by numerous pinnacle reefs now found within a range of water depths between 1–2 km many of which, despite the high rates of sediment supply, are not buried by sediment. Alluvial fan deposits on land in the Togian Islands (Cottam et al. 2011) are separated from their equivalents on the East Arm by water depths up to 1.5 km (Jablonksi et al. 2007) and 25 km further south elevations on land exceed 2 km. Offshore east of East Arm are probable platform carbonates with no sediment cover at water depths of more than 1 km. They are likely to be Middle and/or

Upper Miocene by comparison with limestones beneath the Celebes Molasse in the Togian Islands, implying subsidence of 1 km or more probably in less than 5 Myr. I interpret the subsidence in Gorontalo Bay, and the extension occurring on land today forming core complexes (Spencer 2010), to be driven by rollback at the North Sulawesi trench as the Celebes Sea is subducted.

Seismic lines across Bone Bay also show thick sediments and a similar subsidence history although this subsidence was driven by Banda rollback. However, west of Bone Bay in South Sulawesi is the Tonasa–Tacipi platform/shelf, where there has been carbonate deposition since the Eocene (e.g. Wilson and Bosence 1996, Ascaria 1997, Ascaria et al. 1997). As with Borneo, there is a difference from north to south from young and spectacular uplift and subsidence to areas that appear to have remained stable and close to sea level during much of the Cenozoic.

3.7 Miocene to Recent: Pacific arcs and northern Australia

At the eastern edge of Wallacea are the Halmahera and Sangihe volcanic arcs, which are the only example of an arc–arc collision in the world. Both of the currently active arcs formed during the Neogene. The Sangihe Arc can be traced from Sulawesi northwards into the Philippines. In biogeographic terms it is within Wallacea but in geological terms is arguably part of Neogene Sundaland. It was constructed on Eocene oceanic crust of the Celebes Sea but in the Neogene formed a continuation of the North Arm of Sulawesi, which had collided with the Sula Spur. The modern Halmahera Arc is relatively recently arrived in Wallacea. It is constructed on older intra-oceanic arcs formed in the Pacific, which were part of the Philippine Sea Plate. Between 45 and 25 Ma the Philippines–Halmahera Arc developed above a north-dipping subduction zone where there was subduction of oceanic lithosphere as Australia moved north (Hall 1996, 2002, Hall and Spakman 2002). After collision of this arc with northern Australia (*c.* 25 Ma) it moved west along the New Guinea margin north of the left-lateral Sorong fault zone (Hall et al. 1995a, b).

3.7.1 North Moluccas

The history of the Sorong Fault in New Guinea is not well known. It juxtaposes arc and ophiolitic rocks of Pacific/Philippine Sea Plate origin against Australian continental crust in the New Guinea mobile belt (Pieters et al. 1983, Dow and Sukamto 1984). The fault is an irregular wide zone of strands within which there was transpression and transtension, meaning local uplift and subsidence along the fault. Most of the rocks within the fault zone are oceanic and fragments from the Pacific, although some small slices of continental crust were caught up in fault strands. All these rocks have been carried westwards along the fault zone.

The islands of the North Moluccas are part of the mobile belt and include both arcs and potential arks. The group of islands northwest of the Bird's Head between Halmahera and Waigeo have Middle to Upper Miocene limestones above arc and oceanic basement, a widespread feature of northern New Guinea, indicating that for most of the Miocene this region was close to sea level but mainly submerged (Figs 3.9 to 3.11).

Initiation of east-directed Halmahera subduction probably resulted from locking of strands of the left-lateral Sorong fault zone. The present-day Molucca Sea double subduction system was initiated at about 15 Ma, and the oldest Neogene volcanic rocks in the Halmahera Arc have ages of about 11 Myr (Baker and Malaihollo 1996). Active volcanism would have formed islands similar to the present offshore islands of the Halmahera arc such as Ternate and Tidore. Since 11 Ma the Molucca Sea has since been eliminated by subduction at both its eastern and western sides. The two arcs first came into contact at about 3 Ma and began to build the central Molucca Sea accretionary complex as the two forearcs collided (Hall 2000). Since then, Halmahera has become emergent and the areas of land in the North Moluccas have increased significantly.

3.7.2 New Guinea

New Guinea is not part of Wallacea, but in geological terms the northern part certainly belongs with the North Moluccas, and the Bird's Head has an important connection with the Banda Arc in Seram. In New Guinea there is now a substantial chain of mountains, the New Guinea Highlands, up to 5 km high running from west to east with a coastal plain and tropical swamps to the north and the south. These mountains have risen, and the island has acquired its present form, only in the last 5 Myr (Struckmeyer et al. 1993, Hill and Raza 1999, Hill and Hall 2003, Cloos et al. 2005, van Ufford and Cloos 2005). For most of the Cenozoic the New Guinea Limestones, which now form the high mountains, were part of a wide and long-lived carbonate shelf north of the Australian land mass (Dow 1977, Pieters et al. 1983). Before about 5 Ma, to the north of this shallow marine shelf, it is likely that there were emergent islands in the northern New Guinea mobile belt (Pigram and Davies 1987, Audley-Charles 1991, Crowhurst et al. 1996) as the Philippine Sea Plate moved west, which may have permitted dispersal of faunas and floras. It is likely for most of the Neogene that the palaeogeography resembled the region of small islands currently within splays of the Sorong Fault zone east and west of the Bird's Head but separated from land to the south by a wide, albeit shallow, marine gap. However, the New Guinea Limestone sequences (e.g. Pieters et al. 1983, Carman and Carman 1990, 1993, Brash et al. 1991, Buchanan

1996, Buchanan et al. 2000) include numerous minor unconformities, hard grounds, minor clastic intervals, and signs of karstic alteration suggesting intermittent emergence. It is impossible at present to map the palaeogeography in detail; it seems improbable that the entire shelf was ever completely emergent, but it is likely that the shelf was an area of numerous low islands during most of the Cenozoic.

3.7.3 Bird's Head

The Bird's Head of New Guinea is in a critical position between Australia and Wallacea. At present it is largely aseismic but surrounded by shallow seismicity (hypocentres shallower than 70 km). GPS studies indicate it is moving southwest towards Seram relative to Australia. The Bird's Head has been relatively little studied since regional mapping by Australian–Indonesian teams in the early 1980s (Pieters et al. 1983, Dow and Sukamto 1984, Dow et al. 1986). The northern Bird's Head is crossed by strands of the Sorong Fault system. To the north is the mobile belt and to the south is Australian continental crust. East of the Bird's Head is a triangular embayment, Cenderawasih Bay, with a thick sediment cover above a basement of unknown type (Dow and Hartono 1982). On the east side is a zone of young deformation on the seafloor adjacent to the Bird's 'body', and to the east is the Wandaman peninsula, which includes high grade metamorphic rocks with very young metamorphic ages, and west of that the west-directed Lengguru fold and thrust belt of Late Miocene–Pliocene age (Dow et al. 1985, Bailly et al. 2009, Sapin et al. 2009).

The northeast part of the Bird's Head includes mountains, which supplied sediment to the southwest from at least the Late Miocene. On its southwest side is a topographic ridge, the Misool–Onin–Kumawa Ridge (Fraser et al. 1993, Pairault et al. 2003, Sapin et al. 2009), which is a broad anticline. This was an emergent feature in the Early Pliocene and a widespread unconformity suggests much of the Bird's Head was emergent at times during and since the Late Miocene. In the offshore region near to Misool, the importance of this unconformity was recognised from seismic data only a few years ago (Pairault et al. 2003). New seismic and detailed bathymetric information, as yet unpublished, from the offshore areas including Cenderawasih Bay to the east, the Seram Trough and the Misool–Onin–Kumawa Ridge to the south, and the Sorong fault zone to the west is now adding to our knowledge and changing our picture of the Bird's Head development, which is still not well understood. It is too early to be certain but recent work is highlighting important palaeogeographic changes including more, and more persistent, areas of land during the late Neogene. It is possible that emergent terrestrial dispersal routes from Australia, via the Bird's Head and Sula Spur into Wallacea, were more important and more enduring than previously suggested.

3.8 Miocene to Recent: Sundaland collision, uplift and subsidence

While the most important geological changes were in Wallacea and northern New Guinea as Australia moved north, there were also significant events in Sundaland, notably, but not only, in Borneo.

3.8.1 Borneo collision

Subduction ceased in the Early Miocene when the thinned passive margin of the Dangerous Grounds underthrust northern Borneo causing deformation, uplift and crustal thickening (Hutchison et al. 2000, Hall and Wilson 2000). Deep marine sedimentation stopped and the sediments of the Crocker Fan were exposed on land. In parts of central Borneo there was some Miocene magmatism, but volcanic activity largely ceased following arc-continent collision after complete subduction of the proto-South China Sea. The collision resulted in uplift of much of the interior of Borneo. Most of the deep Cenozoic basins around Borneo are filled by sediments (van Hattum et al. 2006) carried from Borneo by large rivers as the island emerged and land area increased through the Neogene. However, there are differences between the areas north and south of the Paternoster–Lupar lineaments.

3.8.2 Southern Borneo

The area south of the Paternoster–Lupar lineaments has a continental basement rifted from the Australian margin in the Late Jurassic inferred to have a thick strong and cold lithosphere. With the exception of the Meratus Mountains this area has remained close to sea level since the Middle Eocene.

In east Kalimantan, gold mineralisation is associated with Early Miocene magmatism and hydrothermal alteration in volcanic settings similar to those of the present-day North Island of New Zealand (Davies et al. 2008). The gold belt appears never to have been deeply buried (T. van Leeuwen pers. comm. 2008) and was exhumed in the Pleistocene.

Southeast of the gold belt, the Meratus Mountains are a narrow zone of deformation, which may be a reactivated strike-slip suture in the basement, probably elevated between the Late Miocene to Pleistocene. West of the Meratus Mountains is a broad downwarp, the Barito Basin, filled by Eocene to Miocene terrestrial to marginal marine clastic sediments and shallow marine limestones, whereas to the east is the Asem-Asem Basin and the long-lived Eocene to Miocene Paternoster–Tonasa carbonate platform. Both areas are still largely undeformed. The Asem-Asem Basin was the eastern part of the Barito Basin until uplift of the Meratus Mountains. Seismic lines across the Paternoster Platform and field studies on land show that Eocene to Recent largely shallow marine carbonates are of the order of 1–2 km in thickness and record vertical movements relative to sea level of much smaller amounts over 40 Myr. Apart from the Meratus Mountains, southern Borneo has subsided to allow sediment to accumulate, but the surface has always been close to sea level so expansion of the area of shallow marine carbonates, notably during the Oligocene, records eustatic sea-level rises.

3.8.3 Northern Borneo

In contrast, the rest of Borneo has supplied large amounts of sediment to basins on land and offshore. Early Miocene rise of the Central Borneo Mountains shed sediment first into deltas of the Kutai and Sandakan Basins and later into the Tarakan and Baram Basins. In the Kutai Basin sediment was derived from erosion of the Borneo highlands and inversion of older parts of the basin. Inversion, emergence and erosion migrated from west to east during the Early and Middle Miocene. A similar progressive outward movement of coastline is typical of most of Borneo north of the Paternoster–Lupar lineaments.

In Sabah the history is more complicated than a simple and continuous rise of mountains. At the end of the Early Miocene, Sabah was emergent but soon afterwards most of west and east Sabah subsided below sea level (Balaguru et al. 2003) probably leaving a narrow band of highlands along the present Crocker Ranges. Subsidence may be related to break-off of the subducted Proto-South China Sea slab, the load of a lithospheric root formed during the collision, or could be connected to Middle Miocene (Rangin and Silver 1991) opening of the Sulu Sea.

It is possible that this may have been driven by subduction rollback. In south Sabah there was a Middle and Late Miocene volcanic arc in the Dent and Semporna peninsulas (Kirk 1968) formed by north-directed subduction of the Celebes Sea (Chiang 2002). Plio-Pleistocene volcanism can be traced offshore into the Sulu arc but is basaltic and has an ocean island character (Macpherson et al. 2010) suggesting an important tectonic change in the Late Miocene.

During the Middle and Late Miocene in east Sabah, a large river system, flowing northeast, in a similar position to the present Kinabatangan River, deposited sediment in a delta and coastal plain complex extending across most of the area as far as Sandakan. Subsidence broadly kept pace with sediment supply and several kilometres of sediment accumulated. Northwest of the Crocker Ranges there was also deposition of thick sediments in deltas and coastal plains of north Sabah and Brunei by rivers flowing to the north or northwest. The Crocker Ranges were narrow at 15 Ma and the shelf edge moved northwards since then.

At the northern end of the Crocker Ranges is Mount Kinabalu, which is a granite pluton that crystallised several kilometres beneath the surface at about 7–8 Ma (Cottam et al. 2010), and is now exposed at the summit 4 km above sea level. Kinabalu has since been exhumed at an average rate of about 0.5 km/Myr. As Kinabalu and the Crocker Ranges grew, most of east Sabah emerged above sea level, and the former large sedimentary basin to the east was reduced in area by erosion leaving only circular remnants. In northern Sabah and Brunei, the shelf edge continued to move north. Although much of Sabah is topographically relatively low, it has risen from below sea level, the area of emergent land has increased and several kilometres of cover rocks have been removed by erosion. Kinabalu is one of the few mountains between New Guinea and the Himalayas that was capped by ice during the Pleistocene. It is likely that the summit was ice-covered and ice-free several times in the last 2 Myr. Carbon dating of wood in glacial tills shows that ice was present at about 1500 m at 35 ka, whereas dating of plant material in cores from small lakes shows the summit was free of ice by about 9.2 ka.

3.8.4 Dangerous Grounds

Northwest of Sabah between the coast and the Dangerous Grounds is the linear northeast–southwest-trending Northwest Borneo Trough with water depths of up to 3 km. It has been interpreted as an extinct subduction trench, an active trench and the site of thrusting but not subduction. The Dangerous Grounds are a rifted microcontinental fragment from the South China margin capped by locally emergent carbonate reefs, shoals and atolls that give the area its name. Between the Sabah coast and the trough is a thick Neogene sediment wedge in an offshore northward-directed fold and thrust belt.

Offshore hydrocarbon exploration studies have shown that there have been repeated gravitational failures near the shelf edge that are effectively huge submarine landslides. Steep fault scarps can be mapped, and there are large debris fields on the deep sea floor north of Brunei and Sabah that cover hundreds of square kilometres and include blocks up to 1 km across and 150 m high. Debris flows of similar size and character can be recognised in the subsurface, showing that these failures have been repeated numerous times during the last few millions of years. Furthermore, the sequences of debris flows and interbedded deep-water sediments, probably mainly turbidites, are folded and thrust northwards away from Sabah. New exploration work in the Sulu Sea indicates similar structures suggesting transport of material away from land to the northeast.

Seismic lines (Hutchison 2010) show elevated features within the Northwest Borneo Trough, now at water depths close to 3 km, capped by carbonates and pinnacle reefs that must have formed at sea level, indicating major subsidence. There is almost no seismicity associated with the trough, no volcanic activity on land, and nothing to indicate southward subduction; nor is there evidence for converging plates to produce the fold and thrust belt. Recent studies show thin crust beneath the offshore Neogene sediment wedge (Franke et al. 2008), implying thinning of crust previously thickened during the Early Miocene North Borneo collision.

I suggest all these observations indicate a link between young and rapid uplift on land, evidenced by exhumation of the 7–8 Ma Kinabalu granite now exposed at 4 km above sea level, and young and rapid subsidence offshore. Their importance from a biogeographic perspective is the demonstration of significant vertical movements within a few million years, and within the largely seismically and volcanically inactive interior of Sundaland. We do not yet understand their causes. This does not justify a return to mythical land bridges, which were discredited by the plate tectonic revolution (Tarling 1982), but it should alert geologists and biologists to the possibilities of faster palaeogeographical changes than previously expected, possibly driven by subduction rollback or deep crustal movements. Evidence from the biota, particularly freshwater organisms, and molecular techniques, could contribute in some areas to a better mapping of the past distribution of land where the geological record is silent for reasons explained in Hall (2001).

3.8.5 Sumatra and Java

From the Middle Miocene, Sumatra gradually and steadily became the large island of today as the Barisan Mountains rose and widened. A number of factors have contributed and Barber et al. (2005) attribute most of the changes in palaeogeography to regional processes. Sumatra is underlain by continental crust and the processes that led to basin formation never caused substantial and widespread subsidence below sea level. Global sea level has fallen since the Early Miocene. Volcanic activity and strike-slip faulting have contributed to the elevation of the Barisan Mountains. Inversion and uplift may also be due to region-wide Sundaland deformation following Australian collision in east Indonesia. Features on the subducting Indian Plate may have contributed to elevation of Sumatra and its forearc. Thick sediments to the west of the Investigator Fracture which were part of the huge Bengal–Nicobar fan have been subducted beneath northern Sumatra, and the Investigator Ridge itself is likely to have played a role, as suggested by the coincidence in the inferred position of the subducted ridge and the Toba volcanic centre. There are several elevated linear features broadly parallel to the Investigator Ridge on the downgoing plate, and it is plausible that they too are implicated in the elevation of the forearc islands, certainly from Simeulue to the Mentawai Islands. From Nias and Siberut the forearc south of the Sumatran coast is extremely shallow and there is almost no deep basin between the forearc high and the coast, as there is to the northwest and southeast. Several of the large islands, such as Nias and Siberut, were probably connected at times to the Sumatran mainland although this is less likely for the islands further southeast, and improbable for the small island of Enggano at the south end of the forearc high.

In contrast to Sumatra, Java became the large island of today more abruptly and more recently. The Paleogene Southern Mountains volcanic arc ceased activity at about 20 Ma, possibly with a Toba-scale explosive eruption not far from present-day Merapi. Northern Java and the Java Sea was a shallow shelf that was close to sea level and intermittently emergent, whereas further south there was widespread carbonate deposition on the eroded Paleogene arc. Volcanic activity resumed in the Late Miocene at about 10 Ma in a chain of volcanic islands. The position of the volcanic arc moved abruptly northwards at about 7 Ma, and this movement was associated with a widespread episode of thrusting throughout Java, with the most displacement in the west leading to the emergence of most of West Java. The deformation and shift in the arc position is interpreted as a result of the arrival of a buoyant basaltic plateau, similar to the Roo Rise, at the subduction trench, which caused volcanic activity to cease for an interval. Unusual K-rich volcanoes erupted near the edge of the Sunda Shelf during this interval and normal arc volcanism resumed only in the last 1–2 Myr and broadly coincides with the emergence of East Java, which has made Java the elongate large island of today. This picture of the late Neogene emergence of the large island of Java is consistent with palaeogeographic inferences from mammals (van den Bergh et al. 2001).

3.9 Pleistocene change

There were major changes in sea level during the Pleistocene but it was primarily the shelf areas of Sunda and Australia that changed. Simply using a modern bathymetric map and changing sea level is a good guide to what happened (Voris 2000). The present islands around the Sunda Shelf were connected at times and the entire shelf was emergent. The same was true for the Sahul Shelf. However, at the edge of Sundaland the Makassar Straits remained an important topographic barrier although it was narrower, which, because it is the main passageway for Pacific waters, could have made it more difficult for biota to cross because currents could have been much faster.

Glacially driven sea-level change would have had little effect on Wallacea where shelves surrounding islands are narrow, and shelf edges very steep. Modelling the consequences of sea-level change using modern bathymetry is problematical because tectonic change has been substantial on geologically short time scales. In Wallacea, and possibly in north Borneo, average rates of vertical movements of the order of 1.5 km/Myr have been recorded in areas such as Timor (Audley-Charles 1986), Sumba (Fortuin et al. 1997), Savu and Roti (Roosmawati and Harris 2009), with locally higher rates for some intervals. Such movements indicate the possibility of substantial palaeogeographical change on the scale of 1 Myr. The geography of Wallacea has changed very substantially in the last 5 Myr, and the overall trend has been a significant increase in areas of land, and an increase in elevation (3 km above sea level is not unusual). However, recent work in and around Sulawesi suggests surprisingly rapid subsidence of inter-arm bays such as Gorontalo Bay. It still seems unlikely that there were continuous connections between most east Indonesian islands, and most of the islands of the Banda Arc have emerged in the last few million years, but it is probable that there were substantial areas of land between the Bird's Head and Sulawesi since the first collision in the Early Miocene. The palaeogeographical maps (Figs 3.8 to 3.12) may underestimate the extent of land.

3.10 Conclusions

Southeast Asia is an unusual region. Plate tectonics provide a first order description of the region's history, and a microplate model does help explain how the region has grown by the addition of large and small fragments that include rifted continental crust and island arcs. However, Sundaland has a complex internal structure, with strong areas of old continental fragments and oceanic crust within a much weaker crust. It is not a craton or shield. The term Sunda Shield has been inherited from the period before the plate tectonic revolution and is less commonly used today, but the term craton continues to be misused by geologists. Except for an extensive flat area of the Sunda Shelf, Sundaland has nothing in common with the cratons and shields of South and North America, the Baltic, Africa, India and Australia, which are ancient and strong parts of the continents.

Sundaland was not a rigid plate and Wallacea is part of the complex plate boundary zone between this overall weak continental region and the strong surrounding plates of the Pacific, Philippine Sea, Australia and India. The shape of the Australian margin, in particular the Banda embayment, has been an important influence on the development of the region and subduction rollback into the embayment led to formation of young ocean basins at the heart of the collision zone. The heterogeneous character of the Sundaland continent has influenced the way in which different parts of the region have responded to plate tectonic forces. In the weaker areas, the upper crust appears to be deforming almost independently of the deeper lithosphere, and in eastern Indonesia there have been exceptionally high rates of vertical movements, both up and down.

We need to look at the region in new ways to understand its palaeogeography and the consequences for biogeography. The microplate or terrane approach, when applied to smaller and smaller areas, is less and less useful and biologists need to exercise caution in applying such models to explain their patterns. The concept of slicing fragments from New Guinea followed by multiple collisions is increasingly implausible, and new models (Spakman and Hall 2010) interpret the microcontinental fragments to be the result of a larger coherent continental spur that has been fragmented by extension driven by subduction rollback. Different parts of Borneo and East Indonesia have risen and subsided since the Early

Miocene following Australia's initial contact with Southeast Asia. Very substantial change in relief began at about 8 Ma in northern Borneo, and at about 5 Ma in West, North and East Sulawesi. In Timor, the Halmahera Islands, and probably in Seram, significant relief changes are even younger. Subduction rollback has caused some of this rapid subsidence and uplift, but some of the causes of vertical movements appear to be related to movements within the crust not easily linked to specific tectonic mechanisms, and rates of changes are greater than would be expected from conventional geological models. An important conclusion is that the palaeogeography has changed more quickly in and around Wallacea, particularly in the last 15 Myr, than would be expected from geological models developed in other tectonic settings elsewhere in the world. Biogeographically, although terranes may not have behaved as arks, dispersal may have been possible by other means. Rapid vertical movements may mean there was more land than previously expected and perhaps more short-lived stepping stones for terrestrial dispersal and biogeographical convergence. Conversely, for other organisms, or at other times, such rapid changes could have disrupted populations, habitats and biogeographical continuity, resulting perhaps in local extinctions and biogeographical divergence by vicariance. In their discussion of different biogeographical models of Southeast Asia, Bellwood et al. (this volume, Chapter 9) consider that one possible cause of the region's high biodiversity was an ongoing combination of such frequent convergences and divergences in different organisms at different times (their 'dynamic mosaic').

The collision of Australia with the Southeast Asia margin has closed an important major marine passage but water still passes through from the Pacific to the Indian Ocean through a complex gateway that reflects the geological history of Wallacea. The Indonesian Throughflow is now a major topic of research because of its importance in the global oceanographic system and its likely contribution to global climate. The present distribution of plants and animals in the region is the result of many factors. Geology is at the base of pyramid, something Wallace recognised many years ago, and can explain some of the most obvious biogeographic patterns, but to understand the diversity and complexity of Wallacea we need to unravel the interplay and feedbacks between tectonics, palaeogeography, ocean and atmospheric circulation, and the record of life – something that has only just begun (e.g. Renema et al. 2008, Bellwood et al. this volume, Chapter 9).

Acknowledgements

Work by the Southeast Asia Research Group has been funded over many years by a consortium of oil companies who have been generous and open-minded in their support. In addition, our work has been funded at times by the University of

London Central Research Fund, the Natural Environment Research Council and the Royal Society. I thank Pusat Survei Geologi Bandung, Lemigas, Indonesian Institute of Sciences and Institut Teknologi Bandung, for assistance and many colleagues, friends and students in the UK, Europe and Southeast Asia for help and discussion. I am especially grateful to Bob Morley and Duncan Witts for comments on the palaeogeographical maps. I thank Brian Rosen and two anonymous reviewers for comments that assisted my attempt to make the geology comprehensible to non-geologists.

References

Abbott, M. J. and Chamalaun, F. H. (1981). Geochronology of some Banda Arc volcanics. In *The Geology and Tectonics of Eastern Indonesia*, ed. A. J. Barber and S. Wiryosujono. *Geological Research and Development Centre, Bandung, Special Publication*, 2, 253–68.

Abdullah, C. I., Rampnoux, J.-P., Bellon, H., Maury, R. C. and Soeria-Atmadja, R. (2000). The evolution of Sumba Island (Indonesia) revisited in the light of new data on the geochronology and geochemistry of the magmatic rocks. *Journal of Asian Earth Sciences*, **18**, 533–46.

Abdullah, N. T. (2009). Mesozoic stratigraphy. In *Geology of Peninsular Malaysia*, ed. C. S. Hutchison and D. N. K. Tan. Kuala Lumpur, Malaysia: University of Malaya and Geological Society of Malaysia, pp. 87–131.

Aitchison, J. C., Ali, J. R. and Davis, A. M. (2007a). When and where did India and Asia collide? *Journal of Geophysical Research*, **112**, B05423, doi:10.1029/2006JB004706.

Aitchison, J. C., McDermid, I. R. C., Ali, J. R., Davis, A. M. and Zyabrev, S. V. (2007b). Shoshonites in southern Tibet record Late Jurassic rifting of a Tethyan intraoceanic island arc. *Journal of Geology*, **115**, 197–218.

Ali, J. R. and Aitchison, J. C. (2008). Gondwana to Asia: plate tectonics, paleogeography and the biological connectivity of the Indian sub-continent from the Middle Jurassic through latest Eocene (166–35 Ma). *Earth Science Reviews*, **88**, 145–66.

Ascaria, A. N. (1997). *Carbonate facies development and sedimentary evolution of the Miocene Tacipi Formation, South Sulawesi, Indonesia*. PhD Thesis, University of London.

Ascaria, N. A., Harbury, N. and Wilson, M. E. J. (1997). Hydrocarbon potential and development of Miocene Knoll-Reefs, south Sulawesi. In *Proceedings of the International Conference on Petroleum Systems of SE Asia and Australia, Jakarta, Indonesia*, ed. J. V. C. Howes and R. A. Noble. Jakarta: Indonesian Petroleum Association, pp. 569–84.

Audley-Charles, M. G. (1986). Rates of Neogene and Quaternary tectonic movements in the Southern Banda Arc based on micropalaeontology. *Journal of the Geological Society of London*, **143**, 161–75.

Audley-Charles, M. G. (1991). Tectonics of the New Guinea area. *Earth and Planetary Sciences, Annual Review*, **19**, 17–41.

Audley-Charles, M. G. (2004). Ocean trench blocked and obliterated by Banda forearc collision with Australian proximal continental slope. *Tectonophysics*, **389**, 65–79.

Audley-Charles, M. G. (2011). Tectonic post-collision processes in Timor. In *The SE Asian Gateway: History and Tectonics of Australia–Asia Collision*, ed. R. Hall, M. A. Cottam and M. E. J. Wilson. *Geological Society of London Special Publication*, 355, 241–66.

Bailly, V., Pubellier, M., Ringenbach, J.-C., de Sigoyer, J. and Sapin, F. (2009). Jumps of deformation zones in a young convergent setting; the Lengguru fold-and-thrust belt, New Guinea Island. *Lithos*, **113**, 306–17.

Baker, S. and Malaihollo, J. (1996). Dating of Neogene igneous rocks in the Halmahera region: arc initiation and development. In *Tectonic Evolution of SE Asia*, ed. R. Hall and D. J. Blundell. *Geological Society of London Special Publication*, 106, 499–509.

Balaguru, A., Nichols, G. and Hall, R. (2003). Tertiary stratigraphy and basin evolution of southern Sabah: implications for the tectono-stratigraphic evolution of Sabah, Malaysia. *Bulletin of the Geological Society of Malaysia*, **47**, 27–49.

Barber, A. J. (2000). The origin of the Woyla Terranes in Sumatra and the Late Mesozoic evolution of the Sundaland margin. *Journal of Asian Earth Sciences*, **18**, 713–38.

Barber, A. J. and Crow, M. J. (2009). The structure of Sumatra and its implications for the tectonic assembly of Southeast Asia and the destruction of Paleotethys. *Island Arc*, **18**, 3–20.

Barber, A. J., Crow, M. J. and de Smet, M. E. M. (2005). Chapter 14: Tectonic evolution. In *Sumatra: Geology, Resources and Tectonic Evolution*, ed. A. J. Barber, M. J. Crow and J. S. Milsom. *Geological Society of London Memoir*, 31, 234–59.

Bellier, O., Sébrier, M., Seward, D. et al. (2006). Fission track and fault kinematics analyses for new insight into the Late Cenozoic tectonic regime changes in West-Central Sulawesi (Indonesia). *Tectonophysics*, **413**, 201–20.

Bowin, C., Purdy, G. M., Johnston, C. et al. (1980). Arc-continent collision in the Banda Sea region. *American Association of Petroleum Geologists Bulletin*, **64**, 868–918.

Brash, R. A., Henage, L. F., Harahap, B. H., Moffat, D. T. and Tauer, R. W. (1991). Stratigraphy and depositional history of the New Guinea Limestone Group, Lengguru, Irian Jaya. In *Proceedings of the Indonesian Petroleum Association 20th Annual Convention*. Jakarta: Indonesian Petroleum Association, pp. 67–84.

Briais, A., Patriat, P. and Tapponnier, P. (1993). Updated interpretation of magnetic anomalies and sea floor spreading stages in the South China Sea: implications for the Tertiary tectonics of Southeast Asia. *Journal of Geophysical Research*, **98**, 6299–328.

Buchanan, P. G. ed. (1996). *Petroleum Exploration, Development and Production in Papua New Guinea: Proceedings of the Third PNG Petroleum Convention, Port Moresby.* Port Moresby, Papua New Guinea: PNG Chamber of Mines and Petroleum.

Buchanan, P. G., Grainge, A. M. and Thornton, R. C. N. eds. (2000). *Papua New Guinea's Petroleum Industry In The 21st Century. Based On The Proceedings Of The Fourth PNG Petroleum Convention Organised by the PNG Chamber of Mines and Petroleum in Port Moresby, Papua New Guinea, 29–31 May, 2000*. Port Moresby, Papua New Guinea: PNG Chamber of Mines and Petroleum.

Burollet, P. F. and Salle, C. (1981). A contribution to the geological study of Sumba (Indonesia). In *Proceedings of the Indonesian Petroleum Association 10th Annual Convention*. Jakarta: Indonesian Petroleum Association, pp. 331–44.

Calvert, S. J. (2000). *The Cenozoic geology of the Lariang and Karama Regions, Western Sulawesi, Indonesia*. PhD Thesis, University of London.

Calvert, S. J. and Hall, R. (2007). Cenozoic evolution of the Lariang and Karama regions, North Makassar Basin, western Sulawesi, Indonesia. *Petroleum Geoscience*, **13**, 353–68.

Carlson, R. L. and Melia, P. J. (1984). Subduction hinge migration. *Tectonophysics*, **102**, 399–411.

Carman, G. J. and Carman, Z. eds. (1990). *Petroleum Exploration in Papua New Guinea. Based on the Proceedings of the First PNG Petroleum Convention Organised by the PNG Chamber of Mines and Petroleum in Port Moresby, Papua New Guinea, 12–14 February, 1990*. Port Moresby, Papua New Guinea: PNG Chamber of Mines and Petroleum.

Carman, G. J. and Carman, Z. eds. (1993). *Petroleum Exploration and Development in Papua New Guinea. Based on the Proceedings of the Second PNG Petroleum Convention Organised by the PNG Chamber of Mines and Petroleum in Port Moresby, Papua New Guinea, 31 May–2 June 1993*. Port Moresby, Papua New Guinea: PNG Chamber of Mines and Petroleum.

Chiang, K. K. (2002). *Geochemistry of the Cenozoic igneous rocks of Borneo and tectonic implications*. PhD Thesis, University of London.

Clements, B., Burgess, P. M., Hall, R. and Cottam, M. A. (2011). Subsidence and uplift by slab-related mantle dynamics: A driving mechanism for the Late Cretaceous and Cenozoic Evolution of Continental SE Asia? In *The SE Asian Gateway: History and Tectonics of Australia–Asia Collision*, ed. R. Hall, M. A. Cottam and M. E. J. Wilson. *Geological Society of London Special Publication*, 355, 37–51.

Cloos, M., Sapiie, B., Quarles van Ufford, A. et al. (2005). Collisional delamination in New Guinea: the geotectonics of subducting slab breakoff. *Geological Society of America Special Paper*, **400**, 1–51.

Corneé, J.-J., Tronchetti, G., Villeneuve, M. et al. (1995). Cretaceous of eastern and southeastern Sulawesi (Indonesia): new micropalaeontological and biostratigraphical data. *Journal of Southeast Asian Earth Sciences*, **12**, 41–52.

Cottam, M. A., Hall, R., Sperber, C. and Armstrong, R. (2010). Pulsed emplacement of the Mount Kinabalu Granite, North Borneo. *Journal of the Geological Society of London*, **167**, 49–60.

Cottam, M. A., Hall, R., Forster, M. and BouDagher-Fadel, M. B. (2011). Basement character and basin formation in Gorontalo Bay, Sulawesi, Indonesia: new observations from the Togian Islands. In *The SE Asian Gateway: History and Tectonics of Australia–Asia Collision*, ed. R. Hall, M. A. Cottam and M. E. J. Wilson. *Geological Society of London Special Publication*, 355, 177–202.

Crow, M. J. (2005). Chapter 8: Tertiary volcanicity. In *Sumatra: Geology,*

Resources and Tectonic Evolution, ed. A. J. Barber, M. J. Crow and J. S. Milsom. *Geological Society London Memoir*, 31, 98–119.

Crowhurst, P. V., Hill, K. C., Foster, D. A. and Bennett, A. P. (1996). Thermochronological and geochemical constraints on the tectonic evolution of northern Papua New Guinea. In *Tectonic Evolution of SE Asia*, ed. R. Hall and D. J. Blundell. *Geological Society of London Special Publication*, 106, 525–37.

Currie, C. A. and Hyndman, R. D. (2006). The thermal structure of subduction zone back arcs. *Journal of Geophysical Research*, **111**, B08404, doi:10.1029/2005JB004024.

Davies, A. G. S., Cooke, D. R., Gemmell, J. B. et al. (2008). Hydrothermal breccias and veins at the Kelian gold mine, Kalimantan, Indonesia: genesis of a large epithermal gold deposit. *Economic Geology*, **103**, 717–57.

Davies, I. C. (1990). Geology and exploration review of the Tomori PSC, eastern Indonesia. In *Proceedings of the Indonesian Petroleum Association 19th Annual Convention*. Jakarta: Indonesian Petroleum Association, pp. 41–68.

Dickerson, R. E. (1928). *Distribution of Life in the Philippines*, Bureau of Science, Manila, Monographic Publication 21, pp. 1–322.

Dow, D. B. (1977). A geological synthesis of Papua New Guinea. *Bureau of Mineral Resources, Australia, Geology and Geophysics Bulletin*, **201**, 1–41.

Dow, D. B. and Hartono, U. (1982). The nature of the crust underlying Cenderawasih (Geelvink) Bay, Irian Jaya. *Bulletin Geological Research and Development Centre, Bandung*, **6**, 30–36.

Dow, D. B. and Sukamto, R. (1984). Western Irian Jaya: the end-product of oblique plate convergence in the late Tertiary. *Tectonophysics*, **106**, 109–39.

Dow, D. B., Robinson, G. P. and Ratman, N. (1985). New hypothesis for formation of Lengguru Foldbelt, Irian Jaya, Indonesia. *American Association of Petroleum Geologists Bulletin*, **69**, 203–14.

Dow, D. B., Robinson, G. P., Hartono, U. and Ratman, N. (1986). *Geological Map of Irian Jaya, Indonesia. Scale 1:1 000 000*. Bandung, Indonesia: Geological Survey of Indonesia, Directorate of Mineral Resources, Geological Research and Development Centre.

Elsassar, W. (1971). Sea-floor spreading as convection. *Journal of Geophysical Research*, **76**, 1101–12.

Engdahl, E. R., van der Hilst, R. and Buland, R. (1998). Global teleseismic earthquake relocation with improved travel times and procedures for depth determination. *Bulletin of the Seismological Society of America*, **88**, 722–43.

Fortuin, A. R., van der Werff, W. and Wensink, G. (1997). Neogene basin history and paleomagnetism of a rifted and inverted forearc region, on- and offshore Sumba, Eastern Indonesia. *Journal of Asian Earth Sciences*, **15**, 61–88.

Franke, D., Barckhausen, U., Heyde, I., Tingay, M. and Ramli, N. (2008). Seismic images of a collision zone offshore NW Sabah/Borneo. *Marine and Petroleum Geology*, **25**, 606–24.

Fraser, T. H., Bon, J. and Samuel, L. (1993). A new dynamic Mesozoic stratigraphy for the West Irian micro-continent Indonesia and its implications. In *Proceedings of the Indonesian Petroleum Association 22nd Annual*

Convention. Jakarta: Indonesian Petroleum Association, pp. 707–761.

Fullerton, L. G., Sager, W. W. and Handschumacher, D. W. (1989). Late Jurassic–Early Cretaceous evolution of the eastern Indian Ocean adjacent to northwest Australia. *Journal of Geophysical Research*, **94**, 2937–54.

Garrard, R. A., Supandjono, J. B. and Surono (1988). The geology of the Banggai-Sula microcontinent, eastern Indonesia. In *Proceedings of the Indonesian Petroleum Association 17th Annual Convention*. Jakarta: Indonesian Petroleum Association, pp. 23–52.

GEBCO (2003). IHO-UNESCO, *General Bathymetric Chart of the Oceans*, Digital Edition 2003, www.ngdc.noaa.gov/mgg/gebco.

Godfrey, J. S. (1996). The effect of the Indonesian throughflow on ocean circulation and heat exchange with the atmosphere: a review. *Journal of Geophysical Research*, **101**, 12217–38.

Gordon, A. L. (2005). Oceanography of the Indonesian Seas and their throughflow. *Oceanography* **18**, 14–27.

Gordon, A. L., Giulivi, C. F. and Ilahude, A. G. (2003). Deep topographic barriers within the Indonesian seas. *Deep-Sea Research*, **50**, 2205–28.

Haile, N. S. (1974). Borneo. In *Mesozoic–Cenozoic Orogenic Belts*, ed. A. M. Spencer. *Geological Society of London Special Publication*, 4, 333–47.

Hall, R. (1996). Reconstructing Cenozoic SE Asia. In *Tectonic Evolution of SE Asia*, ed. R. Hall and D. J. Blundell. *Geological Society of London Special Publication*, 106, 153–84.

Hall, R. (1998). The plate tectonics of Cenozoic SE Asia and the distribution of land and sea. In *Biogeography and Geological Evolution of SE Asia*, ed. R. Hall and J. D. Holloway. Leiden, The Netherlands: Backhuys, pp. 99–131.

Hall, R. (2000). Neogene history of collision in the Halmahera Region, Indonesia. *Proceedings of the Indonesian Petroleum Association, 27th Annual Convention*. Jakarta: Indonesian Petroleum Association, pp. 487–93.

Hall, R. (2001). Cenozoic reconstructions of SE Asia and the SW Pacific: changing patterns of land and sea. In *Faunal and Floral Migrations and Evolution in SE Asia–Australasia*, ed. I. Metcalfe, J. M. B. Smith, M. Morwood and I. D. Davidson. Lisse, The Netherlands: A. A. Balkema (Swets and Zeitlinger), pp. 35–56.

Hall, R. (2002). Cenozoic geological and plate tectonic evolution of SE Asia and the SW Pacific: computer-based reconstructions, model and animations. *Journal of Asian Earth Sciences*, **20**, 353–434.

Hall, R. (2009a). SE Asia's changing palaeogeography. *Blumea*, **54**, 148–61.

Hall, R. (2009b). Hydrocarbon basins in SE Asia: understanding why they are there. *Petroleum Geoscience*, **15**, 131–46.

Hall, R. (2009c). The Eurasian SE Asian margin as a modern example of an accretionary orogen. In *Accretionary Orogens in Space and Time*, ed. P. A. Cawood and A. Kröner. *Geological Society of London Special Publication*, 318, 351–72.

Hall, R. (2011). Australia–SE Asia collision: plate tectonics and crustal flow. In *The SE Asian Gateway: History and Tectonics of Australia–Asia Collision*, ed. R. Hall, M. A. Cottam and M. E. J. Wilson. *Geological Society of London Special Publication*, 355, 75–109.

Hall, R. and Holloway, J. D., eds. (1998). *Biogeography and Geological Evolution of SE Asia*. Leiden, The Netherlands: Backhuys Publishers.

Hall, R. and Morley, C. K. (2004). Sundaland Basins. In *Continent–Ocean Interactions within the East Asian Marginal Seas*, ed. P. Clift, P. Wang, W. Kuhnt and D. E. Hayes. *Geophysical Monograph*, 149, 55–85.

Hall, R. and Smyth, H. R. (2008). Cenozoic arc processes in Indonesia: identification of the key influences on the stratigraphic record in active volcanic arcs. In *Formation and Applications of the Sedimentary Record in Arc Collision Zones*, ed. A. E. Draut, P. D. Clift and D. W. Scholl. *Geological Society of America Special Paper*, 436, 27–54.

Hall, R. and Spakman, W. (2002). Subducted slabs beneath the eastern Indonesia–Tonga region: insights from tomography. *Earth and Planetary Science Letters*, **201**, 321–36.

Hall, R. and Wilson, M. E. J. (2000). Neogene sutures in eastern Indonesia. *Journal of Asian Earth Sciences*, **18**, 787–814.

Hall, R., Ali, J. R. and Anderson, C. D. (1995a). Cenozoic motion of the Philippine Sea Plate: paleomagnetic evidence from eastern Indonesia. *Tectonics*, **14**, 1117–32.

Hall, R., Ali, J. R., Anderson, C. D. and Baker, S. J. (1995b). Origin and motion history of the Philippine Sea Plate. *Tectonophysics*, **251**, 229–50.

Hall, R., van Hattum, M. W. A. and Spakman, W. (2008). Impact of India-Asia collision on SE Asia: the record in Borneo. *Tectonophysics*, **451**, 366–89.

Hall, R., Clements, B. and Smyth, H. R. (2009a). Sundaland: basement character, structure and plate tectonic development. In *Proceedings of the Indonesian Petroleum Association, 33rd Annual Convention*. Jakarta: Indonesian Petroleum Association, pp. IPA09-G134.

Hall, R., Cloke, I. R., Nur'Aini et al. (2009b). The North Makassar Straits: what lies beneath? *Petroleum Geoscience*, **15**, 147–58.

Hall, R., Cottam, M. A. and Wilson, M. E. J. eds. (2011). *The SE Asian Gateway: History and Tectonics of Australia–Asia Collision. Geological Society of London Special Publication*, 355.

Hamilton, W. B. (1979). Tectonics of the Indonesian Region. *US Geological Survey Professional Paper*, **1078**, 345 pp.

Hamilton, W. B. (2007). Driving mechanism and 3-D circulation of plate tectonics. *Geological Society of America Special Paper*, **433**, 1–25.

Hasan, K. (1990). *The Upper Cretaceous flysch succession of the Balangbaru Formation, Southwest Sulawesi, Indonesia*. PhD Thesis, University of London.

Hazebroek, H. P. and Tan, D. N. K. (1993). Tertiary tectonic evolution of the NW Sabah continental margin. *Geological Society of Malaysia Bulletin*, **33**, 195–210.

Herrington, R. J., Scotney, P. M., Roberts, S., Boyce, A. J. and Harrison, D. (2010). Temporal association of arc-continent collision, progressive magma contamination in arc volcanism and formation of gold-rich massive sulphide deposits on Wetar Island (Banda arc). *Gondwana Research*, **19**, 583–93.

Hill, K. C. and Hall, R. (2003). Mesozoic–Cenozoic evolution of Australia's

New Guinea margin in a west Pacific context. In *Evolution and Dynamics of the Australian Plate*, ed. R. R. Hillis and R. D. Müller. *Geological Society of Australia Special Publication*, 22, and *Geological Society of America Special Paper*, 372, 265–89.

Hill, K. C. and Raza, A. (1999). Arc-continent collision in Papua Guinea: constraints from fission track thermochronology. *Tectonics*, **18**, 950–66.

Hinschberger, F., Malod, J.-A., Dyment, J. et al. (2001). Magnetic lineations constraints for the back-arc opening of the Late Neogene South Banda Basin (eastern Indonesia). *Tectonophysics*, **333**, 47–59.

Hinschberger, F., Malod, J.-A., Réhault, J.-P. et al. (2000). Origine et évolution du bassin Nord-Banda (Indonésie): apport des données magnétiques. *Comptes Rendus de l'Académie des Sciences, Paris*, **331**, 507–14.

Honthaas, C., Rehault, J.-P., Maury, R. C. et al. (1998). A Neogene back-arc origin for the Banda Sea basins: geochemical and geochronological constraints from the Banda ridges (East Indonesia). *Tectonophysics*, **298**, 297–317.

Honthaas, C., Maury, R. C., Priadi, B., Bellon, H. and Cotten, J. (1999). The Plio-Quaternary Ambon arc, Eastern Indonesia. *Tectonophysics*, **301**, 261–81.

Hutchison, C. S. (1996). *South-East Asian Oil, Gas, Coal and Mineral Deposits*. Oxford Monographs on Geology and Geophysics, 36. Oxford: Clarendon Press.

Hutchison, C. S. (2005). *Geology of North-West Borneo*. Amsterdam: Elsevier.

Hutchison, C. S. (2010). The North-West Borneo Trough. *Marine Geology*, **271**, 32–43.

Hutchison, C. S., Bergman, S. C., Swauger, D. A. and Graves, J. E. (2000). A Miocene collisional belt in north Borneo: uplift mechanism and isostatic adjustment quantified by thermochronology. *Journal of the Geological Society of London*, **157**, 783–93.

Huxley, T. H. (1868). On the classification and distribution of the Alectoromorphae and Heteromorphae. *Proceedings of the Zoological Society of London*, **1868**, 294–319.

Hyndman, R. D., Currie, C. A. and Mazzotti, S. P. (2005). Subduction zone backarcs, mobile belts, and orogenic heat. *GSA Today*, **15**, 4–10.

Jablonski, D., Priyono, P., Westlake, S. and Larsen, O. A. (2007). Geology and exploration potential of the Gorontalo Basin, central Indonesia: eastern extension of the North Makassar basin? *In Proceedings of the Indonesian Petroleum Association 31st Annual Convention*. Jakarta: Indonesian Petroleum Association, pp. 197–224.

Khan, S. D., Walker, D. J., Hall, S. A. et al.(2009). Did the Kohistan-Ladakh island arc collide first with India? *Geological Society of America Bulletin*, **121**, 366–84.

Kirk, H. J. C. (1968). The igneous rocks of Sarawak and Sabah. *Geological Survey of Malaysia, Borneo Region, Bulletin*, **5**, 1–210.

Klompé, T. H. F. (1954). The structural importance of the Sula Spur (Indonesia). *Indonesian Journal of Natural Sciences*, **110**, 21–40.

Kudrass, H. R., Wiedecke, M., Cepeck, P., Kreuser, H. and Muller, P. (1986). Pre-Quaternary rocks dredged from the South China Sea (Reed Bank area) and Sulu Sea, during Sonne Cruises in 1982–1983. *Marine and Petroleum Geology*, **3**, 19–30.

Kündig, E. (1956). Geology and ophiolite problems of East Celebes. *Verhandelingen Koninklijk Nederlands Geologisch en Mijnbouwkundig Genootschap, Geologische Serie*, **16**, 210–35.

Lydekker, R. (1896). *A Geographical History of Mammals*. Cambridge: Cambridge University Press.

Lyle, M., Gibbs, S., Moore, T. C. and Rea, D. K. (2007). Late Oligocene initiation of the Antarctic Circumpolar Current: evidence from the South Pacific. *Geology*, **35**, 691–94.

Lyle, M., Barron, J., Bralower, T. J. et al. (2008). Pacific Ocean and Cenozoic evolution of climate. *Reviews of Geophysics*, **46**, RG2002, doi:10.1029/2005RG000190.

McCourt, W. J., Crow, M. J., Cobbing, E. J. and Amin, T. C. (1996). Mesozoic and Cenozoic plutonic evolution of SE Asia: evidence from Sumatra, Indonesia. In *Tectonic Evolution of SE Asia*, ed. R. Hall and D. J. Blundell. *Geological Society of London Special Publication*, 106, 321–35.

McKenna, M. C. (1973). Sweepstakes, filters, corridors, Noah's Arks, and beached Viking funeral ships and palaeogeography. In *Implications of Continental Drift to the Earth Sciences*, ed. D. H. Tarling and S. K. Runcorn. London: Academic Press, pp. 295–308.

Macpherson, C. G., Chiang, K. K., Hall, R. et al.(2010). Plio-Pleistocene intra-plate magmatism from the southern Sulu Arc, Semporna peninsula, Sabah, Borneo: implications for high-Nb basalt in subduction zones. *Journal of Volcanology and Geothermal Research*, **190**, 25–38.

Mayr, E. (1944). Wallace's Line in the light of recent zoogeographic studies. *The Quarterly Review of Biology*, **19**, 1–14.

Metcalfe, I. (1990). Allochthonous terrane processes in Southeast Asia. *Philosophical Transactions Royal Society of London A*, **331**, 625–40.

Metcalfe, I. (1996). Pre-Cretaceous evolution of SE Asian terranes. In *Tectonic Evolution of SE Asia*, ed. R. Hall and D. J. Blundell. *Geological Society London Special Publication*, 106, 97–122.

Metcalfe, I. (1998). Palaeozoic and Mesozoic geological evolution of the SE Asian region: multidisciplinary constraints and implications for biogeography. In *Biogeography and Geological Evolution of SE Asia*, ed. R. Hall. and J. D. Holloway. Leiden, The Netherlands: Backhuys Publishers, pp. 25–41.

Metcalfe, I. (2011). Palaeozoic–Mesozoic History of SE Asia. In *The SE Asian Gateway: History and Tectonics of Australia–Asia Collision*, ed. R. Hall, M. A. Cottam and M. E. J. Wilson. *Geological Society of London Special Publication*, 355, 7–35.

Mitchell, A. H. G. (1984). Post-Permian events in the Zangbo suture zone, Tibet. *Journal of the Geological Society, London*, **141**, 129–36.

Mitchell, A. H. G. (1992). Late Permian-Mesozoic events and the Mergui Group Nappe in Myanmar and Thailand. *Journal of SE Asian Earth Sciences*, **7**, 165–78.

Mitchell, A. H. G. (1993). Cretaceous–Cenozoic tectonic events in the western Myanmar (Burma)-Assam region. *Journal of the Geological Society of London*, **150**, 1089–102.

Molnar, P. and Atwater, T. (1978). Interarc spreading and Cordilleran tectonics as alternates related to the age of subducted oceanic lithosphere. *Earth and Planetary Science Letters*, **41**, 330–40.

Moss, S.J. (1998). Embaluh group turbidites in Kalimantan: evolution of a remnant oceanic basin in Borneo during the Late Cretaceous to Palaeogene. *Journal of the Geological Society*, **155**, 509–24.

Muller, J. (1968). Palynology of the Pedawan and Plateau Sandstone Formations (Cretaceous-Eocene) in Sarawak, Malaysia. *Micropaleontology*, **14**, 1–37.

Pairault, A. A., Hall, R. and Elders, C. F. (2003). Structural styles and tectonic evolution of the Seram Trough, Indonesia. *Marine and Petroleum Geology*, **20**, 1141–60.

Parkinson, C. D., Miyazaki, K., Wakita, K., Barber, A. J. and Carswell, D. A. (1998). An overview and tectonic synthesis of the pre-Tertiary very high-pressure metamorphic and associated rocks of Java, Sulawesi and Kalimantan, Indonesia. *Island Arc*, **7**, 184–200.

Pieters, P. E., Pigram, C. J., Trail, D. S. et al. (1983). The stratigraphy of western Irian Jaya. *Bulletin Geological Research and Development Centre, Bandung*, **8**, 14–48.

Pigram, C. J. and Davies, H. L. (1987). Terranes and the accretion history of the New Guinea orogen. *BMR Journal of Australian Geology and Geophysics*, **10**, 193–212.

Pigram, C. J. and Panggabean, H. (1984). Rifting of the northern margin of the Australian continent and the origin of some microcontinents in eastern Indonesia. *Tectonophysics*, **107**, 331–53.

Polvé, M., Maury, R. C., Bellon, H. et al. (1997). Magmatic evolution of Sulawesi (Indonesia): constraints on the Cenozoic geodynamic history of the Sundaland active margin. *Tectonophysics*, **272**, 69–92.

Powell, C. M., Roots, S. R. and Veevers, J. J. (1988). Pre-breakup continental extension in East Gondwanaland and the early opening of the eastern Indian Ocean. *Tectonophysics*, **155**, 261–83.

Rangin, C. and Silver, E. A. (1991). Neogene tectonic evolution of the Celebes Sulu basins; new insights from Leg 124 drilling. *Proceedings of the Ocean Drilling Program, Scientific Results*, **124**, 51–63.

Renema, W., Bellwood, D. R., Braga, J. C. et al. (2008). Hopping hotspots: global shifts in marine biodiversity. *Science*, **321**, 654–7.

Rigg, J. W. D. and Hall, R. (2011). Structural and stratigraphic evolution of the Savu Basin, Indonesia. In *The SE Asian Gateway: History and Tectonics of Australia–Asia Collision*, ed. R. Hall, M. A. Cottam and M. E. J. Wilson. *Geological Society of London Special Publication*, 355, 225–40.

Robb, M. S., Taylor, B. and Goodliffe, A. M. (2005). Re-examination of the magnetic lineations of the Gascoyne and Cuvier Abyssal Plains, off NW Australia. *Geophysical Journal International*, **163**, 42–55.

Roosmawati, N. and Harris, R. (2009). Surface uplift history of the incipient Banda arc-continent collision: geology and synorogenic foraminifera of Rote and Savu Islands, Indonesia. *Tectonophysics*, **479**, 95–110.

Sandwell, D. T. and Smith, W. H. F. (2009). Global marine gravity from retracked Geosat and ERS-1 altimetry: ridge segmentation versus spreading rate. *Journal of Geophysical Research*, **114**, B01411, doi:10.1029/2008JB006008.

Sapin, F., Pubellier, M., Ringenbach, J.-C. and Bailly, V. (2009). Alternating thin versus thick-skinned decollements, example in a fast tectonic setting: the Misool-Onin-Kumawa Ridge (West Papua). *Journal of Structural Geology*, **31**, 444–59.

Sarasin, P. and Sarasin, S. (1901). *Materialen zur Naturgeschichte der Insel Celebes. Vierter Band. Entwurf Einer Geografisch-Geologischen Beschreibung Der Insel Celebes*. Wiesbaden, Germany: C. W. Kreidel.

Schneider, N. (1998). The Indonesian throughflow and the global climate system. *Journal of Climate*, **11**, 676–89.

Scotese, C. R. (2010). *Paleomap Project*, http://www.scotese.com/.

Scotney, P. M., Roberts, S., Herrington, R. J., Boyce, A. J. and Burgess, R. (2005). The development of volcanic hosted massive sulfide and barite gold orebodies on Wetar Island, Indonesia. *Mineralium Deposita*, **40**, 76–99.

Siebert, L. and Simkin, T. (2002). Volcanoes of the World: an illustrated catalog of Holocene volcanoes and their eruptions. In *Smithsonian Institution, Global Volcanism Program Digital Information Series, GVP-3*, http://www.volcano.si.edu/world/.

Silver, E. A., McCaffrey, R., Joyodiwiryo, Y. and Stevens, S. (1983). Ophiolite emplacement and collision between the Sula platform and the Sulawesi island arc, Indonesia. *Journal of Geophysical Research*, **88**, 9419–35.

Simons, W. J. F., Socquet, A., Vigny, C. et al. (2007). A decade of GPS in Southeast Asia: resolving Sundaland motion and boundaries. *Journal of Geophysical Research*, **112**, B06420, doi:10.1029/2005JB003868.

Simpson, G. G. (1977). Too many lines; the limits of the Oriental and Australian zoogeographic regions. *Proceedings of the American Philosophical Society*, **121**, 107–20.

Smyth, H. R., Hamilton, P. J., Hall, R. and Kinny, P. D. (2007). The deep crust beneath island arcs: inherited zircons reveal a Gondwana continental fragment beneath East Java, Indonesia. *Earth and Planetary Science Letters*, **258**, 269–82.

Spakman, W. and Hall, R. (2010). Surface deformation and slab-mantle interaction during Banda Arc subduction rollback. *Nature Geoscience*, **3**, 562–6.

Spencer, J. E. (2010). Structural analysis of three extensional detachment faults with data from the 2000 Space-Shuttle Radar Topography Mission. *GSA Today*, **20**, doi: 10.1130/GSATG59A.

Stampfli, G. M. and Borel, G. D. (2002). A plate tectonic model for the Paleozoic and Mesozoic constrained by dynamic plate boundaries and restored synthetic oceanic isochrons. *Earth and Planetary Science Letters*, **196**, 17–33.

Struckmeyer, H. I. M., Yeung, M. and Pigram, C. J. (1993). Mesozoic to Cainozoic plate tectonic and palaeogeographic evolution of the New Guinea region. In *Petroleum Exploration and Development in Papua New Guinea. Based on the Proceedings of the Second PNG Petroleum Convention Organised by the PNG Chamber of Mines and Petroleum in Port Moresby, Papua New Guinea, 31 May–2 June 1993*, ed. G. J. Carman and Z. Carman. Port Moresby, Papua New Guinea: PNG Chamber of Mines and Petroleum. pp. 261–90.

Surono (1995). Sedimentology of the Tolitoli Conglomerate Member of the Langkowala Formation, Southeast Sulawesi, Indonesia. *Journal of Geology and Mineral Resources, Geological Research and Development Centre Bandung, Indonesia*, **5**, 1–7.

Suwitodirdjo, K. and Tjokrosapoetro, S. (1975). *Geologic Map of the Atambua Quadrangle, Timor*. Bandung, Indonesia: Geological Research and Development Centre, Bandung.

Tan, D. N. K. and Lamy, J. M. (1990). Tectonic evolution of the NW Sabah continental margin since the late Eocene. *Geological Society of Malaysia Bulletin*, **27**, 241–60.

Tarling, D. H. (1982). Land bridges and plate tectonics. *Geobios*, **15**, 361–74.

Taylor, B. and Hayes, D. E. (1983). Origin and history of the South China Sea Basin. In *The Tectonic and Geologic Evolution of Southeast Asian Seas and Islands, Part 2*, ed. D. E. Hayes. *Geophysical Monographs Series*, 27, 23–56.

Tomascik, T., Mah, A. J., Nontji, A. and Moosa, M. K. (1997). *The Ecology of the Indonesian Seas*. The Ecology of Indonesia Series, Oxford: Periplus Editions, Oxford University Press.

Tongkul, F. (1991). Tectonic evolution of Sabah, Malaysia. *Journal of Southeast Asian Earth Sciences*, **6**, 395–406.

van Bemmelen, R. W. (1949). *The Geology of Indonesia*. The Hague, The Netherlands: Government Printing Office.

van den Bergh, G. D., de Vos, J. and Sondaar, P. Y. (2001). The Late Quaternary palaeogeography of mammal evolution in the Indonesian Archipelago. *Palaeogeography, Palaeoclimatology, Palaeoecology*, **171**, 385–408.

van Hattum, M. W. A. (2005). *Provenance of Cenozoic sedimentary rocks of Northern Borneo*. PhD Thesis, University of London.

van Hattum, M. W. A., Hall, R., Pickard, A. L. and Nichols, G. J. (2006). SE Asian sediments not from Asia: provenance and geochronology of North Borneo sandstones. *Geology*, **34**, 589–92.

van Leeuwen, T. M., Allen, C. M., Kadarusman, A. et al. (2007). Petrologic, isotopic, and radiometric age constraints on the origin and tectonic history of the Malino Metamorphic Complex, NW Sulawesi, Indonesia. *Journal of Asian Earth Sciences*, **29**, 751–77.

van Leeuwen, T. M., Susanto, E. S., Maryanto, S. et al. (2010). Tectonostratigraphic evolution of Cenozoic marginal basin and continental margin successions in the Bone Mountains, Southwest Sulawesi, Indonesia. *Journal of Asian Earth Sciences*, **33**, 233–54.

van Ufford, A. Q. and Cloos, M. (2005). Cenozoic tectonics of New Guinea. *American Association of Petroleum Geologists Bulletin*, **89**, 119–40.

Veevers, J. J. (1988). Morphotectonics of Australia's northwestern margin: a review. In *The North West Shelf. Based on the Proceedings of the North West Shelf Symposium Sponsored by the Western Australian Branch of the Petroleum Exploration Society of Australia Limited and held in Perth, Western Australia, August 10–12, 1988*, ed. P. G. Purcell and R. R. Purcell. Perth,

Australia: Petroleum Exploration Society of Australia (PESA), Western Australian Branch. pp. 19–27.

von der Heydt, A. and Dijkstra, H. A. (2006). Effect of ocean gateways on the global ocean circulation in the late Oligocene and early Miocene. *Paleoceanography*, **21**, DOI 10.1029/2005PA001149.

Voris, H. K. (2000). Maps of Pleistocene sea levels in Southeast Asia: shorelines, river systems and time durations. *Journal of Biogeography*, **27**, 1153–67.

Wakita, K. and Metcalfe, I. (2005). Ocean plate stratigraphy in East and Southeast Asia. *Journal of Asian Earth Sciences*, **24**, 679–702.

Wakita, K., Munasri and Widoyoko, B. (1994). Cretaceous radiolarians from the Luk-Ulo Melange Complex in the Karangsambung area, Central Java, Indonesia. *Journal of Southeast Asian Earth Sciences*, **9**, 29–43.

Wallace, A. R. (1869). *The Malay Archipelago. The Land of the Orang-Utan and the Bird of Paradise: A Narrative of Travel, with Studies of Man and Nature*, 2 Volumes. London: Macmillan and Co. [Hong Kong: Periplus Editions (facsimile)]

Wanner, J. (1910). Einige geologische Ergebnisse einer im Jahre 1909 ausgefuhrten Reise dürch den ostlichen Teil des indoaustralischen Archipels. Neues über die Perm, Trias und Jura Formation des indoaustralischen Archipel. *Centralblatt für Geologie und Paläontologie, Beil.*, **22**, 736–41.

Watkinson, I. M. (2011). Ductile flow in the metamorphic rocks of central Sulawesi. In *The SE Asian Gateway: History and Tectonics of Australia–Asia Collision*, ed. R. Hall, M. A. Cottam and M. E. J. Wilson. *Geological Society of London Special Publication*, 355, 157–76.

Watkinson, I. M., Hall, R. and Ferdian, F. (2011). Tectonic re-interpretation of the Banggai-Sula–Molucca Sea margin, Indonesia. In *The SE Asian Gateway: History and Tectonics of Australia-Asia Collision*, ed. R. Hall, M. A. Cottam and M. E. J. Wilson. *Geological Society of London Special Publication*, 355, 203–24.

Whittaker, J. M., Müller, R. D., Sdrolias, M. and Heine, C. (2007). Sunda–Java trench kinematics, slab window formation and overriding plate deformation since the Cretaceous. *Earth and Planetary Science Letters*, **255**, 445–57.

Whitten, T., Whitten, J., Goettsch, C., Supriatna, J. and Mittermeier, R. A. (1999a). Sundaland. In *Biodiversity Hotspots of the World*, ed. R. A. Mittermeier, P. R. Gil and C. Goettsch-Mittermeier. Prado Norte, Mexico: Cemex, pp. 279–90.

Whitten, T., Whitten, J., Goettsch, C., Supriatna, J. and Mittermeier, R. A. (1999b). Wallacea. In *Biodiversity Hotspots of the World*, ed. R. A. Mittermeier, P. R. Gil and C. Goettsch-Mittermeier. Prado Norte, Mexico: Cemex, pp. 297–307.

Williams, P. R., Johnston, C. R., Almond, R. A. and Simamora, W. H. (1988). Late Cretaceous to early Tertiary structural elements of West Kalimantan. *Tectonophysics*, **148**, 279–98.

Wilson, M. E. J. and Bosence, D. W. J. (1996). The Tertiary evolution of South Sulawesi: a record in redeposited carbonates of the Tonasa Limestone Formation. In *Tectonic Evolution of SE Asia*, ed. R. Hall and

D. J. Blundell. *Geological Society of London Special Publication*, 106, 365–89.

Wolfenden, E. B. (1960). *The Geology and Mineral Resources of the Lower Rajang Valley and Adjoining Areas, Sarawak, British Territories Borneo Region Geological Survey Department, Memoir*, **11**, 1–167.

4

A review of the Cenozoic palaeoclimate history of Southeast Asia

ROBERT J. MORLEY

4.1 Introduction

Southeast Asia experienced a complex geological history during the Cenozoic that had a major impact on the climate and vegetation of the region. During the early Tertiary, the area was profoundly modified following the Paleocene–Eocene collision of the Indian Plate with Asia and the subsequent southeastward extrusion of parts of Indochina (Tarponnier et al. 1982, 1986) that coincided with the development of major fracturing and of multiple pull-apart basins across the Sunda region. This resulted in the closure of the Tethys Ocean, affecting ocean currents throughout the region, and also the dispersal into Southeast Asia of diverse elements of the Indian flora (Morley 2000). The region was subsequently further modified by compression as a result of the combined effects of the westward drift of the Pacific Plate and northern drift of Australasia (Hall 1996, 1998, 2002, this volume, Chapter 3). The Australian Plate collided with Sunda at the end of the Oligocene, resulting in the separation of the Indian and Pacific oceans, the development of the Indonesian Throughflow and the limited interchange of floristic elements between Sunda and Australasia. This complex history of changing land masses and ocean currents, coupled with global climate change, is likely to have had a profound effect on climate change across the region.

Biotic Evolution and Environmental Change in Southeast Asia, eds D. J. Gower et al. Published by Cambridge University Press.

This chapter summarises the Cenozoic climatic and environmental history of the broad Sunda region, from Sulawesi to southern Vietnam, based mainly on the palynological record, but also taking into account lithological indicators of climate, especially the occurrence of coals, which form only during periods of ever-wet climate, and palaeosols. The climatic record for each period is discussed in relation to trends in global temperatures as implied by the Cenozoic oxygen isotope curve of Zachos et al. (2001), which is based on the analysis of deep marine benthonic foraminifera. Previous preliminary reviews of past climates of the area are by Morley and Flenley (1987) and Morley (2000).

Most of the examples used to reconstruct the palaeoclimate history of the Southeast Asian region have been taken from unpublished routine biostratigraphic studies undertaken during the course of hydrocarbon exploration across the region. This work has involved many palynologists over the course of the last 30 years, with intensive studies from the Malay, Natuna and Nam Con Son Basins from the South China Sea, from Tarakan and Kutei Basins in the Makassar Straits, and from East and West Java Sea Basins with additional studies from South and Central Sumatra and Sulawesi. These studies mostly were not undertaken with a view of specifically determining palaeoclimate, except during the last 7 years or so, during which time it has become clear within the petroleum industry that especially for non- and marginal marine stratigraphic successions which are poorly dated, an understanding of the palaeoclimate succession, especially when viewed within a sequence stratigraphic context, helps considerably in unravelling the regional stratigraphy (Morley 1996, Morley et al. 2003). As a result, many of the palaeoclimate inferences brought to attention are fortuitous, and details of the localities and datasets discussed are yet to be published.

The palaeoclimate history is discussed under headings of epochs, which roughly coincide with the main divisions of the 'greenhouse' and 'icehouse' climates of the Tertiary. The Paleocene and Eocene epochs are discussed first, reflecting the thermal maximum of the early Tertiary, followed by Oligocene mainly 'icehouse' climates. The Neogene is then divided into the Early and Middle Miocene period of 'greenhouse' climates, followed by the Late Miocene, and Plio-Pleistocene, during which time global climates underwent profound cooling, culminating in the glacial cycles of the Late Pliocene and Pleistocene.

Climatic inferences are made mainly by attributing pollen and spore types to modern taxa for which the ecology is known. For the Paleocene and Eocene, only a limited number of pollen types can be referred to living taxa, but from the base Oligocene onward, the majority can be attributed to extant genera. Interpretations are thus more confident in younger sediments. Palynological data are presented in a simplified form using summary diagrams, with pollen and spores being placed into groups according to which vegetation type they are considered mainly to have been derived. Although predominantly pollinated by animals, tropical forests

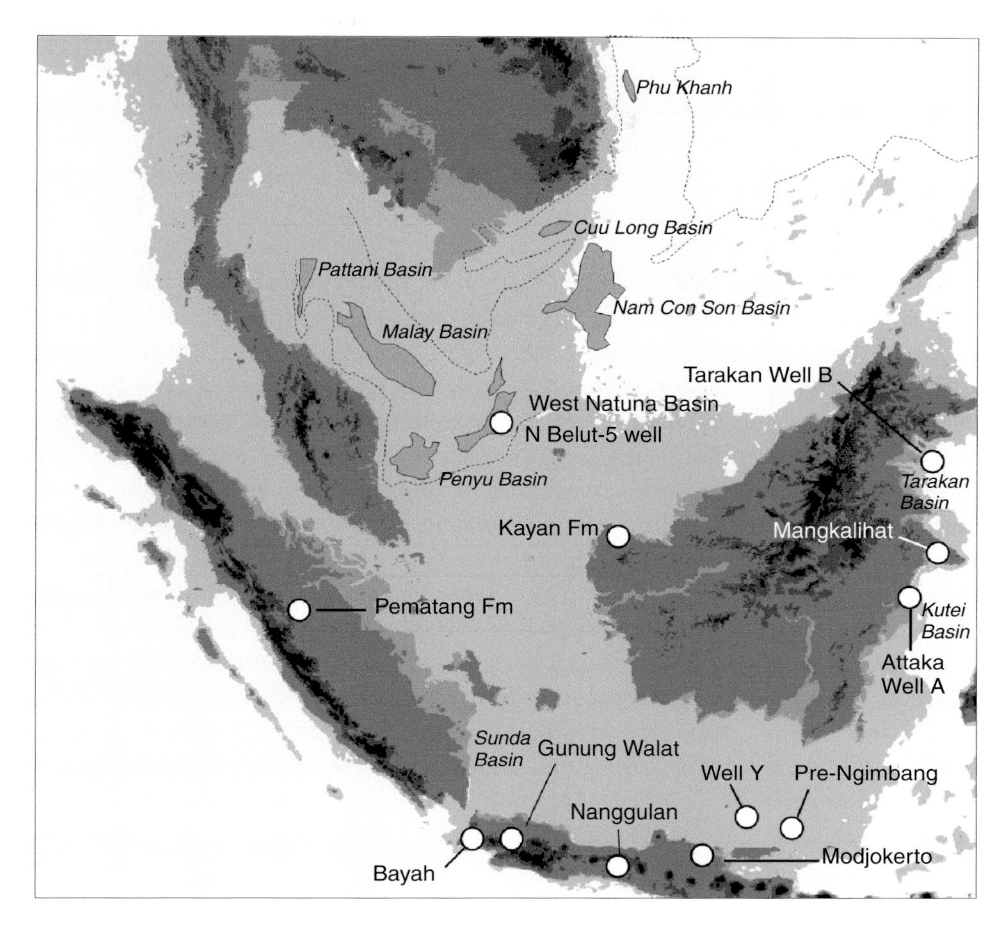

Figure 4.1 Locations mentioned in the text and positions of South China Sea Oligocene lacustrine basins.

actually produce large amounts of pollen, not incomparable to the productivity of temperate forests (e.g. Flenley 1973, Bush 1995).

4.2 Paleocene

There are very few Paleocene sediments reported from the Southeast Asian region, suggesting a period of widespread emergence. During the Paleocene, the Sunda region is thought to have formed a peninsula of the Asian land mass extending to present day southeast Borneo and East Java (Hall 1998). Two localities are noteworthy from the perspective of palaeoclimate, firstly the Kayan Formation in western Sarawak (Fig 4.1), analysed for palynology by Muller (1968) and subsequently reinterpreted by Morley (1998, 2000) and second, coaly sediments of the

Pre-Ngimbang Formation (Bransden and Matthews 1992) from the East Java Sea (Morley unpublished).

The pollen flora of the Kayan Formation was illustrated graphically and examined using correspondence analysis by Morley (2000: 178). The pollen flora has a low diversity, with most pollen types having no modern analogue, but by comparing the distributions of pollen types that can be compared to modern taxa with abundance variations of taxa for which affinity is uncertain using correspondence analysis, an approximate idea of the main pollen associations and suggestions regarding main vegetation types can be obtained. Mangroves, dominated by Nypoidae, were extensive, as were freshwater swamps, with *Ilex* (Aquifoliaceae), rattans, *Anacolosa* (Olacaceae) and Apocynaceae. An upland or open woodland element is suggested by the occurrence of diverse bisaccates and also regular *Ephedra* pollen (Gnetales), which suggest a subhumid climate. The Pre-Ngimbang Formation, on the other hand, penetrated by petroleum exploration wells in the East Java Sea is thought to be of Early Paleocene (or possibly Late Maastrichtian) age (Morley unpublished), based on the presence of trichotomosulcate palm pollen, and *Proteacidites* spp. with affinity to West African Senonian/Paleocene taxa, and with the incoming in the upper part of *Dicolpopollis* spp. It is characterised by coals in some localities suggesting an everwet climate. The coals yield rich assemblages of a trichotomosulcate palm pollen type that most closely resembles the Neotropical cocoid palms *Acrocomia* and *Ceroxylon*. There thus appears to have been a climatic gradient across Sundaland in the Paleocene, with wetter climates toward the southeast.

4.3 Eocene

Evidence for Eocene vegetation and climate mainly comes from palynological studies from outcrops and from the subsurface of Java and Sulawesi (Fig 4.1), which during the Middle Eocene, was attached to Borneo (Hall 1998). By the earliest Middle Eocene, the flora of the region had changed dramatically compared to that of the Paleocene, following the collision of the Indian Plate with Asia, and the dispersal of many taxa of Indian origin into the region, which resulted in the extinction of older, Paleocene pollen types. The Indian element, summarised by Morley (1998, 2000) is indicated by the presence of pollen of palms, especially of the morphogenus *Palmaepollenites* (attributed to the subtribe Iguanurinae by Harley and Morley 1995) and also probable *Eugeissona*, *Alangium* (Alangiaceae), aff. *Beauprea* (Proteaceae), Sapindaceae, *Ctenolophon* (Ctenolophonaceae), aff. *Durio* (Bombacaceae), *Gonystylus* (Gonystylaceae), *Ixonanthes* (Ixonanthaceae) and Polygalaceae. Based on subsurface data from southern Sulawesi where the most complete Eocene succession has been observed (Morley unpublished),

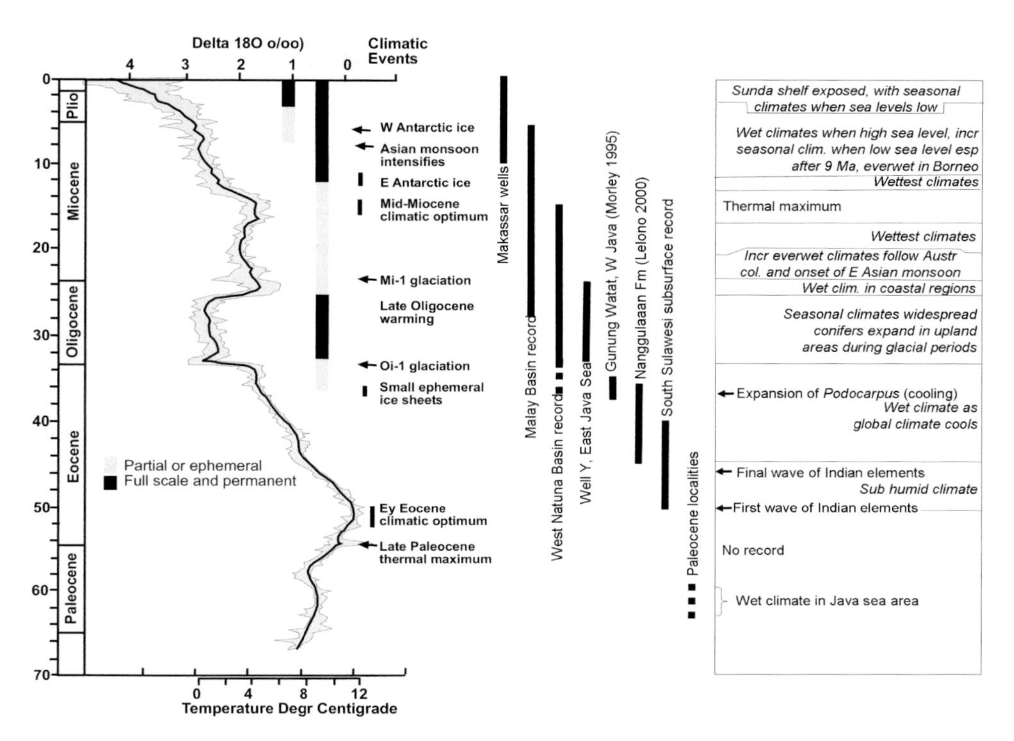

Figure 4.2 Oxygen isotope curve of Zachos et al. (2001), compared to palynological record in basins across Southeast Asia compared to major tectonic and climatic events.

these taxa seem to have dispersed in three phases, with the parent plant of *Palmaepollenites* arriving at the beginning of the Middle Eocene at about 50 Ma, followed by *Alangium*, aff. *Eugeissona* and aff. *Durio* at about 48 Ma, and subsequently with aff. *Beauprea*, *Ctenolophon* and Polygalaceae by about 46 Ma.

The southern Sulawesi record suggests a change in climate through the Middle Eocene. Palynomorph assemblages from the Middle Eocene are of low diversity, with pollen of Restionaceae as one of the dominants, possibly suggesting a subhumid climate. However, during the later Middle Eocene assemblages are of higher diversity and are dominated by *Palmaepollenites* spp. suggesting a period of wetter climate. This succession corresponds with the appearance of coals in the stratigraphic record and is contemporaneous with the Middle Eocene Nanggulan Formation from Central Java (Fig 4.1), studied for palynology by Morley (1982), Takashi (1982) and intensively by Lelono (2000). The interval with common Restionaceae pollen occurs just after the period of warmest global temperatures as suggested by oxygen isotope data of Zachos et al. (2001), whereas by the later Middle Eocene, global temperatures had reduced (Fig 4.2). It is suggested that the subhumid climate inferred by the dominance of Restionaceae pollen in the earlier Middle Eocene may relate to the time of the Eocene thermal maximum, with everwet climates developing as the global climate cooled.

An everwet climate is also suggested for the Late Eocene of the Mangkalihat Peninsula in Kalimantan (Fig 4.1), where coals yield abundant *Palmaepollenites* spp. and *Meyeripollis nayarkotensis* (Morley 2000: 188), and also in West Java, where coals from Bayah and Gunung Walat are also dated palynologically as Late Eocene from the overlap of *Meyeripollis nayarkotensis* and *Proxapertites operculatus (*Morley 1995, unpublished). The presence of regular pollen of Podocarpaceae from Late Eocene sediments at Nanggulan (Lelono 2000) suggests further cooling compared to the Middle Eocene. It is thought that Podocarpaceae dispersed to Java via the Indian Plate (Morley 2010).

4.4 Oligocene

The Oligocene has been extensively studied for palynology from the West Natuna Basin (Morley et al. 2003; Morley 2006), the Malay Basin (Yakzan et al. 1996, Morley and Shamsuddin 2006; Shamsuddin and Morley 2006), and the Java Sea (Lelono and Morley 2011). Most studies have been undertaken in pull-apart rift basins, in which marine influence is absent or minimal and marine index fossils absent. The Oligocene of the Malay Basin and West Natuna is an important hydrocarbon-bearing succession that has proved difficult to date using age-restricted microfossils and is subdivided using a sequence-biostratigraphic approach (Fig 4.3) based on the recognition of climatostratigraphic cycles using palynology (Morley et al. 2003, Morley and Morley 2011).

Oligocene palynomorph assemblages differ markedly from those of the Eocene, with pollen of Poaceae and Pinaceae becoming prominent, and the rich palm pollen assemblages of the Eocene disappearing from the record. The presence of conifer pollen in the Oligocene of the South China Sea region was first brought to attention by Muller (1966), and discussed in more detail by Muller (1972), who suggested that intervals with abundant conifer pollen reflected periods with elevated terrain in the hinterland. The likelihood is that during the Oligocene the South China Sea region, which underwent widespread rifting at the time, was characterised by a horst and graben topography and that conifer pollen was derived from montane forests growing on higher altitude horsts in areas possibly affected by some degree of water stress. This is suggested because in more equatorial areas around Borneo, where montane climates were more likely to be everwet, there is only a poor relationship between conifer pollen and the pattern of palaeoelevation as proposed by Hall (1998 and this volume, Chapter 3), suggesting that conifers were minor elements of montane forests until the diversification of montane podocarps by immigration from Australasia in the Late Neogene, (see below).

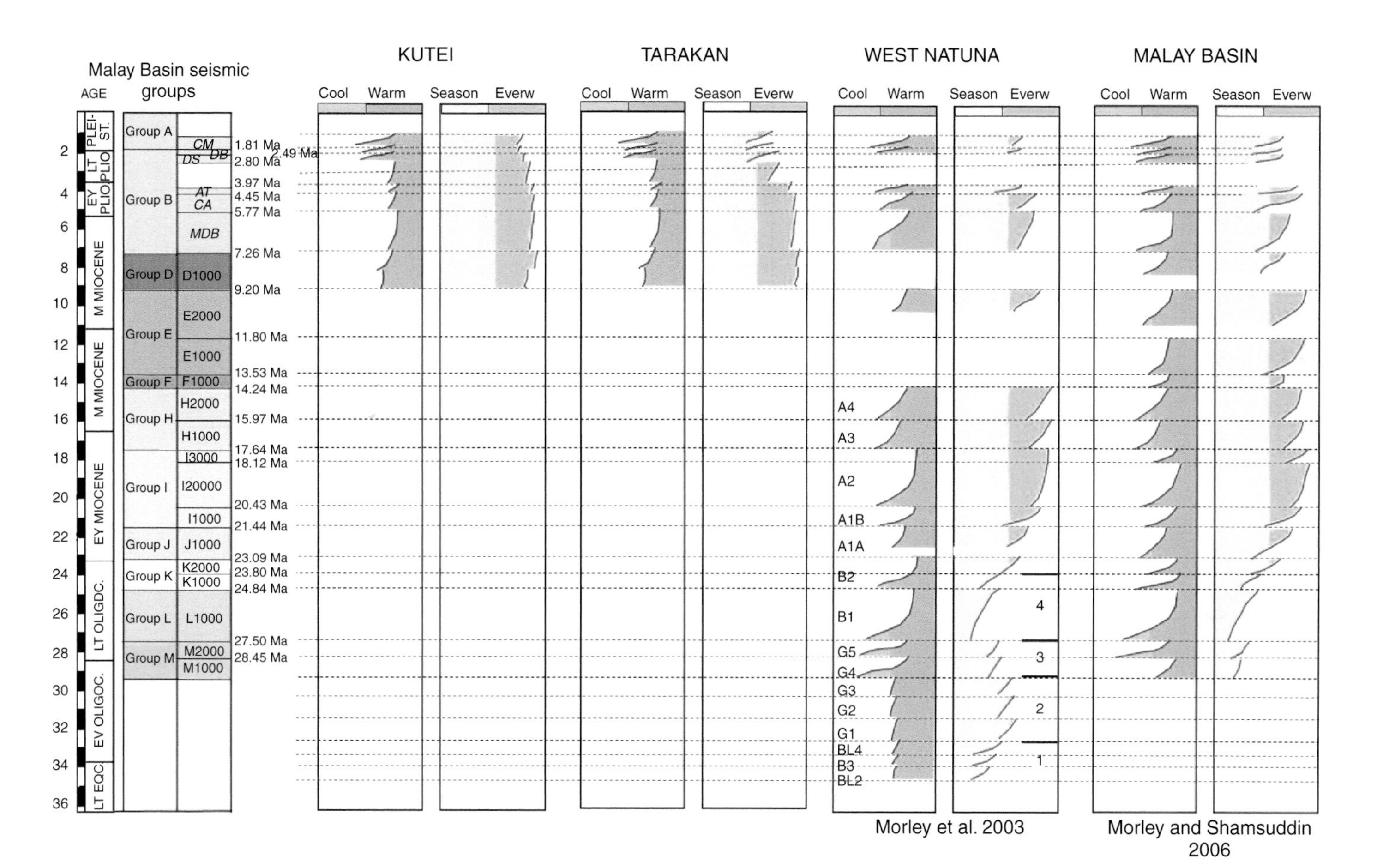

Figure 4.3 Climate trends in the Malay (Morley and Shamsuddin 2006) and Natuna Basins (Morley et al. 2003, Morley 2006) compared to schematic climate history for the Borneo margin of the Makassar Straits. Alpha-numeric terminology for West Natuna basins indicates biostratigraphically defined sequences of Morley et al. (2003). See plate section for colour version.

Songtham et al. (2003) have also proposed that the southeastern shift of Indochina into the tropical zone following the Indian collision with Asia may have been a contributing factor (see below).

The Pinaceae include common pollen of the genera *Abies*, *Picea* and *Tsuga*, accompanied also by pollen of *Alnus* (Betulaceae), which, when occurring in acmes may reflect periods of cooler climate, and probably also reflect freezing temperatures in the mountains. *Pinus* pollen is also common, and is often coupled with the common occurrence of Poaceae pollen, and probably relates to periods of drier climate with open pine savanna, as occurs today in parts of Thailand. It is noteworthy that coals are essentially absent from the Natuna and Malay Basin Oligocene, also supporting the suggestion of subhumid climates in this area at this time, although coals are common in the Late Oligocene of the Nam Con Son Basin in Vietnam, suggesting a more equable maritime climate in coastal regions.

4.4.1 Oligocene climate history of South China Sea pull-apart basins

The detailed pattern of climate change within this region of interlinked intercontinental basins ties closely with the phases of basin development of synrift (sediments deposited during the period of rifting) and postrift (sediments deposited after rifting ceased) and inversion (period of compression by subsequent tectonic movement) (Fig 4.4).

From the perspective of palaeoclimate, the Oligocene can be divided into four broad packages (Fig 4.3), the first two of which relate to the synrift phase. The earliest Oligocene or possibly latest Eocene interval, coinciding with the Benua and Lama Formations and seen only in the West Natuna Basin, has common *Barringtonia* (Lecythidaceae) pollen, suggesting a seasonally inundated freshwater swamp setting (the Tonle Sap 'foret inondee' in Cambodia would provide an analogue), and common Poaceae and *Celtis* pollen without bisaccate conifer pollen, suggesting open woodland and a period of seasonally dry climate The interval corresponds to palynological Zones PIA-C of Morley et al. (2003) (Fig 4.4). The second package, coinciding with the Belut Formation, is characterised mainly by pollen of open woodland trees such as *Celtis* (Ulmaceae) and Combretaceae but without bisaccates and just low frequencies of Poaceae pollen. It also contains common specimens of the pollen morphotaxon *Malvacipollis diversus*, which compares closely with pollen of the Australian Euphorbiaceae *Austrobuxus swainii* and *Dissiliaria baloghioides* (Martin 1974), which are both trees of drier rainforests/warm temperate sclerophyll communities in southeast Queensland/northwest New South Wales (Pickett et al. 2004). This assemblage suggests a more closed canopy sclerophyll woodland (Zones PII to PIVA) and a slightly wetter but still distinctly seasonal dry climate.

The third and fourth packages coincide with the postrift phase. The third coincides with the Gabus Formation in West Natuna and Seismic Group M in Malay

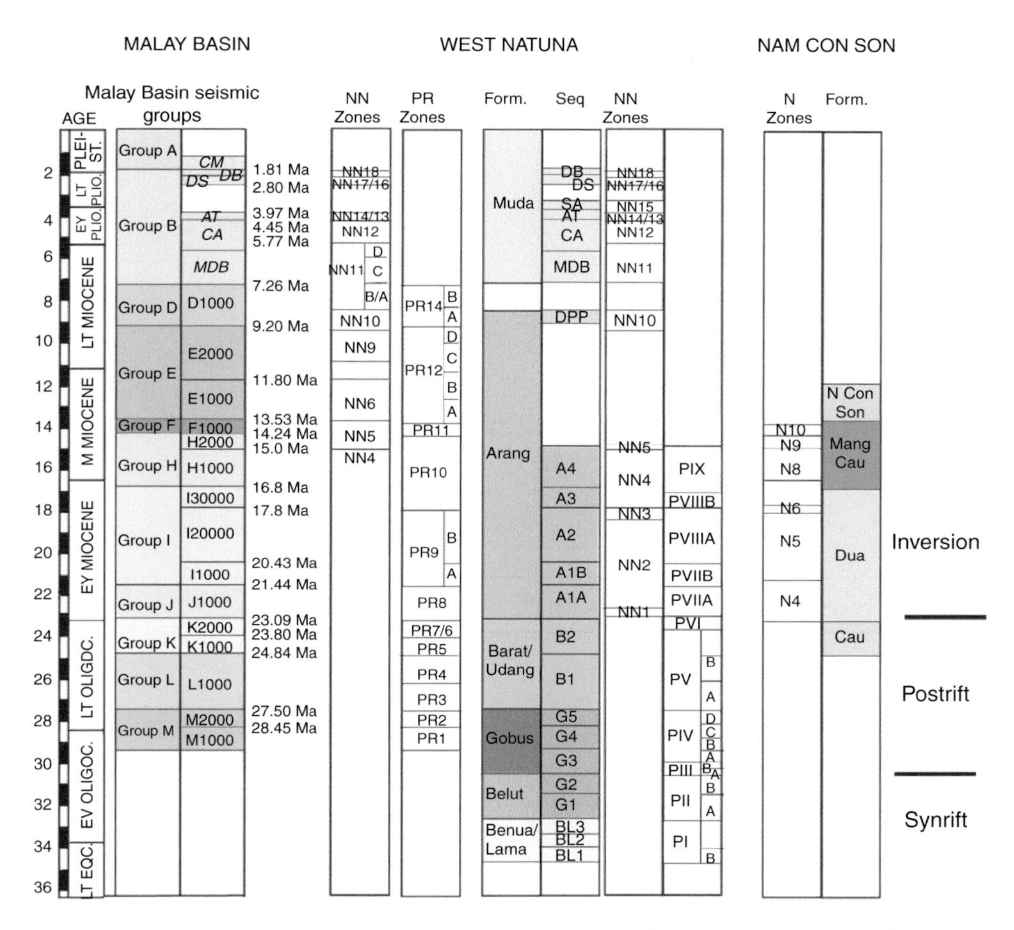

Figure 4.4 Stratigraphic relationships in the Malay Basin (Shamsuddin and Morley 2006), West Natuna (Morley et al. 2003), and Nam Con Son Basins, showing correlations to palynological, nannofossil and planktonic foraminiferal zones and the main phases of basin development across the region. See plate section for colour version.

Basin. It is has the same sclerophyll woodland assemblage as for the previous package, but also contains common *Pinus* pollen and the low representation of Poaceae pollen (Zones PIVB-D, and PR1–2 of Yakzan et al. 1996 in Malay Basin), suggesting seasonally dry pine-dominated forest, but acmes of temperate taxa such as *Picea*, *Abies* and *Tsuga* suggest intermittent cooler periods.

The final interval corresponds to the Barat Formation in West Natuna and to Seismic groups L and K in the Malay Basin. This interval is noteworthy because it is characterised by an acme of *Shorea* type pollen and common Poaceae, suggesting monsoonal dipterocarp forest (Morley 2000: 284), but it also contains common bisaccates, which are dominated by *Pinus*. *Ephedra* pollen is also regularly present

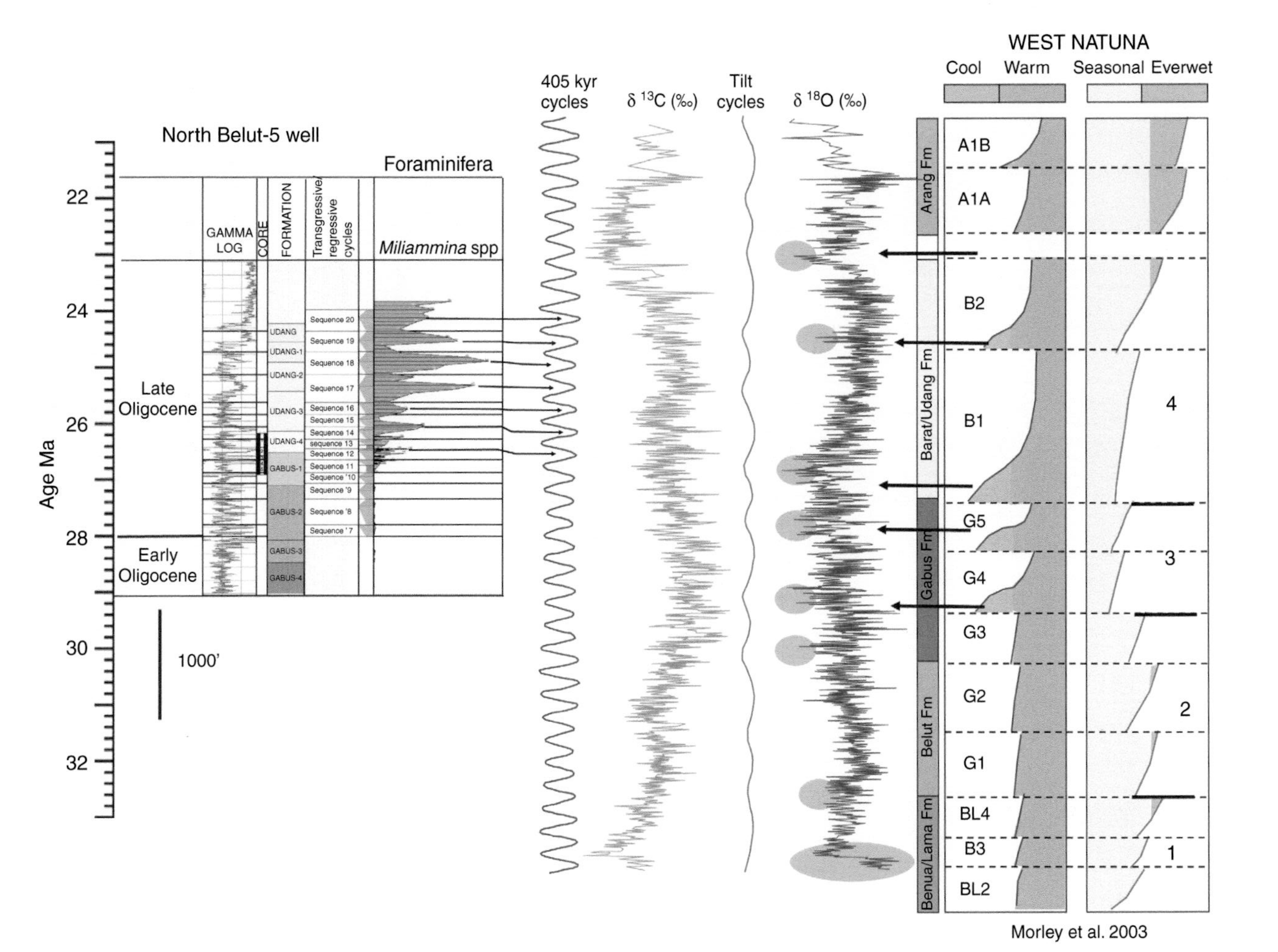

North Belut-5 well
Age Ma
22
24
26
28
30
32
Late Oligocene
Early Oligocene
1000'
GAMMA LOG
CORE
FORMATION
Transgressive/regressive cycles
Foraminifera
Miliammina spp
UDANG
UDANG-1
UDANG-2
UDANG-3
UDANG-4
GABUS-1
GABUS-2
GABUS-3
GABUS-4
Sequence 20
Sequence 19
Sequence 18
Sequence 17
Sequence 16
Sequence 15
Sequence 14
sequence 13
Sequence 12
Sequence 11
Sequence '10
Sequence '9
Sequence '8
Sequence '7
405 kyr cycles
δ ¹³C (‰)
Tilt cycles
δ ¹⁸O (‰)
WEST NATUNA
Cool
Warm
Seasonal
Everwet
Arang Fm
Barat/Udang Fm
Gabus Fm
Belut Fm
Benua/Lama Fm
A1B
A1A
B2
B1
G5
G4
G3
G2
G1
BL4
B3
BL2
4
3
2
1
Morley et al. 2003

Figure 4.5

together with diverse rainforest elements. The interval therefore appears to contain a mixture of wet and seasonal climate elements. It is also characterised by short-lived intervals with pollen of the temperate conifers *Abies*, *Picea* and *Tsuga*, suggesting periods of cooler climate allowing montane forests to expand to lower altitudes in upland areas. This interval coincides with the Late Oligocene thermal maximum as recorded by Zachos et al. (2001) and corresponds to palynological Zones PV–VII for West Natuna or zones PR3–7 in the Malay Basin.

This period also witnesses the first evidence of marine influence across the region, in the form of the brackish water foraminifer *Miliammina*, which appears suddenly and ubiquitously without other foraminiferal elements in most lacustrine facies, occurring at the same stratigraphic horizon across the Con Son to Malay Basins, and subsequently occurs in successive blooms (Fig 4.5) (Morley et al. 2007). The regularity with which the appearance of *Miliammina* appears across the region suggests that it relates to a sudden sea-level rise over extensive low-lying topography which bore numerous freshwater lakes.

An explanation of the cyclical *Miliammina* blooms, and indeed many aspects of the Oligocene climate system in this area, is tentatively forthcoming from comparison of cyclicity patterns with detailed isotope data from ODP Site 1218 in the equatorial Pacific (Pälike et al. 2006), where the climate system response to intricate orbital variations suggests a fundamental interaction of the carbon cycle, solar forcing, and glacial events. *Miliammina* acmes show virtually the same cyclicity as the ^{13}C cycles from ODP 1218 (Fig 4.5), which correlate to the 405 kyr Earth's eccentricity cycles and are likely to reflect repeated marine transgressions driven by global temperature change, whereas conifer pollen acmes correlate to periods with strongly positive ^{18}O signals, and to 1.2 Myr oscillations of the Earth's tilt cycles. The conifer pollen acmes therefore most likely reflect periods of glacial expansion and low sea level, with lowland areas experiencing a distinctly seasonal climate.

4.4.2 Oligocene climate records from the Java Sea

A clear record of palaeoclimate change from a palynological study has recently been published from the East Java Sea (Lelono and Morley 2011). The record spans the whole of the Oligocene, with most detail from the Late Oligocene (Fig 4.6).

Caption for Figure 4.5 North Belut-5 Well, West Natuna Basin, *Miliammina* spp. acmes through latest Oligocene reflecting successive sea-level rises during the latest Oligocene thermal maximum (Morley et al. 2007) correlated against 405 kyr eccentricity cycles, and ^{13}C and ^{18}O data from Pälike et al. (2006). Cool climate episodes based on acmes of conifer pollen are also correlated with periods with highest ^{18}O values. Grey ellipses indicate glacial periods. See plate section for colour version.

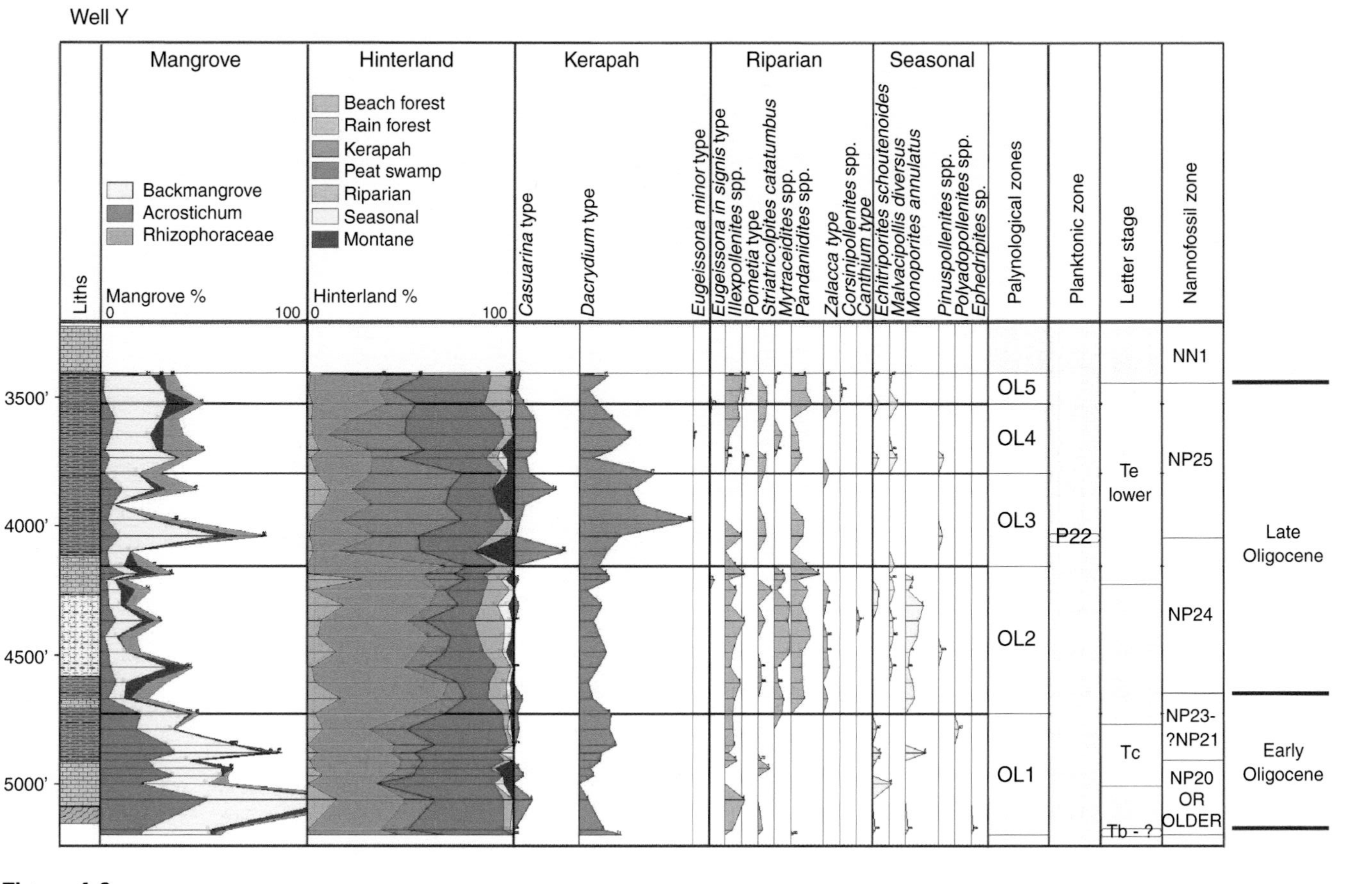

Well Y
Liths
Mangrove
Backmangrove
Acrostichum
Rhizophoraceae
Mangrove %
0
100
Hinterland
Beach forest
Rain forest
Kerapah
Peat swamp
Riparian
Seasonal
Montane
Hinterland %
0
100
Kerapah
Casuarina type
Dacrydium type
Eugeissona minor type
Riparian
Eugeissona in signis type
Ilexpollenites spp.
Pometia type
Striatricolpites catatumbus
Mytraceidites spp.
Pandaniidites spp.
Zalacca type
Corsinipollenites spp.
Canthium type
Seasonal
Echitriporites schoutenoides
Malvacipollis diversus
Monoporites annulatus
Pinuspollenites spp.
Polyadopollenites spp.
Ephedripites sp.
Palynological zones
Planktonic zone
Letter stage
Nannofossil zone
3500'
4000'
4500'
5000'
OL5
OL4
OL3
OL2
OL1
P22
Te
lower
Tc
Tb - ?
NN1
NP25
NP24
NP23-
?NP21
NP20
OR
OLDER
Late
Oligocene
Early
Oligocene

Figure 4.6

The Early Oligocene is characterised by the dominance of rainforest pollen suggesting a wet climate. Within the early part of the Late Oligocene, corresponding to nannofossil zone NP24, palynomorph assemblages contain significant abundances of Poaceae pollen (*Monoporites annulatus*) together with that of *Schoutenia* or *Echitriporites schoutenioides* (Tiliaceae; *Schoutenia* is an element of semi-evergreen rainforest) suggesting a more seasonal climate aspect. This interval is also characterised by the regular occurrence of *Malvacipollis diversus*, as noted above, derived from Australian *Austrobuxus/Dissiliaria*). Similar assemblages, but with more common *Malvacipollis diversus*, are found through the time-equivalent 'freshwater' Talang Akar Formation in the Sunda Basin, and again, seasonality of climate is suggested (Lelono and Morley 2011).

Assemblages from the Late Oligocene, equivalent to nannofossil zone NP25 and also from the time-equivalent 'coaly' Talang Akar Formation in the Sunda Basin contain common *Dacrydium* (Podocarpaceae) and *Casuarina* type pollen (Casuarinaceae; the same pollen type occurs in the peat swamp genus *Gymnostoma*). These taxa are of Gondwanan origin, and following dispersal into the Southeast Asian region found niches in peat swamps and kerangas vegetation (*Dacrydium*, *Gymnostoma*), montane forests (*Dacrydium* spp. and *Casuarina junghuhniana*) and also along beaches (*Casuarina equisitifolia*). The occurrence of abundant *Casuarina* type and *Dacrydium* pollen from the Java Sea Late Oligocene suggests a peat swamp derivation, since these pollen types are common in coals from the 'Coaly' Talang Akar Formation in the Sunda Basin (Morley 2000). *Gymnostoma* and *Dacrydium* do not occur in the phasic communities of the domed topotrophic/ombrotrophic peat swamps which are widespread along the coasts of Kalimantan and Sumatra (e.g. Anderson 1963), but characterise the less well known ombrotrophic (blanket bog type) 'kerapah' or watershed peats which occur locally in Borneo on poorly drained podsolic soils (Brunig 1990, Morley 2000) in areas of 'superwet' climate (Richards 1996). The occurrence of

Caption for Figure 4.6 Pollen record and palynological zonation from Well Y, East Java Sea, Palynological zone OL3 reflects a period of seasonally dry climate, whereas zones OL4 and OL5 reflect a very wet climate (Lelono and Morley 2011). Main components of palynomorph assemblage groupings are as follows: 'Backmangrove', Sonneratioid and *Brownlowia* pollen; *Nypa* (in blue) 'Coastal', *Terminalia*, *Thespesia*, *Barringtonia* pollen; 'Rainforest' pollen of Dipterocarpaceae, Burseraceae, Leguminosae and many others; 'Kerapah', *Casuarina* and *Dacrydium* pollen; 'Peat swamp', *Blumeodendron*, *Cephalomappa*, *Calophyllum*, *Durio* and Sapotaceae pollen; 'Seasonal', Poaceae pollen and *Malvacipollis diversus*; 'Montane', mainly *Podocarpus* pollen and Pinaceae. The groups reflect the main vegetation types from which the fossil pollen is thought to be derived – the taxa are not wholly restricted to the vegetation types indicated. See plate section for colour version.

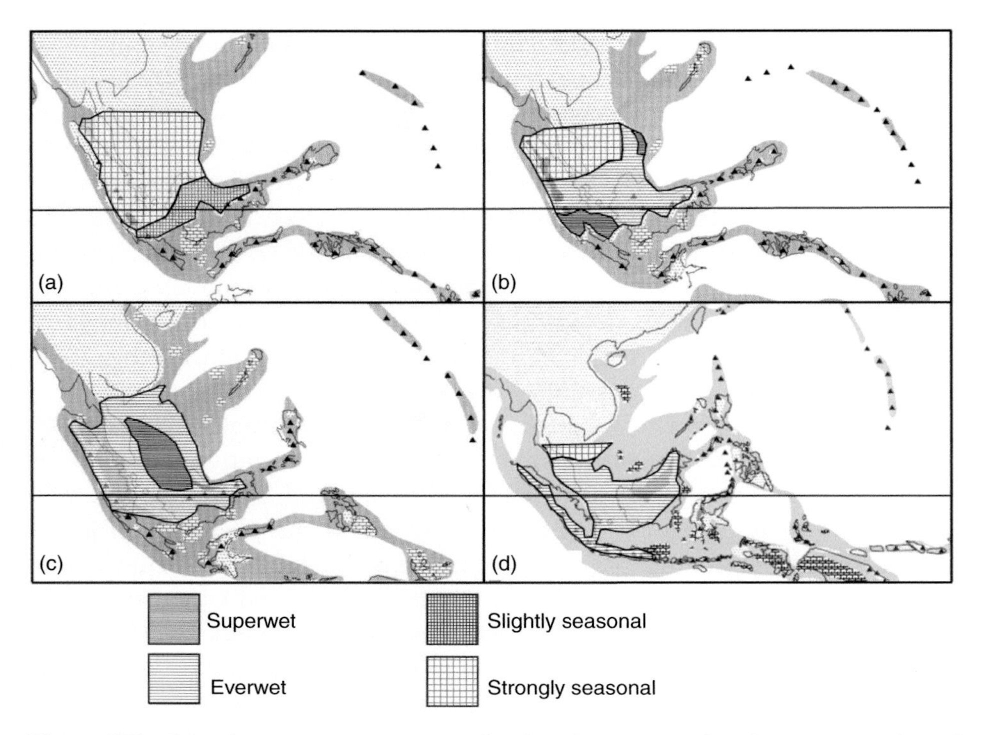

Figure 4.7 Palaeoclimate reconstructions for the Oligocene and earliest Miocene based on subsurface data from the Java and south China Seas: (a) 28 Ma, OL2; (b) 25 Ma OL 3–4; (c) 20 Ma PR9 in Malay Basin; (d) 10 Ma, PR 14 in Malay Basin. (a–c) updated from Lelono and Morley (2011). Base maps from Hall (1998). See plate section for colour version.

'kerapah' swamps is thus convincing evidence for an everwet climate and the pollen record indicates that kerapah peats bear one of the oldest plant communities of Southeast Asia.

Evidence for seasonal climates is also forthcoming from the Oligocene of Sumatra, from the Pematang Formation, which is often characterised by Poaceae pollen (Pribatini 1993) and also sometimes includes red beds, suggesting a subhumid climate (e.g. Carnell et al. 1998), but dating is less satisfactory and it is difficult to place the Pematang succession into that recorded from the Malay and West Natuna Basins.

4.4.3 Summary of Oligocene climate trends

The main pattern was of wetter climates along the southern and eastern fringe of Sundaland, with more seasonal climates to the north and west (Fig 4.7). During the mid Oligocene, subhumid seasonal climates may have extended across much of the region, but by the late Oligocene, an everwet rainforested belt was probably present along the equatorial zone and also along the east coast, with a super-wet climate allowing the development of widespread kerapah swamps along the southern margin.

4.5 Early and Middle Miocene

The collision of the Australian Plate with Sunda at the end Oligocene allowed some degree of intermingling of the Australasian and Southeast Asian floras, but these dispersals had little effect on the overall floristic composition of the Sunda region, unlike the invasion following the collision of India with the Asian Plate in the early Tertiary; just a minor increase in diversity. Even typically 'Australian' elements such as Casuarinaceae and *Dacrydium*, have now been shown to have dispersed into the region during, or possibly before, the Early Oligocene (Lelono and Morley 2011), with *Dacrydium* dispersing into the area via the Indian Plate (Morley 2010). The collision is thought to have had a major impact on the vegetation and climate of the region, however, by disrupting the Indonesian Throughflow, with the result that moisture which previously would have been carried from the Pacific Warm Pool was subsequently shed on Sundaland from this time onward (Morley 2003, 2006), rather than being carried by marine currents to India and East Africa.

For the Early and Middle Miocene, the record is mainly from the Malay Basin including its extension into Vietnam territorial waters, West Natuna and Sarawak/Brunei. Minor studies have also been undertaken from South and Central Sumatra. This review focuses on sequence stratigraphically restrained depositional cycles from the Malay and West Natuna Basins; most other regional evidence is reviewed in Morley (2000) and will not be repeated here.

4.5.1 Malay Basin and West Natuna climate history

The Early and Middle Miocene succession in the Malay, West Natuna and associated basins such as Con Son and Penyu corresponds to the sag or inversion phase of basin development across the region, with sedimentation being much more widespread rather than being confined to the rifts (Fig 4.4).

The pollen record from these basins is very similar, with successions dominated mostly by rainforest and peat swamp pollen. However, due to differences in the representation of mangrove pollen in the two basins during the Early Miocene, different palynological zonations have been developed for the two areas, with the 'PR' scheme of Yakzan et al. (1996) for the Early Miocene, based mainly on evolution of mangrove pollen, applying to the Malay Basin, and the 'P' scheme, of Morley et al. (2003) based more on the record of pollen from freshwater swamps applying to West Natuna. Both schemes, however, can be used and cross-correlated in the southern Malay Basin (Fig 4.4).

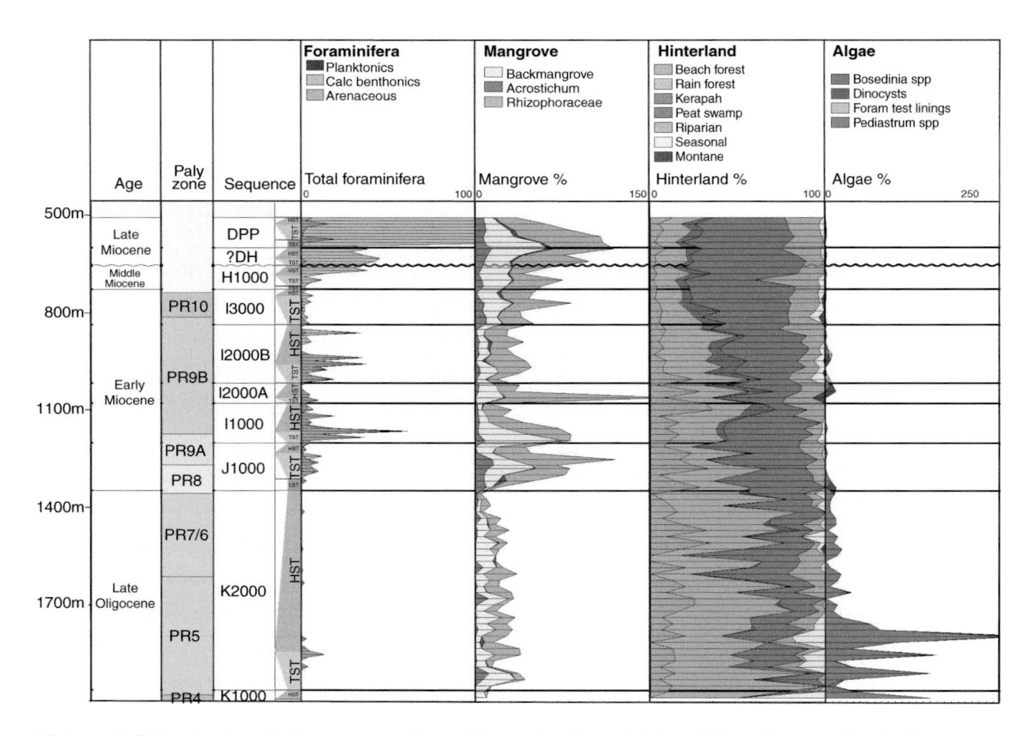

Figure 4.8 Example of climatostratigraphic cycles from Malay/West Natuna Basin from an unnamed well. The Late Oligocene Sequence K1000 is characterised by an initial interval with abundant freshwater algae and common seasonal climate pollen (mainly Poaceae), followed by abundant rainforest pollen. Miocene sequences are mainly characterised by an initial mangrove pollen acme coinciding with, or followed by, a foraminiferal abundance acme with hinterland pollen dominated by rainforest elements but with seasonal climate taxa often represented at the base of a cycle (e.g. Sequences I2000, I3000). See Fig 4.6 for details of palynological groups. See plate section for colour version.

The Early and Middle Miocene succession has been assessed by the integration of biostratigraphic and seismic data in the Malay Basin (Morley and Shamsuddin 2006, Wong et al. 2006); by characterising depositional cycles using palynology and benthonic foraminifera (Fig 4.8). For each depositional cycle, deposition typically begins when sea levels were low, then with rising sea levels mangrove swamps undergo a phase of expansion, followed by an acme of shallow water foraminifera at the time of highest sea level within the cycle. Climate changes during the cycle are reflected by changes in the hinterland pollen record. The typical pattern is an acme of montane or Poaceae pollen at the base of each package when sea levels were low, which is replaced by rainforest pollen as sea levels rise and mangrove swamps become widespread (e.g. Grindrod et al. 1999), with climates becoming warmer and wetter. Assemblages may also be characterised by pollen derived from freshwater swamps, which provide useful climate signals, with peat swamp pollen assemblages indicating an everwet climate, and pollen from herbaceous swamps, dominated by Cyperaceae and also with *Potomogeton* (Potomogetonaceae) or *Trapa* (Trapaceae) suggesting a more seasonal climate. Some intervals are dominated by abundant spores of *Ceratopteris* (Parkeriaceae), an annual floating swamp fern, which suggest the presence of seasonally inundated swamps.

The Early and Middle Miocene successions are best discussed in terms of Malay Basin sequence biostratigraphic units, equivalents of which can be carried across the region (Fig 4.4). The earliest Early Miocene deposition is represented by Sequence J1000 and its stratigraphic equivalents (Fig 4.3), and is characterised by abundant sands, which may reflect erosion at the time of the sea-level fall associated with the M1 glaciation (Zachos et al. 2001). In the West Natuna Basin, this interval is characterised by a change to dominance by rainforest pollen elements, although seasonal climate elements such as *Pinus* and Poaceae pollen often occur. Toward the northern end of the Malay Basin, seasonal swamps are suggested from the common occurrence of herbaceous swamp elements, such as *Potomogeton*, whereas in the Cuu Long Basin in southern Vietnam, *Lagerstroemia* (Lythraceae) swamps were widespread, reminiscent of the Myaing seasonal swamps of the Irawaddy (Yamada 1998). Some seasonality of climate is also suggested for Sequence I1000, except in the West Natuna Basin, where peat swamp elements generally dominate.

In parts of the Early Miocene, coals are often common, even in lowstand deposits, suggesting an everwet climate. Often the initial part of each cycle is characterised by pollen of *Dacrydium* and *Casuarina* types, suggesting kerangas or kerapah swamps, also indicating a very wet climate. This is especially the case for Sequences I2000 and I3000. In the West Natuna and Malay Basins (several Early Miocene incised valleys contain thick coals, and these yield common *Casuarina* type and *Dacrydium* pollen indicating kerapah type valley peats). The Early Miocene is therefore interpreted as the period with the wettest climates and the most extensive development of rainforests across the area. In the northern Malay Basin, Cyperaceae swamps were widespread within Sequence I1000, but within Sequence I2000, abundant spores indicate that swamps were overwhelmingly dominated by terrestrial swamp ferns, with the scrambling fern *Stenochlaena palustris* (Blechnaceae) being prominent.

The most widespread extent of tree-dominated peat swamps was during sequence I3000, during which time peat swamp elements, such as *Austrobuxus, Blumeodendron, Neoscortechinia* (Euphorbiaceae) and *Melanorrhoea* (Anacardiaceae) were common across the whole region from Penyu to North Malay Basin. Peat swamps were also widespread during sequences H1000 and H2000 at the time of the Middle Miocene thermal maximum during which climates

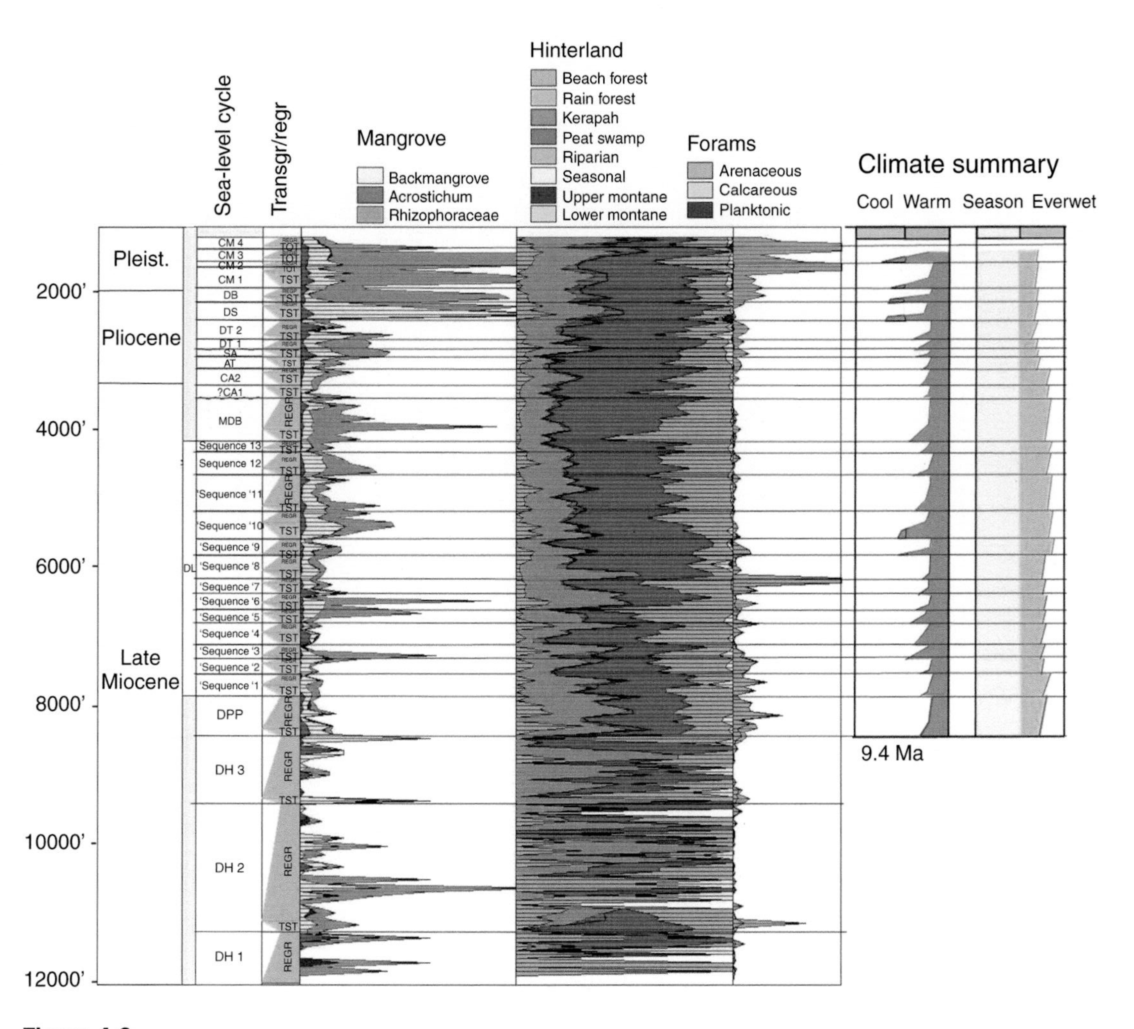

Sea-level cycle
Transgr/regr
Mangrove
Backmangrove
Acrostichum
Rhizophoraceae
Hinterland
Beach forest
Rain forest
Kerapah
Peat swamp
Riparian
Seasonal
Upper montane
Lower montane
Forams
Arenaceous
Calcareous
Planktonic
Climate summary
Cool
Warm
Season
Everwet
Pleist.
Pliocene
Late Miocene
2000'
4000'
6000'
8000'
10000'
12000'
CM 4
CM 3
CM 2
CM 1
DB
DS
DT 2
DT 1
SA
AT
CA2
?CA1
MDB
Sequence 13
Sequence 12
Sequence '11
Sequence '10
Sequence '9
Sequence '8
Sequence '7
Sequence '6
Sequence '5
Sequence '4
Sequence '3
Sequence '2
Sequence '1
DL
DPP
DH 3
DH 2
DH 1
TST
REGR
9.4 Ma

Figure 4.9

remained everwet, in contrast to the end Oligocene thermal maximum, when sub-humid climates prevailed. Based on the record of coals, some of the most extensive peat swamps developed across the Malay Basin during the latest Middle Miocene Sequence E1000, coinciding with the initial stage of Late Neogene cooling. Within the Middle Miocene, indicators of seasonality are also of limited occurrence, but short-lived acmes of conifer pollen suggest intermittent periods of cooler climate at times of low sea level.

4.6 Late Miocene and Pliocene

The most comprehensive overview of Late Miocene and Pliocene climate change is a study of petroleum exploration wells from the Kutei and Tarakan Basins, by Morley (2010, 2011). The Attaka Well B was key to this study (Fig 4.9) and is located close to the modern Mahakam Delta. It has yielded a particularly noteworthy succession because, due to very high sedimentation rates recorded during the mid part of the Late Miocene, it has been possible to characterise in some detail the palynological response of high resolution eustatic cycles, which probably represent the Earth's eccentricity cycles with a typical duration of about 100 ka per cycle.

There are at least 13 transgressive/regressive cycles within the 'mid' Late Miocene interval that are accurately dated from 8.5 to 7.0 Ma using nannofossils and by cross-correlation to well dated wells on the slope, giving a period of 107 ka per cycle. Each cycle is characterised by an initial acme of riparian pollen followed by mangrove pollen, and generally followed by an acme of foraminifera at the maximum flood, and with peat swamp and rainforest elements predominating during the latter part of the cycle. The cycles show a strong rhythmical pattern within mangrove and hinterland pollen groups and these are mainly thought to reflect changes in lowland and coastal vegetation associated with oscillating sea levels and patterns of coastal progradation.

During the later Late Miocene and Early Pliocene, about six palynological cycles are recorded, with each cycle representing about 600 ka. Each cycle shows abundant riparian pollen at the base, with rainforest elements increasing up section. Within the Late Pliocene and Pleistocene, the succession is mainly

Caption for Figure 4.9 Palynological and foraminiferal record for Attaka Well B, located to the northwest of Mahakam Delta, showing age, nannofossil zones, sedimentation rates, (black line shows actual rates uncorrected for compaction), mangrove pollen acmes, hinterland trends, foraminiferal abundance and palaeoenvironments (courtesy of Chevron IndoAsia and Migas). See Fig 4.6 for details of palynological groups. See plate section for colour version.

characterised by high-resolution transgressive/regressive cycles, with strong mangrove pollen acmes reflecting the high amplitude sea-level oscillations of the Quaternary.

There is minimal evidence for change in seasonality of climate through individual cycles irrespective of stratigraphic position within each cycle; rainforest elements dominate all cycles from base to top, from the base of the Late Miocene, up to the Quaternary. The main conclusion, therefore, with considerable implications with respect to the long term continuity of rainforest communities in Borneo, is that everwet climates have occupied the equatorial Makassar Straits in general, and the Mahakam catchment area in particular, from at least basal Late Miocene onward, irrespective of the global climate scenario, and during intervals characterised by both high ('interglacial') and low ('glacial') sea levels.

A similar palynological succession has been obtained from a well from the Tarakan Basin (Fig 4.10). This succession shows less resolution than the Attaka Well, but for the Late Miocene and Early Pliocene, essentially shows the same picture, with everwet climates irrespective of the sea-level scenario.

The most notable features of the Tarakan Basin well, however, is the gradual increase in abundance of seasonal climate elements, mainly of grass and Asteraceae pollen, within the Late Pliocene and Pleistocene. The abundance of seasonal climate elements varies through each sequence, suggesting alternating more seasonal and more everwet climates, but resolution is insufficient to show clearly detailed changes in every package. The presence of seasonal climate pollen is a persistent feature of each of the Neogene successions so far studied in the Tarakan Basin and suggests that during the Quaternary and Late Pliocene, seasonal climates, probably supporting open canopy woodland, or perhaps seasonal swamps along river valleys, were a persistent feature of low sea-level periods in the area of Northeast Borneo. The current evidence suggests that equatorial Borneo has experienced everwet climates without interruption from the topmost Middle Miocene.

The climate history through the Late Miocene and Pliocene is also well reflected in data from the Malay Basin, but less so for Natuna where the Late Miocene is often represented by a hiatus. Most notable is that from the D1000 sequence onward (i.e. from about 9 Ma), seasonality of climate becomes significant, especially during periods of low sea level, during which periods acmes of Poaceae and montane conifers are well represented. Foraminiferal analyses of the Late Miocene and Early Pliocene of the Solo River succession in East Java (Fig 4.1) also suggest intermittent periods of cooler climate during the Late Miocene, based on the distribution of temperature-sensitive planktonic foraminifera (Van Gorsel and Troelstra 1981). The effect of the global climate temperature decline, summarised by Zachos et al. (2001) is thus clearly apparent across the region.

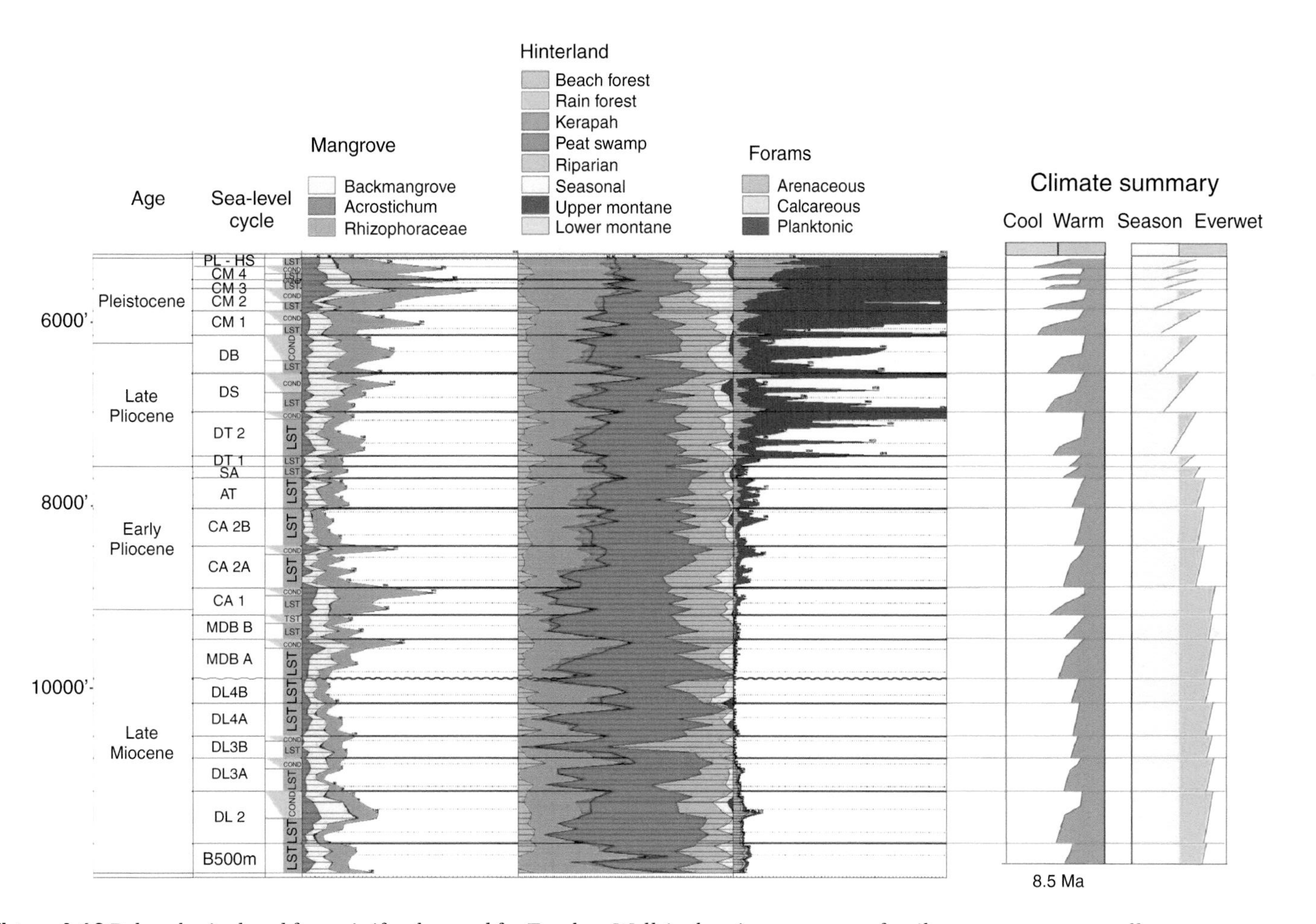

Figure 4.10 Palynological and foraminiferal record for Tarakan Well A, showing age, nannofossil zones, mangrove pollen acmes, hinterland pollen trends and foraminiferal abundance (courtesy of Migas). This well is from the mid/lower continental slope. See Fig 4.6 for details of palynological groups. See plate section for colour version.

4.7 Late Pliocene and Pleistocene

The effect of cooling global climates and increased seasonality, especially at subequatorial latitudes, is clearly shown by palynological analyses of the Late Pliocene or earliest Pleistocene Java Man locality near Modjokerto in East Java. (Morley unpublished in Huffman 2003). In this locality, the overwhelming dominance of Poaceae pollen accompanied by charred Poaceae cuticle fragments suggests the development of an open, nearly treeless savanna subject to burning, with montane rainforests with *Dacrycarpus imbricatus* and *Podocarpus* spp. growing on volcanoes in the hinterland, similar to the setting seen today on the drier islands of Nusa Tenggara. The increased seasonality of climate in the area of Java and Nusa Tenggara from the Pliocene onward relates to the development of the Australian or Javanese monsoon resulting from a combination of the northward drift of the Australian Plate into the southern hemisphere high pressure zone and Late Neogene global cooling (Morley 2000).

The effect of Late Neogene global temperature decline is also clearly marked by the immigration of temperate taxa into the area, especially within montane floras. This is particularly clearly shown by the immigration of *Dacrycarpus imbricatus* and *Phyllocladus* into the region from Australasia. *Dacrycarpus* (Fig 4.11) dispersed into Borneo and Java from New Guinea between about 3.4 and 4 Ma. It reached Sumatra and the Malay Peninsula at about 1.8 Ma, and is probably still extending its range into Indochina (Morley 2010). *Phyllocladus*, on the other hand, dispersed to Borneo from New Guinea at the beginning of the Pleistocene, but has not extended its range further.

Sundaland underwent a fundamental change in character at the beginning of the Quaternary during which time the amplitude of sea-level changes increased (Zachos et al. 2001), resulting in the previously mainly submerged Sunda Shelf being exposed during successive periods of low sea levels. Over the last million years the Sunda region effectively doubled in size during 'glacial' episodes and during the current highstand exhibits its smallest geographical area for that period, with the flora presently being in a state of refuge (Fig 4.12). The Sundaland vegetation dynamics of fragmentation and contraction followed by expansion, driven by glacial cycles, occur in the opposite phase to those in the northern hemisphere and Africa, indicating that Sundaland evergreen rainforest communities are currently in a refugial stage (Cannon et al. 2009).

The probable nature of vegetation growing on the exposed Sunda Shelf has received widespread attention (e.g. Morley and Flenley 1987, Heaney 1991, Voris 2000, Kershaw et al. 2002, Meijaard 2003, Bird et al. 2005, Cannon et al. 2009, Wurster et al. 2010) with differing opinions as to the extent of seasonal climate vegetation. The suggestion of a 'savanna corridor' from Indochina to Java has been widely suggested at periods of sea-level lowstand, initially to facilitate dispersal of

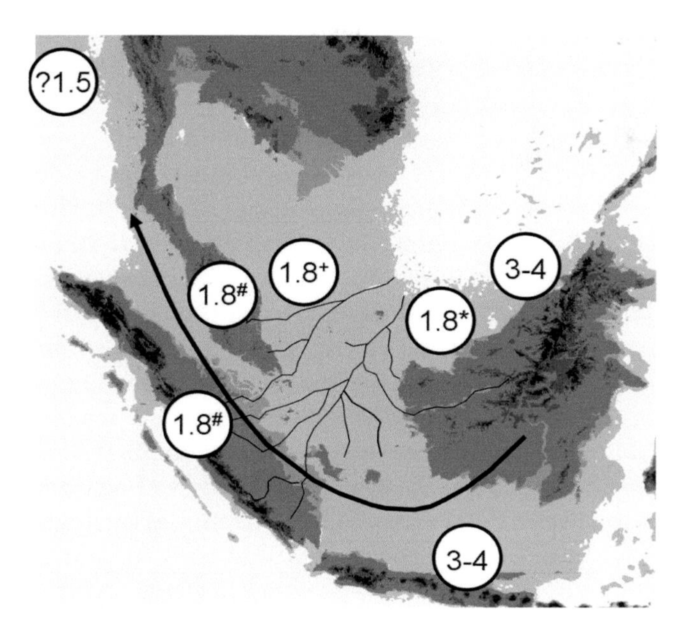

Figure 4.11 Dispersal of *Dacrycarpus imbricatus* across the Sunda region during the Pliocene and Pleistocene based on pollen records. Dates in Ma. The positions of the Proto-Musi, Hari, Kampar and Pahang Rivers at the time of the Last Glacial Maximum (and older sea-level lowstands) are shown following Voris (2000). These rivers carried pollen that was deposited in the Malay Basin (+) and in deltas offshore Northwest Sarawak (*) dating dispersal from Borneo/Java to the mountains of Sumatra and the Malay Peninsula (#).

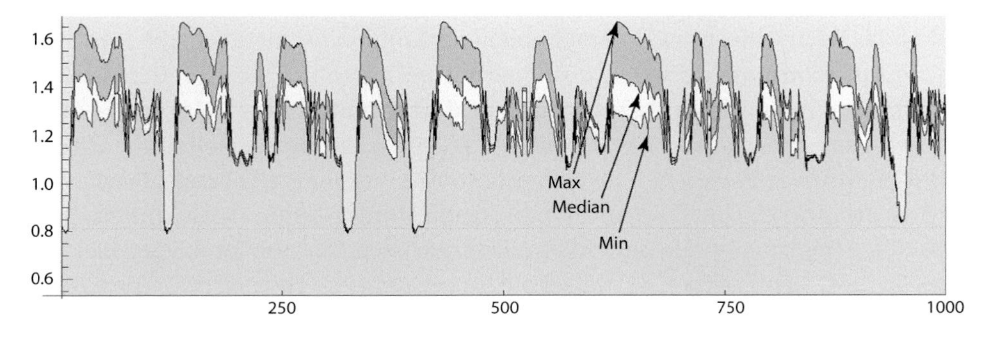

Figure 4.12 Land area of tropical lowland evergreen broadleaf forest over the last million years. The three lines represent the three model conditions (maximal, median and minimal), horizontal scale age in ka, vertical scale in $km^2 \times 10^6$ (from Cannon et al. 2009).

browsing and grazing mammals to Java (van den Bergh 2001), and subsequently to justify a floristic and faunal barrier between Sumatra/Malay Peninsula floras and those of Borneo (e.g. Heaney 1991, Gaythorne Hardy et al. 2002, Gorog et al. 2004, Woodruff 2010, Wurster et al. 2010). Cannon et al. (2009) emphasised that although there is evidence for drier climates at the time of glacial maxima, there is no botanical evidence to suggest that there was a continuous seasonal climate corridor from north to south.

Most discussion on Sundaland 'glacial' climate and vegetation has focused on the last glacial maximum (LGM), when sea levels were about 120 m below present day levels and climates were considerably cooler and also drier. This has misled many into thinking that such conditions apply to all 'glacial' intervals, but for most of the Quaternary, sea levels fluctuated between 40–60 m below present day levels and the climate would have been neither so dry, nor cool, as during the LGM. Based on the predicted nature of soil conditions from the comparison of a large floristic database from Borneo, Malay Peninsula and Sumatra, Slik et al. (2011) suggest that for the period with intermediate sea levels, rainforests were most likely continuous from west to east at equatorial latitudes, but that there may have been a dispersal barrier for many wet-adapted taxa from west to east due to the predicted widespread occurrence of kerangas and kerapah forests on low lying, suspected coarse-textured soils in central Sundaland between Sumatra and Borneo. It is thus necessary to visualise three contrasting sea-level/climate scenarios to understand Quaternary biotic distributions (Fig 4.13).

The most persistent scenario (approx. 55% of last million years based on the marine isotope record) would have been of periods with sea levels 40–50 m below current levels. With perhaps half of the current Sunda Shelf emergent, and with evergreen rainforests extending from Borneo to Sumatra, but with extensive kerangas forests and kerapah swamps, the latter principally on watersheds and poorly drained areas (Fig 4.13b). Seasonal climate vegetation, with common grasses, would have been widespread in emergent areas around, for instance, the Java Sea, based on the common occurrence of Poaceae pollen in the Sangkarang-16 marine core from offshore southern Sulawesi (Morley and Morley 2011), but would not have been so extensive in the Malay Peninsula, based on pollen data from Sungei Remis near Penang which suggests continuation of rainforests, *Pandanus* and peat swamps during Oxygen isotope stages 3 and 5a (Kamaludin and Yakzan 1998). Since the last interglacial, this would have been the case for oxygen isotope stages 5a–d, 4, 3 and early 1. ^{13}C data from guano deposits suggest that rainforests extended to Palawan at this time (Wurster et al. 2010).

The second most persistent scenario (37% of last million years) would be of periods with very low sea levels, such as during the last glacial maximum, when seasonal climate vegetation would have been very widespread, although may not have formed a continuous corridor from north to south (Fig 4.13c), for each glacial maximum. For the Malay Peninsula, the occurrence of *Pinus* and Poaceae-dominated palynomorph assemblages from an undated 'mid' Quaternary tin mine deposit near Kuala Lumpur, suggesting pine savanna (Morley 1998), has

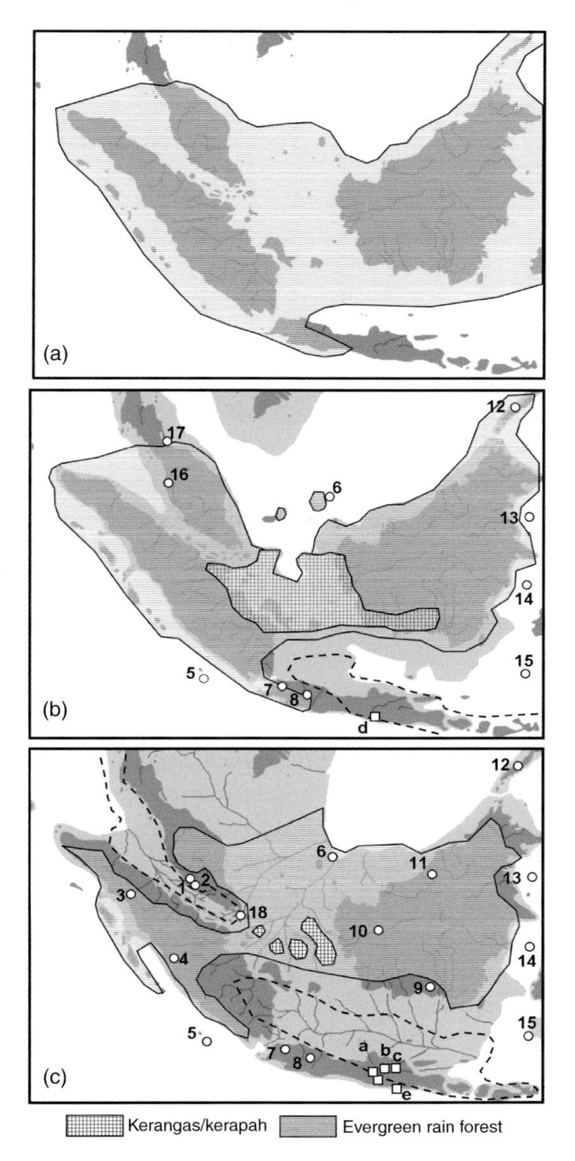

Figure 4.13 Three Quaternary climate-sea-level scenarios: (a) 'interglacial' high sea level, Marine isotope stage (MIS) 1 and 5e; (b) intermediate sea level, MIS stages early 1, 3, 4, 5a–d; (c) low sea level glacial maximum, MIS stage 2. Data for sites from: 1, Morley (1998); 2, Wurster et al. (2010); 3, Maloney and McCormac (1995); 4, Newsome and Flenley (1988); 5, van der Kaars (2010); 6, Wang et al. (2009); 7, van der Kaars et al. (2001); 8, van der Kaars and Dam (1995); 9, Morley (1982); 10, Anshari et al. (2004); 11, 12, Wurster et al. (2010); 13, 14, 15, Morley and Morley (2010); 16, Kamaludin and Yakzan (1998); 17, Rugmai et al. (2008); 18, Taylor et al. (2001). Dashed line indicates more open vegetation with common grasses.

formed the main basis for suggesting seasonality of climate for Quaternary glacial maxima, However a ^{13}C record from bat guano from nearby Batu Caves suggests a similar climate, and has been dated to the LGM (Wurster et al. 2010). Increased seasonality of climate is also suggested from South China Sea cores, from east of Natuna Island (Wang et al. 2009), and from Singapore, Nee Soon swamp (Taylor et al. 2001), based on the greatly increased presence of Poaceae pollen and charcoal compared to the Holocene and to equatorial LGM sites. Even at equatorial latitudes, climates would have been less humid, since peat growth intermittently ceased in equatorial Borneo (Anshari et al. 2004).

The succession of mammalian faunas recorded from the island of Java provides a critical perspective of the evolution of Sundaland vegetation during the Quaternary. The diverse, and well-dated faunas were largely adapted to open woodland vegetation (van den Bergh et al. 2001), and had affinities with Asian mainland faunas (India and China) suggesting that intermittently (on at least seven separate occasions), a belt of more open vegetation must have stretched from the Asian mainland to Java through which they could migrate; the so called 'Savanna Corridor' of Heaney (1991). However, compared with Asian fossil faunas, the Javanese faunas are depauperate, and lack the 'arid adapted' or 'open savanna' elements of time-equivalent Siwalik faunas from India (Medway 1972), suggesting that at no time during the Quaternary was the Sunda Shelf traversed by true savanna, despite the presence of open savanna in East Java during the early Pleistocene (Morley unpublished in Huffman 2001); thus open woodland and riverine grasslands were likely to have been the main open habitats. Similarly, seasonal evergreen forests in West Java and Myanmar contain many taxa in common, whereas Javanese and Indo-Burmese deciduous forests share few species, implying periodic Pleistocene connection of seasonal evergreen forests.

The first faunas (Satir and Ci-Saat) were characterised by dwarfing, indicating the late Pliocene insular nature of Java. However, from 1 to 1.5 (or ?1.8) Ma onward (van den Bergh et al. 2001, Morwood 2003, Swisher et al. 1994), over which time hominins formed part of the fauna, and especially from about 900 ka onward, with the change from 40 ka to increased amplitude 100 ka glacioeustatic cycles, diverse large browsing and grazing mammalian faunas intermittently migrated from the Asian mainland at times of low sea level and appropriate open woodland habitat (van den Bergh et al. 2001),

The first phases of immigration, were represented by the Jetis and Trinil faunas, followed by the Kedung Brubus Fauna, dated at 800–700 ka, whereas the succeeding Ngandong fauna (which may reflect more than one immigration) characterised the Pleistocene up until about 125 ka (van den Bergh et al. 2001). The succeeding Punung Fauna, which has been dated at 80–60 ka, is essentially a rainforest fauna with the first record of *Pongo* for Java. The likelihood is that immigration was during MIS stage 5a-d, when sea levels were lower than for the present day and there would have been a direct forested land connection to Sumatra (van den Bergh et al. 2001).

Javanese faunas from the last glacial period contain no novel woodland mammals (van den Bergh et al. 2001), only the forest dwelling *Elephas maximus* suggesting that opportunities for migration across the Sunda Shelf during the last glacial were inhibited. Thus a 'closed corridor' setting (*sensu* Cannon et al. 2009) may have been the case for the LGM, but a more 'open corridor' setting for earlier glacial maxima needs to be considered.

The third scenario would be represented by the present day with high sea levels and evergreen rainforests extending from the Kra Isthmus to West Java (Fig 4.13a), but significantly, occurring for just 7% of the last million years emphasising the 'refugial' nature of present day rainforests (Cannon et al. 2009).

4.8 Controls on climate change

Attempting an explanation of the changes of climate revealed by the Southeast Asian palynological record is not straightforward. It appears likely that controls were by different forcing mechanisms acting at different times. For the Paleocene and Eocene, it seems possible that global temperatures provided the main control on climate, with the hottest period of the Middle Eocene possibly experiencing a subhumid climate, and with the development of typically everwet climates following further global cooling during the later Middle Eocene. The significance of the closure of the Indonesian Throughflow on regional climates following the collision of the Australian Plate with Sunda at the end of the Oligocene has also been brought to attention and is thought to be responsible for the dramatic change to generally wetter climates, seemingly across the Sunda region, from the earliest Miocene onward (Morley, 2003, 2006), which interestingly coincides with the time of onset of the Asian monsoon across China (Sun and Wang 2005), and thus the possibility needs to be considered that the Indonesian Throughflow closure initiated the East Asian monsoon.

The trend to cooler and more seasonal climates during the later Late Miocene and Pliocene, and the immigration of cool-climate podocarps into the region, is also thought to relate to a combination of continued global cooling associated with the development of the Northern Hemisphere ice sheets, and to shifting plate positions and tectonic uplift. For the northern Sunda Shelf, Himalayan uplift, which initiated the Indian monsoon (Wang et al. 2003) and which also affected the intensity of the East Asian monsoon, (Sun and Wang 2005) may have been partly responsible for increased climatic seasonality seen in the Malay Basin from the Late Miocene onward. For the southern Sunda Shelf, including Java, the northern drift of the Australian Plate into the Southern Hemisphere subtropical high pressure zone, coupled with global cooling, is thought to have been the main control with respect to the Plio-Pleistocene strengthening of the Australian or Javanese monsoon.

An explanation of the remaining trends in climate change during the Oligocene and Early Miocene for the Malay and West Natuna Basins has proved challenging. Previously, Morley et al. (2003) made an attempt to correlate Oligocene climate change with the global climate curve of Abreu and Anderson (1998), but the improved isotope curve of Zachos et al. (2001) would require a different correlation.

Based on comparison of datasets from West Natuna with detailed isotope records from the equatorial Pacific (Pälike et al. 2006), it seems likely that the South China Sea climate responded to both 405 kyr and 1.2 Myr astronomical cycles, with successive marine transgressions relating to periods of climate amelioration correlating to 405 kyr cycles, but with the coolest climates correlating to periods of minimum tilt, with a cyclicity of 1.2 Myr (Fig 4.5).

The main enigma through the Late Oligocene to Middle Miocene is that the two global climate maxima, in the latest Oligocene and basal Middle Miocene, exhibit quite different climate scenarios, with the latest Oligocene being characterised by more seasonal and drier climates (although wetter in coastal regions), but with regionally much wetter climates in the Middle Miocene. The key to understanding the reasons for their differences is thought to relate to the closure of the Indonesian Throughflow in the latest Oligocene following the collision of the Australian Plate with Sunda, and the subsequent development of the East Asian monsoon. Thus prior to the development of this monsoon, during the latest Oligocene thermal maximum, the Sunda region was likely to have experienced a more 'equatorial' rather than monsoonal climate, possibly with two wet seasons corresponding to the equinoxes, and reduced annual rainfall (P. S. Ashton pers. comm. 2010). The East Asian monsoon subsequently dominated the region's climate during the Early Miocene, resulting in a widespread increase in rainfall, and subsequently the basal Middle Miocene thermal maximum was characterised by much wetter climates. Alternatively, continentality may have played the major role, with the greatest extent of subhumid climates coinciding with the Oligocene, when Sundaland showed its greatest extent of land.

For the Miocene of the Malay, West Natuna and associated basins, there also seems to be a possible correlation with the distribution of land and sea irrespective of the global climate scenario, with everwet climates approximately mirroring the extent of marine transgression across the Sunda Shelf. Thus everwet climates were particularly widespread following the initial phase of marine transgression, which culminated during Sequences I2000 and I3000, and continued until Sequence E1000, prior to the onset of global cooling in the Late Miocene. On the other hand, the wettest climates occurred at times of intermediate rather than at the time of highest global temperatures (Fig 4.4).

Songtha et al. (2003) discuss Oligo-Miocene changes in pollen assemblages from northern Thailand, close to the present day northern boundary of tropical rainforest. They attribute assemblage changes from those suggesting a warm temperate climate from the Oligocene/Early Miocene to assemblages suggestive of tropical rainforest from the Middle Miocene as due to the southeast shift of Indochina following the Indian collision with Asia (e.g. Taponnier et al. 1986, Hall 1998). This plausible suggestion raises the question as to whether the southeastern shift of Indochina may have influenced the representation of Laurasian conifers in the South China Sea region over the same period, which display a parallel but less pronounced decline at the same time? The likelihood is that it may have contributed to their greater representation prior to the Middle Miocene, but due to the more southerly position of the Malay and West Natuna Basins, would be unlikely to have been the only cause for their decline.

4.9 Summary and conclusions

The Southeast Asian region shows complex palaeogeographic/climatic interactions through the Tertiary period. Some sort of climatic gradient appears to have been in place during the Paleocene, with a subhumid climate in the region of Sarawak, and a wetter climate in Java. It is suggested that during the Eocene thermal maximum, high ambient temperatures resulted in a subhumid climate in southern Sulawesi, and that with global temperature decline during the later Middle Eocene, the opportunity developed for more typically everwet climates, indicated also from lithologies from the presence of coals.

The Oligocene climate across Sundaland was mainly subhumid, especially during the mid Oligocene, but wet climates developed along the southern and eastern margins of the region during the latest Oligocene. The northern part of Indochina experienced a warm temperate climate during the Oligocene and Early Miocene due to its more northerly position at that time.

Understanding climate changes at the time of the end Oligocene and Early Miocene thermal maxima is problematic, and it is suggested that the Late Oligocene subhumid climate in the South China Sea region may be a response to the thermal maximum. Also at a high-resolution scale, it is suggested that the period of the thermal maximum was characterised by rapid temperature fluctuations resulting in successive eustatic sea-level changes in West Natuna.

The collision of the Australian Plate with Sundaland at the end of the Oligocene is thought to have had a profound effect on regional climate by disrupting the Indonesian Throughflow, with the result that moisture which previously would have been carried from the Pacific Warm Pool was probably shed on the Sunda region from this time onward (Morley 2003, 2006), rather than being carried by marine currents to the west and possibly initiating the East Asian monsoon across Sundaland and also in China (Sun and Wang 2005). Thus, from the basal Miocene onward, the same moist monsoonal climate regime characterised much of the Sunda and East Asian region. However, it is suggested that the climate perturbations recorded from the Oligo-Miocene of the Malay and West Natuna Basins probably also relate to the Early Miocene opening of the South China Sea into the Sunda Shelf, the marine incursion bringing a more oceanic climate to the interior of Sundaland.

For the Late Miocene and Pliocene, most climate trends can be explained in relation to global cooling associated with the expansion of northern hemisphere ice sheets, tectonic uplift in the Himalayan region and also with the continued movement of the Australian Plate in the Southern Hemisphere subtropical high pressure zone resulting in the establishment of the Australian or Javanese monsoon.

Quaternary climates fluctuated rapidly in parallel with global climate perturbations, with three climate scenarios depending on global temperature, which also determined sea levels. At periods of glacial maxima, the Sunda region experienced much cooler and less humid climates, but it is unlikely that a continuous savanna corridor extended from north to south. During periods of intermittent sea level with temperatures intermediate between the LGM and Holocene, some degree of climate seasonality persisted in the region of Java, but the Malay Peninsula experienced a mainly wet climate. Predominantly everwet climates similar to that of today would have characterised interglacial periods with high sea levels.

A major new development in the clarification of the palaeoclimate record of the area is the demonstration that the island of Borneo, at least at equatorial latitudes, has experienced everwet climates without interruption from the basal Late Miocene. Such climates would have been implemental in maintaining rich and diverse rainforests which would have facilitated the diversification and maintenance of all elements of the Bornean rainforest fauna and flora which characterise the region today, possibly helping to explain why some areas of Borneo, such as the Lambir forest Reserve in Sarawak, contain the world's most floristically diverse rainforests (Wright 2001, Lee et al. 2002, Ashton 2010).

Acknowledgements

The two reviewers and editor, Peter Ashton, Carlos Jaramillo and James Richardson, have all made useful suggestions to improve this chapter. Peter Ashton has been particularly helpful in clarifying the discussion of seasonal climates and in understanding the history of monsoons, which has led to a more rational interpretation of events over the Oligo-Miocene boundary.

References

Abreu, V. S. and Anderson, J. B. (1998). Glacial eustacy during the Cenozoic: sequence stratigraphic implications. *American Association of Petroleum Geologists Bulletin*, **82**, 1385–400.

Anderson, J. A. R. (1963). The flora of the peat swamp forests of Sarawak and Brunei, including a catalogue of all recorded species of flowering plants, ferns and fern allies. *Gardens Bulletin, Singapore*, **20**, 131–238.

Anshari, G., Kershaw, A. P. and van der Kaars, W. A. (2004). Environmental change and peatland forest dynamics in the Lake Sentarum area, West Kalimantan, Indonesia. *Quaternary Science*, **19**, 637–55.

Ashton, P. S. (2010). Conservation of Borneo biodiversity: do small lowland parks have a role, or are big inland sanctuaries sufficient? Brunei as an example. *Biodiversity and Conservation*, **19**, 243–56.

Bird, M. I., Taylor, D. and Hunt, C. (2005). Palaeoenvironments of insular Southeast Asia during the Last Glacial Period: a savanna corridor in Sundaland? *Quaternary Science Reviews*, **24**, 2228–42.

Bransden, P. J. E. and Matthews, S. J. (1992). A stratigraphic and structural evolution of the East Java Sea, Indonesia. *Proceedings of the 21st Indonesian Petroleum Association Convention*, 418–53.

Brunig, E. F. (1990). Oligotrophic forested wetlands in Borneo. In *Ecosystems of the World, Vol 15, Forested Wetlands*, ed. A. E. Lugo, M. Brinson and S. Brown. Amsterdam: Elsevier, pp. 299–344.

Bush, M. B. (1995). Neotropical plant reproductive strategies and fossil pollen representation. *The American Naturalist*, **145**, 594–609.

Cannon C. H, Morley R. J. and Bush A. B. G. (2009). The current refugial rainforests of Sundaland are unrepresentative of their biogeographic past and highly vulnerable to disturbance. *Proceedings of the National Academy of Sciences, USA*, **106**, 11188–93.

Carnell, A., Butterworth, P., Hamid, B. et al.(1998). The brown shale of Central Sumatra, A detailed geological appraisal of a shallow lacustrine source rock. *Proceedings of the 28th Indonesian Petroleum Association*, 51–69.

Flenley, J. R. (1973). The use of modern pollen grain samples in the study of the vegetational history of tropical regions. In *Quaternary Plant Ecology*, ed. H. J. B. Birks and R. G. West. London: Blackwell Scientific Publications, pp. 131–41.

Gaythorne-Hardy, F. J., Davies, F. G., Eggleton, P. and Jones, D. T. (2002). Quaternary rainforest refugia in Southeast Asia: using termites (Isoptera) as indicators. *Biological Journal of the Linnean Society* **25**, 453–66.

Gorog, A. J., Sinaga, M. H., and Engstrom, M. S. D. (2004). Vicariance or dispersal? Historical biogeography of three Sunda Shelf murine rodents (*Maxomys surifer, Leololdamis sabanus* and *Maxomys whiteheadii*). *Biological Journal of the Linnean Society*, **81**, 91–106.

Grindrod, J., Moss, P., and van der Kaars, S. (1999). Late Quaternary cycles of mangrove development and decline on the north Australian continental shelf. *Journal of Quaternary Science*, **14**, 465–70.

Hall, R. (1996). Reconstructing Cenozoic SE Asia. In *Tectonic Evolution of Southeast*

Asia, ed. R. Hall and D. J. Blundell. *Geological Society Special Publication*, 106, 152–84.

Hall, R. (1998). The plate tectonics of Cenozoic SE Asia and the distribution of land and sea. In *Biogeography and Geological Evolution in SE Asia*, ed. R. Hall and J. Holloway. Amsterdam: Bukhuys Publishers, pp. 99–131.

Hall, R. (2002). Cenozoic geological and plate tectonic evolution of SE Asia and the SW Pacific: computer-based reconstructions, model and animations. *Journal of Asian Earth Sciences* **20**, 353–431.

Harley, M. M. and Morley, R. J. (1995). Ultrastructural studies of some fossil and extant palm pollen, and the reconstruction of the biogeographical history of the subtribes Iguanurinae and Calaminae. *Review of Palaeobotany and Palynology*, **85**, 153–82.

Heaney, L. R. (1991). A synopsis of climatic and vegetational change in Southeast Asia. *Climatic Change*, **19**, 53–61.

Huffman, O. L. (2003). Geological context and age of the Perning/Mojokerto *Homo erectus*, East Java. *Journal of Human Evolution*, **40**, 353–62.

Kamaludin, H. and Yakzan, M. Y. (1998). Interstadial records of the last glacial period at Pantai Remis, Malaysia. *Journal of Quaternary Science*, **12**, 410–34.

Kershaw, P., van der Kaars, S., Moss, P. and Wang, K. (2002). Quaternary records of vegetation, biomass burning, climate and possible human impact in the Indonesian-northern Australian region. In *Bridging Wallace's Line: The Environmental and Cultural History and Dynamics of the SE Asian-Australian Region*, ed. P. Kershaw, B. David, N. Tapper, D. Penny and J. Brown. *Advances in Geoecology*, **34**, 97–118.

Lee, H. S., Davies, S. J., LaFrankie, J. V. et al. (2002). Floristic and structural diversity of mixed dipterocarp forest in Lambir Hills National Park, Sarawak, Malaysia. *Journal of Tropical Forest Science*, **14**, 379–400.

Lelono, E. B. (2000). *Palynology of the Nanggulan Formation, Central Java, Indonesia*. Unpublished PhD thesis, Royal Holloway, University of London.

Lelono, E. B. and Morley, R. J. (2011). Oligocene palynological succession from the East Java Sea. In *The SE Asian Gateway: History and Tectonics of Australia–Asia Collision*, ed. R. Hall, M. A. Cottam and M. E. J. Wilson. *Geological Society of London Special Publication*, 355, 333–45.

Maloney, B. K. and McCormac, F. G. (1995). A 30 000 year pollen and radiocarbon record from highland Sumatra as evidence for climatic change. *Radiocarbon*, **37**, 181–90.

Martin H. A. (1974). The identification of some Tertiary pollen belonging to the family Euphorbiaceae. *Australian Journal of Botany*, **22**, 271–91.

Medway Lord, (1972). The Quaternary mammals of Malesia: a review. In *The Quaternary Era in Malesia*, ed. P. S. Ashton and H. M. Ashton. *Geography Department University of Hull, Miscellaneous Series* **13**, 63–398.

Meijaard, E. (2003). Mammals of Southeast Asian islands and their Late Pleistocene environments. *Journal of Biogeography*, **30**, 1245–57.

Morley, R. J. (1982). Fossil pollen attributable to *Alangium* Lamarck (Alangiaceae) from the Tertiary of Malesia. *Review of Palaeobotany and Palynology*, **36**, 65–94.

Morley, R. J. (1995). Palynology of Walat Formation, Java, Indonesia. Unpublished poster distributed to participants of field trip to Gunung Walat, following Indonesian Petroleum Association, International Symposium on Sequence Stratigraphy in SE Asia. May 1995.

Morley, R. J. (1996). Biostratigraphic characterisation of systems tracts in Tertiary sedimentary basins. *Indonesian Petroleum Association, Proceedings of International Symposium on Sequence Stratigraphy in SE Asia*, 49–70.

Morley, R. J. (1998). Palynological evidence for Tertiary plant dispersals in the Southeast Asian region in relation to plate tectonics and climate. In *Biogeography and Geological Evolution of SE Asia*, ed. R. Hall and J. D. Holloway. Leiden, The Netherlands: Backhuys, pp. 211–34.

Morley, R. J. (2000). *Origin and Evolution of Tropical Rainforests*. London: Wiley and Sons.

Morley, R. J. (2003). Interplate dispersal routes for megathermal angiosperms. *Perspectives in Plant Ecology, Evolution and Systematics*, **6**, 5–20.

Morley, R. J. (2006). Cretaceous and Tertiary climate change and the past distribution of megathermal rainforests In *Tropical Rainforest Responses to Climatic Change*, ed. M. Bush and J. R. Flenley. Chichester, UK: Praxis-Springer, pp. 1–54.

Morley, R. J. (2010). Palaeoecology of tropical podocarps. In *Ecology of the Podocarpaceae in Tropical Forests*, ed. B. L. Turner and L. M. Cernusak. Smithsonian Institution Contributions to Botany no. 95, Washington DC: Smithsonian Institution Scholarly Press.

Morley, R. J. and Flenley, J. R. (1987). Late Cainozoic vegetational and environmental changes in the Malay Archipelago. In *Biogeographical Evolution of the Malay Archipelago*, ed. T. C. Whitmore. Oxford Monographs on Biogeography 4, Oxford: Oxford Scientific Publications, pp. 50–59.

Morley, R. J. and Morley, H. P. (2010). Neogene climate history of the Makassar Straits with emphasis on the Attaka Field. *Proceedings of the 34th Indonesian Petroleum Association*, IPA10-G-208.

Morley, R. J. and Morley, H. P. (2011). Neogene climate history of the Makassar Straits. In *The SE Asian Gateway: History and Tectonics of Australia–Asia Collision*, ed. R. Hall, M. A. Cottam and M. E. J. Wilson. *Geological Society of London Special Publication*, 355, 319–32.

Morley, R. J. and Shamsuddin Jirin. (2006). The sequence biostratigraphy and chronostratigraphy of the Malay Basin. *Petroleum Geology Conference and Exhibition Malaysia Proceedings*, 77.

Morley, R. J., Morley, H. P. and Restrepo-Pace, P. (2003). Unraveling the tectonically controlled stratigraphy of the West Natuna Basin by means of palaeo-derived Mid Tertiary climate changes. *29th Indonesian Petroleum Association Proceedings*, 1, IPA03-G-118.

Morley, R. J., Salvador, P., Challis, M. L., Morris, W. R. and Adyaksawan, I. R. (2007). Sequence biostratigraphic evaluation of the North Belut Field, West Natuna Basin. *Proceedings of the 31st Indonesian Petroleum Association Convention*, IPA07-G-120.

Morwood, M. J., O'Sullivan, P., Susanto, E. E. and Aziz, F. (2003). Revised age for Mojokerto 1, an early *Homo erectus* cranium from East Java, Indonesia. *Australian Archaeology*, **57**, 1–4.

Muller, J. (1966). Montane pollen from the Tertiary of N.W. Borneo. *Blumea*, **14**, 231–5.

Muller, J. (1968). Palynology of the Pedawan and Plateau Sandstone Formations (Cretaceous–Eocene) in Sarawak. *Micropalaeontology*, **14**, 1–37.

Muller, J. (1972). Palynological evidence for change in geomorphology, climate and vegetation in the Mio-Pliocene of Malesia. In *The Quaternary Era in Malesia*, ed. P. S. Ashton and H. M. Ashton. *Geography Department, University of Hull, Miscellaneous Series*, **13**, 6–34.

Newsome, J. and Flenley, J. R. (1988). Late Quaternary vegetational history of the Central Highlands of Sumatra. II Palaeopalynology and vegetational history. *Journal of Biogeography*, **15**, 555–78.

Pälike, H., Norris, R. D., Herrle, J. O. et al. (2006). The heartbeat of the Oligocene climate system. *Science*, **314**, 1894–8.

Pickett, E. J., Harrison, S. P., Hope, G. et al.(2004). Pollen-based reconstructions of biome distributions for Australia, Southeast Asia and the Pacific (SEAPAC region) at 0, 6000 and 18 000 ^{14}C yr BP. *Journal of Biogeography*, **31**, 1381–1444.

Pribatini, H. (1993). *Geologi dan analysis geokimia dalam explorasi penhahuluan mineral logam dasar daerah Tanjung Balit Kecamatan Pangkalan kotabaru Kabupaten Limapuluhkoto Propinsi Sumatera Barat*. Unpublished PhD thesis, Trisakti University, Jakarta, Indonesia.

Richards, P. W. (1996). *The Tropical Rainforest*, 2nd edn. Cambridge: Cambridge University Press.

Rugmai, W., Grote, P. J., Chongpan Chonglakmani, Zetter, R. and Ferguson, D. K. (2008). A Late Pleistocene palynoflora from the coastal area of Songkla Lake, southern Thailand. *Science Asia*, **34**, 137–45.

Shamsuddin Jirin and Morley, R. J. (2006). Integration of biostratigraphy with seismic for sequence stratigraphic interpretation in the Malay Basin. *Petroleum Geology Conference and Exhibition Malaysia Proceedings*, 101–102.

Slik, F. J. W., Shin-Ichiro Aiba, Bastian, M. et al. (2011). Soils on exposed Sunda Shelf shaped biogeographic patterns in the equatorial forests of Southeast Asia. *Proceedings of the National Academy of Sciences of the United States of America*, **108**, 12343–7.

Songtham, Wickanet, Benjavun Ratanasthien, Mildenhall, D. C., Sampan Singharajwarapana and Wittaya Kandharosaa (2003). Oligocene–Miocene Climatic Changes in Northern Thailand Resulting from Extrusion Tectonics of Southeast Asian Land mass. *Science Asia*, **29**, 221–33.

Sun, X, and Wang, P. (2005). How old is the Asian monsoon system? Palaeobotanical records from China. *Palaeogeography, Palaeoclimatology, Palaeoecology*, **222**, 181–222.

Swisher, C. C. III, Curtis, G. H., Jacob, T. and Getty, A. G. (1994). Age of the earliest known hominids in Java, Indonesia. *Science*, **263**, 1118–21.

Takahashi, K. (1982). Miospores from the Eocene Nanggulan Formation in the Yogyakarta region, Central Java. *Transactions of the Proceedings of the Palaeontological Society of Japan, N.S.*, **126**, 303–26.

Tarponnier, P., Peltzer, G., LeDain, A., Armijo, R. and Cobbold, P. (1982). Propagating extrusion tectonics in Asia; new insights with simple experiments with plasticine. *Geology*, **10**, 611–16.

Tarponnier, P., Peltzer, G. and Armijo, R. (1986). On the mechanics of the collision between India and Asia. *Geological Society Special Publication*, **19**, 115–57.

Taylor, D. Hwee Yen, O., Sanderson, P.G. and Dodson, J. (2001). Late Quaternary peat formation and vegetation dynamics in lowland tropical swamp: Nee Soon, Singapore. *Palaeogeography, Palaeoclimatology, Palaeoecology*, **171**, 269–87.

van den Bergh, G. D., de Vos, J. and Sondaar, P. Y. (2001). The Late Quaternary palaeogeography of mammal evolution in the Indonesian Archipelago. *Palaeogeography, Palaeoclimatology, Palaeoecology*, **171**, 385–408.

van Gorsel, J. T. and Troelstra, S. R. (1981). Late Neogene planktonic foraminiferal biostratigraphy and climatostratigraphy of the Solo River section (Java, Indonesia) *Marine Micropaleontology*, **6**, 183–209.

van der Kaars, W. A. and Dam, M. A. C. (1995). A 135 000-year record of vegetational and climatic change from the Bandung area, West-Java, Indonesia. *Palaeogeography Palaeoclimatology, Palaeoecology*, **117**, 55–72.

van der Kaars, S., Penny, D., Tibby, J. et al. (2001). Late Quaternary palaeoecology, palynology and palaeolimnology of a tropical lowland swamp: Rawa Danau, West Java, Indonesia. *Palaeogeography, Palaeoclimatology, Palaeoecology*, **171**, 185–212.

van der Kaars, S., Bassinot, F., De Deckker, P. and Guichard, F. (2010). Changes in monsoon and ocean circulation and the vegetation cover of southwest Sumatra through the last 83 000 years: The record from marine core BAR94-42c. *Palaeogeography, Palaeoclimatology, Palaeoecology*, **296**, 52–78.

Voris, H. K. (2000). Maps of Pleistocene sea levels in Southeast Asia: shorelines, river systems and time durations. *Journal of Biogeography*, **27**, 1153–67.

Wang, B., Clemens, S. C. and Liu, Pl. (2003). Contrasting the Indian and East Asian monsoons. *Marine Geology*, **201**, 5–21.

Wang, X. M., Sun, X. J., Wang, P. X. and Stattegger, K. (2009). Vegetation on the Sunda shelf, South China Sea, during the Last Glacial Maximum. *Palaeogeography, Palaeoclimatology, Palaeoecology*, **278**, 88–97.

Wong, R., Ahmad, M., Boyce, B. et al. (2006). Sequence stratigraphy and chronostratigraphy of the Malay Basin unravels new exploration plays. *American Association of Petroleum Geologists Bulletin*, **91**, 368–98.

Woodruff, D. S. (2010). Biogeography and conservation in Southeast Asia: how 2.7 million years of repeated environmental fluctuations affect today's patterns and the future of the remaining refugial-phase biodiversity. *Biodiversity and Conservation*, **19**, 919–41.

Wright, S. J. (2001). Plant diversity in tropical forests: a review of mechanisms of species coexistence. *Oecologia*, **130**, 1–14.

Wurster, C. M., Bird, M. I., Bull, I. D. et al. (2010). Forest contraction in north equatorial Southeast Asia during the last Glacial Maximum. *Proceedings of the National Academy of Sciences, USA*, **107**, 15508–11.

Yakzan, A. M., Awalludin, H., Bahari, M. N., and Morley, R. J. (1996). Integrated biostratigraphic zonation for the Malay Basin. *Geological Society of Malaysia*, **39**, 157–84.

Yamada, I, 1998. *Tropical Rainforests of Southeast Asia, A Forest Ecologist's View* (transl. P. Hawkes). Hawaii: University of Hawaii Press.

Zachos, J. C, Pagini, M., Sloan, L., Thomas, E. and Billups, K. (2001). Trends, rhythms and aberrations in global climate 65 Ma to Present. *Science*, **292**, 686–93.

5

Quaternary dynamics of Sundaland forests

Charles H. Cannon

5.1 Introduction

The historical dynamics and distribution of Sundaland forests during the Quaternary Period involved the interaction between two complex systems: (1) global climate change through numerous glacial cycles and their associated effects on land area and local climate, and (2) the ecological evolution of rainforest trees. Over the past decade, our understanding of the spatial dynamics of tropical rainforests in Sundaland has improved substantially, through detailed geographic (Voris 2000, Woodruff and Turner 2009) and modelling work (Cannon et al. 2009). Spatial modelling of historical land area and climate suggests that Sundaland's wet evergreen forests are currently experiencing a brief transitional refugial stage. Given probable forest area over the last 2.6 Myr, total extent and overall connectivity are now at millennial lows. In fact, current distribution and geographic location stabilised less than 10 ka at the end of the last glacial cycle (Hanebuth et al. 2000, Bintanja et al. 2005, Hanebuth et al. 2009). Clearly, the current and familiar archipelago setting of Sundaland rainforests, with the major land masses divided among the islands of Sumatra, Java and Borneo, is actually an unusual biogeographic setting. This statement is true whichever climate and vegetation scenario is favoured by the reader. We must strive to understand these historical forests to gain any insight into present forests (Woodruff 2010).

Biotic Evolution and Environmental Change in Southeast Asia, eds D. J. Gower et al. Published by Cambridge University Press.

Interpreting the effect of these dramatic cyclical changes on the ecology and evolution of forest trees remains difficult, because our general understanding of the evolutionary behaviour of tropical trees is quite poor. Many aspects of their life history, genetic diversity and reproductive behaviour are known only from a few scattered empirical studies, which frequently focus on responses to disturbance. We also lack a convincing general theory for their diversification and speciation given that sympatry of numerous closely related species is a common feature of these communities. Additionally, what is known about the evolutionary and ecological characteristics of trees (Petit and Hampe 2006) does not easily fit into standard evolutionary models or analytical approaches, often derived using zoological subjects or plant crop models.

In this chapter, I first review our understanding of the historical biogeography of Sundaland forests through the Quaternary Period. I highlight the power of spatial modelling to generate testable hypotheses (Graham et al. 2006, Carnaval et al. 2009), particularly for the comparative analysis of different major forest types. Because some key elements of the historical biogeography of these forest types are robust to a wide range of model parameters, Sundaland exemplifies a marvellous natural laboratory for testing assumptions about community assembly processes, historical population size and the natural formation of refugia. This review also pinpoints significant outstanding questions related to the geomorphology of the shelf and the interaction of forests and soils.

Second, I outline several important aspects of rainforest tree life history, emphasising the spectacular 'non-model' nature of their evolutionary behaviour. These elements require a novel approach to the study of speciation and phenotypic adaptation in forest trees. We should move quickly to employ new and powerful genomic technologies. Pulling these two major themes together, I discuss how the unique biogeographic history of Sundaland forests and the evolutionary characteristics of rainforest trees have interacted to produce the forests of today. On a final note, I stress the importance of capturing the natural spatial patterns of community composition and genetic variation prior to their complete erasure by human activities (Petit et al. 2008).

5.2 Old wine in new bottles: palaeoforests of Sundaland in the present archipelago

The beginning of the Quaternary Period was recently recognised by the International Union of Geological Sciences as 2.58 Ma (Gibbard et al. 2010), to coincide with the cyclical glaciations, which became increasingly prevalent in the last 800 kyr. These glacial cycles were broken by approximately eight brief periods of warm interglacial conditions, each typically lasting only a few thousand years.

We are currently enjoying one of these unusual interglacial periods, following an extensive and widespread glaciation that lasted roughly 120 kyr. The current warm period began roughly 12 ka (Hanebuth et al. 2009), at which point the percentage of lowland rainforest in 'core' areas of 10 km radius or more dropped from >70% to less than 50% (Cannon et al. 2009). Given all model parameters, the current extent of evergreen lowland rainforests is among the smallest 10% in terms of total land area over the last million years. Ultimately, the current geographic distribution and connectivity of Sundaland forests are unrepresentative of the vast majority of their history through the Quaternary Period.

While trees can generally migrate fast enough to track the movement of appropriate climate (Clark 1998) and can have high rates of gene flow (Konuma et al. 2000, Lee et al. 2000, Lee et al. 2002, Ng et al. 2006, Petit and Hampe 2006), the environmental conditions of the interglacial and their current geographic distribution were established quite recently. Given the slow pace of macro-evolution in trees, this rapid transition over roughly 10 kyr would have provided little opportunity for significant phenotypic adaptation, given current models of phenotypic adaptation and gene flow. Current interglacial dynamics may have played little role in shaping Sundaland forests through the Quaternary Period. Instead, we should assume that phenotypic traits have evolved primarily in response to historical selection pressures. Potentially, no present analogy exists for the prevalent combination of historical selection pressures.

As discussed above, climatic conditions over Sundaland during the glacial period were probably slightly drier and cooler, while remaining appropriate to support rainforest. Additionally, the spatial relationships between these altered environmental conditions with soils and landforms would have been different, requiring communities to shift geographically to track their preferred conditions. The soils exposed by lowered sea levels would have been substantially different in their structure and nutrient properties in comparison to terrestrial soils. Very little is known about the rate of soil evolution after sea levels recede. Obviously, the palaeomorphology and sedimentation patterns of the shelf have been complex (Steinke et al. 2003). Ultimately, given the slow pace of community assembly and evolutionary processes in trees, current Sundaland rainforests probably exist at an ecological and evolutionary disequilibrium. Potentially, the present ecological fit of species characteristics and their phenotypes with specific environmental conditions and community composition may be fairly poor.

Given all hypothetical scenarios during the Quaternary Period, we can safely assume that rainforest communities on Sundaland would have been substantially different just 10 ka in terms of community composition, species population densities and species ranges. Species that are rare today might have been dominant in glacial period communities, with their slightly cooler and drier conditions, while today's dominant species were rare. Given the repeated and unstable nature of dramatic change in Sundaland, both in climate and land area, with no episode lasting more than 200 kyr, 'change' has been relatively constant, rapid, and profound for Sundaland tree species. Given the slow evolutionary dynamics of trees, species may evolve mechanisms to maximise their phenotypic plasticity and focus on bet-hedging strategies instead of evolving towards a phenotypic optimum for a specific environment (Borges 2005, 2009).

5.2.1 Historical distribution of Sundaland rainforest

The dramatic change in land area on Sundaland through the Quaternary Period, caused by lowered sea levels, has long been recognised (Mollengraaff 1921, Heaney 1991, Voris 2000). The major land masses on the shelf, including the major islands of Borneo, Sumatra and Java, have been connected to the Asian mainland through land bridges for the majority of the Quaternary Period. This connectivity should have provided abundant opportunity for dispersal throughout the region and yet a large amount of evidence suggests that strong biogeographic barriers exist (Zhi et al. 1996, Fernando et al. 2003, Hirai et al. 2005, Steiper 2006), particularly between Sumatra and Borneo, or east to west. Apparently, the barrier was environmental and not geographic, with several authors (Heaney 1991, Bird et al. 2005) proposing a 'savanna corridor' running north–south through the western portion of Sundaland. This corridor would have linked seasonal open vegetation communities along the southern margin of the shelf (in Java) with similar communities at higher latitudes in Thailand and northern peninsular Malaysia. The width and severity of this 'savanna' corridor is difficult to assess from the palaeontological record because much of Sundaland is now submerged. Several palaeontological studies have indicated that Sundaland rainforests might have been larger at the Last Glacial Maximum (LGM) with relatively little seasonality (Kershaw et al. 2001, Hu et al. 2003, Kershaw et al. 2007). Recently, a series of sea cores in the South China Sea has indicated that tropical rainforest was predominant on the shelf at the LGM, even though the LGM palaeoclimate was slightly cooler and drier (Sun et al. 2002, Wang et al. 2009).

Numerical simulations of the palaeoclimate at the Last Glacial Maximum (Bush and Fairbanks 2003) indicate that the majority of Sundaland was well within the climatic envelope for tropical everwet forest (Cannon et al. 2009). The modelled LGM climate agrees with the majority of existing palaeontological evidence, with pronounced cooler and drier conditions along the southern margin of Sundaland and on the islands of Sulawesi and the Philippines. The modelled LGM climate does not include a north–south corridor of seasonal or 'savanna' climate. This discrepancy is probably caused by the coarse spatial and topographic resolution in the global climate model. These results suggest that the corridor was probably fairly narrow and probably due to rain-shadow effects. Additionally, while conditions were more seasonal during glacial periods, the presence of a strongly seasonal 'savanna' climate on Sundaland seems unlikely. The vegetation in this seasonal corridor was probably similar to open structured, seasonally dry tropical forest. This difference in vegetation type might have been an effective biogeographic barrier to the dispersal of core rainforest species (Meijaard and van der Zon 2003). The potential impact of fires and hunting by humans in this seasonal forest cannot be underestimated, because hominids have been present in the region throughout the Quaternary Period (van den Bergh et al. 2001). The impact of simple hunting technologies has been shown to be devastating to megafaunal rainforest populations (Marshall et al. 2006). Human activity, concentrated in the more open seasonal forests over the last two or three glacial cycles, could have reinforced a mild environmental biogeographic barrier between the eastern and western blocks of core everwet forest habitat.

Given the median model of lowland rainforest distribution at the LGM, Sundaland rainforests probably covered a substantially larger area than they do at present through more than 90% of the Quaternary Period. Lowland forests would have existed as a large contiguous block in the Sunda basin, which would have drained two major river systems – the Siam originating from the now Bay of Thailand and the North Sunda originating from the interior of Sundaland, including Borneo and Sumatra (Voris 2000). This lowland palaeoforest, potentially twice as large as current forests, would have been similar in its topographic setting to the forests of the Amazon basin. Bornean forests possess equivalent levels of biodiversity to those found in the much larger Amazon basin (Condit et al. 2000, Pitman et al. 2001). Our historical model (Cannon et al. 2009) provides a simple explanation for this rather anomalous pattern. On the other hand, if savanna and open seasonal forests had covered a large proportion of Sundaland during glacial periods, with rainforests isolated in small refugia around the margins of the shelf (Brandon-Jones 1996, Abegg and Thierry 2002, Meijaard and van der Zon 2003, Louys and Meijaard 2010, Wurster et al. 2010), overall biodiversity on the Greater Sunda islands would be expected to be substantially lower than Amazonian forests, as seen in equatorial Africa where rainforests were reduced to several small refugia through glacial periods (Bonnefille 2007). The Sundaland savanna hypothesis poses numerous difficult and unresolved questions about the maintenance and diversification of tree species in these mega-diverse forests.

Significant questions about historical distribution

Several major questions remain about the composition and structure of the Sundaland palaeoforests, primarily related to geomorphological processes and plant–soil interactions.

How quickly did marine sediments evolve into terrestrial soils on the exposed shelf?

The exposed soils, after the retreat of the ocean, would have been substantially different from terrestrial soils. The action of the ocean would have sorted the soils, probably making them much more compact and nutrient poor at the surface. A successional process, perhaps somewhat similar to the process of invasion after volcanic eruptions (Whittaker et al. 1989, Bush et al. 1995), would have occurred prior to the development of typical organic soil conditions and nutrient cycles. The oceanic soils would probably have been much poorer in nutrients and with lower water retention capacity than terrestrial soils. The pace of this successional process would probably have been relatively rapid, on the scale of centuries, and mature rainforest would probably have become established soon after the ocean retreat, although initial stages would have probably resembled stunted peat forests on sands and heavy clays. The pace and impact of these geomorphological processes are worth exploring.

Do current population sizes correlate with past population sizes?

If rare species were always rare while dominant species were always dominant, the significant difference in the historical population sizes of these species should be evident in their genetic diversity. A simple comparison between the observed genetic diversity and current population density of a community of trees would provide insight into whether population density has been constant through time or not. If a rare species was actually quite common during glacial periods, populations should harbour a large amount of genetic diversity with little geographic structure (but see discussion below about refugia formation, which could be a largely stochastic process, generating random but strong patterns across the landscape). If at the community level, current genetic diversity did not correlate with current population density, historical population sizes would not be correlated with current population sizes.

How do refugia form naturally? What are the community assembly processes of contraction?

At the transition from glacial to interglacial conditions, land area and forest distribution was reduced from its greatest extent, with the Sundaland coast lying beyond the continental margin, to a historical low in just over 7 kyr. Because glacial forests at higher latitudes are interpreted to be at the refugial stage followed by rapid expansion during the interglacial, we have a detailed understanding of the expansion and migration routes of the biota across Europe (Petit et al. 2003, Hewitt 2004) and its consequences for population genetics and community structure. On the other hand, very little is known about the natural process of population and community retreat into a small refugium (Channell and Lomolino 2000, Sagarin and Gaines 2002). The best-studied natural experiment involved the creation of an island archipelago after the damming of a Venezuelan lake (Terborgh et al. 2001), where long-term species loss seems to occur with large scale fragmentation of forests (Feeley and Terborgh 2005). Because the contraction happened so quickly, due to the loss of land area to rising seas, the process must have imposed a tremendous migration pressure on forest communities and probably generated similar stochastic effects on community composition as observed in Venezuela. Could established forests in relatively stable environmental conditions have resisted the invasion of species from communities being forced inland? What happens when an entire community must shift geographically into a smaller location where environmental conditions are slightly different than previously?

Did a large inland lake exist on Sundaland, in the position of the Bay of Thailand?

The spatial model suggests that the upper portion of the Bay of Thailand was below or near sea level for a large proportion of the last glacial cycle. While this model does not incorporate the uplift of the shelf (Lambeck and Purcell 2005) expected after the removal of the water load caused by the South China Sea, the potential for a very large freshwater inland lake exists, larger than the Tonle Sap of Cambodia. Evidence for the presence of such a large lake could potentially be found in the genetic diversity of flora and fauna specialising in these types of habitats due to very large historical population sizes.

How did the El Niño-Southern Oscillation (ENSO) cycle differ during the glacial period? Does general or supra-annual mast fruiting require a special hypothesis?

Global models indicate that the El Niño-Southern Oscillation (ENSO) could have been substantially different during glacial periods, potentially weaker and more frequent (Bush 2007, Zheng et al. 2008). These cycles have a major impact on the community biology of Sundaland forests and are hypothesised to be the cue for supra-annual mast fruiting (SMF) (Sakai et al. 2006). The length of time between ENSO events would play an important role in the evolution of SMF, if predator-satiation processes are important (Janzen 1974). If SMFs occurred more frequently during glacial periods, once every 2 years or so and were the strongest and clearest community-level environmental cue, then no special hypothesis is actually necessary to explain their evolution. The long period of time between SMF events we currently observe is simply due to the lengthening of time between ENSO events.

Was the Sunda Shelf covered by vast peat swamps?

The exposed soils on the shelf after the oceans receded would have been quite similar to the soils inhabited by current peat forests (Morley 1981, Newbery 1991). They currently cover a small fraction of the land area of lowland forests, although they represent some of the largest peat formations in the tropics (Rieley and Page 1995). Several different families dominate these forests and several genera are quite species rich, even within one community (Anderson 1963), and the species composition is substantially different from other lowland forest types (Cannon and Leighton 2004). Although peat forests do generally have lower species richness than other forest types (Connell and Lowman 1989, Davies and Becker 1996, Cannon and Leighton 2004), they do possess considerable tree diversity. The flat Sunda Shelf was near sea level through much of the Quaternary Period and the cyclical rise and fall of the ocean, exposing marine sediments and sandy beaches, would have made conditions favourable for the establishment of peat forests. If species confined to peat forests harbour substantial genetic diversity, historical population sizes might have been substantially larger. If little geographic structure in genetic diversity exists among now isolated populations, then historical biogeographic conditions might have presented fewer barriers to gene flow.

5.2.2 Variance in historical biogeography among forest types is predictable

We included three major forest types in our spatial model (Cannon et al. 2009): (1) coastal mangrove (0–10 m above sea level); (2) lowland wet evergreen forests (10–450 m above sea level); and (3) upland wet evergreen forests (1000–2200 m above sea level). We chose these three types because of their relative taxonomic distinctiveness. The distribution of these forest types was estimated at the LGM, given a conservative definition for appropriate rainforest climate and adjusting the elevational zonation according to a range of lapse rates. We then interpolated the distribution of these types through the 1 Myr in 500 year time steps, allowing sea level and temperature level to change independently, adjusted for the equator given a global dataset (Bintanja et al. 2005). Animations of this model are available at the PNAS website.

One of the more useful things to emerge from the model was the distinct and predictable differences among the forest types in their Quaternary dynamics at the regional scale. These predictable differences allow us to generate testable hypotheses about community assembly and historical population size among the types. These large differences in the biogeographic distribution of forest types provide a unique perspective on the selective advantage of various phenotypic traits. Ecologists typically assume that community composition and population structure is shaped by current biotic interactions and edaphic constraints. The different historical dynamics experienced by each forest type should have imposed significant and distinct biogeographic filters on phenotypic composition and community assembly. For example, tree species found only in coastal forests would have to be able to persist through rapid and repeated geographic relocation while montane tree species would have experienced little geographic relocation through the Quaternary Period. Phenotypic trait composition should differ predictably between these two forest types, with traits associated with high dispersal and fecundity (r strategy) concentrated in coastal forests versus those associated with low dispersal and fecundity (K strategy) in montane forests. The selective advantage of these traits would not be detectable by measuring current environmental conditions and selection pressures.

Mangrove forests

The forest type with the most dynamic historical distribution is obviously coastal mangrove forest. These forests have gone through repeated and almost complete geographic relocations at each glacial–interglacial transition. The forced migration, from the outer margins of the continental shelf to the inner margins of the South China Sea, of mangrove species would have occurred over just a few thousand years. Because of the relatively flat topography of the Shelf, a large area would have been at or near sea level at the beginning and ending of each glacial cycle. The model predicts that during these brief periods, mangrove forests would have gone through a brief but substantial population explosion, particularly at the centre of Sundaland but also along the eastern edge of Borneo (Fig 5.1a). These two effects should be relatively easy to detect in the population genetic variation among the dominant mangrove species (Hewitt 1996, Templeton 1998). The dramatic biogeographic filter imposed by these two historical factors (complete geographic relocation and explosive population growth followed by rapid contraction), might partially explain the low species diversity found in these forests (Duke et al. 1998, Ellison et al. 1999), as only taxa with appropriate seed dispersal strategies and life history characteristics would have been able to persist and dominate. By examining the temporal depth of mangrove forests on Sundaland (see legend for Fig 5.2 for a detailed description of temporal depth maps), the current distribution of mangrove forests corresponds weakly with the dominant historical distribution given a linear decay in historical effect (Fig 5.2).

Lowland forests

The spatial distribution of lowland rainforest varies considerably across Sundaland through the last glacial cycle (Fig 5.1b). In eastern and central Borneo, the percentage of the area occupied by lowland forest has been relatively stable during much of the glacial period with a small reduction in area during the interglacials. This pattern contrasts sharply with the central basin of Sundaland, which was largely covered by lowland rainforest during the latter half of the glacial cycle. This area is now completely submerged. In contrast, northwest Borneo, largely corresponding to Sarawak, was largely occupied by lowland forests during the early phase of the glacial cycle but experienced a major dip at the end of the last glacial maximum but then recovered again at the interglacial. In the western portion of Sundaland, lowland forest distribution dropped dramatically during the latter half of the glacial period. This part of the model is largely due to the imposed corridor of seasonal conditions. A narrow corridor of seasonal forest agrees with recently published evidence from four caves in Southeast Asia (Wurster et al. 2010).

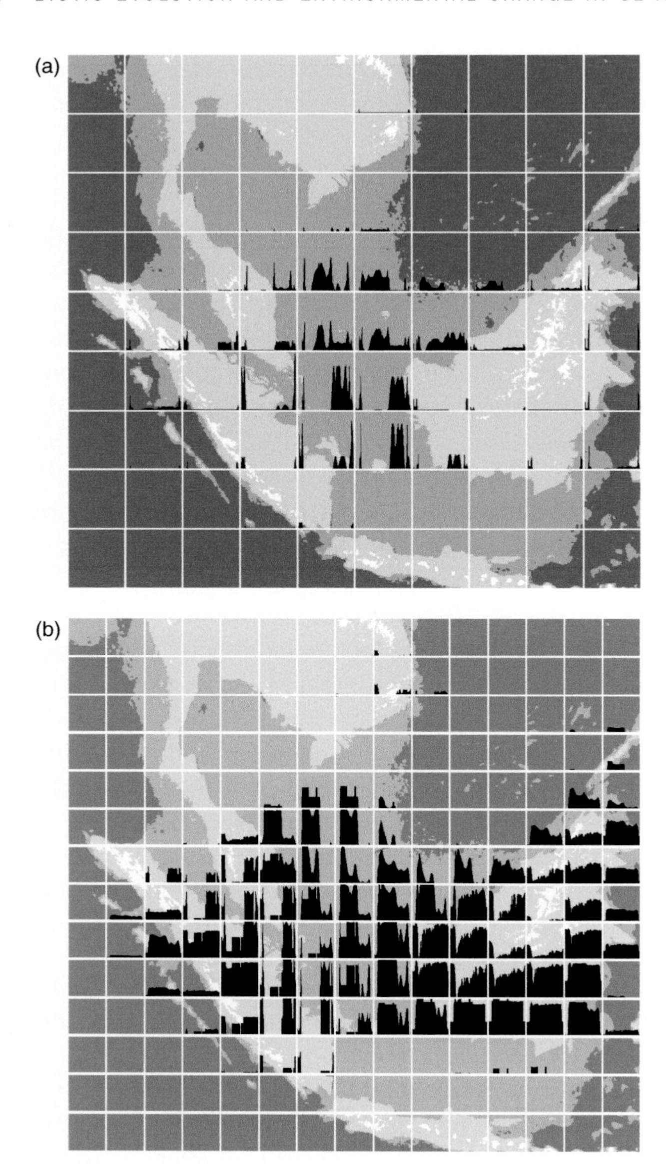

Figure 5.1 Percentage occupancy of three forest types in Sundaland over the last glacial cycle (0–120 ka) for specific geographic gridcells: (a) coastal mangrove areas within 0–10 m sea level and (b) lowland areas. The black area under the line for each grid cell indicates the percentage occupancy of each grid cell, bounded by white margins, through time. The x-axis in each gridcell indicates time before present, starting with 0 at the left and 120 ka at the right. For mangrove forests in (a), the vertical scale runs from 0–45%, while for lowland forests, the vertical scale runs from 0–100%. Land masses at current sea level are shown in light grey and white while the exposed shelf at the Last Glacial Maximum is shown in mid-grey. Ocean is dark grey.

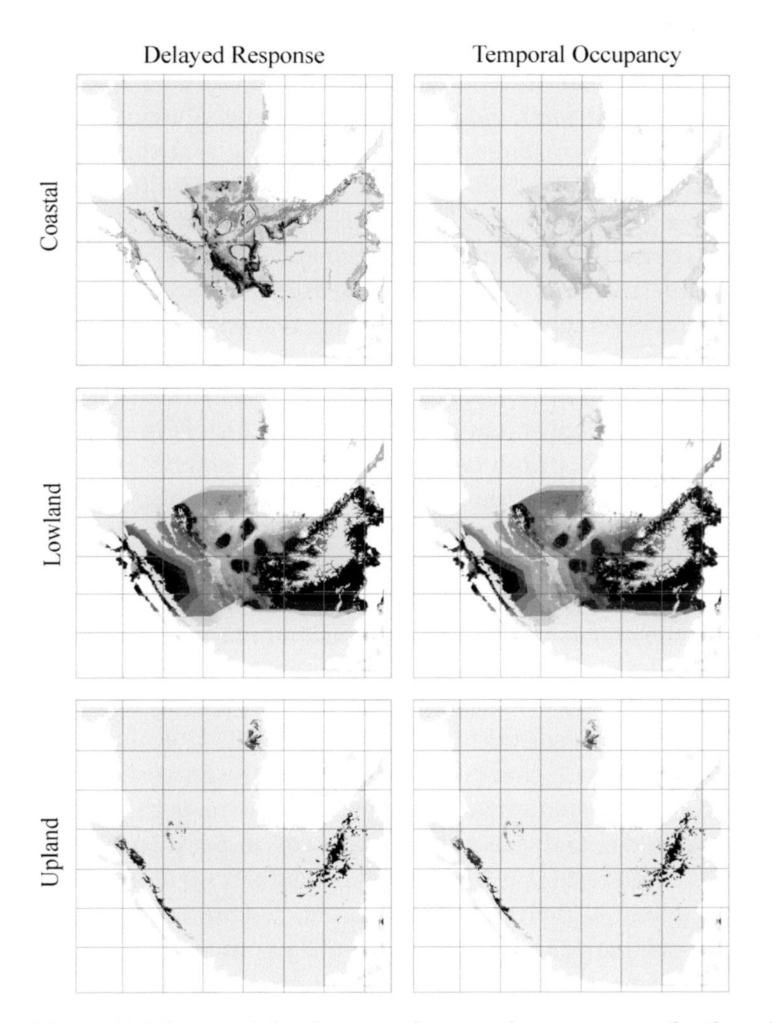

Figure 5.2 Temporal depth maps of geographic occupancy for three forest types in Sundaland over the last glacial cycle (0–120 ka in 500 year time steps). The cumulative occupancy in each pixel is indicated by the grey scale value; black is equal to 100% occupancy throughout the glacial cycle while white indicates no occupancy. 'Delayed Response' maps, in the left column, indicate the temporal depth assuming a lag time in response, equal to the summation multiplied by the reciprocal of the number of thousands of years into the past. 'Temporal Occupancy' simply indicates the number of time steps each pixel was occupied divided by the total number of time steps.

The biogeographic districts recognised on Sundaland within lowland forests (Slik et al. 2003) are largely apparent on the temporal depth maps (Fig 5.2), although their historical geographic locations are different from their current positions. Through the majority of the Quaternary Period, the peninsular Malaysian forests probably resided in the region off the eastern coast with a large community in the northern portion of Sundaland, near present day Vietnam and Cambodia, while the Sumatran forests would have been largely confined to the flat alluvial plains to the east of the central mountain range. Another surprising and somewhat depressing result is that the small islands in the middle of the South China Sea, particularly Natuna, have a deep and central position in the history of lowland Sundaland rainforests but these islands almost vanished and now have very little natural forest remaining. These islands would have been excellent natural laboratories to study the formation of refugia, similar to ongoing research in Venezuela on the creation of islands in human-made reservoirs (Feeley and Terborgh 2005), but on a continental scale. On Borneo, a division between the northwestern corner and the northeastern is somewhat apparent with a reduced history in the area between the two, as noted in the distribution of species on the island (Slik et al. 2003). The central and eastern half of the island represents the largest and deepest block of forest history.

The extent and distribution of lowland rainforest on Sundaland are also highly sensitive to the model parameters, unlike mangrove or montane forests. The decline in temperature at the equator, the degree of seasonality in precipitation, and the response in elevational zonation of vegetation types to these climate changes strongly affect the expected distribution of lowland rainforest. The minimal model for lowland rainforest suggests a highly variable distribution and at the last glacial maximum, assuming a -3° change in temperature, the greatest reduction in overall precipitation suggested by the coupled atmosphere–ocean global circulation palaeoclimate model, and a response rate of 207 m per degree temperature decline, lowland rainforest is pushed almost completely off of the shelf. Coastal forests are highly dynamic, their distribution is affected primarily by sea-level change and therefore the patterns are robust to model parameters. Montane forests are confined to high elevations and their distribution only responds to the vegetation lapse rate with no effect of sea-level change. Only the lowland rainforest responds to all of the model parameters. This high level of variability reduces confidence in the output from a single set of parameters and also indicates high variability among the eight glacial cycles that have occurred since the beginning of the Quaternary Period. The distribution of lowland rainforest probably varied considerably among the different cycles while the distribution of coastal and montane forests was basically repeated with each cycle.

Upland forests

Upland forests have always been relatively small and isolated across Sundaland, with three major cores: central Borneo, western Sumatra and Indochina. These areas were persistent throughout the Quaternary Period, although the model suggests that at the LGM, precipitation in the northern Indochinese portion of

Sundaland might have decreased below the threshold for rainforest leading to the widespread loss of everwet tropical forests. This result seems unlikely given the high levels of diversity and unique taxonomic composition found in these forests (Slik et al. 2009). Overall, there is much less variability in the spatial distribution of upland forests. A slight expansion of area can be seen through the glacial cycle, peaking at the last glacial maximum with a fairly substantial reduction at the transition to the interglacial phase. Given the diversity of montane forests on the island of Java and in Indochina, small isolated refugia probably existed throughout Sundaland on the various mountain ranges found in the region.

Outstanding questions about specific forest types

The different and predictable patterns modelled for these three forest types generate fundamental questions about community assembly processes and the effects of historical population size.

Do different biogeographic histories lead to predictable differences in the dominant phenotypic traits and seed dispersal characteristics?

Mangrove forests are known to be particularly species poor (Duke et al. 1998). This lack of diversity is thought to be caused by strong selection pressures due to the harsh saline environment. Peat forests also persist in an extreme environment, with poor acidic soil conditions, and yet maintain some of their diversity. While species diversity is generally lower in peat forests in comparison to other lowland forests (Cannon and Leighton 2004), mangrove forests are substantially less species diverse than other forest types and strongly dominated by a relatively small number of families (Anderson 1963, Duke et al. 1998, Ellison et al. 1999). Additionally, obligate mangrove species have evolved in several different plant families indicating that adaptation to the edaphic conditions is not restricted to a single group with the appropriate ancestral traits. The highly dynamic biogeographic history of mangroves imposed a dramatic filter on the species and phenotypic traits that can track and dominate this forest type. Given the model of sea-level change, coastal mangrove areas around northern Borneo and along the coast of Vietnam have maintained their geographic positions throughout the Quaternary Period. These gradients in forced migration could be examined to measure the impact of differing histories on community composition: taxonomic, genetic and phenotypic.

On the other hand, upland forests have largely occupied the same geographic locations through the Quaternary Period with little forced migration. While rapid seed dispersal would provide a major selective advantage to coastal forest trees, dispersal ability in upland trees would have evolved to exploit the greater stability of the montane climate and its island archipelago setting. In upland forests with strong dispersal limitation, competitive exclusion and local adaptation to strict and stable environmental conditions should be more frequent. In contrast, mangrove species would benefit from enhanced ability to rapidly colonise newly formed habitat, including fast growth and strong dispersal ability. Given the sensitivity of lowland rainforest distribution on the shelf to model parameters, such as precipitation, and unknown parameters, such as marine soil evolution, a mixed strategy was probably the most advantageous. Some areas of rainforest have been entirely stable throughout the Quaternary, providing a haven for species with poor dispersal ability, while forests on the shelf would have needed greater dispersal ability.

Testable hypotheses about historical population size can be generated

Mangrove populations found along the eastern coast of Borneo should bear a genetic signature of the predicted explosive population growth and range expansion that occurred at the beginning and end of the glacial period. The historical population sizes of several locations around the current coastline of the Sunda Shelf islands are predicted to be substantially different (Fig 5.1a). Further development of spatial models that include likely migration paths would be informative, as the predicted rate of migration also varies considerably, from the western coast of Borneo (which should have required rapid, long-distance dispersal) to the eastern coast of peninsular Malaysia (relatively short-distance dispersal) through to points along the northwestern coast of peninsular Malaysia, where historical population sizes were probably small and almost no migration would have been necessary.

Likewise, lowland forest communities that now exist in different regions vary considerably in their predicted historical population sizes (Fig 5.1b). As stated above, the central and eastern populations have been remarkably stable through the last glacial cycle and, by extrapolation, through the majority of the Quaternary Period. This pattern is in marked contrast to the highly dynamic population sizes predicted in the northwest corner of Borneo, where large populations were sustained for tens of thousands of years but punctuated by near local extinction. Again, a spatial migration model would be informative in comparing the migration distances and landscape level turnover of different current lowland populations around Sundaland. The three main upland centres differ in their predicted historical population sizes, with central Borneo being relatively stable but marked by a steady, slow expansion with the progression of the ice age with a dramatic reduction with the onset of the interglacial. Upland forests in Indochina on the other hand were considerably more unstable, with potential local extinction at the LGM because of increasing rainfall seasonality and cooling of temperatures. Upland forests in Sumatra have been substantially smaller in area and slightly more unstable than those in central Borneo. All of these patterns could be examined using currently available genetic techniques.

5.3 Macro-evolutionary dynamics of tropical trees

Petit and Hampe (2006) provided an excellent review of the evolutionary consequences of being a tree and the many unusual characteristics of their life history strategy. Here, I expand briefly upon certain aspects of these characteristics, focusing particularly on tropical trees. Unfortunately, relatively little empirical data exist in relation to many of these topics. Some reasonable speculation seems appropriate in the development of models and questions specifically relevant to these enigmatic organisms.

5.3.1 Life history and phenotypic plasticity

Determining the age of tropical rainforest trees is notoriously difficult, because the structural wood in their stem does not possess clear seasonal growth patterns and stem size does not correlate well with age. Small poles can be quite old (Vieira et al. 2005), persisting in the shaded understory until an opening in the canopy appears, while growth rates can vary substantially across a population, depending on the immediate environmental setting. Various techniques have been used to age trees, including examining turnover rates over long periods of time (Laurance et al. 2004) and radiocarbon dating of structural tissue (Chambers et al. 1998). These studies indicate that many large trees in tropical communities are >400 years old with several individuals obtaining ages greater than 1000 years. The survivorship curve for most tree species is strongly convex, having very high mortality among the smallest size classes but once individuals obtain a certain size threshold, they enjoy high survivorship and fecundity (Van Valen 1976) and under certain conditions may only die as a result of stochastic events and not due to intrinsic age-related performance decline (Seymour and Doncaster 2007).

The size class distributions for the populations of many species contain a few exceptionally large individuals (Condit et al. 2006). These 'super-trees' can potentially contribute enormous numbers of offspring across numerous reproductive events spaced out over several years. Their general longevity and the presence of super-trees in the population means that local climate conditions may be substantially different over the span of relatively few generations, both on millennial and glacial scales (Borges 2009). The simple definition of 'generation time' for rainforest trees is rather problematic because of the plasticity of life history within a population and between communities. Demographic structure of rainforest tree populations, dominated by a few super-trees but also containing many large, mature, but substantially younger individuals, must be extremely complex, with highly overlapped generations.

The lifetime genetic potential and reproductive capacity of trees can be enormous, each individual producing hundreds of thousands of flowers and seeds

(Curran and Leighton 2000). This reproductive effort multiplied over one hundred flowering events would allow mature trees to perform a stunning amount of genetic experimentation. Given the quality and variety of pollination received, a tree may be able to produce a wide range of phenotypes in a single reproductive episode through mixed mating strategies (Holsinger 1991, Goodwillie et al. 2005), from selfing to interspecific hybridisation (Page et al. 2010). Trees have a very limited ability to manipulate pollinator behaviour after visiting their flowers, meaning that individuals possess little opportunity for assortative mating, at least not physically (Marshall and Folsom 1991). The main control would thus be post-pollination genetic pre-zygotic and post-zygotic developmental barriers. The control of resource investment in these developing zygotes could be exercised through selective abortion, supporting those that are most fit developmentally (Bawa and Webb 1984, Burd 1998). Either way, intraspecific assortative mating might be a weak force in tree populations, particularly if individuals are pollinator limited (Ashman et al. 2004, Vamosi et al. 2006).

Generally speaking, the possibility of generating interspecific hybrids is very high, with dispersal limitation leading to clumped populations, generalised pollination systems and widespread and variable sympatry among numerous closely related taxa. Given the lifetime fecundity of trees and their iteroparous reproductive behaviour, interspecific hybrid crosses could play a small but substantial role in the evolution of local populations, even with low hybrid viability. Likewise, trees have no control over the physical destination of their seeds and exercise relatively weak control over the type of disperser. Seed dispersal limitation thus creates neighbourhoods of offspring, all produced by the same mother tree (Condit et al. 2000). These groves of closely related offspring would further erode the ability of trees for assortative mating. If local environmental conditions are predictable and stable over sufficiently long periods of time for adaptation and competitive exclusion to occur, this homing strategy, where successful trees establish outcrossed offspring in their immediate vicinity, could be effective.

Overall, many evolutionary characteristics of trees limit their ability to adapt to their immediate surroundings – long lifespan, weak assortative mating, stochastic choice of germination site. Because of these slow evolutionary dynamics, selection pressures imposed by the environment and the surrounding community are probably always relatively unstable. Given the dramatic change in the biogeographic distribution of rainforest across Sundaland, over the course of just a few thousand years, the selective landscape must have always been in flux from the perspective of tree populations. The present distribution of species and individuals represents the integration of numerous selective forces, both past and present, varying rapidly and unpredictably in time and space. The large impact of stochastic forces on the distribution of species and populations should not be surprising. Tree species and their phenotypic traits may match narrow environmental niches but their historical 'extinct' niches may have played a more important role in their evolution. The phenotypic identity of species should also be expected to change through time. Combining the effect of the 'ghosts of soils–climate–community past' with the relatively shallow and spatially unpredictable environmental gradients present in Sundaland forests, the macro-evolutionary objective of tree species may not be the optimisation of specific phenotypes, fit perfectly to a refined niche, but instead the maintenance of a large capacity for phenotypic variability in their offspring.

5.4 The natural historical record: going, going, gone?

Humans now completely dominate the planet and this explosion in population and resource consumption has occurred within a few lifetimes, after tens of thousands of years of existence in more 'primitive' conditions. Some scientists suggest that we have entered a new era of Earth's history, called the Anthropocene (Steffen et al. 2007). While the potential for massive species extinction has been frequently discussed in the literature (Sodhi et al. 2004), the transition into the Anthropocene is also erasing our rich natural history, underlying the Earth's great wealth of biodiversity and its natural resources. This historical record is found in the natural geographic distribution of species, phenotypic traits and genetic diversity. These patterns are the result of millennial processes and represent the integration of numerous ecological and evolutionary forces. Human activities threaten to largely destroy this record in a century or less. Understanding the forests of the past is essential to understanding current forests and for making viable predictions about their future (Petit et al. 2008). Eventually, even if we manage to conserve and sustain a substantial fraction of biodiversity through the transition into the Anthropocene, we may yet have lost the necessary information to thoroughly comprehend the deep evolutionary and ecological forces that created that biodiversity.

The Sundaland forests provided Alfred Russel Wallace a key inspiration for the development of his theories of natural selection and biogeography (Wallace 1869: 114):

> We are now enabled to trace out with some probability the course of events. Beginning at the time when the whole of the Java sea, the Gulf of Siam, and the Straits of Malacca were dry land, forming with Borneo, Sumatra, and Java, a vast southern prolongation of the Asiatic continent, the first movement would be the sinking down of the Java sea and the Straits of Sunda, consequent on the activity of the Javanese volcanoes. [...] As the volcanic belt of Java and Sumatra increased in activity, more and more of the land was submerged,

> till first Borneo, and afterwards Sumatra, became entirely severed. Since the epoch of the first disturbance, several distinct elevations and depressions may have taken place, and the islands may have been more than once joined with each other or with the mainland, and again separated.

Interestingly enough, this vivid and fairly accurate description of the history of the region invokes rapid geological change without a hint that sea level or climate might have changed. We now know that climate-driven factors change much more rapidly than geology. Perhaps this bias on Wallace's part reveals something about human perception of the environment. Physical evidence of earthquakes and volcanoes and wave action can be found everywhere, so geological change is obvious. Climate change on the other hand leaves a much more subtle record, one that requires more diligence and imagination to visualise and demonstrate. Because climate change caused by the glacial cycles had a direct physical impact on Sundaland forests, through the repeated flooding and exposure of the shelf, we have a unique opportunity to test directly a number of fundamental aspects of tropical community ecology and evolution. Hopefully, we will capture enough evidence about the past before it becomes too fragmented to fully interpret.

References

Abegg, C. and Thierry, B. (2002). Macaque evolution and dispersal in insular south-east Asia. *Biological Journal of the Linnean Society*, **75**, 555–76.

Anderson, J. A. R. (1963). The flora of the peat swamps of Sarawak and Brunei, including a catalogue of all recorded species of flowering plants, ferns and fern-allies. *Gardens Bulletin of Singapore*, **120**, 131–228.

Ashman, T. L., Knight, T. M., Steets, J. A. et al. (2004). Pollen limitation of plant reproduction: Ecological and evolutionary causes and consequences. *Ecology*, **85**, 2408–21.

Bawa, K. S. and Webb, C. J. (1984). Flower, fruit and seed abortion in tropical forest trees: implications for the evolution of paternal and maternal reproductive patterns. *American Journal of Botany*, **71**, 736–51.

Bintanja, R., van de Wal, R. S. W. and Oerlemans, J. (2005). Modelled atmospheric temperatures and global sea levels over the past million years. *Nature*, **437**, 125–8.

Bird, M. I., Taylor, D. and Hunt, C. (2005). Environments of insular Southeast Asia during the Last Glacial Period: a savanna corridor in Sundaland? *Quaternary Science Reviews*, **24**, 2228–42.

Bonnefille, R. 2007. Rainforest responses to past climatic change in tropical Africa. In *Tropical Rainforest Responses to Climatic Change*, ed. M. B. Bush and J. R. Flenley. Chichester, UK: Springer-Praxis, pp. 117–70.

Borges, R. M. (2005). Do plants and animals differ in phenotypic plasticity? *Journal of Biosciences*, **30**, 41–50.

Borges, R. M. (2009). Phenotypic plasticity and longevity in plants and animals:

cause and effect? *Journal of Biosciences*, **34**, 605–11.

Brandon-Jones, D. (1996). The Asian Colobinae (Mammalia: Cercopithecidae) as indicators of Quaternary climatic change. *Biological Journal of the Linnaean Society*, **59**, 327–50.

Burd, M. (1998). 'Excess' flower production and selective fruit abortion: a model of potential benefits. *Ecology*, **79**, 2123–32.

Bush, A. B. G. (2007). Extratropical influences on the El Niño-Southern Oscillation through the late Quaternary. *Journal of Climate*, **20**, 788–800.

Bush, A. B. G. and Fairbanks, R. G. (2003). Exposing the Sunda shelf: tropical responses to eustatic sea level change. *Journal of Geophysical Research-Atmospheres*, **108**(D15), art. 4446.

Bush, M. B., Whittaker, R. J. and Partomihardjo, T. (1995). Colonization and succession on Krakatau: an analysis of the guild of vining plants. *Biotropica*, **27**, 355–72.

Cannon, C. H. and Leighton, M. (2004). Tree species distributions across five habitats in a Bornean rainforest. *Journal of Vegetation Science*, **15**, 257–66.

Cannon, C. H., Morley, R. J. and Bush, A. B. G. (2009). The current refugial rainforests of Sundaland are unrepresentative of their biogeographic past and highly vulnerable to disturbance. *Proceedings of The National Academy of Sciences of the United States of America*, **106**, 11188–93.

Carnaval, A. C., Hickerson, M. J., Haddad, C. F. B., Rodrigues, M. T. and Moritz, C. (2009). Stability predicts genetic diversity in the Brazilian Atlantic Forest Hotspot. *Science*, **323**, 785–9.

Chambers, J. Q., Higuchi, N. and Schimel, J. P. (1998). Ancient trees in Amazonia. *Nature*, **391**, 135–6.

Channell, R. and Lomolino, M. V. (2000). Trajectories to extinction: spatial dynamics of the contraction of geographical ranges. *Journal of Biogeography*, **27**, 169–79.

Clark, J. S. (1998). Why trees migrate so fast: confronting theory with dispersal biology and the paleorecord. *American Naturalist*, **152**, 204–24.

Condit, R., Ashton, P. S., Baker, P., Bunyavejchewin, S. et al. (2000). Spatial patterns in the distribution of tropical tree species. *Science*, **288**, 1414–18.

Condit, R., Ashton, P., Bunyavejchewin, S. et al. (2006). The importance of demographic niches to tree diversity. *Science*, **313**, 98–101.

Connell, J. H. and Lowman, M. D. (1989). Low-diversity tropical rainforests: some possible mechanisms for their existence. *American Naturalist*, **134**, 88–119.

Curran, L. M. and Leighton, M. (2000). Vertebrate responses to spatiotemporal variation in seed production of mast-fruiting Dipterocarpaceae. *Ecological Monographs*, **70**, 101–28.

Davies, S. J. and Becker, P. (1996). Floristic composition and stand structure of mixed dipterocarp and heath forests in Brunei Darussalam. *Journal of Tropical Forest Science*, **8**, 542–69.

Duke, N. C., Ball, M. C. and Ellison, J. C. (1998). Factors influencing biodiversity and distributional gradients in mangroves. *Global Ecology And Biogeography Letters*, **7**, 27–47.

Ellison, A. M., Farnsworth, E. J. and Merkt, R. E. (1999). Origins of mangrove ecosystems and the mangrove biodiversity anomaly. *Global Ecology and Biogeography*, **8**, 95–115.

Feeley, K. J. and Terborgh, J. W. (2005). The effects of herbivore density on soil nutrients and tree growth in tropical forest fragments. *Ecology*, **86**, 116–24.

Fernando, P., Vidya, T. N. C., Payne, J. et al. (2003). DNA analysis indicates that Asian elephants are native to Borneo and are therefore a high priority for conservation. *PLoS Biology*, **1**, 110–15.

Gibbard, P. L., Head, M. J. and Walkers, M. J. C. (2010). Formal ratification of the Quaternary System/Period and the Pleistocene Series/Epoch with a base at 2.58 Ma. *Journal of Quaternary Science*, **25**, 96–102.

Goodwillie, C., Kalisz, S. and Eckert, C. G. (2005). The evolutionary enigma of mixed mating systems in plants: occurrence, theoretical explanations, and empirical evidence. *Annual Review of Ecology Evolution and Systematics*, **36**, 47–79.

Graham, C. H., Moritz, C. and Williams, S. E. (2006). Habitat history improves prediction of biodiversity in rainforest fauna. *Proceedings of the National Academy of Sciences of the United States of America*, **103**, 632–36.

Hanebuth, T., Stattegger, K. and Grootes, P. M. (2000). Rapid flooding of the Sunda Shelf: a late-glacial sea-level record. *Science*, **288**, 1033–5.

Hanebuth, T., Stattegger, K. and Bojanowski, A. (2009). Termination of the Last Glacial Maximum sea-level lowstand: the Sunda-Shelf data revisited. *Global and Planetary Change*, **66**, 76–84.

Heaney, L. R. (1991). A synopsis of climatic and vegetational change in Southeast-Asia. *Climatic Change*, **19**, 53–61.

Hewitt, G. M. (1996). Some genetic consequences of ice ages, and their role in divergence and speciation. *Biological Journal of the Linnaean Society*, **58**, 247–76.

Hewitt, G. M. (2004). Genetic consequences of climatic oscillations in the Quaternary. *Proceedings of the Royal Society B*, **359**, 183–95.

Hirai, H., Wijayanto, H., Tanaka, H. et al. (2005). A whole-arm translocation (WAT8/9) separating Sumatran and Bornean agile gibbons, and its evolutionary features. *Chromosome Research*, **13**, 123–33.

Holsinger, K. E. (1991). Mass-action models of plant mating systems: the evolutionary stability of mixed mating systems. *American Naturalist*, **138**, 606–22.

Hu, J. F., Peng, P., Fang, D. Y. et al. (2003). No aridity in Sunda Land during the Last Glaciation: Evidence from molecular-isotopic stratigraphy of long-chain n-alkanes. *Palaeogeography Palaeoclimatology Palaeoecology*, **201**, 269–81.

Janzen, D. H. (1974). Tropical blackwater rivers, animals, and mast fruiting by the Dipterocarpaceae. *Biotropica*, **4**, 69–103.

Kershaw, A. P., Penny, D., Van der Kaars, S., Anshari, G. and Thamotherampillai, A. (2001). Evidence for vegetation and climate in lowland southeast Asia at the Last Glacial Maximum. In *Faunal and Floral Migrations and Evolution in SE Asia-Australasia*, ed. I. Metcalfe, J. M. B. Smith, M. Morwood and I. Davidson. Lisse, The Netherlands: A. A. Balkema, pp. 227–36.

Kershaw, A., van der Kaars, S. and Flenley, J. (2007). The Quaternary history of Far Eastern rainforests. In *Tropical Rainforest Responses to Climatic Change*, ed. M. B. Bush and J. F. Flenley. Chichester, UK: Springer-Praxis, pp. 77–115.

Konuma, A., Tsumura, Y., Lee, C. T., Lee, S. L. and Okuda, T. (2000). Estimation of gene flow in the tropical-rainforest tree *Neobalanocarpus heimii* (Dipterocarpaceae), inferred from paternity analysis. *Molecular Ecology*, **9**, 1843–52.

Lambeck, K. and Purcell, A. (2005). Sea-level change in the Mediterranean Sea since the LGM: model predictions for tectonically stable areas. *Quaternary Science Reviews*, **24**, 1969–88.

Laurance, W. F., Nascimento, H. E. M., Laurance, S. G. et al. (2004). Inferred longevity of Amazonian rainforest trees based on a long-term demographic study. *Forest Ecology and Management*, **190**, 131–43.

Lee, S. L., Wickneswari, R., Mahani, M. C. and Zakri, A. H. (2000). Genetic diversity of a tropical tree species, *Shorea leprosula* Miq. (Dipterocarpaceae), in Malaysia: implications for conservation of genetic resources and tree improvement. *Biotropica*, **32**, 213–24.

Lee, S. L., Ng, K. K. S., Saw, L. G. et al. (2002). Population genetics of *Intsia palembanica* (Leguminosae) and genetic conservation of virgin jungle reserves in peninsular Malaysia. *American Journal of Botany*, **89**, 447–59.

Louys, J. and Meijaard, E. (2010). Palaeoecology of Southeast Asian megafauna-bearing sites from the Pleistocene and a review of environmental changes in the region. *Journal of Biogeography*, **37**, 1432–49.

Marshall, A. J., Nardiyono, Engstrom, L. M. et al. (2006). The blowgun is mightier than the chainsaw in determining population density of Bornean orangutans (*Pongo pygmaeus morio*) in the forests of East Kalimantan. *Biological Conservation*, **129**, 566–78.

Marshall, D. L. and Folsom, M. W. (1991). Mate choice in plants: an anatomical to population perspective. *Annual Review of Ecology and Systematics*, **22**, 37–63.

Meijaard, E. and van der Zon, A. P. M. (2003). Mammals of south-east Asian islands and their Late Pleistocene environments. *Journal of Biogeography*, **30**, 1245–57.

Mollengraaff, G. A. F. (1921). Modern deep-sea research in the east Indian archipelago. *Geographical Journal*, **57**, 95–121.

Morley, R. J. (1981). Development and vegetation dynamics of a lowland ombrogenous peat swamp in Kalimantan Tengah, Indonesia. *Journal of Biogeography*, **8**, 383–404.

Newbery, D. M. (1991). Floristic variation within kerangas (heath) forest: re-evaluation of data from Sarawak and Brunei. *Vegetatio*, **96**, 43–86.

Ng, K. K. S., Lee, S. L., Saw, L. G., Plotkin, J. B. and Koh, C. L. (2006). Spatial structure and genetic diversity of three tropical tree species with different habitat preferences within a natural forest. *Tree Genetics and Genomes*, **2**, 121–31.

Page, T., Moore, G. M., Will, J. and Halloran, G. M. (2010). Breeding behaviour of *Kunzea pomifera* (Myrtaceae): self-incompatibility, intraspecific and interspecific cross-compatibility. *Sexual Plant Reproduction*, **23**, 239–53.

Petit, R. J. and Hampe, A. (2006). Some evolutionary consequences of being a tree. *Annual Review of Ecology Evolution and Systematics*, **37**, 187–214.

Petit, R. J., Aguinagalde, I., de Beaulieu, J. L. et al. (2003). Glacial refugia: hotspots but not melting pots of genetic diversity. *Science*, **300**, 1563–5.

Petit, R. J., Hu, F. S. and Dick, C. W. (2008). Forests of the past: a window to future changes. *Science*, **320**, 1450–2.

Pitman, N. C. A., Terborgh, J. W., Silman, M. R. et al. (2001). Dominance and distribution of tree species in upper Amazonian terra firme forests. *Ecology*, **82**, 2101–17.

Rieley, J. O. and Page, S. E. (1995). *Biodiversity and Sustainability of Tropical Peatlands*, Cardigan, UK: Samara Publishing Limited.

Sagarin, R. D. and Gaines, S. D. (2002). The 'abundant centre' distribution: to what extent is it a biogeographical rule? *Ecology Letters*, **5**, 137–47.

Sakai, S., Harrison, R. D., Momose, K. et al. (2006). Irregular droughts trigger mass flowering in aseasonal tropical forests in Asia. *American Journal of Botany*, **93**, 1134–9.

Seymour, R. M. and Doncaster, C. P. (2007). Density dependence triggers runaway selection of reduced senescence. *PloS Computational Biology*, **3**, e256.

Slik, J. W. F., Poulsen, A. D., Ashton, P. S. et al. (2003). A floristic analysis of the lowland dipterocarp rainforests of Borneo. *Journal of Biogeography*, **30**, 1517–31.

Slik, J. W. F., Raes, N., Aiba, S. I. et al. (2009). Environmental correlates for tropical tree diversity and distribution patterns in Borneo. *Diversity and Distributions*, **15**, 523–32.

Sodhi, N. S., Koh, L. P., Brook, B. W. and Ng, P. K. L. (2004). Southeast Asian biodiversity: an impending disaster. *Trends in Ecology and Evolution*, **19**, 654–60.

Steffen, W., Crutzen, P. J. and McNeill, J. R. (2007). The Anthropocene: are humans now overwhelming the great forces of nature? *Ambio*, **36**, 614–21.

Steinke, S., Kienast, M. and Hanebuth, T. (2003). On the significance of sea-level variations and shelf paleo-morphology in governing sedimentation in the southern South China Sea during the last deglaciation. *Marine Geology*, **201**, 179–206.

Steiper, M. E. (2006). Population history, biogeography, and taxonomy of orangutans (Genus: Pongo) based on a population genetic meta-analysis of multiple loci. *Journal of Human Evolution*, **50**, 509–22.

Sun, X. J., Li, X. and Luo, Y. L. (2002). Vegetation and climate on the Sunda shelf of the South China Sea during the last glaciation-pollen results from station 17962. *Acta Botanica Sinica*, **44**, 746–52.

Templeton, A. R. (1998). Nested clade analyses of phylogenetic data: testing hypotheses about gene flow and population history. *Molecular Ecology*, **7**, 381–97.

Terborgh, J., Lopez, L., Nunez, P. et al. (2001). Ecological meltdown in predator-free forest fragments. *Science*, **294**, 1923–6.

Vamosi, J. C., Knight, T. M., Steets, J. A. et al. (2006). Pollination decays in biodiversity hotspots. *Proceedings of the National Academy of Sciences of the United States of America*, **103**, 956–61.

van den Bergh, G. D., de Vos, J. and Sondaar, P. Y. (2001). The Late Quaternary palaeogeography of mammal evolution in the Indonesian Archipelago. *Palaeogeography, Palaeoclimatology, Palaeoecology*, **171**, 385–408.

Van Valen, L. (1976). Ecological species, multispecies, and oaks. *Taxon*, **25**, 233–9.

Vieira, S., Trumbore, S., Camargo, P. B. et al. (2005). Slow growth rates of Amazonian trees: consequences for carbon cycling. *Proceedings of the National Academy of Sciences of the United States of America*, **102**, 18502–507.

Voris, H. K. (2000). Maps of Pleistocene sea levels in Southeast Asia: shorelines, river systems and time durations. *Journal of Biogeography*, **27**, 1153–67.

Wallace, A. R. (1869). *The Malay Archipelago*. New York: Harper and Brothers.

Wang, X. M., Sun, X. J., Wang, P. X. and Stattegger, K. (2009). Vegetation on the Sunda Shelf, South China Sea, during the Last Glacial Maximum. *Palaeogeography Palaeoclimatology Palaeoecology*, **278**, 88–97.

Whittaker, R. J., Bush, M. B. and Richards, K. (1989). Plant recolonization and vegetation succession on the Krakatau Islands, Indonesia. *Ecological Monographs*, **59**, 59–123.

Woodruff, D. (2010). Biogeography and conservation in Southeast Asia: how 2.7 million years of repeated environmental fluctuations affect today's patterns and the future of the remaining refugial-phase biodiversity. *Biodiversity and Conservation*, **19**, 919–41.

Woodruff, D. S. and Turner, L. M. (2009). The Indochinese–Sundaic zoogeographic transition: a description and analysis of terrestrial mammal species distributions. *Journal of Biogeography*, **36**, 803–21.

Wurster, C. M., Bird, M. I., Bull, I. D. et al. (2010). Forest contraction in north equatorial Southeast Asia during the Last Glacial Period. *Proceedings of the National Academy of Sciences*, **107**, 15508–511.

Zheng, W., Braconnot, P., Guilyardi, E., Merkel, U. and Yu, Y. (2008). ENSO at 6 ka and 21 ka from ocean–atmosphere coupled model simulations. *Climate Dynamics*, **30**, 745–62.

Zhi, L., Karesh, W. B., Janczewski, D. N. et al. (1996). Genomic differentiation among natural populations of orang-utan (*Pongo pygmaeus*). *Current Biology*, **6**, 1326–36.

6

The Malesian floristic interchange: plant migration patterns across Wallace's Line

JAMES E. RICHARDSON, CRAIG M. COSTION AND ALEXANDRA N. MUELLNER

6.1 Introduction

The convergence of formerly disconnected land masses presents opportunities for plant and animal dispersal and their possible subsequent diversification in new terrain. Examples include the closure of the Isthmus of Panama in the Pliocene, the collision of India with continental Asia and the convergence of the Australian and Eurasian Plates. Perhaps the most widely discussed example of biotic migration is that of the Great American Interchange between the Laurasian North American Plate and the Gondwanan South American Plate that was the subject of George Gaylord Simpson's 'Splendid Isolation' (1980) that focused on migration patterns of animals. Prior to the development of phylogenetic methods, the primary mode of interpreting patterns in migration relied on analysis of fossil and extant distributions. Molecular phylogenetic methods that incorporate the dimension of time are a relatively new and powerful tool in the study of biogeographic history. In an analysis of plant and animal dated phylogenies of taxa that are distributed across the Isthmus of Panama, Cody et al. (2010) demonstrated that plants have a greater capacity for dispersal than animals because they had been dispersing between

Biotic Evolution and Environmental Change in Southeast Asia, eds D. J. Gower et al. Published by Cambridge University Press.

these areas since the Eocene whereas animals only began to disperse in the mid Miocene. Interestingly, this study indicated that many animals also had the capacity for over-water dispersal prior to closure of the Isthmus of Panama but in general they were more reliant on the formation of a direct land connection. One difference between the Panamanian and Indian convergences and that between the Australian and Eurasian Plates is that the former two both culminated in the formation of a direct land bridge, but the latter has not. Wallace (1869) noted the strong faunistic divide around the area of convergence between Australian and Eurasian Plates that would come to be known as 'Wallace's Line'. Wallace was less forthcoming when describing floristic differences and determination of the significance of Wallace's Line for plants has proven far more complex than for the fauna. Any study of biotic interchange between the Eurasian and Australian Plates will ultimately enable better assessment of the capability of over water dispersal of different groups of organisms.

Successful migration depends on a number of variables including dispersal capability, distance between source and sink area, size of source and sink area, biotic interactions with established species and the ability to become established with regard to the abiotic environment. Dated molecular trees support the view that plants readily disperse over oceans, as has been documented in many regions of the world including the Panama Isthmus (Cody et al. 2010), Hawaii (Price and Clague 2002), New Zealand (Winkworth et al. 2002, Perrie and Brownsey 2007), across the Atlantic Ocean (Renner 2004), and across the entire Pacific Ocean (Motley et al. 2005, Muellner et al. 2008); and for several plant genera previously acknowledged as examples of vicariance through continental drift, including *Agathis* (Setoguchi et al. 1998, Knapp et al. 2007), *Cycas* (Keppel et al. 2008), and *Nothofagus* (Cook and Crisp 2005, Knapp et al. 2005). Crisp et al. (2009) reported 226 transoceanic dispersal events in the southern hemisphere, and Renner et al. (2010) reported over 1000 publication hits in Web of Science for 'long distance dispersal'.

The likelihood of transoceanic dispersal should increase with decreased distance between emergent land masses as continental plates converge. The general pattern in Southeast Asia during the Mio-Pliocene has been one of plate convergence leading to an increased likelihood of dispersal between Sundania and Australasia. Houle (1998) demonstrated that potential dispersal time via oceanic currents between Southeast Asian land masses has been decreasing as continental plates converged. Inter-island dispersal success is also governed in part by the size of the receiving island, another feature that has varied through time in the Southeast Asian archipelago. The ability to become established may also be influenced by the existing biota, with the probability of establishment being inversely related to the number of species in the area into which the migration event occurs. Finally, migration success is also dependent on the ability to become established subsequent to dispersal. This ability may in part be controlled by the degree of phylogenetic niche conservatism (PNC) with, for example, representatives of wet-adapted lineages being less likely to become established in dry areas because they cannot adapt across this physiological barrier and are also less likely to compete successfully with resident lineages that have already adapted.

In this chapter we attempt to determine the contribution of Sundanian elements to Australasian land masses and vice versa, and to offer explanations for any bias in the direction of migration. We refer to the origin of a group as being Sundanian or Australasian (i.e. to the west or east of Wallace's Line, respectively) rather than Gondwanan or Laurasian. This is because some Sundanian elements may have had a Gondwanan origin, having arrived in Sundania via the former Gondwanan land mass of India. Our focus will be on the interchange of everwet forest elements and to a lesser extent the montane vegetation. We summarise current vegetation patterns and discuss traditional approaches to assessing biotic interchange based on floristic composition and fossils. We assess the contribution of dated molecular phylogenetic trees to describe these patterns, reassess overall plant migration patterns in the Malesian region, and discuss which variables may be important in controlling successful colonisation.

6.2 Modern vegetation

The forests of Southeast Asia can be separated climatically into three broadly defined groups: everwet tropical rainforest, drought-adapted tropical moist forest (known as monsoon forest in Southeast Asia) and montane forest (Steenis 1950, Whitmore 1984). Numerous vegetation types and associations occur within these three categories and their distribution and floristic composition have been the subject of varying hypotheses explaining how Southeast Asia, one of the most biodiverse regions on earth, was populated by terrestrial plants.

The everwet tropical rainforest, also known as equatorial evergreen rainforest, receives 1750–2000 mm of rain spread evenly throughout the year. Monsoon forest is distinguished from the former by having an average of less than 1270 mm of seasonal rainfall per year and the occurrence of a prolonged period of low rainfall or drought. Monsoon forest is restricted mostly to the Lesser Sunda Islands, Java, Sulawesi, southern New Guinea, parts of mainland Asia and most of Northern Australia, whereas everwet rainforest stretches across Borneo, Sumatra, the Malay Peninsula, parts of Sulawesi and most of New Guinea. The Early Miocene to Pliocene of the Sundanian region was predominantly perhumid with monsoonal vegetation becoming more widespread during the Pliocene (Morley 2000, this volume, Chapter 4). Figure 6.1 shows the degree of seasonality of precipitation (range of annual precipitation) across the region. The darker areas with low values indicate less seasonality and the paler areas with higher values are those

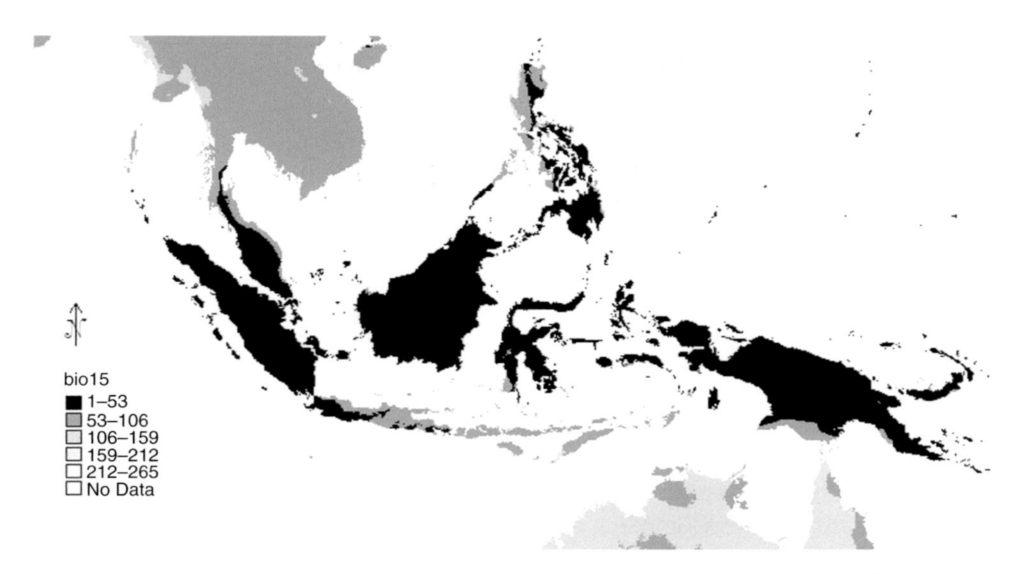

Figure 6.1 Degree of seasonality of precipitation (range of annual precipitation) in Southeast Asia. Darker areas with low values have less seasonality and the paler areas with higher values have a more pronounced dry season. Map generated from Bioclim data (Hijmans et al. 2005).

that have a more pronounced dry season. This map, generated from Bioclim data (Hijmans et al. 2005), agrees with the distribution of what have been traditionally regarded as monsoon forest and everwet forest areas of Southeast Asia. The map lacks fine-scale resolution, and does not indicate some forest areas such as the pockets of everwet forest scattered across the high volcanoes of the Lesser Sundas, or the specific boundaries between everwet forest and monsoon forest. It does however provide a general reflection of where these forest types occur across the entire region.

The montane forests can be subdivided into lower and upper zones that are divided by the treeline. In Malaya, 1500 m in elevation marks the start of the upper montane zone while in New Guinea it marks the start of the lower montane zone. The upper montane zones are largest in New Guinea but also occur in Sumatra, the Malay Peninsula, western Sulawesi, the Mount Kinabulu region of Borneo and other scattered isolated montane areas. Geologically, Mount Kinabalu is less than 8 million years old (Cottam et al. 2010); rapid elevation of the New Guinea highlands seems to date from about 5 Ma – the very end of the Miocene or Pliocene (R. Hall pers. comm. 2009); Sulawesi highlands are estimated to have reached 2500 m in the Miocene in the south, and Plio-Pleistocene in the east, central and western parts (I. Watkinson pers. comm. 2010). The montane vegetation that occupies these formations is therefore expected to be of a similar age, and evolution of montane forms from lowland forms or migration of temperate elements would therefore have occurred during these time frames.

6.2.1 Phytogeography and biogeographic divisions

Traditional approaches to establishing floristic origins in Southeast Asia have been phytogeographic. Much of the earlier work on this topic was focused on finding a botanical equivalent to Wallace's Line or one of the many other faunal lines (Fig 6.2a). Three major floristic regions or provinces were suggested by Steenis (1950) and Balgooy (1971); West Malesia, South Malesia and East Malesia, with major demarcation lines separating them (Fig 6.2b). Minor demarcation lines further separate Wallacea within East Malesia, and the Philippines within West Malesia. This treatment was based on all genera present across the region and has traditionally served as a baseline for inferring biogeographic patterns in the region. Welzen et al. (2005), based on a larger analysis of species treated in Flora Malesiana, defined three different regions (Fig 6.2c): the Sunda Shelf (minus Java), Wallacea (with Java) and the Sahul Shelf. One quarter of the total species predicted for Malesia were included in their analysis and some of the most species-rich families such as Rubiaceae, Myrtaceae and Euphorbiaceae were not included. Thus it is difficult

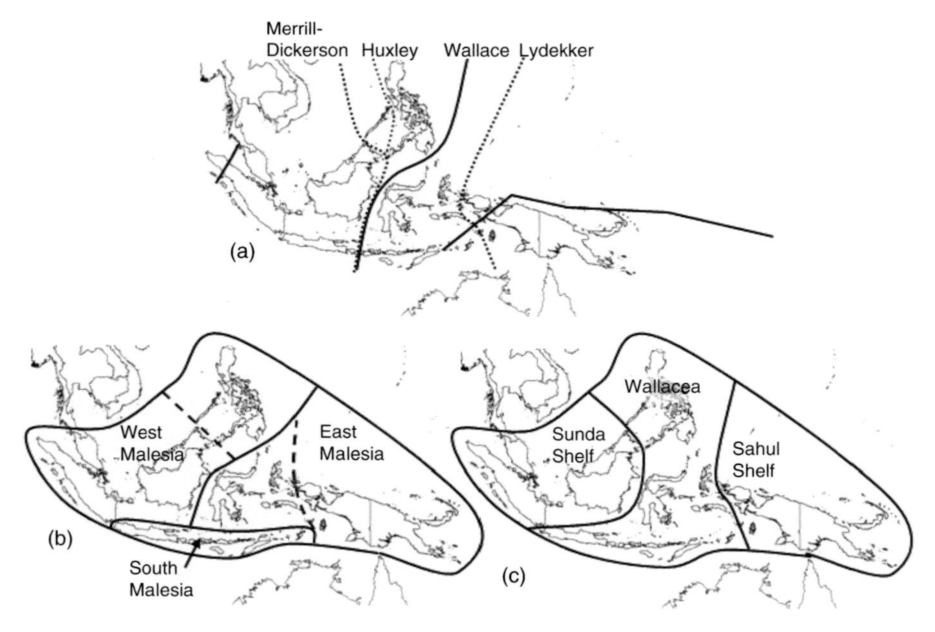

Figure 6.2 (a) Biogeographic divisions in Southeast Asia (largely based on faunal discontinuities). (b) Phytogeographic regions of Southeast Asia as defined by Steenis (1950). (c) Phytogeographic regions of Southeast Asia as defined by Welzen et al. (2005). Major (solid) and minor (dotted) floristic demarcation lines in Southeast Asia are indicated in Figs 6.2b and c.

to directly compare with Steenis' (1950) classification that was based on genera. However, the analysis was statistically robust and produced an obvious pattern that is consistent with the currently accepted geology of the region. A new analysis (P. C. van Welzen pers. comm. 2010) subdivides Wallacea into: (1) the Moluccas plus Sulawesi and Java; (2) the Lesser Sunda Islands; and (3) the Philippines.

6.2.2 Wallace's Line for plants

Kalkman (1955) was one of the first to systematically address the issue of Wallace's Line for plants within a specific region. His study presented quantifiable evidence of floristic exchange between Australia and Southeast Asia occurring in the Lesser Sunda Islands. Kalkman's data showed that the proportions of what he considered to be Malesian taxa or Australian taxa, attenuate gradually in both directions (see Fig 6.3), with Wallace's Line seeming to have no effect.

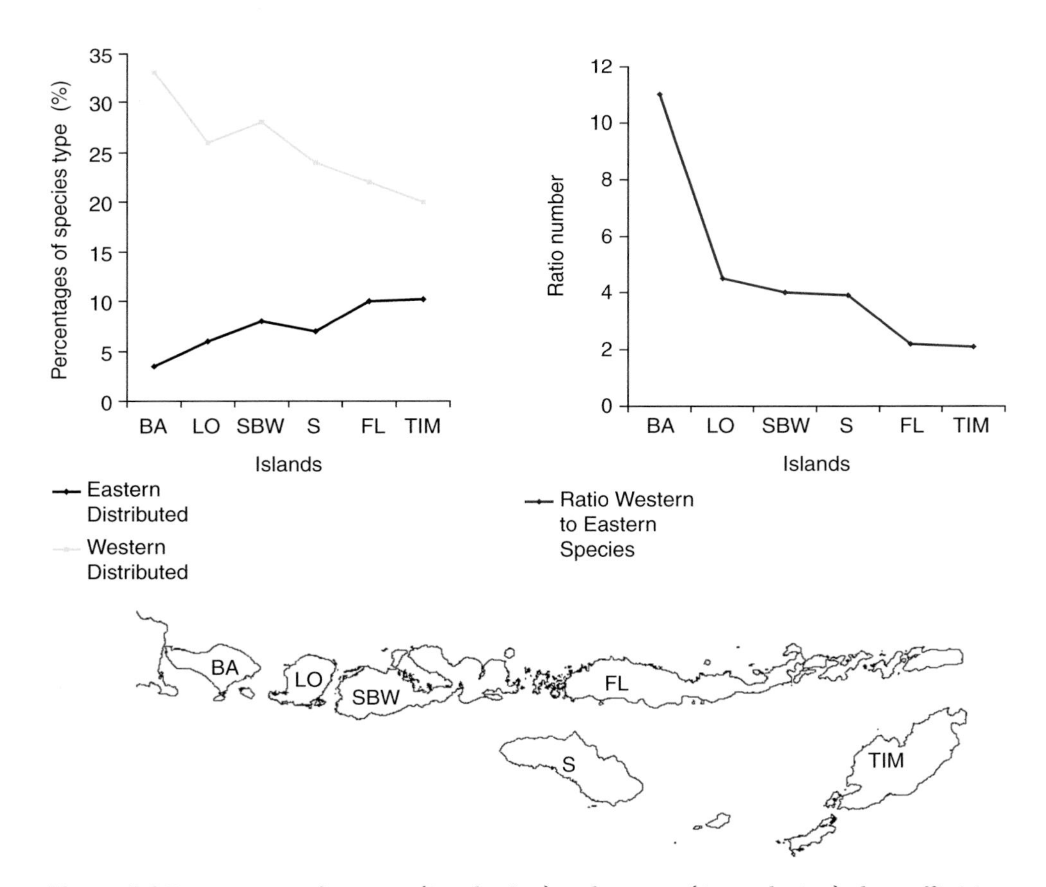

Figure 6.3 Percentages of western (Sundanian) and eastern (Australasian) plant affinities in the Lesser Sunda Islands (redrawn from Kalkman 1955). BA = Bali, LO = Lombok, SBW = Sumbawa, S = Sumba, FL = Flores, TIM = Timor.

Thus he concluded that the southern extremity of Wallace's Line between Bali and Lombok, 'for plants at least has no importance'. Both traditional and recent classifications have supported this (Fig 6.2b, c).

Although opportunities for dispersal across Wallace's Line are expected to have increased as the Australian and Eurasian Plates converged there are examples of lineages that are well represented to one side of the line that are extremely poorly represented on the other, indicating that Wallace's Line still represents a significant barrier (Welzen et al. 2005), for example in Dipterocarpaceae. The following 13 families are found only to the west of Wallace's line: Anisophylleaceae, Nartheciaceae, Caprifoliaceae, Valerianaceae, Gelsemiaceae, Irvingiaceae, Peraceae, Rafflesiaceae, Trigoniaceae, Saururaceae, Berberidaceae, Altingiaceae and Lowiaceae; and the following eight families are found only to the east: Corynocarpaceae, Nothofagaceae, Byblidaceae, Atherospermataceae, Rhipogonaceae, Himantandraceae, Eupomatiaceae and Heliconiaceae (Stevens 2001, pers. comm. 2010, Welzen and Slik 2009). Similar lists could be produced for genera that occur only to the east or west.

Huxley's Line was first used by Steenis (1950) as a minor demarcation line but was later altered by Balgooy (1971) to the Merrill–Dickerson line. Lydekker's Line seems to have more significance for plant species distributions. It has held up through time, first as a minor demarcation line and more recently acknowledged as a major one. It is evident that the lines biologists draw to infer patterns in this complex region are entirely dependent on the criteria and methods employed, specifically in regards to the taxonomic rank assessed. Demarcation lines are useful for inferring broad patterns and categories, but they do not resolve the biogeographic history of specific lineages that we require to fully understand biome assembly. Welzen (2005) proposed that Wallace's line for plants should be likened to that of a zone that spans the entire region of Wallacea, with the exact location of the line changing according to the taxonomic group studied. Because many plant lineages are unaffected by this zone, having dispersal tracks through Wallacea, Wallace's Line may be best considered a filtering zone for plants, one which spans Wallacea and inhibiting distribution of some lineages but not others, as suggested by Morley (2000).

New Guinea is of particular importance and interest to understanding the affinities of the entire region. Zoologists classify New Guinea with the Australian region (Darlington 1957) because it forms part of the Australian Plate and was connected to mainland Australia when sea levels were lower (e.g. during the Last Glacial Maximum). This allowed faunal migrations to occur from Australia to New Guinea. Few of these animals were able to disperse further west and virtually none made it beyond Sulawesi. This faunal discontinuity was the basis for Wallace's Line. The patterns for plants, however, are somewhat different. Good (1960) suggested that 37% of the angiosperm genera of New Guinea had an Indomalayan distribution (that we take to mean origin), the Australian element represented only 4.5% of the genera and 37% had palaeotropical or wider distributions. How many of the latter were of Australasian or Sundanian origin is unclear from that study. Balgooy (1976) reported that West Malesia and New Guinea share more families than New Guinea and Australia, and that twice the number of genera found in New Guinea have distributions centred in Malesia, Asia or other countries to the north and west of New Guinea than in Australia and New Zealand. From these baseline data Balgooy justified a major demarcation line separating Australia and New Guinea and inferred a primarily Malesian derived New Guinean flora. The phytogeography of Australia is also of interest due to its more recent proximity to Sundania. Hooker (1860) commented on the dispersal of plants into Australia from Southeast Asia when he divided Australia's flora into three major affinities: the Indomalayan and Antarctic elements, both considered 'intrusive'; and the 'autochthonous' or endemic element. This classification was the predominant view for over a century until the 1980s when many authors challenged the 'intrusion' hypothesis. These authors contended that the extant flora was a relict of an ancient palaeo-tropical flora (Webb and Tracey 1981, Werren and Sluiter 1984, Truswell et al. 1987) and rejected notions of any large-scale invasion of Indomalayan taxa into Australia. This view has remained pervasive in the literature (Kershaw et al. 2005) until recently. Molecular studies now show ample support for Hooker's original hypothesis with a significant percentage of Australia's northern tropical flora derived from Southeast Asia.

6.3 Floristic dispersal tracks and barriers

6.3.1 Malesian mountain tracks

The occurrence of temperate plants in the montane zones of Southeast Asia has long been acknowledged (e.g. Wallace 1869). Steenis (1935) and Brandon-Jones (2001) note that Mount Kinabalu in Sabah, Malaysian north Borneo, is far richer than Timor in Australian temperate plants. Laumonier (1997) added that the montane flora of Sumatra had less of an affinity with East Malesia, Sulawesi, and the Moluccas, than with Australia and New Guinea. This suggests that a strong westward flow of Australian taxa has taken place between the montane regions of Australasia. Steenis (1934) suggested that plants derived from temperate climates (stenotherms) and also temperate species derived from tropical lineages (eurytherms) occurred in the montane zone of Malesia. He hypothesised three floristic tracks from which the stenotherms originated (see Fig 6.4). The tracks were defined by the assumed affinities of plants within them and their inferred geological histories. The Sumatran track traces the length of Sumatra, which underwent major uplift since the Middle Miocene, and all along the recently emerged Sunda Island

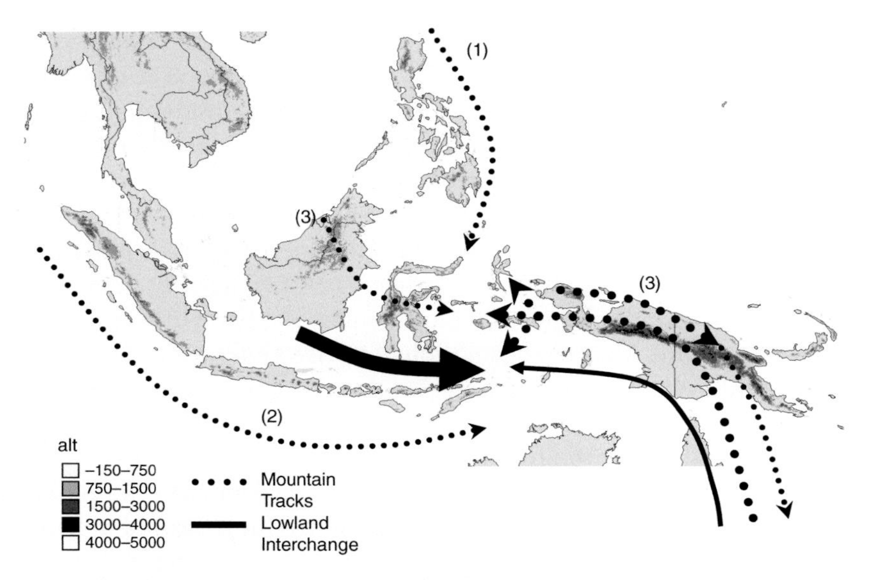

Figure 6.4 Steenis' Mountain Flora Tracks: (1) Luzon, (2) Sumatran and (3) Papuan. Montane areas shaded according to the zonation boundaries of the two extremes following Whitmore (1984).

chain, which reached its present configuration during the Late Neogene. The Luzon track starts in eastern Asia and extends south through the Philippines to southern Sulawesi. The Papuan track extends from Australia across New Guinea, overlapping the Luzon track, and into Northern Borneo.

6.3.2 Monsoon/savannah corridor

The monsoonal region stretching from northern Australia west across the Sunda Islands has been hypothesised to have acted as a corridor for seasonally adapted elements during drier periods. During these periods the rainforest blocks would have retreated to smaller areas (Steenis 1979, Whitmore 1981). This 'savanna corridor' through the Lesser Sunda Islands has been proposed to be one of the main floristic tracks that have enabled interchange between Malesia and Australia (Steenis 1979). Kershaw et al. (2005) state that the existence of this savanna corridor inhibited interchange of rainforest taxa between Australia and West Malesia. If the Savanna Corridor was a barrier to dispersal of wet forest elements, it would have been effective only during the drier periods of the Mid-Oligocene and from the Late Pliocene (Morley this volume, Chapter 4). Recent analyses by Cannon et al. (2009) indicate that glacial periods may in fact have been characterised by greater expanses of lowland evergreen rainforest, and therefore there would not have been a significant barrier to dispersal of wet forest elements during most of the Pleistocene. For the majority of the Tertiary and Quaternary, wet forest seems to have been dominant in Sundania which would have allowed easier movement of wet forest elements throughout the region.

6.4 Palaeobotanical contributions

The contributions of the fossil record to our understanding of the historical assembly of the flora of the Malay Archipelago are documented by numerous publications, for example Morley (1998, 2000) and Truswell et al. (1987), and only a brief summary of these accounts that focuses on the Sundanian–Australasian floristic interchange is presented here. A series of dispersal events into the Sundania region in the mid Eocene are evident from the appearance of elements from older sediments in India that coincided with the collision of Indian and Asian plates when both regions were experiencing everwet climates (Morley, 2000). This invasion was sufficiently strong to result in the Indian and Sundanian regions being considered a single floristic unit at this time. From the Oligocene, continued diversification of the Southeast Asian flora was evident and Morley (2000) associated this with a period of geological and environmental change and in particular the development of drier, more seasonal, monsoonal climates. This period of change was followed by the establishment of more perhumid climates from about 25 Ma. The extent of dry vegetation at various points in time is difficult to assess largely because dry conditions are less conducive to the preservation of organic matter. Dispersal across Wallace's Line as the Australian and Eurasian Plates converged is first evident in the appearance of fossils with Australian (Australasian) Plate affinities in Sundanian areas, e.g. Casuarinaceae in the Early Oligocene and *Phormium* (Hemerocallidaceae) and *Korthalsia* (Arecaceae) between 22 and 21 Ma. Myrtaceae abundance began to increase in Sundania from about 17 Ma (Morley 2000) and representatives of Podocarpaceae such as *Podocarpus* and *Phyllocladus hypophyllus* also migrated west during the Pliocene–Pleistocene boundary (Morley 2000).

Parts of New Guinea must have been emergent during the latest Oligocene because pollen of *Casuarina*, a Gondwanan group according to fossil and molecular evidence, indicated that the genus was a dominant element in a sequence in northern New Guinea at this time (Morley 2000). However, many of the islands to the east of Sulawesi were submerged from the Late Eocene to the end of the Oligocene. The exact time of emergence of the region from East Papua to the Bird's Head of New Guinea is not entirely clear, but Hall (2009) indicated that it was probably no earlier than 10 Ma. Hall (pers. comm. 2009, this volume, Chapter 3) also indicated that the region north of the present highlands is complex and may have had emergent areas during some or all of the Miocene. Morley (pers. comm. 2009) has indicated that there is minimal fossil pollen recorded for New Guinea for the Early Miocene and much of the Middle Miocene, and that the New Guinea record itself shows very little change over the Late Mio/Pliocene. The absence of fossil records from New Guinea for the period up until the mid Miocene is consistent with it being largely submerged and thus unavailable for colonisation by plants. Khan (1974a, 1974b, 1976) revealed that Myrtaceae, *Casuarina*, Rubiaceae and Palmae were important components of the vegetation throughout the late Miocene to Pleistocene, along with mangrove taxa such as Rhizophoraceae, *Sonneratia* and possible *Camptostemon* (Bombacaceae). Khan (1974a, 1976) also demonstrated *Nothofagus* dispersing into New Guinea from the Late Miocene. Playford (1982) indicated the presence of angiosperm groups such as *Barringtonia* (Lecythidaceae) in the Neogene.

Although the Gondwanan heritage of Queensland's wet tropics is well accepted, the number of Sundanian elements present has remained contentious. Truswell et al. (1987) indicated that there was fossil evidence for long distance dispersals from Southeast Asia to Australia during its isolation phase in the Tertiary and the Miocene but no evidence to suggest that this invasion was massive or enhanced by increasing proximity of the Australasian and Sundanian plates. Continuous pollen sequences from northeast Queensland and Southeast Asia are sparse in comparison to southeastern Australia. There are no reliable dated Tertiary sequences older than the late Oligocene from all of Queensland (Macphail et al. 1994). More importantly, pollen cores in the tropical north have been limited to the uplands (Kershaw 1988) and offshore marine deposits (Kershaw et al. 2005). There are no lowland sites in the extant Queensland Wet Tropics to our knowledge, which, as we discuss below, is crucial to understanding the Sundanian component. Kershaw (1988) suggested that the lowlands were the most likely place where interchange would be recorded. Kershaw et al. (1994) and Truswell et al. (1987) also indicated that the invading lineages would most likely have been insect-pollinated taxa and thus difficult to detect in palynosequences.

6.5 Dated phylogenetic trees and historical biogeography

Fossils provide an important record of how floras are assembled in space and time. However, this record is fragmentary and has limitations for providing an adequate picture of developing biomes. Phylogenetic trees that incorporate a temporal dimension have considerably clarified our understanding of historical migration and diversification patterns (reviewed in Pennington et al. 2006). Geological history can be used to make predictions for expected topologies and timing of diversification events within phylogenetic trees. For example, an origin in Sundania followed by dispersal to Australasia across Wallace's Line as the Asian and Australian plates converged would give the phylogenetic tree illustrated in Fig 6.5.

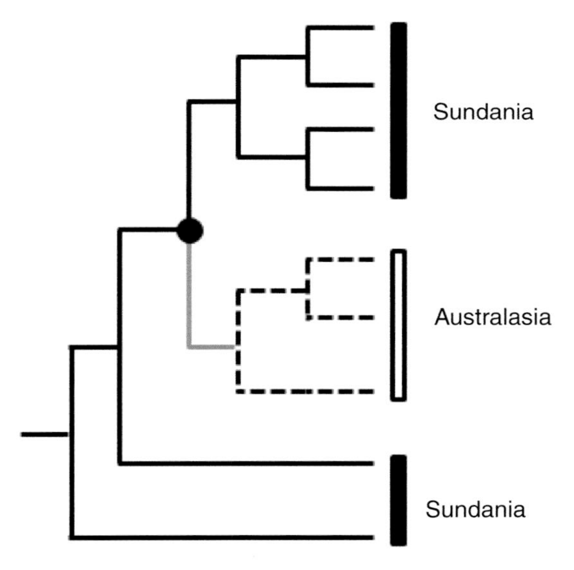

Figure 6.5 Phylogeny expected from a lineage with a Sundanian origin that dispersed across Wallace's Line as the Australian and Eurasian Plates converged. Solid black lines, Sundanian lineages; dashed black lines, Australasian lineages; grey line, equivocal. The black circle represents the basal divergence within the Australasian lineage, expected to be *c.* 20 Ma or younger.

The following examples are all based on fossil-calibrated, dated molecular phylogenetic trees. Muellner et al. (2008) reconstructed ancestral distributions for Aglaieae (Meliaceae) that are distributed throughout the Malay Archipelago, and they determined that the group had an ancestral area that included the Malay Peninsula, Sumatra, Borneo, Palawan, Java and Bali (i.e. Sundania). This ancestral area is also the area with greater species diversity for the group, and additionally fossils are found to the east of Wallace's Line at the Cretaceous/Tertiary boundary in India (Trivedi and Srivastava 1982) but are not, to our knowledge, found to the east of Wallace's Line. Dispersal was an important driving force in diversification in the group and occurred across Wallace's Line on multiple occasions during the Miocene and Pliocene. Australia was colonised by *Aglaia* only during comparably recent times (around *c.* 5 Ma; Muellner et al. 2008, 2009) and the Fijian species of *Aglaia* are of similar age (Muellner et al. 2008).

Australasian lineages of Annonaceae are much less species-rich than Sundanian ones and Richardson et al. (2004) indicated that they were also derived from within Sundanian lineages. This is consistent with the fossil record, with fossils commonly found in areas to the west of Wallace's Line including megafossils attributed to *Annona* from the Oligocene of Yunnan and Tibet; Neogene woods of *Polyalthioxylon stelechocarpides* and *P. platymitroides* (*Polyalthia stelechocarpus*, *P. hypoleuca* and *Platymitra* sp. and *P. macrocarpa*) in Java and Borneo (Kramer

1974) and *Mayogynophyllum paucinervium* leaves (Upper Miocene; Kräusel 1929). We do not know of any fossil records of Annonaceae to the east of Wallace's Line. Su and Saunders (2009) demonstrated that *Pseuduvaria* (Annonaceae) originated in Sundania in the late Miocene and that subsequent migration events were predominantly eastwards towards Australasia, although several back migrations are also evident. Subsequent speciation of *Pseuduvaria* within New Guinea may have occurred after *c.* 6.5 Ma, possibly coinciding with the formation of its Central Range Mountains. Baker and Couvreur (this volume, Chapter 7) also confirm an eastward bias in the migration of palm lineages that had been previously suggested by Dransfield (1987). Although not dated, a phylogenetic tree of *Neonauclea* (Rubiaceae) by Razafimandimbison et al. (2005) also displays a pattern consistent with west to east migration. A common feature of each of the molecular studies discussed above, along with the pattern of west to east migration, is that they are all wet forest groups.

In addition to the above cases, many lineages had been previously considered Gondwanan (e.g. Raven and Axelrod 1974) but dated molecular trees have indicated that they are of Laurasian origin. These include lineages that were distributed at higher latitudes in the Northern Hemisphere during times of higher temperature, particularly during the Paleocene/Eocene thermal maximum (PETM). The PETM (60–49 Ma) permitted the expansion of megathermal groups with a tropical distribution to higher latitudes. Many groups that now have an exclusively tropical distribution may have actually originated at higher latitudes in the North American and Eurasian 'Boreotropical' province during the Late Cretaceous (Morley 2003) or expanded their distributions by migrating through this northern hemisphere megathermal rainforest belt. This is evident from floras at high latitudes during this period that had multiple representatives of lineages that are now restricted to the tropics, such as the London Clay Flora (Reid and Chandler 1933, Chandler 1964), the German Messel oil shales (Wilde 1989, 2004, Manchester et al. 2007) and several German xylofloras (Böhme et al. 2007). Dated molecular phylogenetic trees of multiple tropical wet forest lineages have indicated that they were most probably not old enough to have been affected by Gondwanan break-up but that phylogenetic splits between lineages in the Neotropics, Africa and Asia occurred as a result of climatic deterioration and break-up of a northern hemisphere megathermal rainforest belt following the PETM. Examples of groups that follow this pattern include Annonaceae (Richardson et al. 2004), Melastomataceae (Renner et al. 2001), Meliaceae (Muellner et al. 2006, 2010), Moraceae (Zerega et al. 2005), Burseraceae (Weeks et al. 2005), Malpighiaceae (Davis et al. 2002) and Rubiaceae (Antonelli et al. 2009). Many of these groups would therefore have likely migrated into the Malesian region in the first instance via Sundania, and subsequently crossed Wallace's Line. Again, most of these lineages are predominantly restricted to wet forests.

Migration has not been unidirectional and there is evidence, also from dated molecular phylogenetic trees, of dispersal from east to west. A dated phylogenetic tree of Myrtaceae (Sytsma et al. 2004) is consistent with that group having an Australasian origin with genera with representatives to the west of Wallace's Line being phylogenetically nested within lineages having a distribution to the east. Similarly, dated molecular trees indicate an Australasian origin for other groups spanning Wallace's Line that have been traditionally regarded as Gondwanan, notably Proteaceae (Barker et al. (2007) and Elaeocarpaceae (Crayn et al. 2006), each of which have their earliest fossils recorded from Australia (Greenwood and Christophel 2005). In addition, subfamily Cryptocaryeae of Lauraceae (Chanderbali et al. 2001), Monimiaceae (Renner et al. 2010) and Winteraceae (Marquínez et al. 2009) also have an Australian origin and each of these groups is also more species rich in Australasia than Sundania.

6.6 Reassessing plant migration patterns across Wallace's Line

In the light of recently published molecular phylogenetic studies, we have reassessed the geographic origin of as many groups as possible to determine whether they originated to the west of Wallace's Line in Sundania or east of Wallace's Line in Australasia. Previous assessments of areas of origin were based on those areas that had greatest species diversity or the older fossil record. The dated molecular trees discussed above have confirmed that the area of origin that one would expect for those groups based on these criteria is correct. Therefore, in the absence of molecular data, one could continue to use the traditional approach to determining areas of origin with a greater degree of confidence. However, we acknowledge that there will be instances where areas of lower species diversity or younger fossils may, in the light of molecular phylogenetic evidence, prove to be areas of origin.

We determined areas of origin for two data sets: the first was based on tree checklists for particular islands or regions (we limited our analysis to trees because checklists for all plant species proved too difficult to compile) and the second was based on published Flora Malesiana monographic accounts and a specimen data set for the island of Borneo (courtesy of L. Willemse and P. van Welzen at the Nationaal Herbarium Nederland). For the latter, we were able to make better assessments of area origins because of more complete accounts of species distributions. We could also assess origins of the Philippine flora that was not possible in the first data set because of the lack of a complete checklist of trees. For the first dataset we used the following checklists: Kalimantan (Whitmore et al. 1990); Irian Jaya (Sidiyasa et al. 1997); Conn and Damas's online checklist of the Trees of Papua New Guinea

Table 6.1 Relative contributions of Sundanian/Australasian species on islands on either side of Wallace's Line based on dataset 1 (tree species checklists: see text for sources).

	Kalimantan	Java	Sulawesi	LSI	New Guinea	Queensland wet forest
Sundanian	2007	1462	507	255	1449	477
Australasian	230	270	110	72	842	533
Equivocal/ unknown	744	1277	254	144	660	332
Sundanian percentage	98.3	84.4	82.2	78.0	63.2	47.2

Table 6.2 Relative contributions of Sundanian/Australasian species on islands on either side of Wallace's Line based on dataset 2 (species checklists for Java, Sulawesi, the Lesser Sunda Islands, New Guinea and the Philippines extracted from Flora Malesiana accounts and a checklist of Borneo based on an NHN Leiden specimen database). LSI: Lesser Sunda Islands.

	Borneo	Java	Sulawesi	Philippines	LSI	New Guinea
Sundanian	3112	525	520	821	315	1127
Australasian	437	116	133	221	102	609
Equivocal/ unknown	3345	706	562	804	485	1140
Sundanian percentage	87.7	81.9	79.6	78.8	75.5	64.9

(2006–); Sulawesi (Whitmore 1989a); Bali, Nusa Tenggara and Timor (Whitmore 1989b); The Flora of Java (Backer and Bakhuizen van den Brink, 1963–68) and the wet tropical flora of Queensland (Ford and Metcalfe unpublished). Where possible we classified each species as being either of Australasian or Sundanian in origin. Assignation of origin was performed based on information from molecular trees (dated or not), the fossil record, and/or areas of greatest species diversity. A summary of these data is presented in Tables 6.1 and 6.2 (raw data available on request from the first author).

The results indicate that Sundanian elements outnumber Australasian elements on all of the islands we assessed (Tables 6.1 and 6.2). The Queensland wet tropical flora had a greater percentage of Australasian elements than Sundanian ones. As expected, there was an increase in Australasian elements from western to eastern islands. However, the proportion of Sundanian elements in Australasia is much greater than Australasian elements in Sundania. Particularly striking is the large contribution of Sundanian elements to the floras of Australasian land masses in New Guinea and Queensland. The relative contribution of Australasian elements varies considerably, and this seems to be best explained simply by distance from Australasian land masses. Australia has the highest proportion of Australasian elements, followed by New Guinea, the Lesser Sunda Islands (Bali to Timor in data set 1), Sulawesi, Java and Borneo (Kalimantan in data set 1), respectively. The pattern we demonstrate indicates a clear bias in the direction of movement from west to east rather than in the opposite direction. There are numerous possible explanations for this pattern.

6.6.1 Prevailing oceanic or wind currents

One mode of potential dispersal is via oceanic currents, such that the dominant directions of dispersal would be expected to be the same as those of the prevailing currents. However, the northern equatorial and eastern Australian currents are westward and our results clearly indicate migration against the prevailing oceanic currents although we acknowledge that palaeocurrents may have been flowing in different directions to modern ones. Other vectors for dispersal include wind or animals. Seed dispersal by wind is possible but prevailing wind directions are also from the east, and the seed size of most tropical trees tends to be too great for migration in this manner (although there are exceptions, e.g. *Toona* (Meliaceae)). Animal dispersal may play a major role and may lead to successful colonisation by the plant but not necessarily the animal that carried it given the difference in reproductive strategies between these groups of organisms. It has been suggested that transoceanic dispersal is often the result of unusual events (Higgins et al. 2003, Nathan et al. 2008) that will of course increase in number over time, and these events could be exerting a strong influence on dispersal across Wallace's Line. None of these factors however seem to explain the apparent bias of dispersal from west to east, thus other possible explanations must be considered.

6.6.2 Area effects and geological history

There is a large discrepancy in size between the far larger Sundania and the Australasian islands: the combined area of Peninsular Malaysia, Sumatra, Java, Borneo and the Philippines is 1 779 763 km^2 compared to 983 600 km^2 for Sulawesi, New Guinea and Queensland wet forest. Here we can also consider Australia's wet tropical flora in its present state to be primarily insular in nature, being an island of rainforest within an arid continent (Bowman 2000). This size discrepancy is particularly apparent during glacial periods when sea levels were lower and the land area of Sundania may have been up to twice that of modern times (Cannon 2009). As a result of this difference, Sundania simply has had more lineages available to disperse than Australasia. This is the case even before we take into account the pre-Pleistocene geological history of the region. Sundania already existed and also likely had a well-established flora during the aggregation of the islands of Eastern Malesia. Hall (pers. comm. 2009) suggested that much of New Guinea emerged only from *c.* 10 Ma. At this time, New Guinea land masses were closer to emergent Sundanian land masses than if they had emerged earlier. We might therefore expect that New Guinea should have a higher representation of elements from Sundania than if it had emerged earlier when it would have been further from Sundania. The dated molecular phylogenies of groups that dispersed from Sundania (discussed above) indicate that they began to diversify on New Guinea from the Late Miocene, and this is consistent with the island's recent emergence, as is the absence of endemic families (Good 1960, Balgooy 1976), the absence of otherwise widely distributed families such as Lamiaceae, Amaranthaceae, Phytolaccaceae, Cytinaceae and Balanopaceae (Good 1960), and the absence of fossils up until the mid Miocene (Morley pers. comm. 2009). Recent emergence could therefore partly explain the high percentage of Sundanian elements on New Guinea. Immigrants into Sundania would have had to compete with an already established flora, which would have made successful establishment less likely, whereas immigrants to recently emerged land masses in Australasia or the Philippines would not have had to face such competition.

Wallace drew his line to the east of the Philippines, and this is supported by our findings of a higher percentage of Sundanian elements in the Philippines than Australasian ones. The geological origin of the Philippines is still poorly known, with the reconstructions of Hall (1998, 2009) being somewhat equivocal about whether this group of islands have generally been closer to Sundanian than to Australasian terrains throughout their history. In addition to uncovering the history of its biota, dated molecular trees may help to resolve the geological history of the Philippines. An additional possible explanation for the high Sundanian element in the Philippines in our analysis is the fact that Palawan was derived from part of Continental Asia. Similarly, the high numbers of Sundanian elements in Sulawesi could also be due to the fact that Sulawesi is composed of fragments from both Sundania and Australasia (Hall 1998), and many Sundanian elements may already have been present on the Sundanian fragment that makes up the south-western arm of Sulawesi as the island aggregated.

6.6.3 Habitat compatibility

The concept of phylogenetic niche conservatism suggests that organisms have a low capacity to adapt to conditions outside their ancestral niche, and this may play an important role in the structure and distribution of communities. If immigrants are unable to adapt quickly then they will not be able to compete with the resident flora. Crisp et al. (2009) indicated that the availability of a suitable biome could have substantially influenced the success of a lineage's ability to become established on more than one land mass. Representatives of wet forest adapted lineages will have a greater chance of successfully colonising an area of wet forest than dry forest. Figure 6.1 indicates the modern distribution of everwet and seasonal monsoonal climates in the region. New Guinea is predominantly everwet and therefore might be expected to have a higher proportion of Sundanian elements because the source area of wet forest in Sundania is greater than in Australasia. Australasia is predominantly Australia, whose climate over most of its area is arid. It is likely that few wet adapted lineages have successfully migrated from Australia to the west because at the time of New Guinea's emergence, Australia's source area of wet forest was well in retreat to its current restricted extent as it cooled and aridified.

A wealth of Australian macro- and microfossil data exist that indicate a widespread cooler rainforest flora for much of its history. Macrofossils from southeastern Australia indicate a diverse rainforest flora present during the Eocene (Greenwood and Christophel 2005). These forests persisted until Australia's final separation from Antarctica that lead to the formation of the circum-Antarctic current that resulted in aridification and a massive retraction of rainforest evident from Neogene microfloral and faunal data (Greenwood and Christophel 2005). Rainforest progressively declined on a continental scale but persisted in small refugia along the eastern coast. There were periodic warm-wet phases when rainforest temporarily re-expanded and diversified (Kershaw 1988, Moritz 2005), however the long-term trend of rainforest decline continued as New Guinea was uplifted. By the time New Guinea emerged and other emergent land masses of the Australian Plate neared those of Sundania, the areas of wet forest in Australasia were substantially less than those in Sundania. Dry-adapted taxa from the huge source area for dry-adapted lineages in Australia may well have dispersed across Wallace's Line, but few of these lineages would have been able to successfully colonise the wet forests that currently and historically have occupied much of Sundania.

Previous authors have argued that invading Sundanian lineages were unlikely to have been able to compete with an already established rainforest flora in Australia, one that was gradually declining in suitable habitat (e.g. Adam 1992). However, periodic glacial events forced rainforest to contract and re-expand. The re-expansion periods may have provided opportunities for invading lineages. Another possible mechanism worthy of investigation is the potential role of cyclones in facilitating the establishment of foreign lineages.

6.7 Summary

Our analysis of geographic origins of species indicate a greater plant migration across Wallace's Line from west to east than vice versa, and this supports what has been suggested historically by earlier botanists. We postulate that a combination of the area available for colonisation and the effects of PNC explain, at least in part, this pattern. More studies are necessary to confirm this pattern, including production of more complete checklists of all land plants, based on sound taxonomy, which will then need to be subjected to analyses similar to those presented here.

Dated molecular phylogenetic trees are making a powerful contribution to our understanding of migration patterns in the region but more are needed, particularly focusing on taxa of a broader range of habits and environmental preferences. These dated trees might provide useful information for geologists on islands in the region that have an unclear history, such as the Philippines and New Guinea. If the hypothesised recent emergence of New Guinea proves correct, then this island has experienced a truly spectacular diversification of its biota, with R. Johns (pers. comm. 2010) estimating a total of up to 25 000 flowering plants on the island. The relative positions and sizes of islands have changed dramatically over a short period of time (Hall 2009, this volume, Chapter 3) and the climates of islands have also changed (Morley et al. 2010, this volume, Chapter 4). The incorporation of the dimension of time in phylogenetic studies will also permit the development of models that incorporate these changes. Approaches such as those of Ree et al. (2005), Ree and Smith (2008) and Webb and Ree (this volume, Chapter 8) might be appropriate for such studies.

If we assume PNC, then we can begin to make predictions such as, for example, that dated trees will indicate that dry-adapted lineages migrate from Australasia to Sundania during periods of drier climate in Sundania (i.e. early Oligocene or from the Late Pliocene; see Morley 2010, this volume, Chapter 4). These kinds of studies will allow us to assess the validity of floristic tracks or corridors, or whether patterns for montane elements are different from lowland ones. Molecular data from animal groups will allow us to compare their migration histories with those of plants. We predict that, given the lack of a direct land connection, the difference between plant and animal patterns will be even greater than indicated by Cody et al. (2010) for the interchange across the Isthmus of Panama. Dated phylogenies also have the potential to contribute to geological and climatological studies by providing temporal information on island emergence or the development of different climatic zones. Collaboration among systematists, palaeobiologists, palaeoclimatologists and geologists will provide a clearer picture of both the biotic and abiotic history of the region.

Acknowledgements

Peter van Welzen, Bill Baker, Robert Morley and David Gower provided comments that greatly enhanced the quality of this manuscript. Dan Metcalfe and Andrew

Ford (Queensland Wet Tropics) and Luc Willemse and Peter van Welzen (Malesia) provided checklist data. Thanks to Peter Stevens for help with family distributions. The following people contributed information on the biogeography of various groups: Olivier Maurin (Combretaceae), Santiago Madriñán (Hypericaceae), Chuck Cannon (Fagaceae), Tanya Livschultz (Apocynaceae), Mark Simmons (Celastraceae), Lena Struwe (Gentianaceae), Cynthia Frasier (Loganiaceae), Livia Wanntorp (*Meliosma*, Sabiaceae), Sutee Duangjai (*Diospyros*, Ebenaceae) and Richard Saunders (Annonaceae), Darren Crayn and Mark Harrington (Queensland flora). A.N.M. received financial support from the research funding programme 'LOEWE—Landes-Offensive zur Entwicklung Wissenschaftlich-oekonomischer Exzellenz' of Hesse's Ministry of Higher Education, Research, and the Arts.

References

Adam, P. (1992). *Australian Rainforests*. Oxford: Oxford University Press.

Antonelli, A., Nylander, J. A. A., Persson, C. and Sanmartín, I. (2009). Tracing the impact of the Andean uplift on Neotropical plant evolution. *Proceedings of the National Academy of Sciences of the United States of America*, **106**, 9749–54.

Backer, C. A. and Bakhuizen van den Brink, R. C., Jr. (1963–68). *Flora of Java*, 3 vol. Groningen, The Netherlands: Noordhoff.

Balgooy, M. M. J. van. (1971). Plant geography of the Pacific. *Blumea*, Suppl. **6**

Balgooy, M. M. J. van. (1976). Phytogeography. In *New Guinea Vegetation*, ed. K. Paijmans. Canberra, Australia: CSIRO, pp. 1–22.

Barker, N. P., Weston, P. H., Rutschmann, F. and Sauquet, H. (2007). Molecular dating of the 'Gondwanan' plant family Proteaceae is only partially congruent with the timing of the break-up of Gondwana. *Journal of Biogeography*, **34**, 2012–27.

Böhme M., Bruch, A. A. and Selmeier, A. (2007). The reconstruction of Early and Middle Miocene climate and vegetation in Southern Germany as determined from the fossil wood flora. *Palaeogeography, Palaeoclimatology, Palaeoecology*, **253**, 91–114.

Bowman, D. M. J. S. (2000). *Australian Rainforests: Islands of Green in a Land of Fire*. Cambridge: Cambridge University Press.

Brandon-Jones, D. (2001). Borneo as a biogeographic barrier to Asian–Australasian migration. In *Faunal and Floral Migrations and Evolution in SE Asia–Australasia*, ed. I. Metcalfe, J. M. B. Smith, M. Morwood and I. Davidson. Lisse, The Netherlands: A. A. Balkema Publishers, pp. 365–72.

Cannon, C. H., Morley, R. J. and Bush, A. B. (2009). The current refugial rainforests of Sundaland are unrepresentative of their biogeographic past and highly vulnerable to disturbance. *Proceedings of the National Academy of Sciences of the United States of America*, **106**, 11188–93.

Chanderbali, A. S., van der Werff, H. and Renner, S. S. (2001). Phylogeny and historical biogeography of Lauraceae: evidence from the chloroplast and nuclear genomes. *Annals of the Missouri Botanical Garden*, **88**, 104–34.

Chandler, M. E. J. (1964). The lower Tertiary floras of southern England. IV. A summary and survey of findings in the light of recent botanical observations. *Bulletin of the British Museum (Natural History), Geology Series*, **12**, 1–151.

Cody, S., Richardson, J. E., Rull, V., Ellis, C. and Pennington, R. T. (2010). The Great American Biotic Interchange revisited. *Ecography*, **33**, 1–7.

Conn, B. J. and Damas, K. Q. (2006–) Guide to Trees of Papua New Guinea. http://www.pngplants.org/PNGtrees/.

Cook, L. G. and Crisp, M. D. (2005). Not so ancient: the extant crown group of *Nothofagus* represents a post-Gondwana radiation. *Proceedings of the Royal Society of London, Series B (Biological Sciences)*, **272**, 2535–44.

Cottam, M., Hall, R., Sperber, C. and Armstrong, R. (2010). Pulsed emplacement of the Mount Kinabalu granite, northern Borneo. *Journal of the Geological Society*, **167**, 49–60.

Crayn, D. M., Rossetto, M. and Maynard, D. J. (2006). Molecular phylogeny and dating reveals an Oligo-Miocene radiation of dry-adapted shrubs (Tremandraceae) from rainforest tree progenitors (Elaeocarpaceae) in Australia. *American Journal of Botany*, **93**, 1328–42.

Crisp, M. D., Arroyo, M. T. K., Cook, L. G. et al. (2009). Phylogenetic biome conservatism on a global scale. *Nature*, **458**, 754–6.

Darlington, P. J. (1957). *Zoogeography: The Geographical Distribution of Animals*. New York: Wiley.

Davis, C. C., Bell, C. D., Mathews, S. and Donoghue, M. J. (2002). Laurasian migration explains Gondwanan disjunctions: evidence from Malpighiaceae. *Proceedings of the National Academy of Sciences of the United States of America*, **99**, 6833–7.

Dransfield, J. (1987). Bicentric distributions in Malesia as exemplified by palms. In *Biogeographical Evolution of the Malay Archipelago*, ed. T. C. Whitmore. Oxford: Clarendon Press, pp. 60–72.

Ford A. J. and Metcalfe D. J. (unpublished). *Checklist of the vascular plant species of Queensland's Wet Tropics*. Atherton, Australia: CSIRO.

Good, R. (1960). The geographical relationships of the angiosperm flora of New Guinea. *Bulletin of the British Museum (Natural History) Botany*, **2**, 203–26.

Greenwood, D. R. and Christophel, D. C. (2005). The origins and Tertiary history of Australian Tropical rainforests. In *Tropical Rainforests: Past, Present, and Future*, ed. E. Bermingham, C. Dick and C. Moritz. Chicago, IL: University of Chicago Press, pp. 336–73.

Hall, R. (1998). The plate tectonics of Cenozoic SE Asia and the distribution of land and sea. In *Biogeography and Geological Evolution of SE Asia*, ed. R. Hall and J. D. Holloway. Leiden, The Netherlands: Backhuys, pp. 99–131.

Hall, R. (2009). Southeast Asia's changing palaeogeography. *Blumea*, **54**, 148–61.

Higgins, S. I., Nathan, R. and Cain, M. L. (2003). Are long-distance dispersal events in plants usually caused by nonstandard means of dispersal? *Ecology*, **84**, 1945–56.

Hijmans, R. J., Cameron, S. E., Parra, J. L., Jones, P. G. and Jarvis, A. (2005). Very high resolution interpolated climate surfaces for global land areas. *International Journal of Climatology*, **25**, 1965–78.

Hooker, J. D. (1860). *Flora of Australia, its origin, affinities, and distribution. Botany of the Antarctic Voyage III, Flora of Tasmania 1*. London: Reeve Brothers.

Houle, A. (1998). Floating islands: a mode of long-distance dispersal for small and medium-sized terrestrial vertebrates. *Diversity and Distributions*, **4**, 201–16.

Kalkman, C. (1955). A plant-geographical analysis of the Lesser Sunda Islands. *Acta Botanica Neerlandica*, **4**, 200–25.

Keppel, G., Hodgskiss, P. D. and Plunkett, G. M. (2008). Cycads in the insular Southwest Pacific: dispersal, vicariance? *Journal of Biogeography*, **35**, 1004–15.

Kershaw, A. P. (1988). Australasia. In *Vegetation History*, ed. B. Huntley and T. Webb III. Alphen aan den Rijn, The Netherlands: Kluwer Academic Publishers, pp. 237–306.

Kershaw, A. P. (1994). Pleistocene vegetation of the humid tropics of northeastern Queensland, Australia. *Palaeogeography, Palaeoclimatology, Palaeoecology*, **109**, 399–412.

Kershaw, A. P., Moss, P. T. and Wild, R. (2005). Patterns and causes of vegetation change. In *Tropical Rainforests: Past, Present, and Future*, ed. E. Bermingham C. W. Dick and C. Moritz. Chicago, IL: University of Chicago Press, pp. 374–400.

Khan, A. H. (1974a). Palynology of Tertiary sediment from Papua New Guinea I. New form-genera and species from Upper Tertiary sediments. *Australian Journal of Botany*, **24**, 753–81.

Khan, A. H. (1974b). Palynology of Tertiary sediment from Papua New Guinea II. Gymnosperm pollen from Upper tertiary sediments. *Australian Journal of Botany*, **24**, 783–91.

Khan, A. H. (1976). Palynology of Neogene sediments from Papua New Guinea stratigraphic boundaries. *Pollen et Spores*, **16**, 265–84.

Knapp, M., Stöckler, K., Havell, D. et al. (2005). Relaxed molecular clock provides evidence for long-distance dispersal of *Nothofagus* (southern beech). *PLOS Biology*, **3**, e14.

Knapp, M., Mudaliar, R., Havell, D., Wagstaff, S. J. and Lockhart, P. J. (2007). The drowning of New Zealand and the problem of *Agathis*. *Systematic Biology*, **56**, 862–70.

Kramer, K. (1974). *Die Tertiären hölzer südost-Asiens (unter ausschluss der Dipterocarpaceae)*. Dissertation in Geology–Palaeontology. University of Bonn.

Kräusel, B. (1929). Fossile Pflanzen aus dem Tertiär von Sud Sumatra. *Verhandelingender Geologie en Mijnbouw Genootschap voor Nederland enkolonien, Geologie Serie*, **8**, 329–42.

Laumonier, Y. (1997). *The Vegetation and Physiography of Sumatra*. Alphen aan den Rijn, The Netherlands: Kluwer Academic Publishers.

Macphail, M. K., Alley, N. F., Truswell, E. M. and Sluiter, I. R. K. (1994). Early Tertiary vegetation: evidence from spores and pollen. In *History of the Australian Vegetation, Cretaceous to Recent*, ed. R. S. Hill. Cambridge: Cambridge University Press, pp. 189–261.

Manchester, S. R. Wilde, V. and Collinson, M. E. (2007). Fossil cashew nuts from the Eocene of Europe: biogeographic links between Africa and South America. *International Journal of Plant Science*, **168**, 1199–206.

Marquínez, X. Lohmann, L. G., Faria Salatino, M. F., Salatino, A. and

González, F. (2009). Generic relationships and dating of lineages in Winteraceae based on nuclear (ITS) and plastid (rpS16 and psbA-trnH) sequence data. *Molecular Phylogenetics and Evolution*, **53**, 435–49.

Moritz, C., Dick, C. W. and Bermingham, E. (2005). Overview: rainforest history and dynamics in the Australian Wet Tropics. In *Tropical Rainforests: Past, Present, and Future*, ed. E. Bermingham, C. W. Dick, and C. Moritz. Chicago, IL: University of Chicago Press, pp. 313–21.

Morley, R. J. (1998). Palynological evidence for Tertiary plant dispersals in the SE Asian region in relation to plate tectonics and climate. In *Biogeography and Geological Evolution of SE Asia*, ed. R. Hall and J. D. Holloway. Leiden, The Netherlands: Backhuys, pp. 211–34.

Morley, R. J. (2000). *Origin and Evolution of Tropical Rainforests*. Chichester, UK: Wiley.

Morley, R. J. (2003). Interplate dispersal paths for megathermal angiosperms. *Perspectives in Plant Ecology, Evolution and Systematics*, **6**, 5–20.

Motley, T. J., Wurdack, K. J. and Delprete, P. G. (2005). Molecular systematics of the Catesbaeeae–Chiococceae complex (Rubiaceae): flower and fruit evolution and biogeographic implications. *American Journal of Botany*, **92**, 316–29.

Muellner, A. N., Savolainen, V., Samuel, R. and Chase, M. W. (2006). The mahogany family 'out-of-Africa': divergence time estimation, global biogeographic patterns inferred from plastid *rbcL* DNA sequences, extant and fossil distribution of diversity. *Molecular Phylogenetics and Evolution*, **40**, 236–50.

Muellner, A. N., Pannell, C. M., Coleman, A. and Chase, M. W. (2008). The origin and evolution of Indomalesian, Australasian and Pacific island biotas: insights from *Aglaieae* (Meliaceae, Sapindales). *Journal of Biogeography*, **35**, 1769–89.

Muellner, A. N., Pannell, C. M. and Greger, H. (2009). Genetic diversity and geographic structure in *Aglaia elaeagnoidea* (Meliaceae, Sapindales), a morphologically complex tree species, near the two extremes of its distribution. *Blumea*, **54**, 207–16.

Muellner A. N., Pennington T. D., Koecke A. V., Renner S. S. (2010). Biogeography of *Cedrela* (Meliaceae, Sapindales) in Central and South America. *American Journal of Botany*, **97**, 511–18.

Nathan, R., Schurr, F. M., Spiegel, O., Steinitz, O., Trakhtenbrot, A. and Tsoar, A. (2008). Mechanisms of long-distance seed dispersal. *Trends in Ecology and Evolution*, **23**, 638–47.

Pennington, R. T., Richardson, J. E. and Lavin, M. (2006). Insights into the historical construction of species-rich biomes from dated plant phylogenies, neutral ecological theory and phylogenetic community structure. *New Phytologist*, **172**, 605–616.

Perrie, L. R. and Brownsey, P. (2007). Molecular evidence for long-distance dispersal in the New Zealand pteridophyte flora. *Journal of Biogeography*, **34**, 2028–38.

Playford, G. (1982). Neogene palynomorphs from the Huon Peninsula, Papua New Guinea. *Palynology*, **6**, 29–54.

Price, J. P. and Clague, D. A. (2002). How old is the Hawaiian biota? Geology and phylogeny suggest recent divergence. *Proceedings of the Royal Society of London, Series B (Biological Sciences)*, **269**, 2429–35.

Raven, P. H. and Axelrod, D. I. (1974). Angiosperm biogeography and past continental movements. *Annals of the Missouri Botanical Garden*, **61**, 539–673.

Razafimandimbison, S. G., Moog, J., Lantz, H., Maschwitz, U. and Bremer, B. (2005). Re-assessment of monophyly, evolution of myrmecophytism, and rapid radiation in *Neonauclea* s.s. (Rubiaceae). *Molecular Phylogenetics and Evolution*, **34**, 334–54.

Ree, R. H. and Smith, S. A. (2008). Maximum-likelihood inference of geographic range evolution by dispersal, local extinction, and cladogenesis. *Systematic Biology*, **57**, 4–414.

Ree, R. H., Moore, B. R., Webb, C. O. and Donoghue, M. J. (2005). A likelihood framework for inferring the evolution of geographic range on phylogenetic trees. *Evolution*, **59**, 2299–311.

Reid, E. M. and Chandler, M. E. J. (1933). *The London Clay Flora*. London: British Museum (Natural History).

Renner, S. S., Clausing, G. and Meyer, K. 2001. Historical biogeography of Melastomataceae: the roles of Tertiary migration and long-distance dispersal. *American Journal of Botany*, **88**, 1290–300.

Renner, S. S. (2004). Plant dispersal across the tropical Atlantic by wind and sea currents. *International Journal of Plant Science*, **165**, s23–s33.

Renner, S. S., Strijk, J. S., Strasberg, D. and Thébaud, C. (2010). Biogeography of the Monimiaceae (Laurales): a role for East Gondwana and long-distance dispersal, but not west Gondwana. *Journal of Biogeography*, **37**, 1227–38.

Richardson, J. E., Chatrou, L. W., Mols, J. B. Erkens, R. H. J. and Pirie, M. D. (2004). Historical biogeography of two cosmopolitan families of flowering plants: Annonaceae and Rhamnaceae. *Philosophical Transactions of the Royal Society of London*, **B359**, 1495–508.

Setoguchi, H., Osawa, T. A., Pintaud, J.-C., Jaffré, T. and Veillon, J.-M. (1998). Phylogenetic relationships within Araucariaceae based on *rbcL* gene sequences. *American Journal of Botany*, **85**, 1507–16.

Sidiyasa, K., Whitmore, T. C., Tantra, I. Gusti, M. and Sutisna, U. (1997). *Tree Flora of Indonesia, Checklist for Irian Jaya*. Jakarta, Indonesia: Indonesian Ministry of Forestry.

Simpson, G. G. (1980). *Splendid Isolation: The Curious History of South American Mammals*. New Haven, CT: Yale University Press.

Steenis, C. G. G. J. van. (1934). On the origin of the Malaysian mountain flora 1. Facts and statement of the problem. *Bulletin du Jardin Botanique de Buitenzorg*, **13**, 1–262.

Steenis, C. G. G. J. van. (1935). On the origin of the Malaysian mountain flora. 2. Altitudinal zones, general considerations and renewed statement of the problem. *Bulletin du Jardin Botanique de Buitenzorg*, **13**, 289–417.

Steenis, C. G. G. J. van. (1950). The delimitation of Malaysia and its main plant geographical divisions. In *Flora Malesiana, Vol. I(1)*, ed. C. G. G. J. Van Steenis. Jakarta, Indonesia: Noordhoff, pp. 70–75.

Steenis, C. G. G. J. van. (1979). Plant geography of east Malesia. *Botanical Journal of the Linnean Society*, **79**, 97–178

Stevens, P. F. (2001–). Angiosperm Phylogeny Website. Version 9, June 2008 [and updates]. http://www.mobot.org/MOBOT/research/APweb/.

Su, Y. C. F. and Saunders, R. M. K. (2009). Evolutionary divergence times in the Annonaceae: evidence of a late

Miocene origin of *Pseuduvaria* in Sundaland with subsequent diversification in New Guinea. *BMC Evolutionary Biology*, **9**, 153.

Sytsma, K. J., Litt, A., Zjhra, M. L. et al. 2004. Clades, clocks, and continents: historical and biogeographical analysis of Myrtaceae, Vochysiaceae, and relatives in the southern hemisphere. *International Journal of Plant Sciences*, **165**(4 suppl.), S85–S105.

Trivedi, B. S. and Srivastava, K. (1982). *Aglaioxylon mandlaense* gen et sp. nov. from the Deccan Intertrappean beds of Mandla District (MP), India. In *Phyta, Studies on Living and Fossil Plants, Pant Commemorative Volume*, ed. D. D. Nautiyal. Ranchi, India: Catholic Press, pp. 255–8.

Truswell, E. M., Kershaw, A. P. and Sluiter, I. R. (1987). The Australian–south-east Asian connection: evidence from the paleobotanical record. In *Biogeographical Evolution of the Malay Archipelago*, ed. T. C. Whitmore. Oxford: Clarendon Press, pp. 32–49.

Wallace, A. R. (1869). *The Malay Archipelago*. New York: Harper and Brothers.

Webb, L. J. and Tracey, J. G. (1981). Australian rainforests: patterns and change, In *Ecological Biogeography of Australia*, ed. A. Keast. The Hague, The Netherlands: W. Junk, pp. 607–94.

Weeks, A., Daly, D. C. and Simpson, B. B. (2005). The phylogenetic history and biogeography of the frankincense and myrrh family (Burseraceae) based on nuclear and chloroplast sequence data. *Molecular Phylogenetics and Evolution*, **35**, 85–101.

Welzen, P. C. van and Slik, J. W. F. (2009). Patterns in species richness and composition of plant families in the Malay Archipelago. *Blumea*, **54**, 166–74.

Welzen, P. C. van, Slik, J. W. F. and Alahuhta, J. (2005). Plant distribution patterns and plate tectonics in Malesia. *Biologiske Skrifter*, **55**, 199–217.

Werren, G. L. and Sluiter, I. R. (1984). Australian rainforests: a time for reappraisal. In *Australian National Rainforests Study Report to World Wildlife Fund (Aust.), Vol. 1, Proceedings of a Workshop on the Past, Present, and Future of Australian Rainforests*, ed. G. L. Werren and A. P. Kershaw. Melbourne, Australia: Geography Department, Monash University, pp. 488–500.

Whitmore, T. C. (1981). *Wallace's Line and Plate Tectonics*. Oxford: Clarendon Press.

Whitmore, T. C. (1984). *Tropical Rainforests of the Far East*. Oxford: Oxford Science Publications.

Whitmore, T. C. (1989a). *Tree Flora of Indonesia, Check List for Sulawesi*. Bogor, Indonesia: Indonesian Forest Research and Development Centre.

Whitmore, T. C., Tantra, I. G. M. and Sutisna, U. (1989b). *Tree Flora of Indonesia, Check List for Bali, Nusa Tenggarra and Timor*. Bogor, Indonesia: Indonesian Forest Research and Development Centre.

Whitmore, T. C., Tantra, I. and Sutisna, U. (1990). *Tree Flora of Indonesia, Check List for Kalimantan*. Bogor, Indonesia: Indonesian Forest Research and Development Centre.

Wilde, V. (1989). Untersuchungen zur Systematik der Blattreste aus dem Mitteleozän der Grube Messel bei Darmstadt (Hessen, Bundesrepublik Deutschland). *Courier Forschungsinstitut Senckenberg*, **115**, 1–213.

Wilde, V. (2004). Aktuelle Übersicht zur Flora aus dem mitteleozänen 'Ölschiefer' der Grube Messel bei Darmstadt (Hessen, Deutschland).

Courier Forschungsinstitut Senckenberg, **252**, 109–14.

Winkworth, R. C., Wagstaff, S. J., Glenny, D. and Lockhart, P. J. (2002). Plant dispersal N.E.W.S. from New Zealand. *Trends in Ecology and Evolution*, **17**, 514–20.

Zerega, N. J. C., Clement, W. L., Datwyler, S. L. and Weiblen, G. D. (2005). Biogeography and divergence times in the mulberry family (Moraceae). *Molecular Phylogenetics and Evolution*, **37**, 402–16.

7

Biogeography and distribution patterns of Southeast Asian palms

William J. Baker and Thomas L. P. Couvreur

7.1 Introduction

Southeast Asia is recognised as an outstanding centre of plant diversity (Mittermeier et al. 1999). Malesia, which stretches from the Malay Peninsula to New Guinea (van Steenis 1950, Marsh et al. 2009, Raes and van Welzen 2009, Fig 7.1), alone contains an estimated 42 000 species of vascular plants (Frodin 2001). The rich biodiversity of the region is frequently attributed to the complexity of its geological history, but despite increasingly detailed tectonic models (Hall 1998, 2001, 2002, 2009, this volume, Chapter 3) the origins of the region's flora and its biogeographic patterns are still incompletely understood and remain a focus for ongoing research (e.g. Roos et al. 2004, Brown et al. 2006, Marsh et al. 2009, van Welzen and Slik 2009).

Palms occur throughout the tropical and subtropical regions of the world and contain an estimated 2400 species in 183 genera (Dransfield et al. 2005, Govaerts and Dransfield 2005). Their wide distribution, diversity of distribution patterns coupled with high rates of endemism have prompted several biogeographic analyses (Moore 1973a, 1973b, Dransfield 1981, 1987, Uhl and Dransfield 1987, Dransfield 1988, Baker et al. 1998, Dransfield et al. 2008) although most of these pre-date phylogenetic research in palms. In the past, some authors have argued that palm distribution patterns reflect geological history more closely than other plant families because of the apparent ancientness of palms and their limited dispersal capabilities, inferred from the relatively large size of palm seeds (Dransfield 1981, Uhl and

Biotic Evolution and Environmental Change in Southeast Asia, eds D. J. Gower et al. Published by Cambridge University Press.

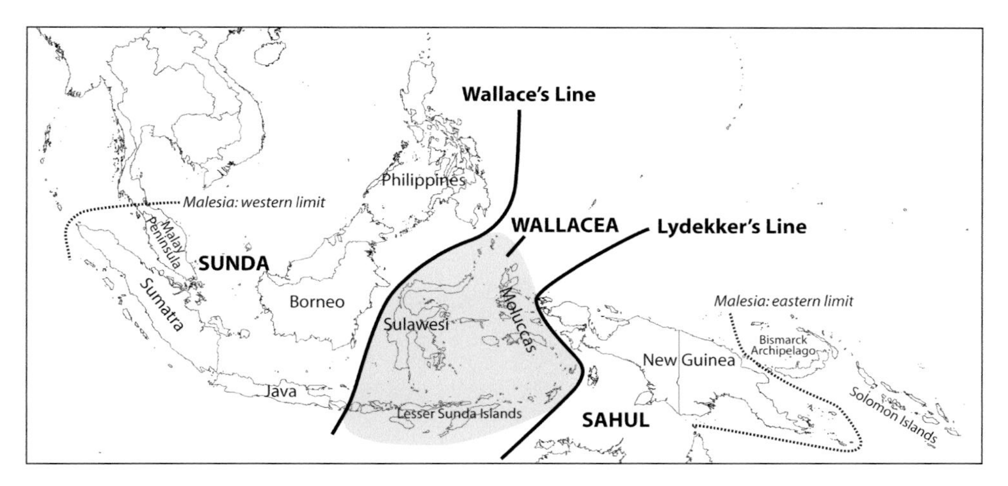

Figure 7.1 Map of Malesian region indicating the key geographical features discussed in this chapter. The Malay Peninsula, Sumatra, Java, Borneo and Palawan are located on the Sunda Shelf, whereas New Guinea and the Aru Islands are on the Sahul Shelf.

Dransfield 1987). We question this viewpoint in the light of well-known examples of long-distance dispersal, such as the coconut (*Cocos nucifera*) and evidence from phylogenetics, molecular dating and the fossil record, which indicates that many other angiosperm lineages are older than palms. The fossil record of palms is however extensive and informative (Harley and Baker 2001, Harley 2006, Dransfield et al. 2008), especially in the Tertiary, with the earliest unequivocal records dating to the Turonian in the Late Cretaceous (Crié 1892, Berry 1914, Kvacek and Herman 2004). Molecular clock estimates for palm origins pre-date the fossil record, suggesting that the family diverged from its closest relatives 91–120 million years ago (Bremer 2000, Wikström et al. 2001, Janssen and Bremer 2004), which is consistent with the earliest accepted monocot fossils (Friis et al. 2004).

Around half of all palm species diversity (*c.* 1200 species in 57 genera) occurs in tropical Asia (extending from the Indian subcontinent to the Solomon Islands) with almost 1000 species (42% of the family total) in 50 genera found in Malesia alone (Dransfield et al. 2008). Although Malesia can be regarded as the greatest palm diversity hotspot, there is considerable differentiation in species richness patterns throughout the region. A strong bimodal distribution occurs across Wallace's Line (Wallace 1860, Fig 7.1), a product of high species richness in both the western end of Malesia (e.g. 302 species in Borneo, 162 species in the Philippines) and the eastern end (e.g. New Guinea in 243 species), and much lower species richness in Wallacea (Fig 7.1), which comprises Sulawesi (*c.* 62 species), the Moluccas (*c.* 40 species) and the Lesser Sunda Islands (6 species; Dransfield 1981, 1987). A glance at generic distributions (Table 7.1) reveals that the lineages comprising the Malesian palm flora are also not distributed uniformly across the region.

Table 7.1 The genera of Malesian palms with their systematic placements (Dransfield et al. 2008) and species richness across the regions of Southeast Asia. Species richness figures are taken from Govaerts et al. (2010) with modifications (Keim 2003, Dransfield et al. 2008, Dowe 2009, Heatubun 2009, Heatubun et al. 2009, Bacon and Baker 2011, Barfod and Saw pers. comm.). *Cocos nucifera* and *Areca catechu* are of uncertain wild origin and have been excluded. Abbreviations: BIS = Bismarck Archipelago, BOR = Borneo, JAV = Java, LSI = Lesser Sunda Islands, MAL = Malay Peninsula, MOL = Moluccas, NWG = New Guinea, PHI = Philippines, SOL = Solomon Islands, SUL = Sulawesi, SUM = Sumatra. The grey bar represents Wallace's Line (Wallace 1860).

				Species Richness											
				Sunda Region and Philippines					Wallacea			Sahul	W. Pacific		Global
Subfamily	Tribe	Subtribe	Genus	MAL	BOR	SUM	JAV	PHI	SUL	MOL	LSI	NWG	BIS	SOL	
Calamoideae	Eugeissoneae		*Eugeissona*	2	4	–	–	–	–	–	–	–	–	–	6
	Calameae	Korthalsiinae	*Korthalsia*	9	15	9	2	5	1	1	–	2	–	–	26
		Salaccinae	*Eleiodoxa*	1	1	1	–	–	–	–	–	–	–	–	1
			Salacca	7	10	4	1	2	–	–	–	–	–	–	20
		Metroxylinae	*Metroxylon*	–	–	–	–	–	–	1	–	1	1	1	7
		Pigafettinae	*Pigafetta*	–	–	–	–	–	1	1	–	1	–	–	2
		Plectocomiinae	*Myrialepis*	1	–	1	–	–	–	–	–	–	–	–	1
			Plectocomia	3	3	3	2	2	–	–	–	–	–	–	16
			Plectocomiopsis	4	3	3	–	–	–	–	–	–	–	–	5
		Calaminae	*Calamus*	62	82	46	18	44	27	11	2	52	4	3	376
			Ceratolobus	2	3	4	2	–	–	–	–	–	–	–	6
			Daemonorops	22	34	33	3	13	6	4	–	1	–	–	101
			Pogonotium	1	3	–	–	–	–	–	–	–	–	–	3
			Retispatha	–	1	–	–	–	–	–	–	–	–	–	1

Nypoideae			*Nypa*	1	1	1	1	1	1	1	1	1	1	1	1
Coryphoideae	Phoeniceae		*Phoenix*	1	–	1	–	1	–	–	–	–	–	–	14
	Trachycarpeae	Rhapidinae	*Maxburretia*	2	–	–	–	–	–	–	–	–	–	–	3
			Rhapis	–	–	1	–	–	–	–	–	–	–	–	8
		Livistoninae	*Johannesteijsmannia*	4	1	1	–	–	–	–	–	–	–	–	4
			Licuala	41	46	6	4	1	3	2	–	32	1	2	170
			Livistona	4	2	–	1	1	–	–	–	2	–	–	27
			Pholidocarpus	2	1	3	–	–	1	1	–	–	–	–	6
			Saribus	–	1	–	–	2	1	1	–	7	–	1	9
	Caryoteae		*Arenga*	6	5	7	2	4	2	–	–	1	–	–	20
			Caryota	2	2	2	2	3	2	1	–	2	1	1	13
	Corypheae		*Corypha*	1	1	1	1	2	1	1	1	1	–	–	6
	Borasseae	Lataniinae	*Borassodendron*	1	1	–	–	–	–	–	–	–	–	–	2
			Borassus	–	–	–	1	–	1	–	1	1	–	–	6
Arecoideae	Oranieae		*Orania*	1	2	1	1	3	1	1	–	19	–	–	25
	Pelagodoxeae		*Sommieria*	–	–	–	–	–	–	–	–	1	–	–	1
	Areceae	Archontophoenicinae	*Actinorhytis*	–	–	–	–	–	–	–	–	1	–	1	1
		Arecinae	*Areca*	4	22	4	1	11	2	1	–	2	2	2	41
			Nenga	3	2	2	1	–	–	–	–	–	–	–	5
			Pinanga	23	39	23	2	24	7	2	1	1	–	–	131

Table 7.1 (cont.)

Subfamily	Tribe	Subtribe	Genus	Species Richness											
				Sunda Region and Philippines					Wallacea			Sahul	W. Pacific		Global
				MAL	BOR	SUM	JAV	PHI	SUL	MOL	LSI	NWG	BIS	SOL	
		Basseliniinae	*Physokentia*	–	–	–	–	–	–	–	–	–	1	3	7
		Linospadicinae	*Calyptrocalyx*	–	–	–	–	–	–	1	–	25	1	–	27
			Linospadix	–	–	–	–	–	–	–	–	2	–	–	9
		Oncospermatinae	*Oncosperma*	2	2	2	1	3	1	–	–	–	–	–	5
		Ptychospermatinae	*Adonidia*	–	1	–	–	1	–	–	–	–	–	–	1
			Brassiophoenix	–	–	–	–	–	–	–	–	2	–	–	2
			Drymophloeus	–	–	–	–	–	–	2	–	2	1	3	8
			Ptychococcus	–	–	–	–	–	–	–	–	2	1	1	2
			Ptychosperma	–	–	–	–	–	–	1	–	27	1	1	29
		Unplaced Areceae	*Clinostigma*	–	–	–	–	–	–	–	–	–	1	2	11
			Cyrtostachys	1	1	1	–	–	–	–	–	6	1	1	7
			Dransfieldia	–	–	–	–	–	–	–	–	1	–	–	1
			Heterospathe	–	–	–	–	12	–	2	–	16	1	6	40
			Hydriastele	–	–	–	–	–	4	4	–	30	3	2	47
			Iguanura	15	13	2	–	–	–	–	–	–	–	–	32
			Rhopaloblaste	1	–	–	–	–	–	1	–	2	1	1	6
			Regional Totals	229	302	162	46	135	62	40	6	243	22	32	

Malesian palm biogeography has been reviewed in detail in three key papers (Dransfield 1981, 1987, Baker et al. 1998). Dransfield (1987) recognised three main types of distributions: (1) distributions to the west of Wallace's Line (West Malesia), (2) distributions to the east of Wallace's Line (East Malesia) and 3) bimodal ('bicentric') distributions. Dransfield concluded 'there are two distinct elements in Malesia, of northern and western Laurasian–Tethyan, and southern and eastern Gondwanic origin respectively.' He also emphasised the need to explain the limited Sulawesi palm flora and transitions across Wallacea in general (Dransfield 1981). Although some of these papers benefited from innovations in palm classification (Uhl and Dransfield 1987), they are all essentially pre-phylogenetic. An overview of global palm biogeography has since been provided by Dransfield et al. (2008) and Southeast Asian groups have been discussed in some focused studies (Hahn and Sytsma 1999, Baker and Dransfield 2000, Loo et al. 2006, Norup et al. 2006, Heatubun et al. 2009, Crisp et al. 2010). These authors draw strongly on tectonic models for Southeast Asia (e.g. Hall 1998, 2001, 2002, 2009), emphasising the likely importance of the Miocene juxtaposition of the Sunda and Sahul regions (Fig 7.1). However, a general review of patterns of palm distribution in Malesia has not been conducted in the context of new phylogenetic evidence for palms.

Important resources are now available to biogeographers due to significant advances in palm systematic and evolutionary biology. Highly resolved and well-supported phylogenetic trees for palms are now available (Uhl et al. 1995, Baker et al. 1999, Asmussen et al. 2000, Asmussen and Chase 2001, Hahn 2002, Asmussen et al. 2006) including, most recently, a supertree that summarises all molecular and morphological data available for all genera (Baker et al. 2009). These and other papers provided the rationale for a phylogenetic classification in which only monophyletic groups are recognised and form the basis of an entirely revised generic monograph of the family (Dransfield et al. 2008). In addition, a complete checklist of accepted palm species names and their distributions are now available (Govaerts and Dransfield 2005) and regularly updated in an on-line database (Govaerts et al. 2010) that can rapidly deliver data on species diversity across the globe. Finally, and for the first time, an absolute timescale for diversification at the palm family level has been obtained through molecular dating of the genus-level supertree using four fossil calibrations and the relaxed molecular clock assumption (Couvreur et al. 2011). Owing to these major advances, opportunities now exist for in-depth analyses of palm diversification in space and time.

In this chapter we review patterns of distribution of palm clades in Malesia in the context of new phylogenetic, temporal and taxonomic frameworks. We document the diversity of distributions exhibited by major palm lineages, comparing them to interpretations in previous studies and exploring the biogeographic implications of our divergence time estimates and the fossil record. In this way, we aim to provide a foundation for in-depth biogeographic studies of individual Malesian palm lineages.

7.2 Phylogenetic composition of the Malesian palm flora

The lineages that make up the Malesian palm flora are highly dispersed across the palm tree of life, indicating that the biogeographic history of Malesian palms has been complex (Fig 7.2). This pattern of high phylogenetic diversity is also found in the palm floras of other major tropical realms like South America and Africa (Dransfield et al. 2008). Eleven major clades (10 tribes listed in Table 7.1 plus subfamily Nypoideae) are native, each potentially representing an independent dispersal into the region and displaying a wide diversity of distribution patterns (see below). The broad-scale patterns of higher-level groups frequently conceal distinct patterns in less inclusive subgroupings, indicative of hierarchically nested biogeographic histories. Most of the 11 clades have diversified extensively within the region, with few exceptions (Corypheae, Nypoideae, Pelagodoxeae, Phoeniceae).

Four of the five palm subfamilies are represented in Malesia, with the fifth (Ceroxyloideae) represented just beyond the region's margins in Queensland (Dransfield et al. 1985, Trenel et al. 2007). The rattan subfamily Calamoideae is predominantly Asian and reaches a peak of diversity at the genus and species levels in Malesia, especially in the Sunda region. Two of its three tribes occur in the region, Eugeissoneae and Calameae, while the third, Lepidocaryeae, is restricted to South America and Africa. Subfamily Nypoideae, which consists of the single species *Nypa fruticans*, the mangrove palm, occurs throughout the region. Five of the eight tribes of the fan-palm subfamily Coryphoideae are found in Malesia, some that are largely restricted to tropical Asia (Caryoteae, Corypheae) whereas others are more widespread, occurring also in Africa, Madagascar, the Americas and the Pacific (Borasseae, Phoeniceae, Trachycarpeae). Finally, the Arecoideae is the largest in terms of species number of the five subfamilies. Its diversity in Malesia, however, is concentrated in three of its 14 tribes, Areceae, Oranieae and Pelagodoxeae. New Guinea is notable for its high diversity of Arecoideae (139 species) compared with other Malesian islands (e.g. Borneo with 82 species). The coconut, *Cocos nucifera* (tribe Cocoseae), is common throughout Malesia and has been viewed as native to the region (Harries 1978, Gruezo and Harries 1984). However, the distribution of the coconut has been heavily modified by humans. Recent phylogenetic evidence indicates that the coconut is most closely related to American genera (Meerow et al. 2009). Tertiary fossil fruits attributed to *Cocos* have been recovered in India and New Zealand (Dransfield et al. 2008), but these

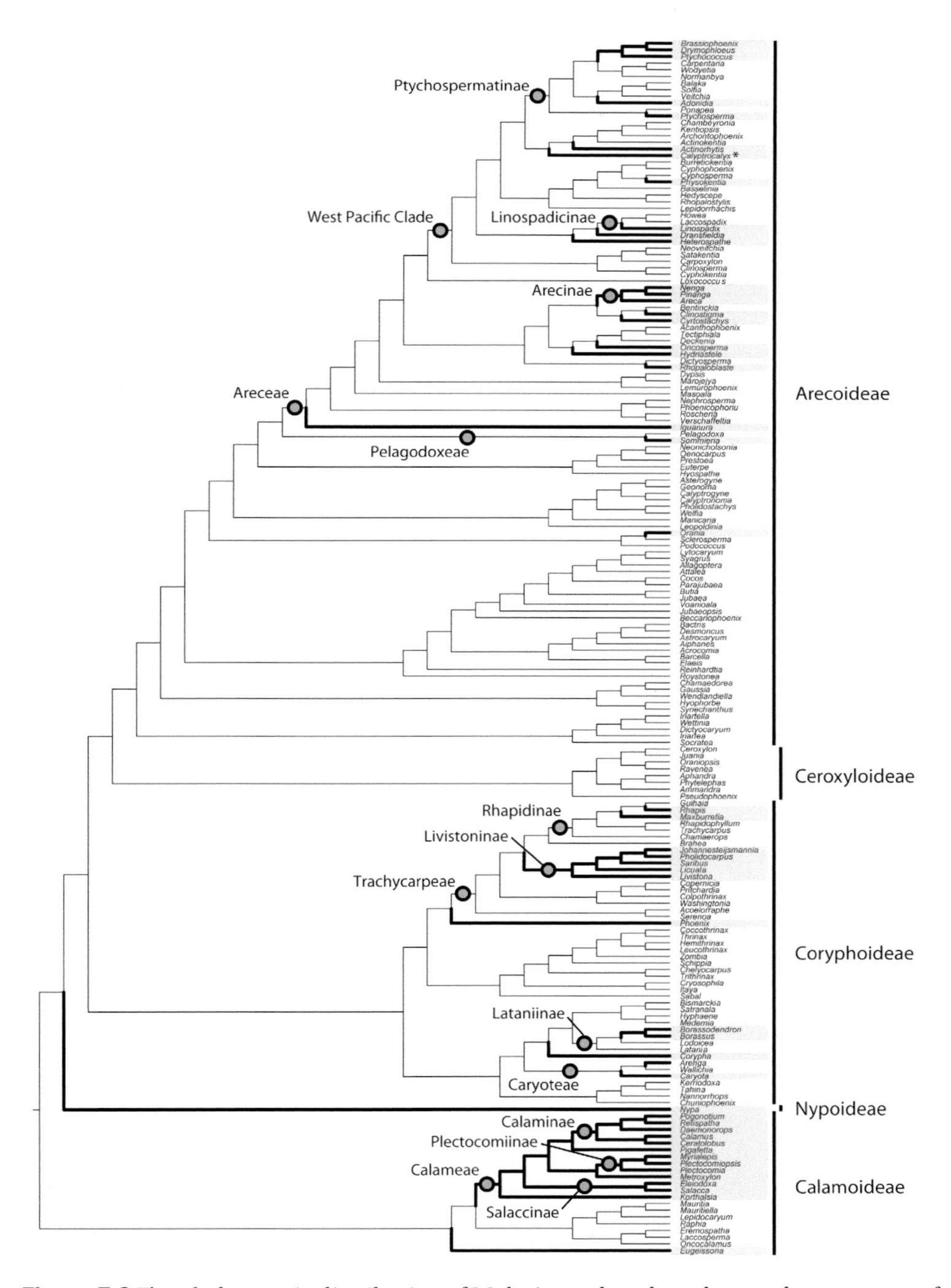

Figure 7.2 The phylogenetic distribution of Malesian palms, based upon the supertree of all palm genera (Baker et al. 2009, Couvreur et al. 2011). Malesian groups are highlighted by bold branches and grey boxes. Subtribe Linospadicinae is not monophyletic in this topology (note placement of *Calyptrocalyx*, indicated with an asterisk).

could equally be attributed to other members of the tribe. For the purposes of this chapter, we disregard *Cocos* due to its equivocal status in Malesia.

7.3 Distributions to the west of Wallace's Line

Malesian palm species diversity is much greater on the western side of Wallace's Line than to the east. Around two thirds (*c.* 650; Table 7.1) of Malesian species occur in this region (termed here West Malesia), few of which are shared with the eastern side of the line (East Malesia). Though species from widespread clades contribute substantially to the skewed distribution of diversity, many groups are strictly or largely endemic to the region (Fig 7.3a).

The most unambiguous West Malesian endemic clades are restricted to the Sunda region, sometimes with extensions into Indochina or the Philippines. Tribe Eugeissoneae, which consists of a single genus of six species, *Eugeissona*, with many unique morphological characters, is a strict Sunda region endemic, occurring only in Borneo and the Malay Peninsula (including southern Thailand). However, pollen evidence (Muller 1981, Phadtare and Kulkarni 1984, Morley 1998, Dransfield et al. 2008) suggests that *Eugeissona* (as the fossil genus *Quillonipollenites*) once occurred outside its current range, notably in the Eocene and Miocene of India, and possibly earlier at the Cretaceous/Tertiary boundary of South America, Africa and India, if the fossil genus *Longapertites* is correctly referred to *Eugeissona*. Morley (1998, 2003) gives evidence for plant dispersals into the Sunda region from India during its collision with Eurasia in the Middle Eocene when a humid, tropical corridor is thought to have existed between the regions. *Eugeissona*-like pollen is widespread in the Oligocene and Miocene of the Sunda region, perhaps as an immigrant from India (Dransfield et al. 2008, R. J. Morley pers. comm.). This, in conjunction with its phylogenetic placement as sister to all remaining Calamoideae, led Baker and Dransfield (2000) to propose a hypothesis that Eugeissoneae diverged from other Calamoideae in Gondwana, later rafting via India (where it later became extinct) and from thence dispersing into the Sunda region during the Tertiary. Extinction in India may be a result of Neogene and Quaternary climate change (Morley 2000). This scenario has since been questioned (Dransfield et al. 2008), primarily due to a mismatch between the timing of Gondwanan break-up through the Early Cretaceous (McLoughlin 2001) and later molecular dates for calamoid palms (Bremer 2000, Wikström et al. 2001, Janssen and Bremer 2004). Our own molecular dates (Couvreur et al. 2011), which indicate a Late Cretaceous divergence time for the tribe, confirm this conflict, although we acknowledge that connectivity between India and other parts of Gondwana may have persisted through the Cretaceous (Ali and Aitchison 2008). Though the biogeographic history remains unresolved it is evident from phylogeny, morphology, palaeopalynology and molecular dating that Eugeissoneae has a long history that is distinct from that of other calamoid palms.

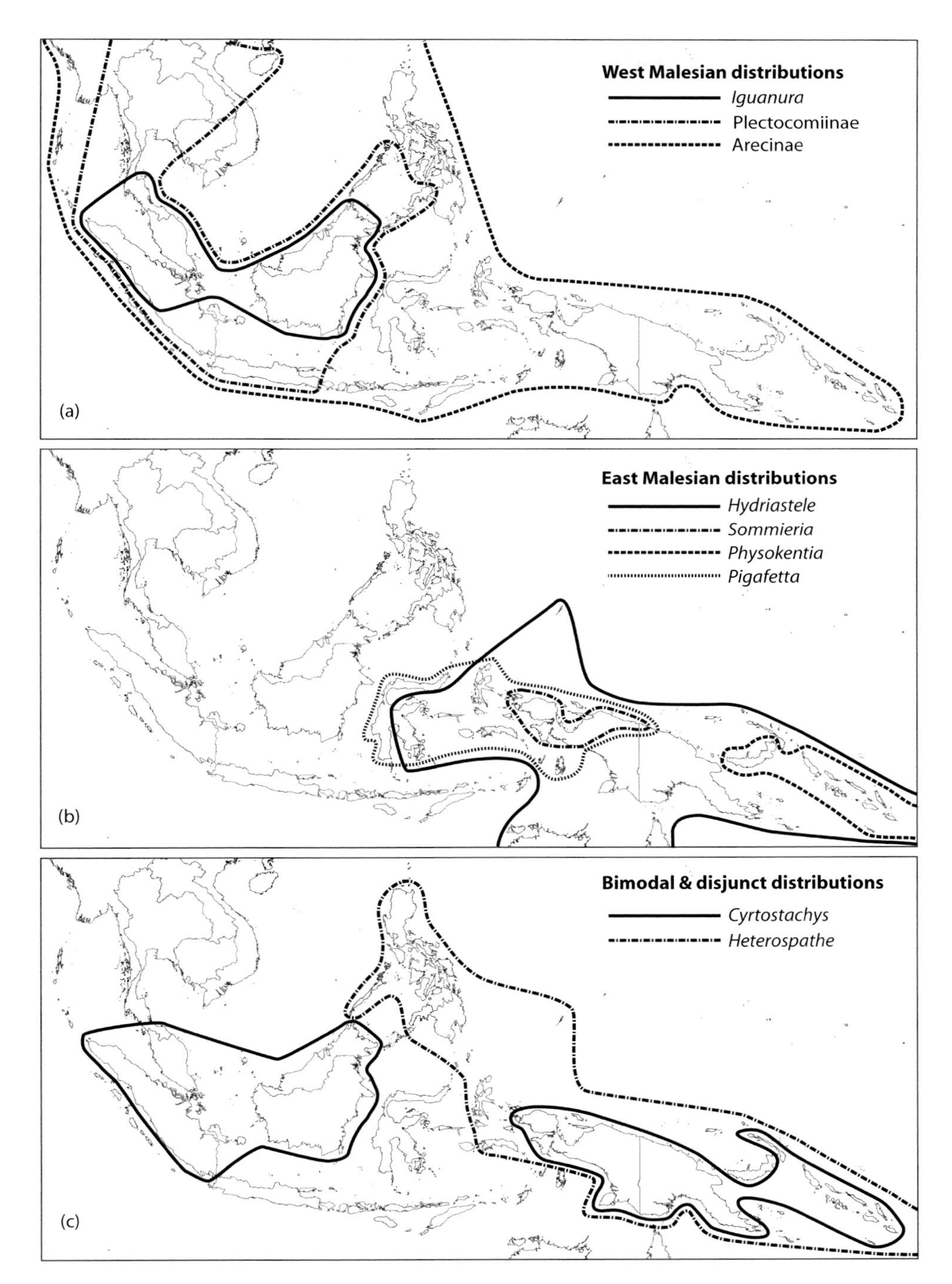

Figure 7.3 Examples of palm distributions in Malesia. (a) Distributions centred to the west of Wallace's Line, (b) distributions east of Wallace's Line, (c) bimodal and disjunct distributions.

Two other subtribes, Salaccinae and Plectocomiinae (Fig 7.3a) from tribe Calameae (Calamoideae) have a strict West Malesia distribution. These two clades, comprising two and three genera respectively, are more widely distributed than Eugeissoneae, ranging from Indochina to the Philippines and Java. Similarly, three genera belonging to more widespread higher groups are strictly endemic to the Sunda region, namely *Borassodendron* (Borasseae: Lataniinae), *Iguanura* (Areceae, Fig 7.3a) and *Johannesteijsmannia* (Trachycarpeae: Livistoninae). Certain rattan genera of subtribe Calaminae (Calameae) also display such a pattern (*Ceratolobus, Pogonotium, Retispatha*). We note that none of the genera adhering strictly to this distribution is particularly species rich. The largest are *Iguanura* (32 species), *Salacca* (20 species) and *Plectocomia* (16 species), the remainder including six species or fewer. Our estimated times of divergence between these groups and their sister taxa (stem node ages) vary widely as follows: Salaccinae – Late Palaeocene, Plectocomiinae – Middle Eocene, *Iguanura* – Late Eocene, *Borassodendron* and *Johannesteijsmannia* – Late Miocene. While the similar distribution patterns of these groups suggest that they may share similar biogeographic histories, the varied ages imply that this may not be so. Nevertheless, they highlight the robustness of Wallace's Line as a biogeographic boundary both before and after the Miocene collision of Sahul with the Sunda region (Hall 2009).

Five other clades display high diversity west of Wallace's Line, but are also present to the east in much lower species numbers. Two of these, *Oncosperma* (Areceae: Oncospermatinae) and *Pholidocarpus* (Trachycarpeae: Livistoninae), do not extend beyond Wallacea, whereas the remainder (Arecinae, Caryoteae, *Korthalsia*) reach New Guinea or beyond. The Arecinae (Areceae, Fig 7.3a), for example, contains three genera that reach their peak of species diversity in the Sunda region, one of which (*Nenga*) is endemic to the west. The remaining two, *Pinanga* and *Areca*, cross into East Malesia, but their species diversity is much reduced. *Pinanga* comprises 39 species in Borneo, but only seven in Sulawesi and just one at its eastern limit in central New Guinea. Similarly, 21 *Areca* species occur in Borneo, with two in each of Sulawesi, New Guinea and the Solomon Islands. The distribution of the Arecinae has previously been misinterpreted as bimodal (Dransfield 1987, Uhl and Dransfield 1987). It has now been shown that the perceived dual peaks of diversity were due to the inclusion of two unrelated lineages (*Hydriastele, Loxococcus*) within the subtribe and taxonomic inflation in New Guinea *Areca* (Loo et al. 2006, Heatubun 2009). The distribution of species diversity in *Korthalsia* (Korthalsiinae) is similar to Arecinae, whereas tribe Caryoteae extends to northern Australia and the West Pacific as far as Vanuatu. The historical biogeography of *Caryota* itself was explored by Hahn and Sytsma (1999) in the context of major biogeographic boundaries. They conclude that *Caryota* underwent a dispersal from west to east, concurring with Dransfield's (1987) general view for palms. However, their conclusions have not been revisited since the discovery of endemic species in New Guinea and Vanuatu.

Like the clades restricted to West Malesia, groups that cross Wallace's Line with a few species to the east have varied temporal origins, diverging from their sister taxa in the Late Maastrichtian (*Korthalsia*), the Middle Eocene (Caryoteae) to the Miocene (Arecinae, *Oncosperma, Pholidocarpus*). These dates are largely consistent with fossil reports attributed to these groups (summarised in Dransfield et al. 2008). We concur with other authors (Dransfield 1981, 1987, Hahn and Sytsma 1999, Loo et al. 2006, Dransfield et al. 2008) that dispersal into East Malesia following the Miocene collision of the Sunda and Sahul regions is the most obvious explanation for the patterns displayed by these groups. Fossil pollen attributed to *Oncosperma* is reported in West Malesia from the Oligocene, but records also appear in the Middle Eocene and Miocene of Sulawesi, which, if accurate, indicate that a pre-Miocene dispersal may have occurred in this group (Muller 1968, R. J. Morley pers. comm.) The phylogenetic relationships, molecular dates and the fossil record of *Oncosperma* and its relatives merit closer scrutiny. An interpretation of *Korthalsia* pollen in the Early Miocene of the Java Sea as an Australian element (Morley 2000) seems unlikely in the light of phylogenetic relationships and modern distributions.

Two further lineages that do not cross to the east of Wallace's Line are discussed here because they reach their distributional limits in West Malesia, rather than being centred on West Malesia, and are otherwise widespread primarily in the northern hemisphere. Tribe Phoeniceae, which contains the date palm genus *Phoenix* alone, is represented in West Malesia by two species, *Phoenix paludosa* and *P. loureiroi*, whereas the 13 remaining species range throughout tropical and subtropical Asia to North Africa, Europe and Macronesia. Our divergence time estimate of latest Early Eocene for Phoeniceae and the wide distribution of both fossil and extant species are consistent with the view of Dransfield et al. (2008) that the lineage has a long history in the northern hemisphere; a robust species-level phylogeny of *Phoenix* is needed to advance the interpretation further. The second group, subtribe Rhapidinae (Trachycarpeae), is represented by *Maxburretia* and *Rhapis*, which occur in West Malesia. The remainder of the subtribe is distributed throughout Eurasia with an outlying genus (*Rhapidophyllum*) in Southeastern North America, an example of the well-known floristic disjunction between eastern Asia and eastern North America (Tiffney 1985, Donoghue and Smith 2004). Our date estimates suggest that the Rhapidinae diverged and diversified through the Miocene in Eurasia. This is consistent with Donoghue and Smith's conclusions that temperate Asia–American disjuncts arose in Asia and dispersed to America during the past 30 Myr.

7.4 Distributions to the east of Wallace's Line

As outlined above, species richness to the east of Wallace's Line is characterised by high species diversity in New Guinea and low diversity in Wallacea, the region comprising Sulawesi, the Moluccas and the Lesser Sunda Islands. A greater diversity of palm distribution patterns can be discerned in this region compared with West Malesia (Fig 7.3b). Most groups are strongly centred on New Guinea and no group restricted to East Malesia is absent from New Guinea (Table 7.1). Palm diversity is remarkably high in New Guinea (243 species) given that much of the island was submerged until around 10 Ma (Hall 2009). Thus New Guinea rivals other large island hotspots for palms such as Borneo (302 species) and Madagascar (188 species). In addition to high species endemism, three genera are endemic to New Guinea, *Brassiophoenix* (Ptychospermatinae: Areceae), *Dransfieldia* (Areceae) and *Sommieria* (Pelagodoxeae, Fig 7.3b), although each contains just one or two species.

Many of the groups that reach their peak of species diversity in Papuasia (New Guinea and the Solomon Islands) show strong biogeographic affinities with the West Pacific. For example, approximately 28 of the 38 species in two of the four genera of subtribe Linospadicinae (Areceae) occur in New Guinea, the remaining taxa occurring in eastern Australia and Lord Howe Island, with a single species in the Moluccas. Similarly, 33 of 63 species (in four of the 11 genera) of subtribe Ptychospermatinae (Areceae) occur in New Guinea, with the majority of the remaining taxa being distributed in Australia, the Solomon Islands, Vanuatu, Fiji and Samoa, and three widespread species in the Moluccas. The Ptychospermatinae also contains a single disjunct monotypic genus west of Wallace's Line, *Adonidia*, in Palawan and northern Borneo. The genus *Hydriastele* (Areceae, Fig 7.3b), with 30 of its 47 species in New Guinea, also penetrates deep into the Pacific, through the Solomon Islands to Vanuatu and Fiji. However, unlike the other groups discussed here, it has also speciated in Wallacea where eight species are recorded. Dransfield (1981) drew attention to the apparent confinement of the genus (as its synonym *Gronophyllum*) to Gondwanic parts of Sulawesi. This distribution has not been verified in the light of subsequent plant exploration in Sulawesi. If correct, it is unlikely to be directly linked to rafting on Gondwanic fragments, which were submerged at various times during their migration towards modern Sulawesi (Hall 2009).

Pacific relationships are also displayed by less species-rich groups in Papuasia. *Actinorhytis* (Areceae: Archontophoenicinae), for example, is endemic to Papuasia, but the remaining four genera of its subtribe are restricted to Australia and New Caledonia. *Physokentia*, which ranges from the Bismarck Archipelago to Fiji, belongs to a subtribe of six genera (Basseliniinae) present also in the Solomon Islands, Vanuatu, New Caledonia and Lord Howe Island (Fig 7.3b). *Clinostigma* (Areceae), which is present in the Bismarck Archipelago, occurs throughout the West Pacific from the Bonin and Caroline Islands to Fiji, Vanuatu and the Solomon Islands. One widespread species of the sago palm genus, *Metroxylon* (Calameae: Metroxylinae), occurs throughout the Moluccas and New Guinea while the remaining six species are scattered through the West Pacific.

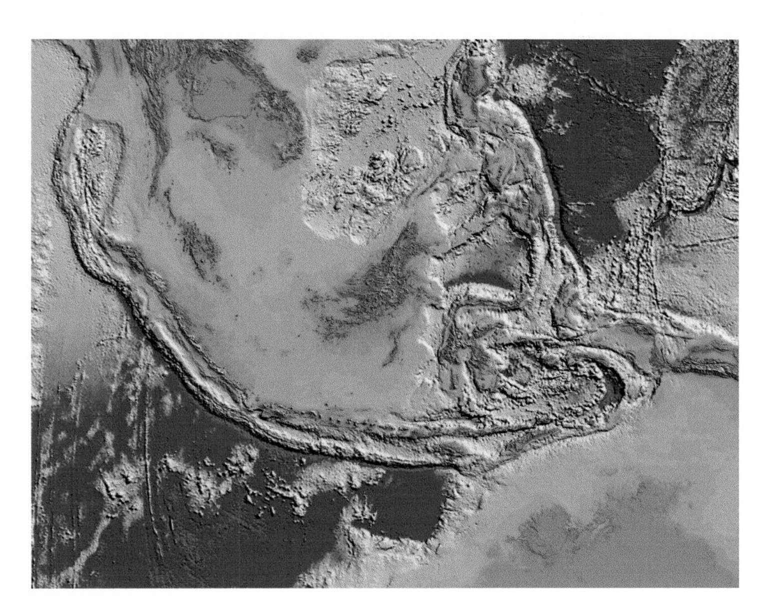

Figure 3.2 Digital elevation model showing satellite gravity-derived bathymetry combined with SRTM (Shuttle Radar Topography Mission) topography (Sandwell and Smith 2009) of the region shown in Fig 3.1. The image highlights the shallow shelves surrounding Southeast Asia and Australia with the complex geology of eastern Indonesia and the Philippines.

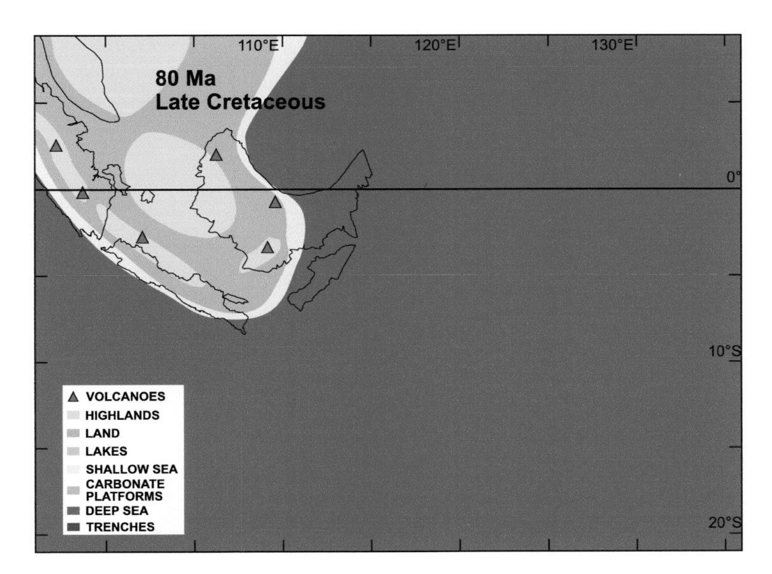

Figure 3.4 Palaeogeography of the region at 80 Ma. The very schematic aspect of this map results from the almost complete absence of rocks of this age from Southeast Asia, because much of the region was emergent. It was probably also significantly elevated judging from the character of sedimentary rocks in Sarawak where there are some of the few rocks of this age that are preserved. They are poorly sorted clastic sediments derived from a granitic source (e.g. Wolfenden 1960) implying deep erosion to expose continental basement rocks from several kilometres depth.

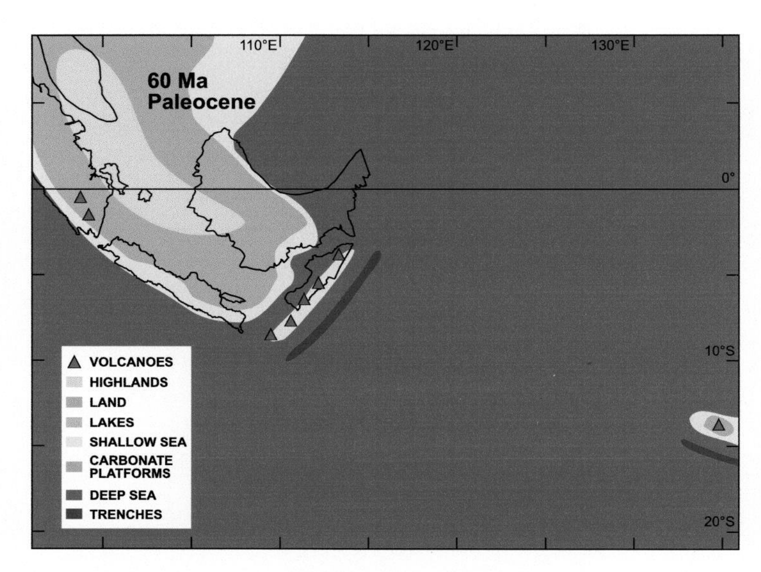

Figure 3.5 Palaeogeography of the region at 60 Ma. Again, there are few rocks of this age in Southeast Asia, because much of the region was emergent. There was a short-lived volcanic arc at the eastern edge of Sundaland in present-day West Sulawesi and Sumba. Other parts of the Sundaland margins were inactive, although there is some uncertainty about the margin in Sarawak and northwards at this time, because much of the region is offshore and covered by younger sediments.

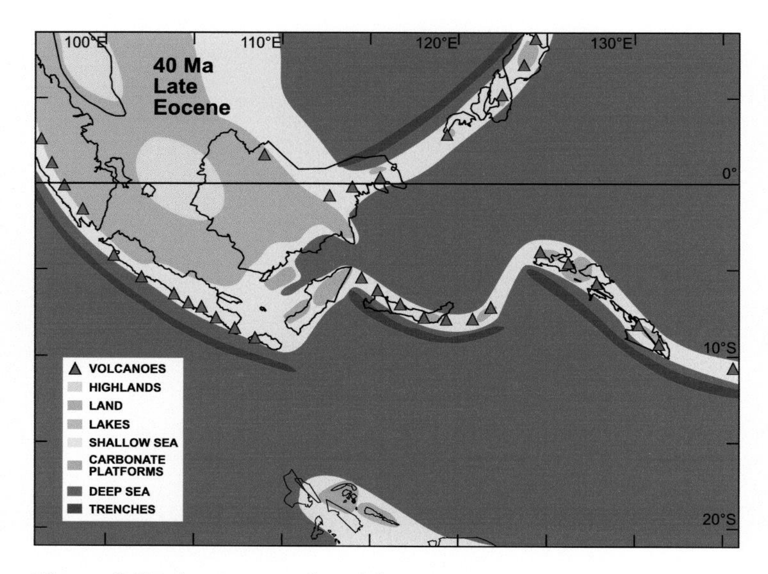

Figure 3.6 Palaeogeography of the region at 40 Ma. Subduction resumed around much of Sundaland in the Middle Eocene and many new sedimentary basins began to form, in which terrestrial sediments were deposited, derived from local sources. The Makassar Straits were already a significant marine gap, especially in the northern straits, at the eastern edge of Sundaland.

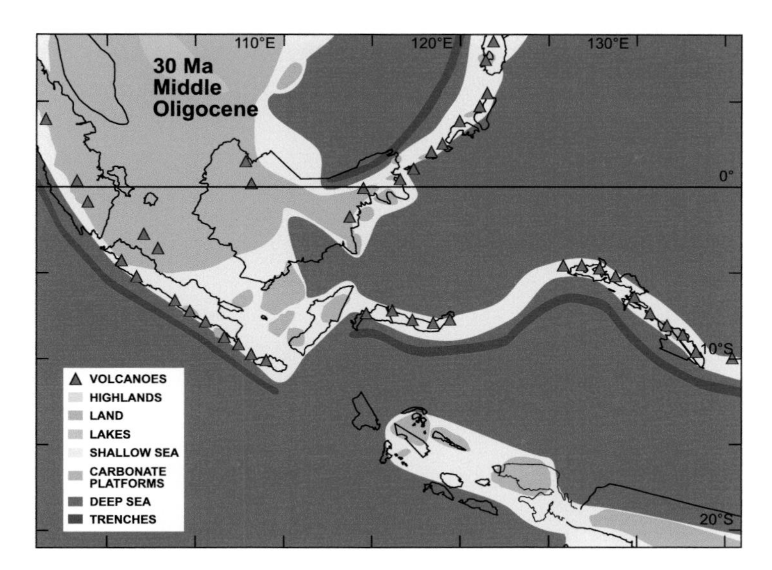

Figure 3.7 Palaeogeography of the region at 30 Ma. Although much of Sundaland was emergent it is likely that topography was significantly lower than earlier in the Cenozoic. Rivers carried recycled clastic sediments to internal basins and the continental margins. On the Sunda Shelf there were large freshwater lakes, not linked to the ocean, which are shown in a different shade from normal salinity seas.

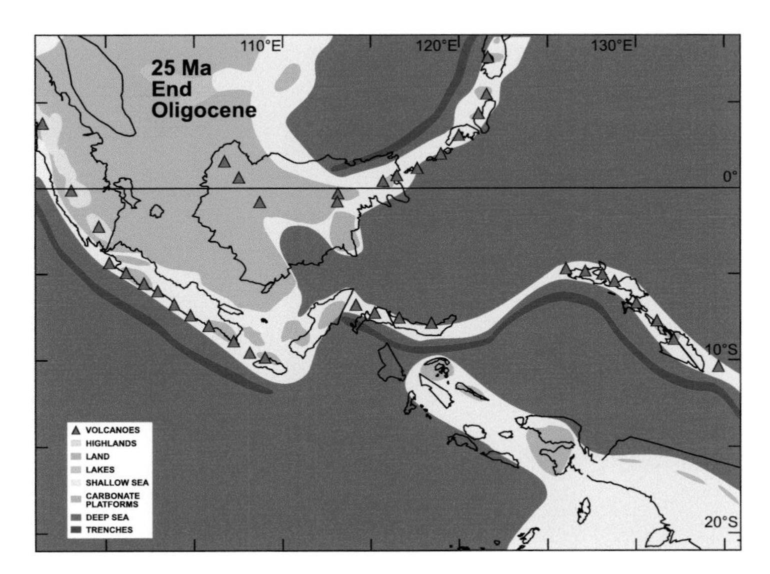

Figure 3.8 Palaeogeography of the region at 25 Ma. On the Sunda Shelf lakes (shown in a different shade) were intermittently connected to the sea and were brackish. The Sula Spur was about to collide with the volcanic arc of Sulawesi's North Arm resulting in ophiolite emplacement and uplift, and becoming the first part of the Australian continent to connect Australia and Southeast Asia, although there was no continuous land connection.

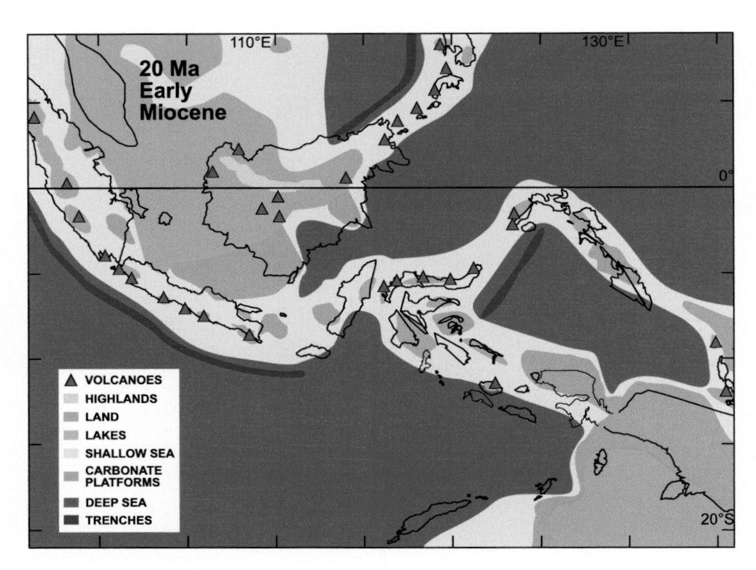

Figure 3.9 Palaeogeography of the region at 20 Ma. There was significant marine incursion onto the Sunda Shelf and extensive areas of carbonate build-ups throughout the region on the wide shallow shelves. Borneo became an important source of clastic sediments which began to pour into the deep offshore basins to the north, east and southeast. It is likely that the flow of water from the Pacific to the Indian Ocean was significantly reduced from 20 Ma until after 10 Ma. Much of Wallacea between Sulawesi and the Bird's Head was the site of shallow marine carbonate deposition and may have been emergent in some areas.

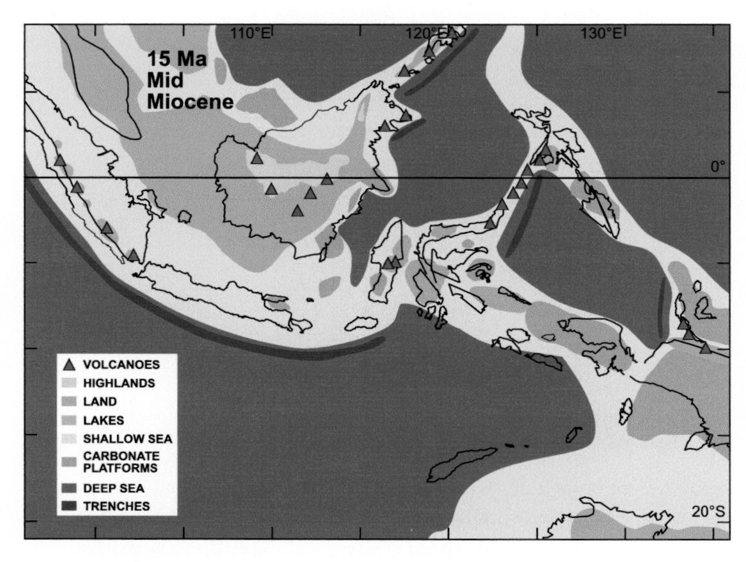

Figure 3.10 Palaeogeography of the region at 15 Ma. Soon after 15 Ma the Java Trench subduction zone began to propagate east as it tore along to the continent ocean boundary south of the Sula Spur and Bird's Head. Rollback of the subduction hinge caused major extension in Sulawesi and fragmented the Sula Spur.

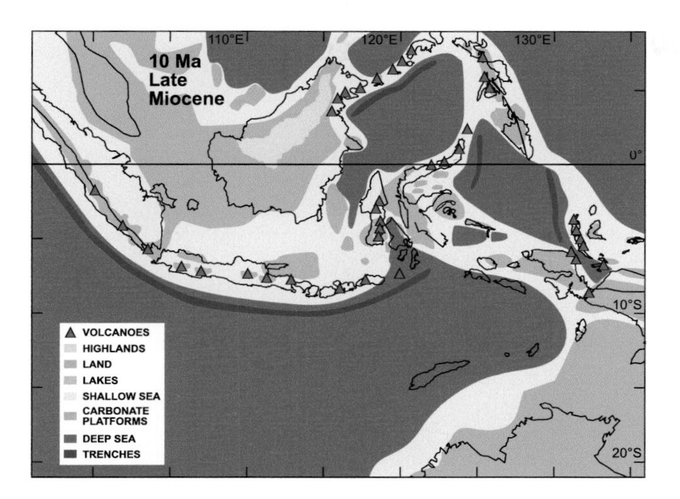

Figure 3.11 Palaeogeography of the region at 10 Ma. Subduction rollback was well under way and extension in Sulawesi led to volcanic activity, subsidence in Bone Gulf and oceanic crust formation in the North Banda Sea. Borneo was now a significantly emergent and elevated area. Wallacea probably had a much more complex palaeogeography than shown as the Sulu Spur was fragmented. Note that this fragmentation was not the result of slicing of fragments from the Bird's Head but driven by rollback of the subduction hinge into the Banda embayment.

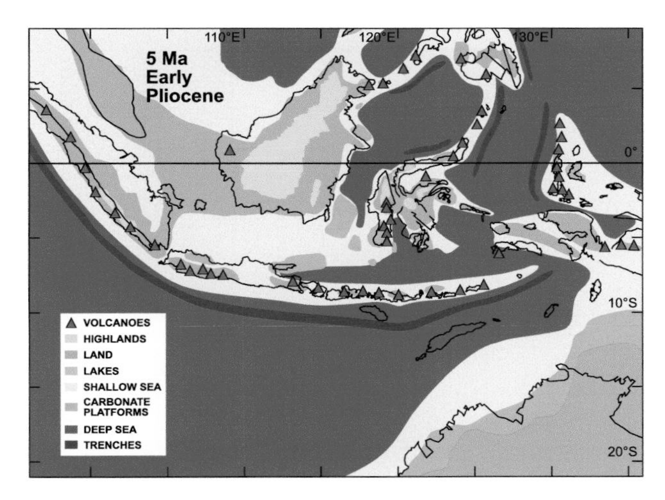

Figure 3.12 Palaeogeography of the region at 5 Ma. New deep oceanic basins had formed throughout Wallacea. Oceanic spreading in the North Banda Sea had ceased but had begun in the South Banda Sea. Deepening of the Flores Sea is inferred. At the same time substantial and rapid elevation began in Sulawesi, as rollback commenced at the North Sulawesi Trench driving core complex formation in the neck and East Arm, but Goontalo Bay was about to subside to present depths of up to 2 km below sea level. Uplift was about to begin in Timor as the Banda volcanic arc collided with the Australian continental margin in Timor. The last 5 Myr has seen significant changes in palaeogeography and dramatic uplift and subsidence. Seram and Timor are the two largest islands in the Banda region to have emerged in this interval.

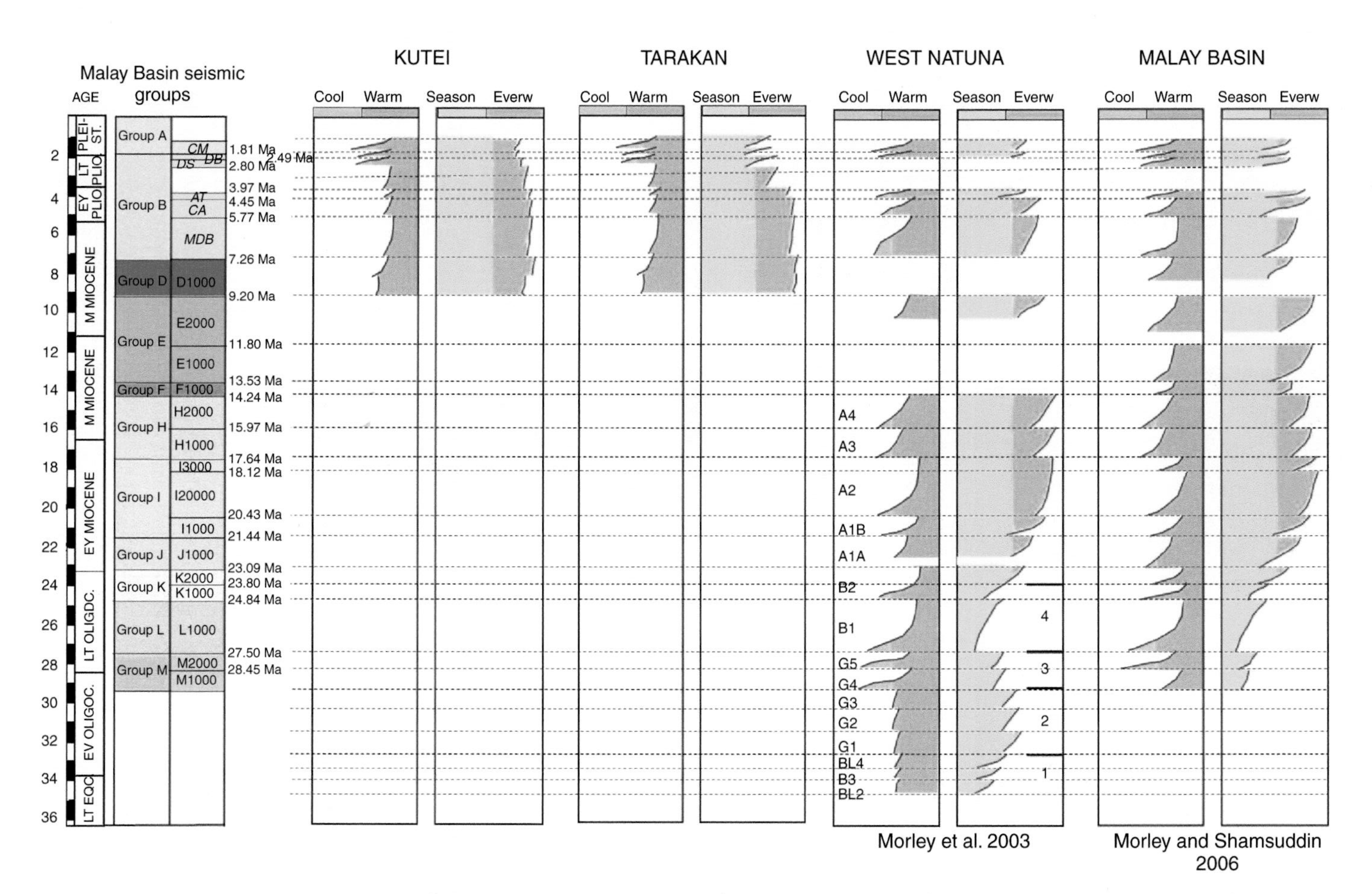

Figure 4.3 Climate trends in the Malay (Morley and Shamsuddin 2006) and Natuna Basins (Morley et al. 2003, Morley 2006) compared to schematic climate history for the Borneo margin of the Makassar Straits. Alpha-numeric terminology for West Natuna basins indicates biostratigraphically defined sequences of Morley et al. (2003).

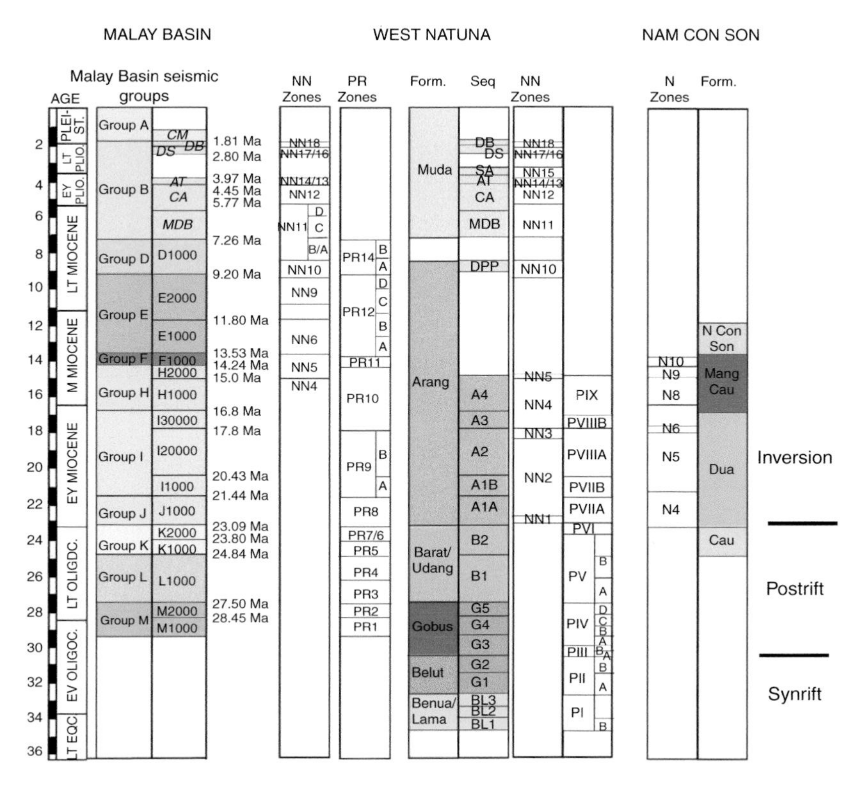

Figure 4.4 Stratigraphic relationships in the Malay Basin (Shamsuddin and Morley 2006), West Natuna (Morley et al. 2003), and Nam Con Son Basins, showing correlations to palynological, nannofossil and planktonic foraminiferal zones and the main phases of basin development across the region.

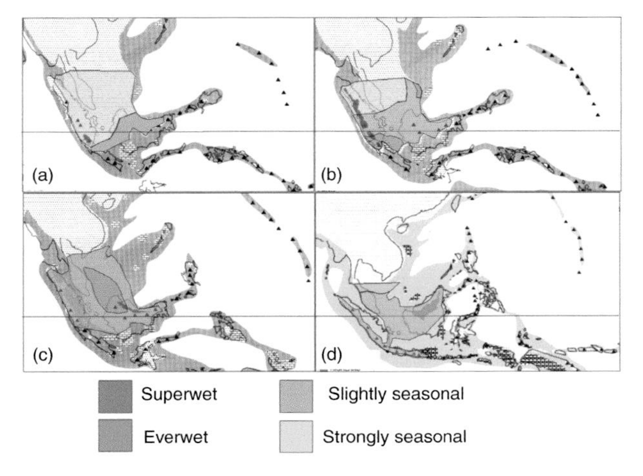

Figure 4.7 Palaeoclimate reconstructions for the Oligocene and earliest Miocene based on subsurface data from the Java and south China Seas: (a) 28 Ma, OL2; (b) 25 Ma OL 3–4; (c) 20 Ma PR9 in Malay Basin; (d) 10 Ma, PR 14 in Malay Basin. (a–c) updated from Lelono and Morley (2011). Base maps from Hall (1998).

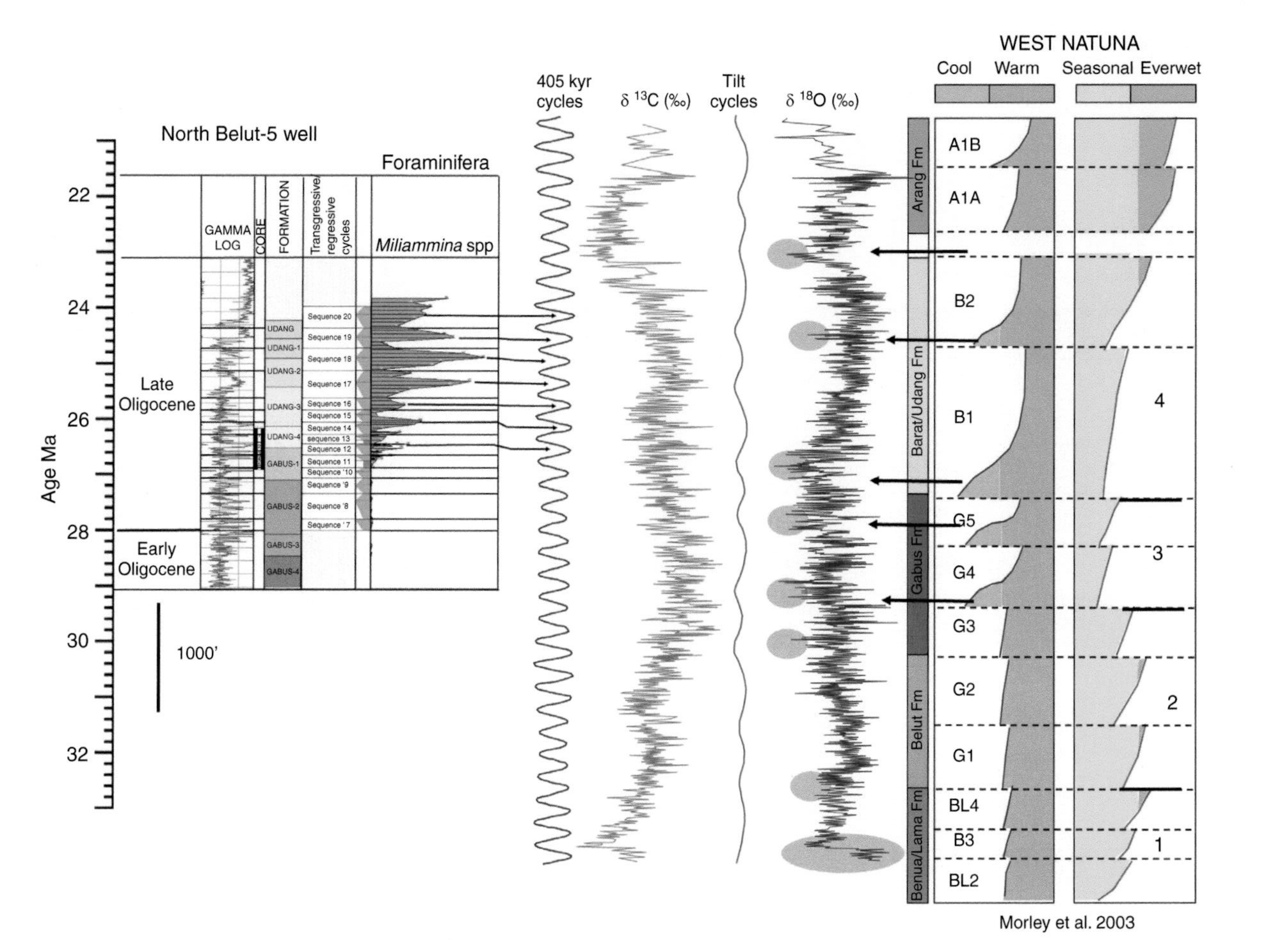

Figure 4.5 North Belut-5 Well, West Natuna Basin, *Miliammina* spp. acmes through latest Oligocene reflecting successive sea-level rises during the latest Oligocene thermal maximum (Morley et al. 2007) correlated against 405 kyr eccentricity cycles, and ^{13}C and ^{18}O data from Pelike et al. (2006). Cool climate episodes based on acmes of conifer pollen are also correlated with periods with highest ^{18}O values. Grey ellipses indicate glacial periods.

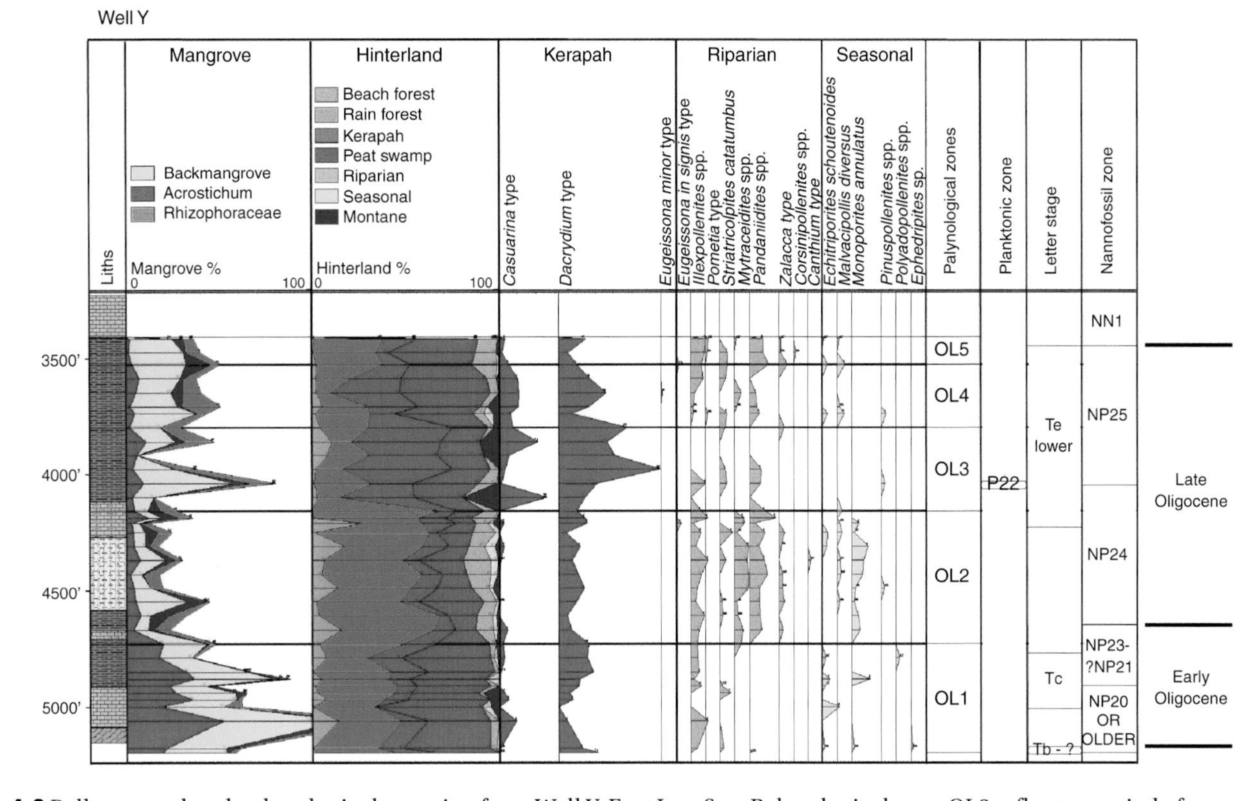

Figure 4.6 Pollen record and palynological zonation from Well Y, East Java Sea, Palynological zone OL3 reflects a period of seasonally dry climate, whereas zones OL4 and OL5 reflect a very wet climate (Lelono and Morley 2011). Main components of palynomorph assemblage groupings are as follows: 'Backmangrove', Sonneratioid and *Brownlowia* pollen; *Nypa* (in blue) 'Coastal', *Terminalia*, *Thespesia*, *Barringtonia* pollen; 'Rainforest' pollen of Dipterocarpaceae, Burseraceae, Leguminosae and many others; 'Kerapah', *Casuarina* and *Dacrydium* pollen; 'Peat swamp', *Blumeodendron*, *Cephalomappa*, *Calophyllum*, *Durio* and Sapotaceae pollen; 'Seasonal', Poaceae pollen and *Malvacipollis diversus*; 'Montane', mainly *Podocarpus* pollen and Pinaceae. The groups reflect the main vegetation types from which the fossil pollen is thought to be derived – the taxa are not wholly restricted to the vegetation types indicated.

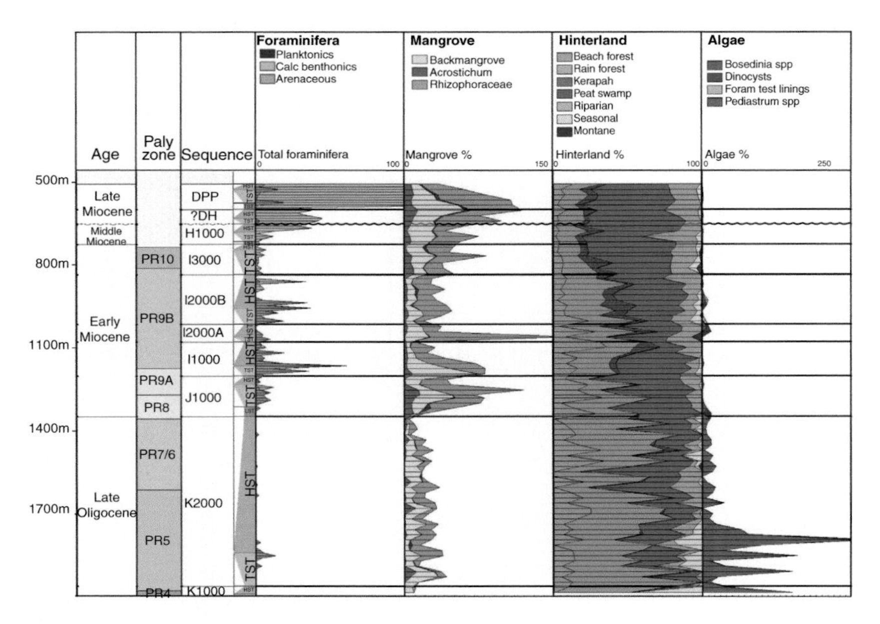

Figure 4.8 Example of climatostratigraphic cycles from Malay/West Natuna Basin from an unnamed well. The Late Oligocene Sequence K1000 is characterised by an initial interval with abundant freshwater algae and common seasonal climate pollen (mainly Poaceae), followed by abundant rainforest pollen. Miocene sequences are mainly characterised by an initial mangrove pollen acme coinciding with, or followed by, a foraminiferal abundance acme with hinterland pollen dominated by rainforest elements but with seasonal climate taxa often represented at the base of a cycle (e.g. Sequences I2000, I3000). See Fig 4.6 for details of palynological groups.

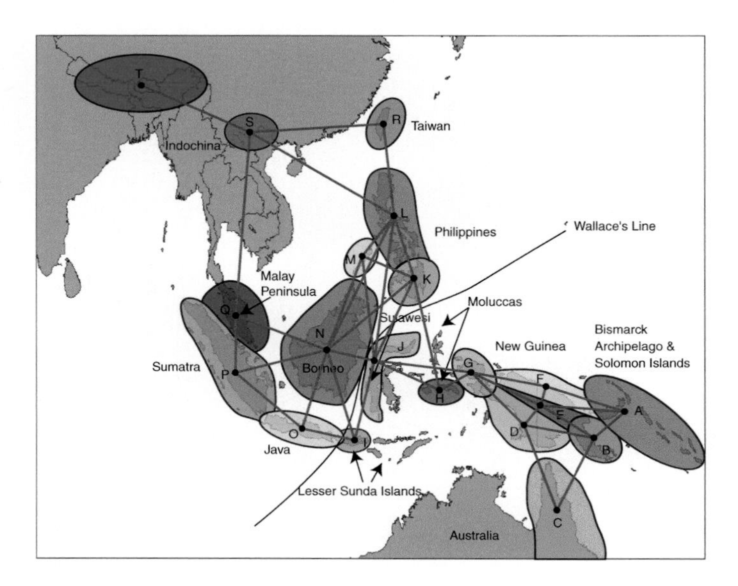

Figure 8.3 Areas of endemism for *Vireya* defined by Brown et al. (2006b), showing adjacency connections used to constrain dispersal routes in reanalysis using LAGRANGE.

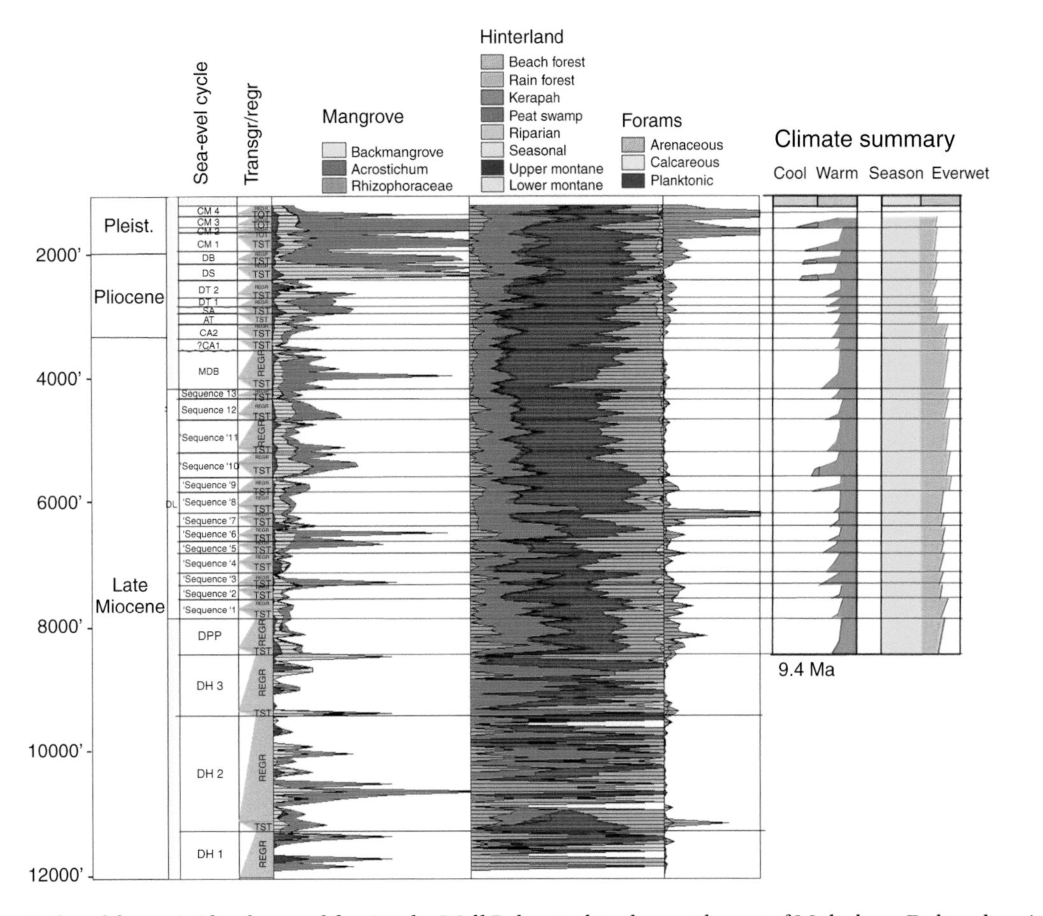

Figure 4.9 Palynological and foraminiferal record for Attaka Well B, located to the northwest of Mahakam Delta, showing age, nannofossil zones, sedimentation rates, (black line shows actual rates uncorrected for compaction), mangrove pollen acmes, hinterland trends, foraminiferal abundance and palaeoenvironments (courtesy of Chevron IndoAsia and Migas). See Fig 4.6 for details of palynological groups.

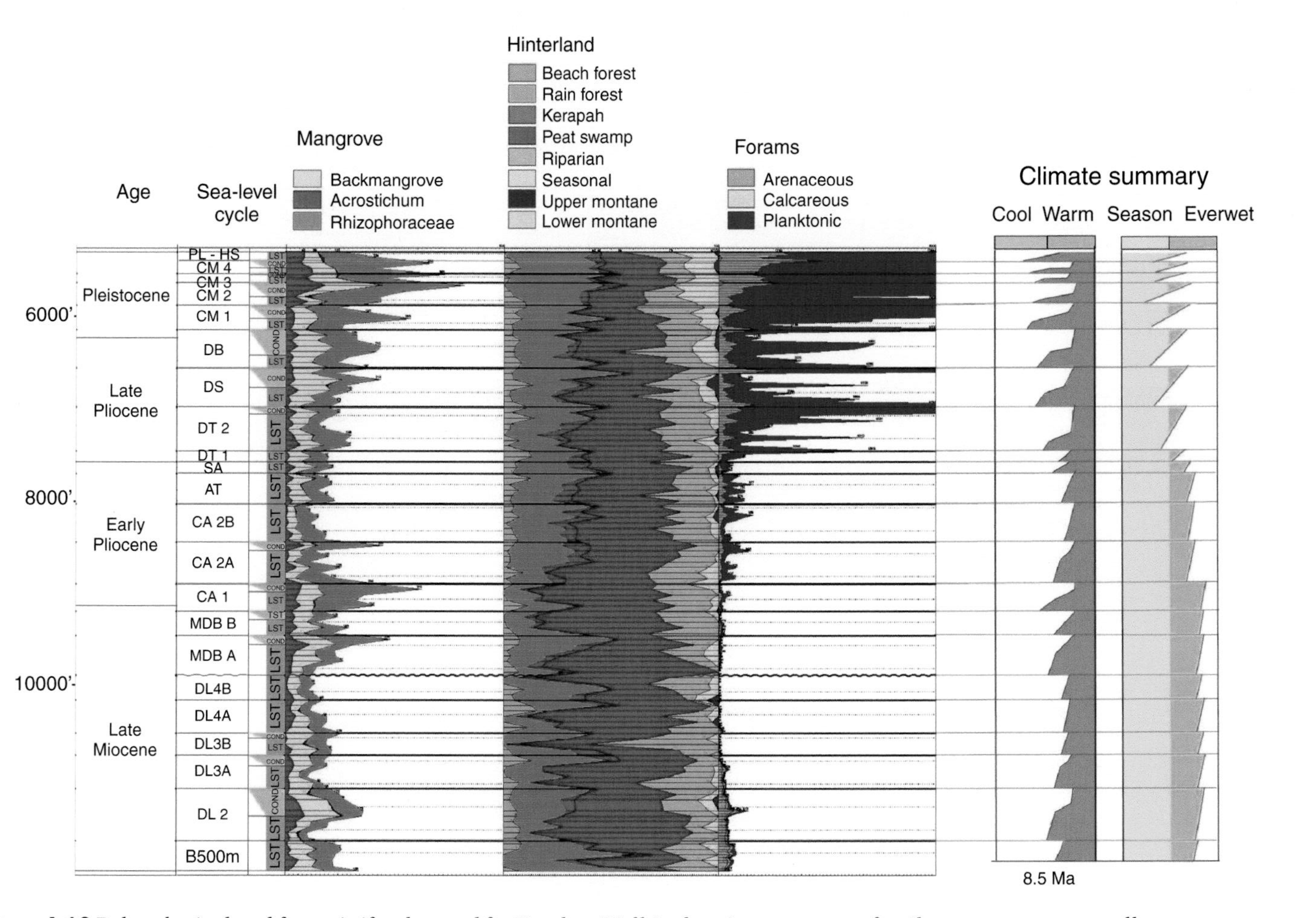

Figure 4.10 Palynological and foraminiferal record for Tarakan Well A, showing age, nannofossil zones, mangrove pollen acmes, hinterland pollen trends and foraminiferal abundance (courtesy of Migas). This well is from the mid/lower continental slope. See Fig 4.6 for details of palynological groups.

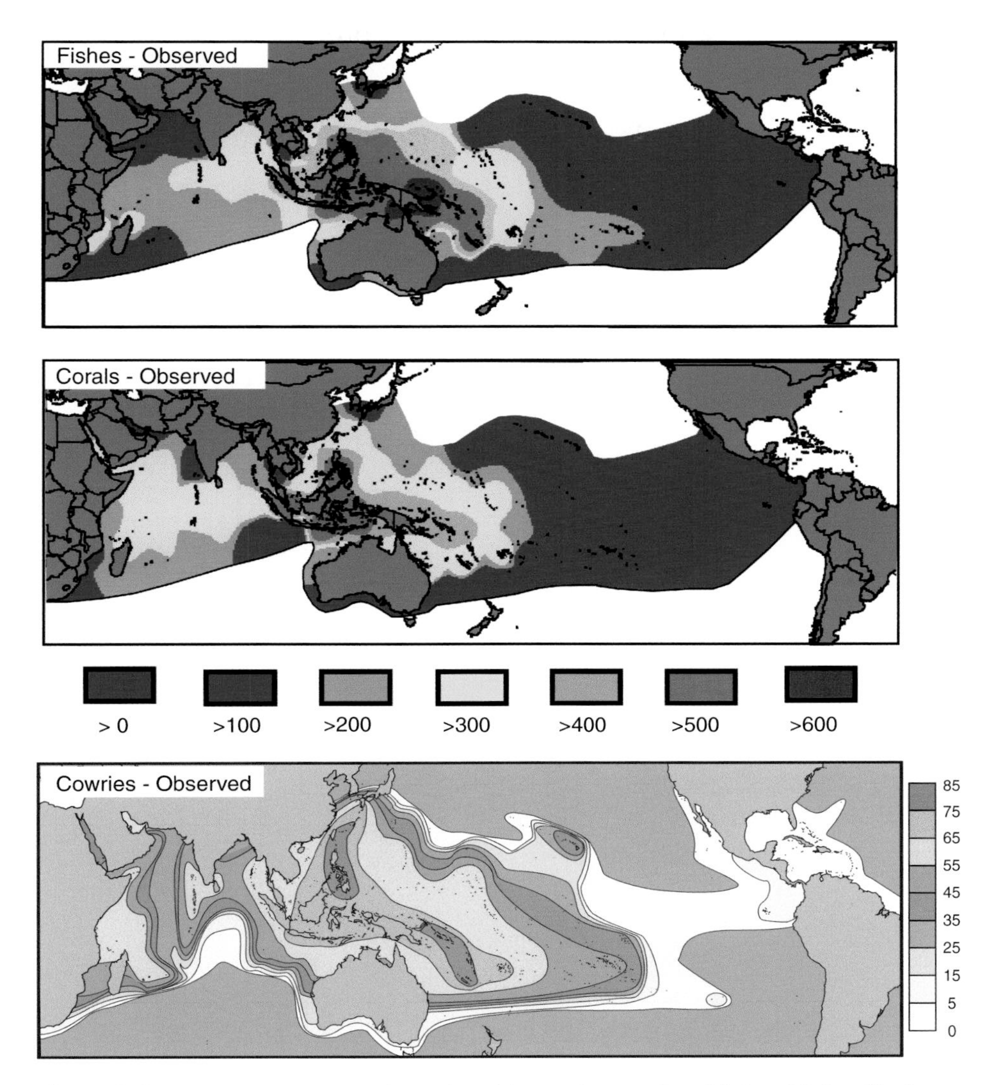

Figure 9.1 Congruent patterns of species richness for coral reef fishes (13 families), corals and cowries, showing the distinctive 'bulls-eye' biodiversity 'hotspot' in the Indo-Australian Archipelago (after Bellwood et al. 2005, Bellwood and Meyer 2009a).

A Bismarck Archipelago & Solomon Islands
B Papuan Peninsula
C northeast Australia
D New Guinea craton
E central New Guinea
F north New Guinea
G Vogelkop Peninsula
H south Moluccas
I Lesser Sunda Islands
J north and west Sulawesi
K south Philippines
L north Philippines
M Palawan
N Borneo
O Java and Bali
P Sumatra
Q Malay Peninsula
R Taiwan
S north Vietnam and south China
T Himalaya
Eastern Malesia
dispersal across Wallace's Line
R. inundatum
R. herzogii
R. rarum
R. baenitzianum
R. leptanthum
R. gracilentum
R. christii
R. rubineiflorum
R. commonae
R. goodenoughii
R. luraluense
R. loranthiflorum
R. konori
R. laetum
R. zoelleri
R. leucogigas
R. carringtoniae
R. multinervium
R. hyacinthosmum
R. spondylophyllum
R. inconspicuum
R. maius
R. solitarium
R. superbum
R. tuba
R. culminicola
R. saxifragoides
R. lochiae
R. rhodopus
R. pudorinum
R. vanvuurenii
R. alternans
R. williamsii
R. zollingeri
R. lagunculicarpum

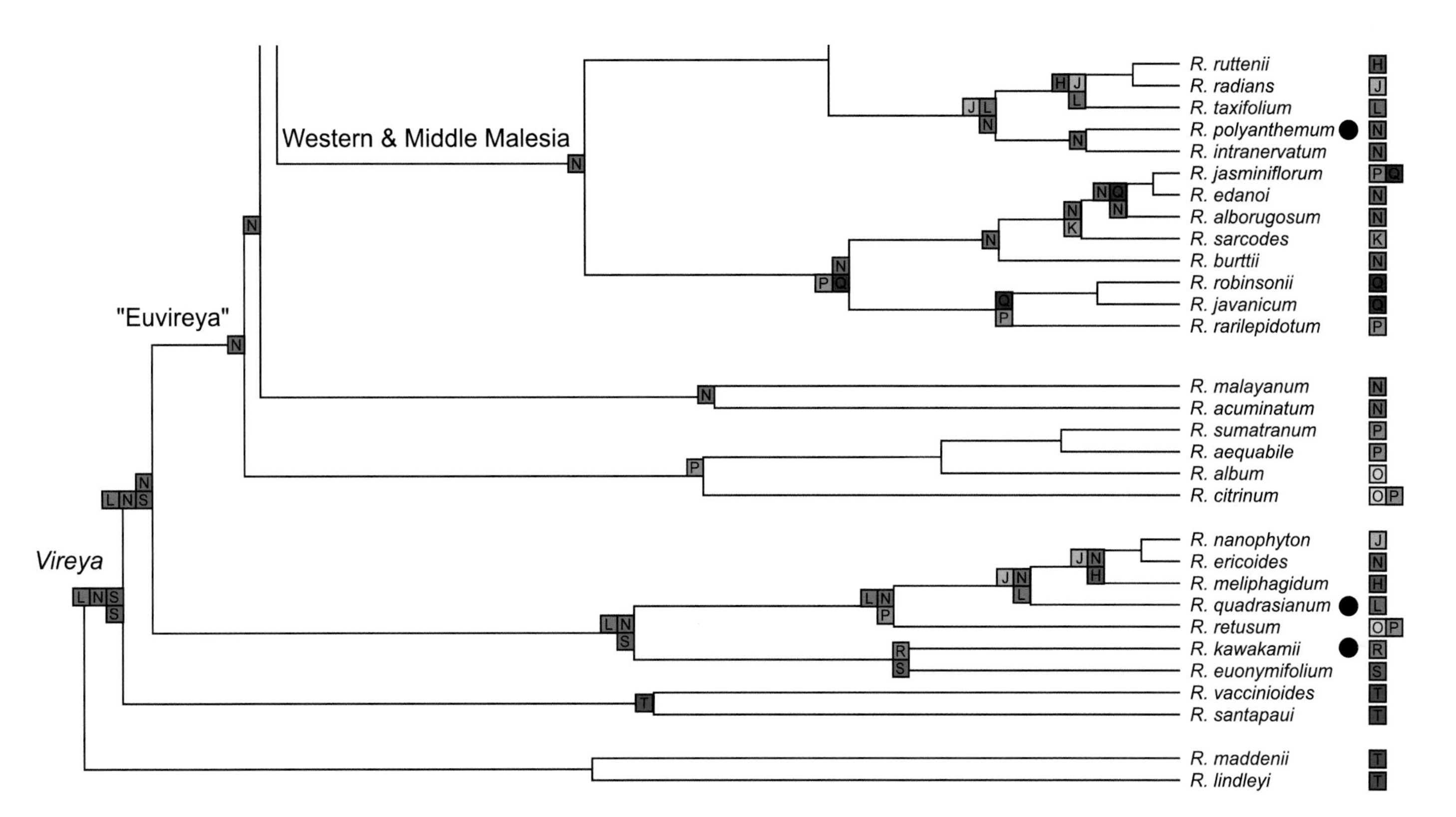

Figure 8.4 Results of LAGRANGE re-analysis of *Vireya* biogeography given 20 areas of endemism in southeast Asia, following Brown et al. (2006b). The relative-time chronogram was produced by applying nonparametric rate smoothing to a maximum likelihood (ML) molecular phylogeny. Nodes are annotated with ML estimates of ancestral geographic ranges. Rates of dispersal and local extinction were estimated subject to constraints implied by the adjacency graph shown in Figure 8.3 (see text for details). Colonisation of eastern Malesia is inferred as dispersal across Wallace's Line. Species used in the SHIBA runs are indicated with a dot.

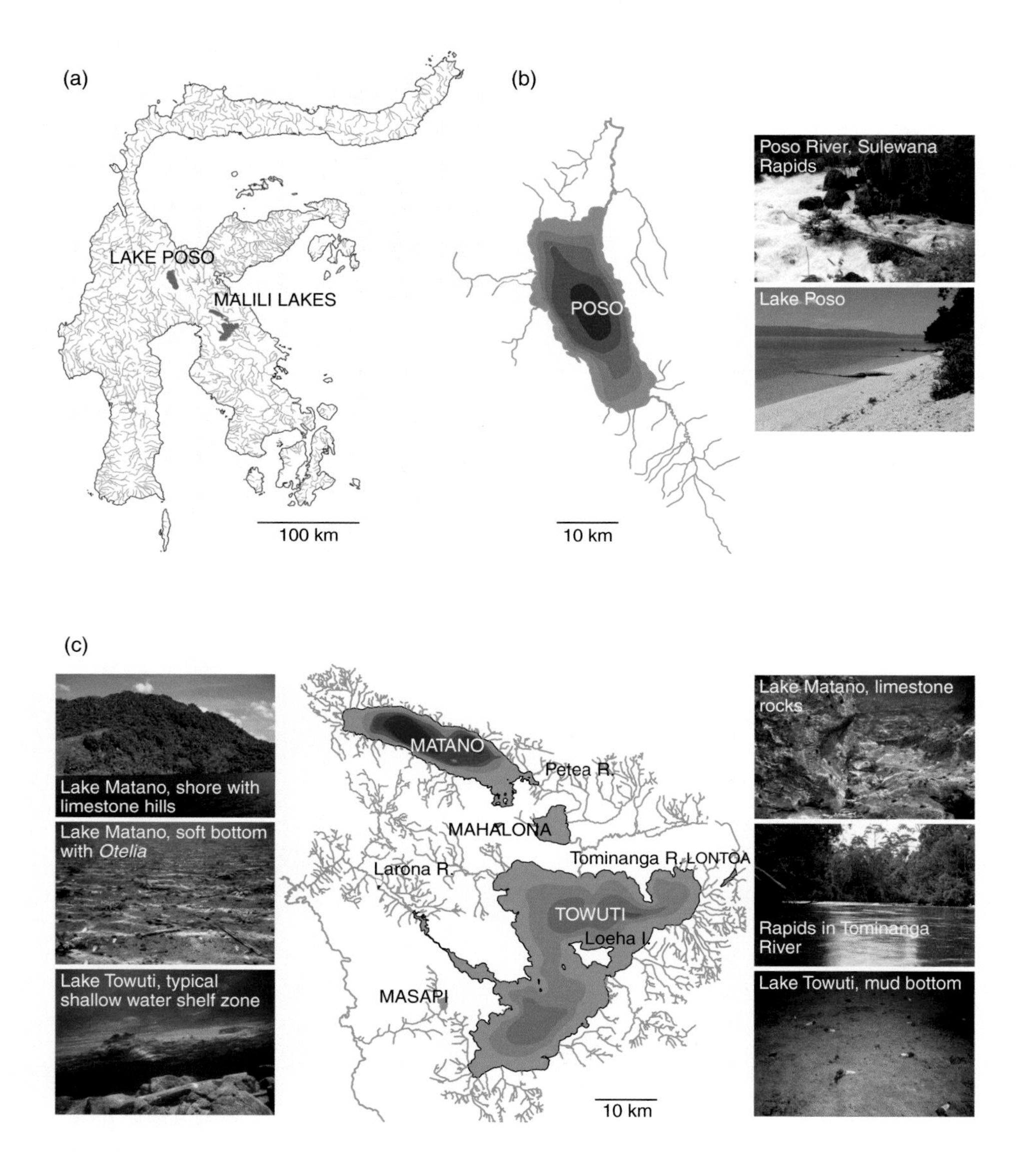

Figure 12.1 Indonesia, Sulawesi and the ancient lakes with characteristic habitats. (a) Sulawesi; the two ancient lake systems are highlighted in red. (b) Lake Poso. (c) Malili lake system. Lake names are printed in capital letters. Abbreviations: R., River; I., Island.

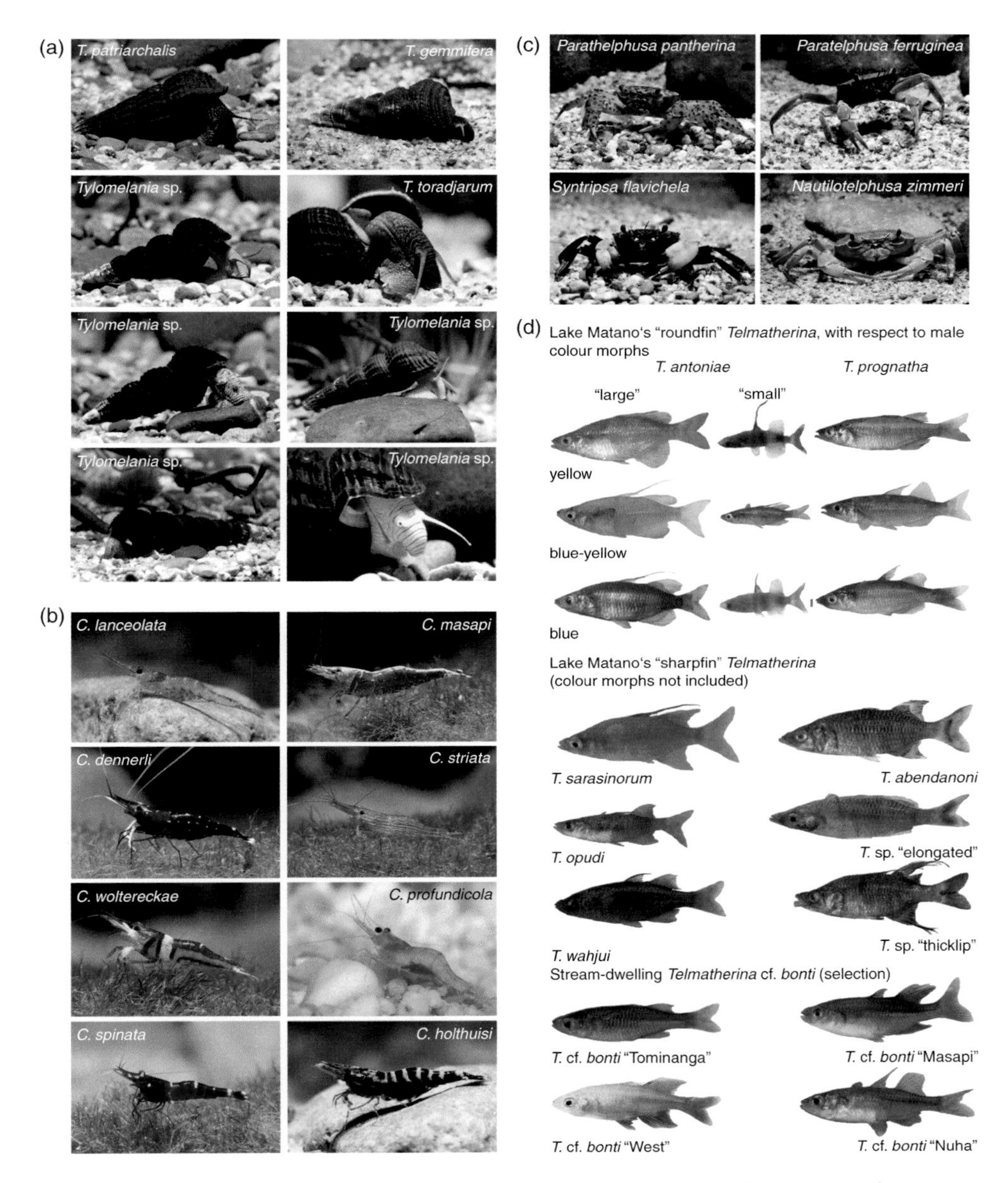

Figure 12.2 Characteristic species of the Sulawesi lakes. (a) Snails (*Tylomelania)* (b) Shrimps (*Caridina*). (c) Crabs (Gecarcinucidae). (d) Fishes (Telmatherinidae). Photographs of living animals (a–c) courtesy Chris Lukhaup.

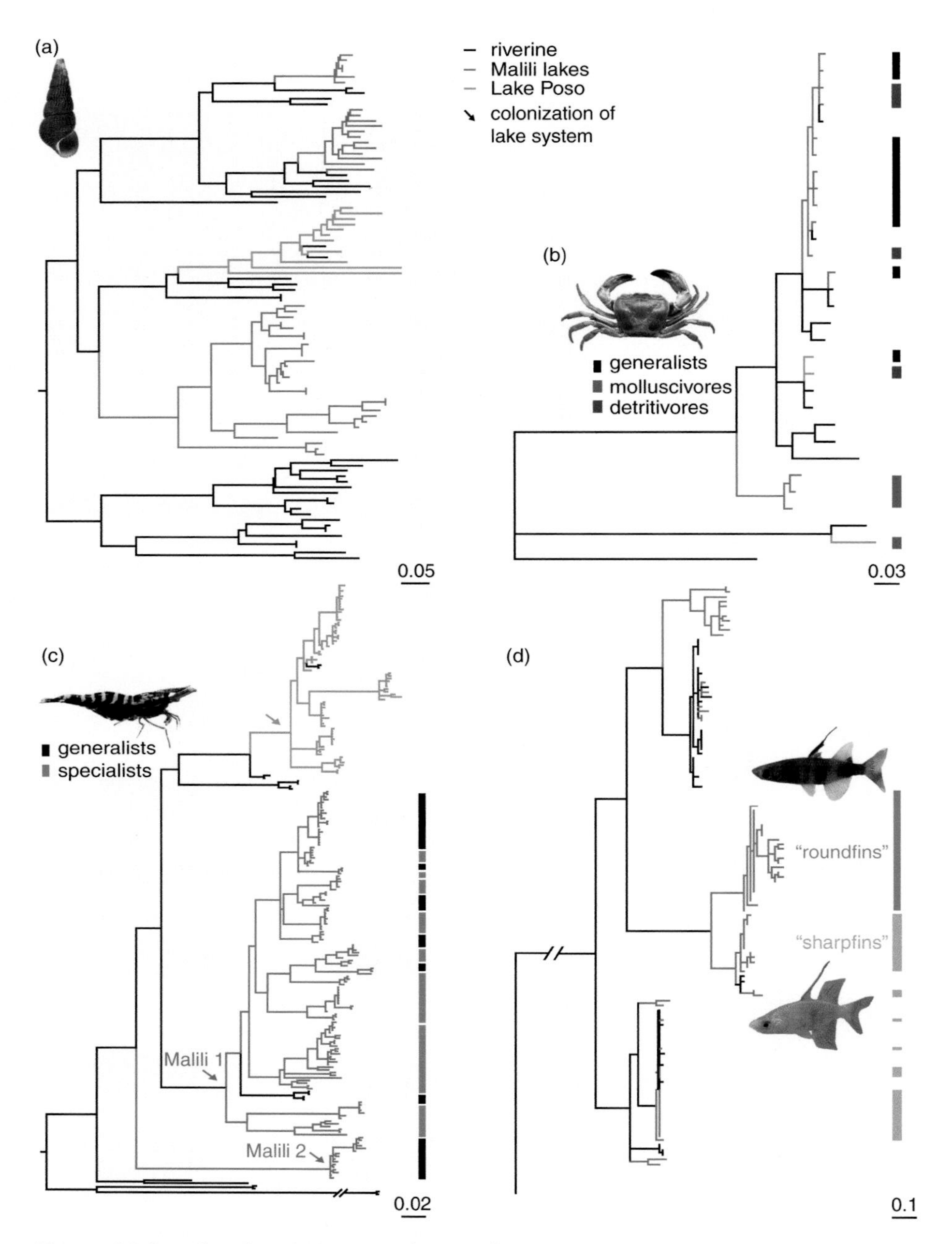

Figure 12.3 Molecular phylogenies (mtDNA) of the four major taxa in the Sulawesi lakes. (a) Snails (*Tylomelania)* (b) Crabs (Gecarcinucidae). (c) Shrimps (*Caridina*). (d) Fishes (Telmatherinidae). Bayesian inference (BI) based on mtDNA sequences. Lacustrine taxa are highlighted in red (Malili lakes) and green (Lake Poso), riverine species in black. The bars on the right highlight selected morphological or ecological groups. Modified or expanded from von Rintelen et al. 2004 (snails), von Rintelen et al. 2010b (shrimps), Schubart and Ng 2008 (crabs) and Herder et al. 2006 (fishes). Refer to these papers for methodological details.

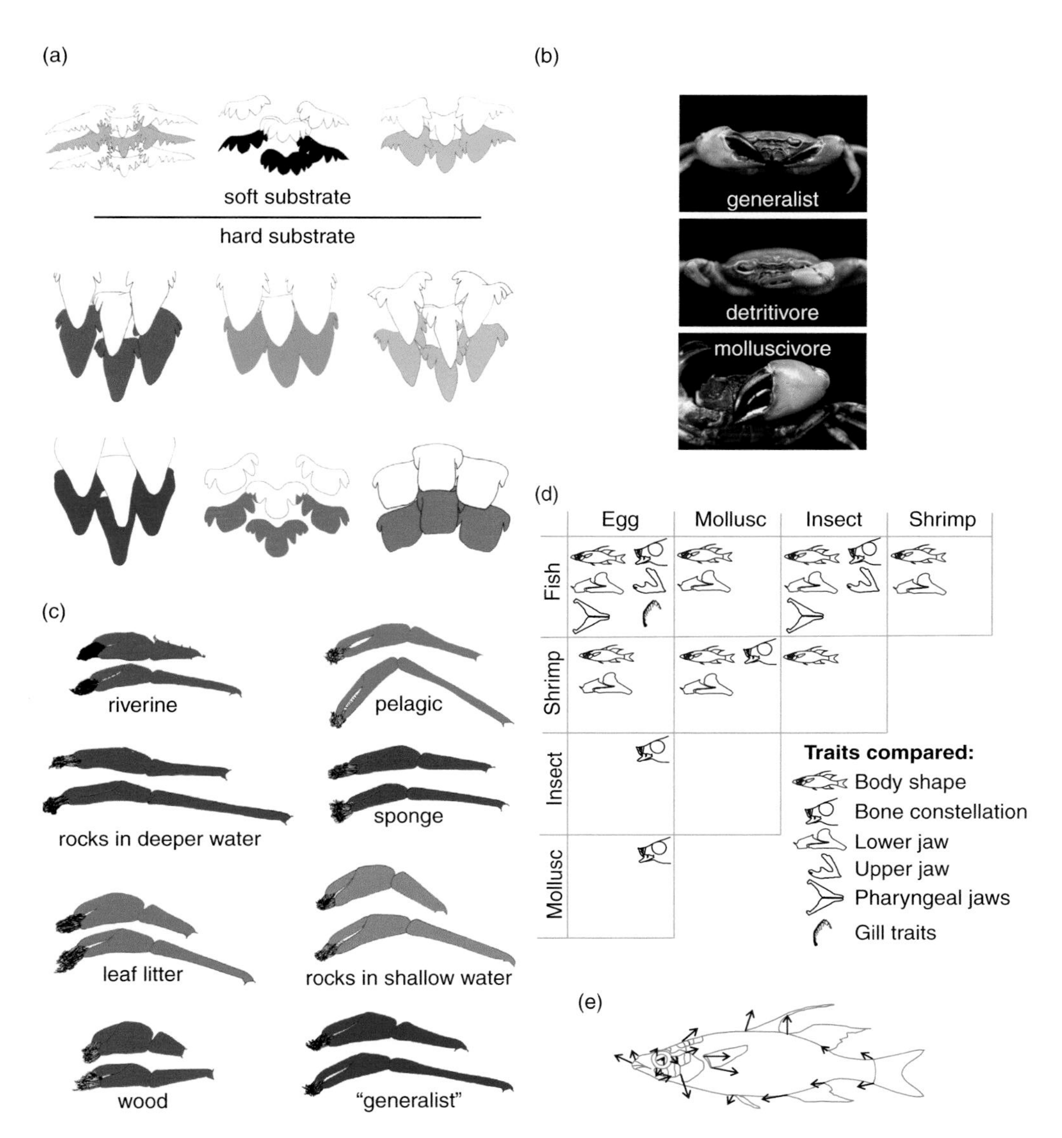

Figure 12.4 Habitat and trophic specialization in the Malili lakes. (a) Snails (*Tylomelania*) – major substrate-specific radula types highlighted by colour. (b) Crabs (Gecarcinucidae) – chela diversification in ecogroups defined by food preference. (c) Shrimps (*Caridina*) – cheliped diversification in substrate specific ecogroups highlighted by colour; modified from von Rintelen et al. (2010). (d) Fishes (Telmatherinidae): significant pairwise differentiation in candidate traits among major trophic groups as defined by stomach contents. (e) Body shape variation along a multivariate axis distinguishing shrimp-feeders from fish-feeders and other trophic specialists. Vector displacements indicate the direction of variation in shape, and show substantial differences in body depth, head shape, fin position and caudal peduncle length (d and e modified from Pfaender et al. 2010).

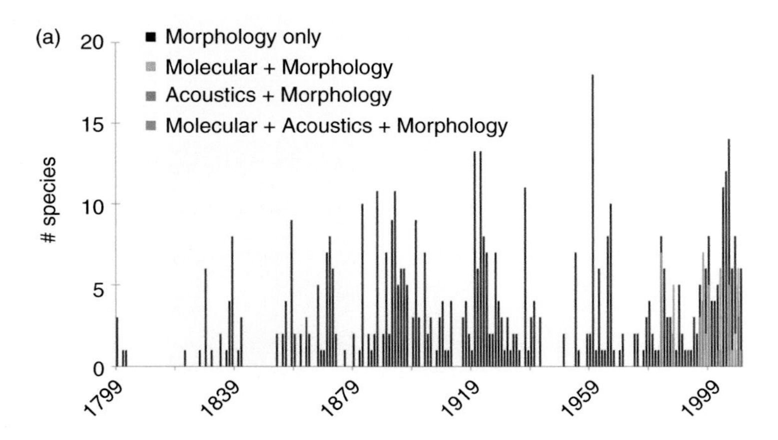

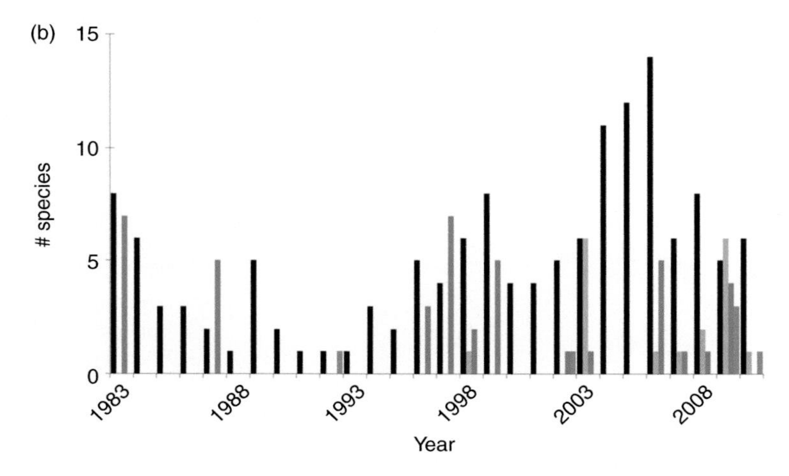

Figure 14.2 (a) Sources of data for recognising Southeast Asian amphibian species over the past two centuries. Note concentration of diversity of data types over the last three decades. (b) Expansion of Fig 15.2a to include descriptions using more than one data type (beginning in 1983 with the publication of Dring's anuran descriptions based on morphology and advertisement calls).

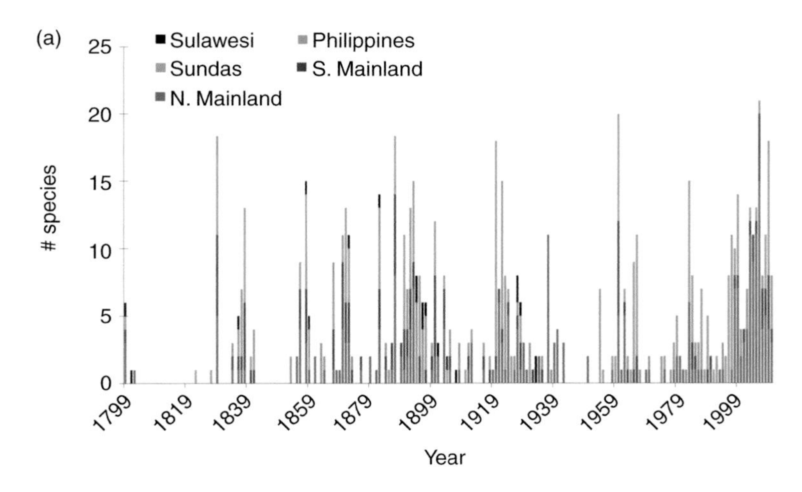

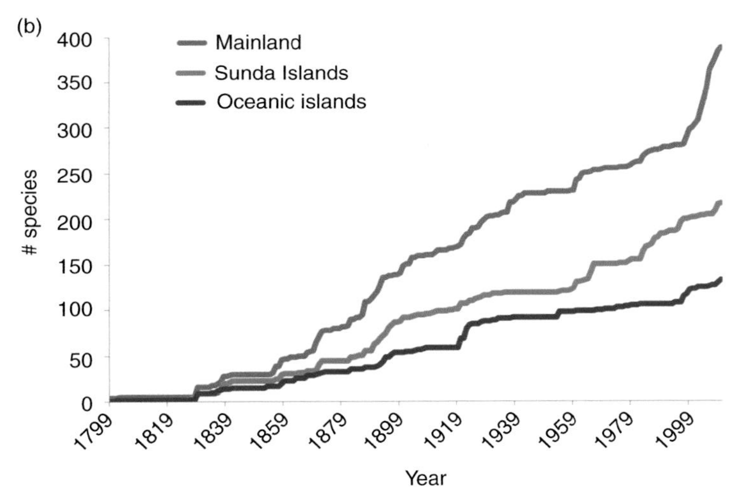

Figure 14.3 (a) Frequency distribution of Southeast Asian amphibian species descriptions over the past two centuries according to geographical region. Note the near absence of activity in Sulawesi after the 1930s and intense activity over the past decade in the northern mainland. (b) Species accumulation curves of Southeast Asian amphibian descriptions in each of three land mass types.

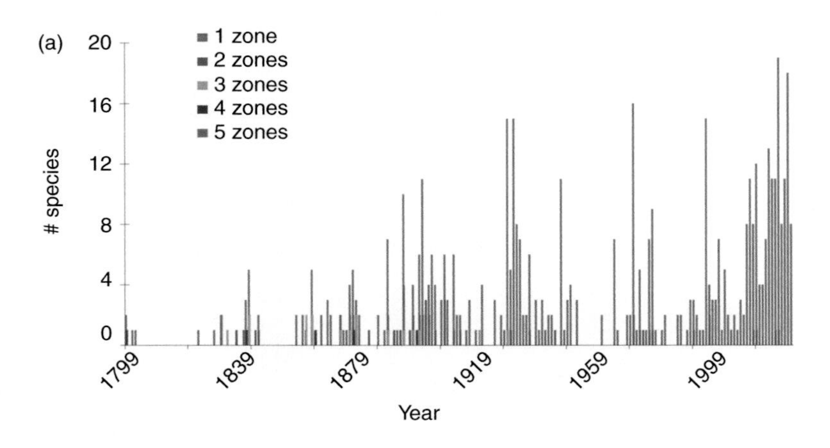

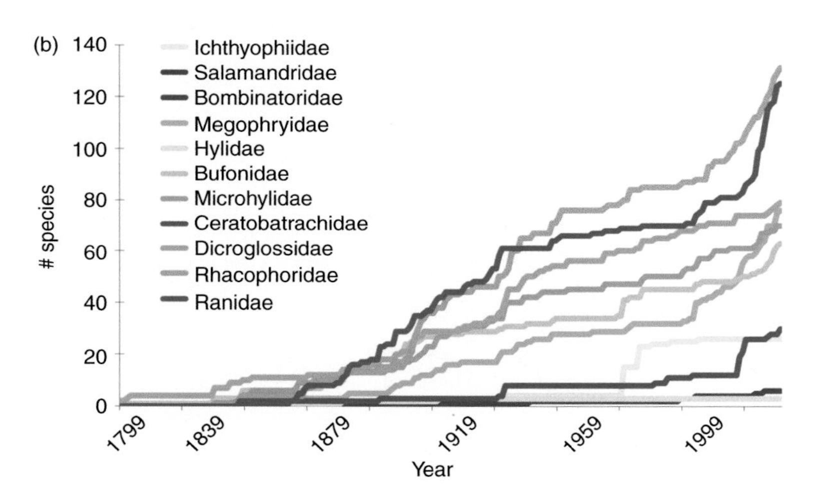

Figure 14.4 (a) Southeast Asian amphibian species descriptions through time categorised by the number of biogeographic zones inhabited (zones defined in main text). Note that widespread species (2–5 zones) were all discovered and described in the 1800s and in most cases, species described in the last century are found in one, or at most two, biogeographic regions. (b) Southeast Asian amphibian species accumulation curves by family.

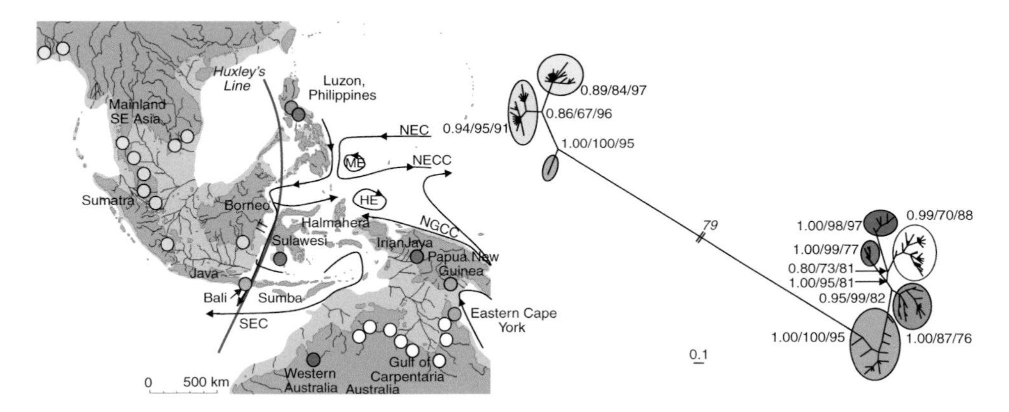

Figure 13.3 Map of sampling locations of *Macrobrachium rosenbergii* showing the distribution of lineages, and Bayesian consensus tree for 93 unique mitochondrial *COI* haplotypes obtained from sampling 861 giant freshwater prawns from 26 locations east and west of Huxley's Line (reproduced from de Bruyn and Mather 2007, with permission John Wiley and Sons). Major surface currents of interest are shown. Dashed line indicates seasonally reversing currents. North Equatorial Current (NEC); North Equatorial Counter Current (NECC); New Guinea Coastal Current (NGCC); Indonesian Throughflow (IT; location of Makassar Strait); Mindanao Eddy (ME); Halmahera Eddy (HE). Pale grey shading on map indicates –120 m sea-level contour, and major fresh watersheds at this time are shown (Voris 2000; reproduced with kind permission The Field Museum Chicago). Values at tree nodes indicate (in order), Bayesian posterior probabilities, neighbour joining bootstrap values, maximum likelihood bootstrap values.

Figure 16.2 View inside undisturbed peat swamp forest in Riau province, Sumatra.

Figure 16.4 One of the authors (CB) standing next to a peat subsidence measurement pole in drained peatland in Johor, southern peninsular Malaysia. Around 2.3 m of peat have been lost over a period of 28 years, a consequence of drainage, compaction and peat oxidation.

Figure 16.5 Fire-impacted peatland in Central Kalimantan. This part of the former Mega Rice Project area (see text for details) has been burnt at least three times since 1997. The vegetation, which was once species-diverse peat swamp forest, is now dominated by ferns and just two or three tree species capable of withstanding the greatly altered environmental conditions.

With the exception of *Metroxylon* (discussed below), our divergence time estimates suggest that the East Malesian taxa listed above all diverged from their nearest living relatives during an 18 Myr window between the latest Oligocene and Late Miocene, consistent with a period of increasing land surface area in Papuasia (Hall 1998, 2009). All but *Sommieria* belong to the major Indo-Pacific tribe Areceae, which is discussed in more detail below.

Of the lineages restricted to the east of Wallace's Line, *Pigafetta* (Calameae: Pigafettinae) stands out because of its lack of a biogeographic link to the Pacific (Fig 7.3b). *Pigafetta elata* is endemic to Sulawesi, whereas *P. filaris* occurs in the Moluccas and western New Guinea. *Pigafetta* and *Metroxylon* are unusual in tribe Calameae because they are absent from West Malesia. Baker and Dransfield (2000) suggested that both these genera could be the result of a recent dispersal into East Malesia. However, our divergence time estimates for both lineages coincide with the opening of the Makassar Strait in the Middle Eocene, which led to separation of West Sulawesi from East Borneo (Hall 2009). On the assumption that their closest relatives in tribe Calameae occurred in the Sunda region, as Baker and Dransfield (2000) suggest, this geological event may provide a vicariance explanation for the distribution of *Pigafetta* at least, and possibly also *Metroxylon* if eastwards dispersal and extinction in Sulawesi occurred.

7.5 Bimodal distributions across Wallace's Line

The rattan subtribe Calaminae (Calameae) displays the most pronounced bimodal distribution among Malesian palm groups. With around 500 species distributed from India to Australia and Fiji and a single outlier in humid tropical Africa (*Calamus deerratus*), it is the most species-rich subtribe in the palm family. *Calamus* itself is the largest of all palm genera (376 species), although it is paraphyletic as currently circumscribed with the remaining genera (*Ceratolobus, Daemonorops, Pogonotium, Retispatha*) nested within it (Baker et al. 2000). The Calaminae reflects palm species richness patterns in Malesia as a whole. It is most diverse to the west of Wallace's Line, especially in the Sunda region, declining sharply through Wallacea before reaching a further peak of species diversity in New Guinea. For example, *Calamus* consists of 82 species in Borneo, 27 in Sulawesi, 11 in the Moluccas and 52 in New Guinea (Table 7.1). This pattern has been attributed to secondary radiation in New Guinea following dispersal across Wallace's Line (Dransfield 1981). Current phylogenetic evidence (Baker et al. 2000) indicates that multiple independent lineages from within the subtribe are present in New Guinea, arguing against a single origin for the group on the island. Macrofossils and fossil pollen data attributed to *Calamus* or its close relatives have been widely reported (Harley and Baker 2001, Harley 2006, Dransfield et al. 2008), although these may in some cases represent other groups in tribe Calameae. Records summarised by Dransfield et al. (2008) indicate that, within its modern distribution, the Calaminae occurs from the Palaeocene (e.g. Borneo) onwards. The earliest records, however, appear in the Late Cretaceous of Somalia (Schrank 1994) with many other subsequent records outside the current distribution range (e.g. Eocene–Miocene of Europe, Eocene of North America). Most of the records in the Asia–Pacific region occur west of Wallace's Line. However, pollen records from the Eocene of Australia (Truswell et al. 1987) and Sulawesi led Morley (1998) to contest Dransfield's hypothesis and suggest that *Calamus* was already present east of Wallace's Line prior to collision. Eocene fossils from New Zealand have recently been associated with *Calamus*, although the evidence is unconvincing (Hartwich et al. 2010). However, if the fossil records of the Calaminae east of Wallace's Line are accurate, it would appear that the group had achieved a very wide distribution encompassing Laurasian and Gondwanic land masses by the Early Tertiary. Thus we might expect that the early history of the group dates to the Early Cretaceous, prior to the full separation of major continents. However, our molecular dates suggest that the Calaminae diverged from their closest relatives in the Middle Eocene, which is in conflict both with Morley's perspective and parts of the fossil record. We acknowledge that our molecular dates can only be treated as minimum age estimates. Nevertheless, we suggest that a critical reappraisal of the fossil record and its taxonomic affinities is as important as thorough new phylogenetic research on this group.

Subtribe Livistoninae (Trachycarpeae) displays a broadly similar distribution to the Calaminae. It is most pronounced in *Licuala* with 170 species in total, the highest diversity occurring in the Sunda region (e.g. around 46 in Borneo, L. G. Saw pers. comm.), very few through Wallacea (3 species in Sulawesi, 2 in the Moluccas) and a further peak of diversity in New Guinea (around 32 species, A. Barfod pers. comm.). Like the Calaminae, the Livistoninae are represented in New Guinea by multiple lineages. For example, *Saribus*, a genus resurrected recently as a result of phylogenetic research (Bacon and Baker 2011, Bacon et al. in press) is centred on New Guinea (six endemic species, one of which also extends to the Solomon Islands), with an endemic species in each of the Philippines and New Caledonia, and a widespread species ranges from Borneo and the Philippines to the Raja Ampat Islands. The genus *Livistona* (discussed below) is also present in New Guinea. In contrast to the Calaminae, the high diversity of the Livistoninae appears to have been achieved relatively recently, according to our molecular dates, with the subtribe diverging from its nearest living relative at the end of the Oligocene and diversifying through the Miocene.

Subgroupings of tribe Areceae have been discussed several times above, but the bimodal distribution of the group as a whole merits further consideration. In total, tribe Areceae consists of more than 660 species in 59 genera and 11 subtribes distributed from Pemba (Tanzania), Madagascar and the Comoro Islands in the eastern Indian Ocean, through tropical Asia to Australia and Samoa in the Pacific. The identification of this major Indo-Pacific clade is one of the most significant findings in recent palm phylogenetic studies (Lewis and Doyle 2002). The oceanic distances spanned by this group are immense and demand numerous dispersal explanations, as demonstrated for other groups of plants and animals with similar distributions (Keppel et al. 2009). The position of the Malesian archipelago at the interface of the two oceans suggests that the region played a significant role in the history of the tribe. Although phylogenies containing all genera of Areceae are now available (Norup et al. 2006, Savolainen et al. 2006, Baker et al. 2009), conflicting resolutions and poor support preclude firm conclusions. Nevertheless, a clade comprising most West Pacific taxa is resolved, including some that occur in the eastern islands of Malesia (New Guinea, Philippines, Moluccas), specifically Archontophoenicinae, Basseliniinae, Linospadicinae, Ptychospermatinae, *Dransfieldia* and *Heterospathe*. The remaining taxa (including all Indian Ocean genera) resolve either as a monophyletic group (excluding *Loxococcus*, Norup et al. 2006) or a paraphyletic group (Baker et al. 2009). Groups with distributions biased towards West Malesia (Arecinae, *Iguanura*, *Oncosperma*) do not resolve in the West Pacific clade, showing closer relationships with Indian Ocean taxa. However, several Malesian genera with a strong Papuasian or West Pacific emphasis (*Clinostigma*, *Cyrtostachys*, *Hydriastele*, *Rhopaloblaste*) also fall outside the West Pacific clade. Our molecular dates indicate that the Areceae diverged from its closest relatives at the end of the Middle Eocene and began to diversify in the Oligocene, results that are broadly consistent with previous dated phylogenies of the tribe (Savolainen et al. 2006). The West Pacific clade then diverged at the end of the Oligocene. A fossil pollen taxon, *Palmaepollenites kutchensis*, distributed between India and Southeast Asia has been linked to members of the West Pacific clade and dates back to the Palaeocene (Harley and Morley 1995). This record is clearly somewhat older than our molecular dates, but its taxonomic affinity should be treated with caution. Despite the ambiguous relationships of the Indian Ocean taxa, to quote Dransfield et al. (2008), 'it appears that tribe Areceae has penetrated the Pacific region on several independent occasions, whereas the same may not be true for Areceae in the Indian Ocean.' As Norup et al. (2006) indicated, due to the complex nature of tectonic changes in Southeast Asia and the West Pacific (Burrett et al. 1991, Hall 2009), simple biogeographic patterns cannot be expected in this group.

We recognise a further type of bimodal distribution across Wallace's Line. Here, the peaks of species diversity are found in New Guinea and the Philippines. The genus *Heterospathe* (Fig 7.3c) exemplifies this pattern with 12 species in the Philippines, 16 in New Guinea and just two species occurring in the intervening Moluccan islands (further species are found from the Solomon Islands to Vanuatu).

Phylogenetic evidence indicates that the Philippine species form a single clade embedded within New Guinea and West Pacific taxa (Norup et al. 2006), a topology consistent with a single dispersal into the Philippines. *Orania*, the sole genus of tribe Oranieae, displays a similar pattern, but less marked, with three species in the Philippines and 19 in New Guinea, one of which is shared with the Philippines, Sulawesi and the Moluccas (Keim 2003). The one remaining Asian species is widespread in southern Thailand and West Malesia (excluding the Philippines). *Orania* also contains a disjunct group of three species in Madagascar, an unusual biogeographic link found in few other angiosperm genera, such as *Nastus* (Poaceae, Baker et al. 1998) and *Langsdorffia* (Balanophoraceae, Mabberley 2008). Links between New Guinea and the Philippines have been reported in other plant groups, for example *Sararanga* (Pandanaceae, Baker et al. 1998) and subgroupings of *Vaccinium* (Ericaceae, Heads 2001). The relationship between New Guinea and the Philippines has been linked to a possible migration route via the Philippine–Halmahera arc that provided dispersal opportunities from 25 Ma onwards through the Miocene (Hall 1998, Norup et al. 2006). Our molecular date estimates are consistent with this, giving an Early Miocene age for the origin of *Heterospathe* and an Early Oligocene origin for *Orania*.

7.6 Disjunct distributions

Several disjunctions are well known in Malesian palm genera (Dransfield 1981, 1987, Baker et al. 1998, Dransfield et al. 2008). The monophyly of these genera has been confirmed (see Dransfield et al. 2008), indicating that the disjunctions are not artefacts of taxonomic delimitation. The Indian Ocean disjunction described above for *Orania* is distinct from the remaining disjuncts, which occur across Wallace's Line. *Cyrtostachys* (Fig 7.3c) and *Rhopaloblaste* (Areceae) display strikingly similar patterns of distribution with six and four species respectively occurring in Papuasia and one species each in the Sunda region. *Cyrtostachys renda* is widespread in Borneo, Sumatra and the Malay Peninsula, while *Rhopaloblaste singaporensis* is restricted to the Malay Peninsula. A further species of *Rhopaloblaste* is found in the Nicobar Islands. A related pattern is seen in subtribe Ptychospermatinae, as discussed above, due to the presence of the disjunct monotypic genus *Adonidia* in the Sunda region in Palawan and northern Borneo. More recently, a new disjunction has been identified (Bacon and Baker 2011) in the genus *Livistona* as a result of the exclusion of species belonging to the re-circumscribed genus *Saribus*. The new monophyletic delimitation of *Livistona* contains nine species in tropical Asia from the Himalayas to Borneo, Sumatra and the Philippines, 18 species in Australia, two of which are shared with seasonal parts of southern New Guinea, and a further disjunct in the Arabian Peninsula and the Horn of Africa. Strong evidence indicates that the Australian species are derived from a single dispersal across Wallace's Line in the Miocene (Crisp et al. 2010, Bacon and Baker 2011, Bacon et al. in press), as suggested previously by Dransfield (1987). Our divergence time estimates are similar for *Cyrtostachys*, *Rhopaloblaste* and *Livistona*, indicating that these genera diverged from their closest relatives around the same time in the Early Miocene. This is consistent with the study of Crisp et al. (2010), which indicated that a single *Livistona* ancestor colonised Australia from the north 10–17 Ma. The divergence time of *Adonidia* falls within the Late Miocene.

Thus, it seems likely that these disjunctions are associated with geological and perhaps climatic changes through the Miocene, such as the terrestrial emergence of New Guinea. If the distributions are a result of long-distance dispersals, it appears that such events have taken place in both directions, west to east in *Livistona*, east to west in the case of *Adonidia*. *Cyrtostachys* and *Rhopaloblaste* are more ambiguous due to their phylogenetic placement among taxa with both Indian and Pacific Ocean affinity (Baker et al. 2009) although Norup et al. (2006) have suggested an Indian Ocean origin for *Rhopaloblaste*. Alternatively, the disjunction may be due to extinction in Wallacea, perhaps as a result of drier climates that developed in the Late Neogene and Quaternary (Dransfield 1981, Morley 1998). Such an explanation could equally account for bimodal patterns, with the disjunct distributions interpreted merely as extreme cases of the bimodal pattern in which species diversity in Wallacea has diminished to zero.

7.7 Widespread distributions

A small number of widespread distributions do not fit the categories listed above. The simplest pattern, perhaps, is found in the sole species of subfamily Nypoideae. Distributions of the mangrove palm *Nypa fruticans* and its fossil records have been reviewed previously (Gee 1990, Baker et al. 1998, Morley 2000, Harley 2006, Pan et al. 2006, Dransfield et al. 2008). This species is currently distributed from India and Sri Lanka through to Australia and the Solomon Islands. However, it is well established that the lineage was once widespread in the Americas, Europe, Asia, Africa and Australia, primarily on account of its distinctive pollen that features widely in the fossil record as early as the Maastrichtian (Late Cretaceous), although our molecular dates suggest the group diverged from other palms around the Turonian. Climatic cooling at the end of the Eocene may explain the retraction of the range of *Nypa*. An in-depth review of the palaeodistribution of this important indicator of humid tropical environments is long overdue.

A further widespread distribution is displayed by subtribe Lataniinae (Borasseae), which is represented within the region by *Borassodendron* in the Sunda region (discussed above) and *Borassus* with three species in Africa and

Madagascar, *B. flabellifer* in South and Southeast Asia and *B. heineanus* endemic to western New Guinea (Bayton 2007). *Borassus flabellifer* is widely cultivated, for food and drink, construction materials, weaving, utensils and other uses, and as a result its natural distribution is not known. It is widespread in western India and Indochina, but in Malesia is reported to be native in Java, the Lesser Sunda Islands and Buton Island off South Sulawesi, reflecting its preference for mores seasonal climates (Govaerts and Dransfield 2005, Bayton 2007). Fossil pollen in the Pliocene–Pleistocene boundary of East Java (R. J. Morley pers. comm.) supports this claim. The Borasseae as a whole displays a strong African–Indian Ocean distribution; the two remaining genera of Lataniinae are *Latania* from the Mascarenes and *Lodoicea* (the double coconut) from the Seychelles, while the second subtribe (Hyphaeninae) is restricted to Africa, Madagascar, the Arabian Peninsula and India. The distribution has, in the past, been attributed to the break-up of Gondwana (Uhl and Dransfield 1987, Whitmore 1998), but divergence time estimates, including our own, indicate that the diversification of the tribe has taken place since the Late Eocene and involved eastwards dispersal into Malesia (Bayton 2005, Couvreur et al. 2011). Fossil records of widespread *Borassodendron* pollen in the Sunda region from the Miocene onwards are consistent with this hypothesis (R. J. Morley pers. comm.).

The distribution of *Corypha*, sole genus of Corypheae, resembles that of Borasseae, although it is most species rich in India and Indochina, with a single species, *C. utan*, occurring through Malesia to New Guinea and Australia, and one endemic in the Philippines (*C. microclada*). Like *Borassus flabellifer*, *C. utan* is not typical of rainforest, occurring in more open habitats. Dransfield et al. (2008) attributed the low diversity of *Corypha* to recent dispersal in the region.

7.8 Discussion

Our review illustrates that while the three broad categories of distribution types identified by Dransfield (1987) are broadly robust to new phylogenetic evidence, they over-simplify a much more complex set of distributions. A re-defined set of distribution patterns of Southeast Asian palms can be summarised as follows:

1. Distributions west of Wallace's Line (West Malesia)
    a. Distributions restricted to West Malesia, with species diversity centred in the Sunda region (e.g. Plectocomiinae, Salaccinae, *Borassodendron*, *Eugeissona*, *Iguanura*, *Johannesteijsmannia*).
    b. Distributions with species diversity centred on the Sunda region, but with few species crossing to the east of Wallace's Line (e.g. Arecinae, Caryoteae, *Korthalsia*, *Oncosperma*, *Pholidocarpus*).

   c. Marginal distributions in West Malesia with most of species diversity situated outside the region (e.g. Rhapidinae, *Phoenix*).
2. Distributions east of Wallace's Line (East Malesia)
   a. Distributions restricted to New Guinea (e.g. *Brassiophoenix*, *Dransfieldia*, *Sommieria*).
   b. Distributions with species diversity centred in New Guinea with affinities in the West Pacific (e.g. Linospadicinae, Ptychospermatinae, *Hydriastele*).
   c. Distributions with low species diversity in Papuasia with affinities in the West Pacific (e.g. *Actinorhytis*, *Clinostigma*, *Metroxylon*, *Physokentia*).
   d. Distributions spanning Wallacea and New Guinea (*Pigafetta*).
3. Bimodal distributions
   a. Bimodal distributions peaking in the Sunda region and New Guinea (e.g. Areceae, Calaminae, Livistoninae).
   b. Bimodal distributions peaking in Philippines and New Guinea (e.g. *Heterospathe*, [*Orania*]).
4. Disjunct distributions
   a. Disjunctions between Papuasia/West Pacific and the Sunda region (e.g. Livistoninae, *Adonidia*, *Cyrtostachys*, *Rhopaloblaste*).
   b. Disjunctions between Malesia and Madagascar (*Orania*).
5. Widespread distributions (e.g. *Corypha*, Lataniinae, *Nypa*).

The similarity between many of these distributions suggests that they may result from common biogeographic drivers. However, our molecular dates indicate that similar patterns can be achieved over very different timescales. It is likely then that different groups have achieved similar ranges via contrasting evolutionary histories.

It is not our intention to construct a general biogeographic narrative for Southeast Asian palms. Rigorous phylogenetic and biogeographic analyses of each independent lineage are required before such a synthesis can be achieved. Nevertheless, some general statements can be provided based on the review delivered here. First, the fossil record and divergence time estimates indicate that the early diversification of palms started long before the Tertiary tectonic evolution of Southeast Asia, apparently around the Middle to Late Cretaceous by which time the break-up of Laurasia and the continents of Gondwana was nearing completion. This suggests that simple scenarios such as that of Dransfield (1987), which proposed that palms originated in Pangaea, diversified independently in Laurasia and Gondwana before entering Southeast Asia from the north and south, are unlikely to apply. While elements of this scenario may hold true, we envisage a more complex situation in which a diverse array of biogeographic histories intersect in Malesia. Second, our analysis does not support the argument that palms are poor dispersers (Dransfield 1981, Uhl and Dransfield 1987). Although many patterns can be linked to geological history, for example in the Sunda region, the distribution of palm diversity across the region can only be explained if significant long-distance dispersals across ocean gaps have occurred. We see no other explanation, for example, for the Indo-Pacific distribution of tribe Areceae. We concur with Dransfield (1981, 1987) that long-distance dispersal across Wallace's Line is consistent with modern distribution patterns, especially from west to east, which is also evident in other dated molecular trees of angiosperm lineages as emphasised by Richardson et al. (this volume, Chapter 6). Morley (1998) considers the evidence for large-scale eastwards dispersal across Wallace's Line in the Miocene to be limited and asserts that many Sundaic elements occurring east of the divide were contributed by the Sundaic fragment of Sulawesi following the opening of the Makassar Straits in the Middle Eocene. However, the origin of some clades (e.g. Arecinae, *Oncosperma*) may in fact post-date this period, reinforcing the eastwards dispersal hypothesis. In any case, taxa stranded in Sulawesi would have to have migrated further eastwards from Sulawesi to reach their current day ranges. We note that East Malesian palm taxa fail to cross Wallace's Line in the manner displayed by some West Malesian taxa; their transgressions are manifest in disjunctions, which we have failed to explain satisfactorily.

The overall bimodal distribution of palms in Malesia can be broken down into two components: (1) the geographical distributions of the clades that occur in Malesia and (2) variation in species richness patterns within each clade. While geological history may strongly influence clade distributions, species richness patterns within lineages remain inadequately explained, especially the *paucity* of palm species in Wallacea. Global analyses of plant species distributions reveal a general pattern of lower species richness through Wallacea relative to the Sunda region and New Guinea, and demonstrate that the strongest determinant of plant species richness on islands is area (Kreft et al. 2008). Preliminary analyses indicate that species–area relationships may also largely explain the bimodal distribution in palms (Kissling et al. in press). Palaeoclimate variation may also have influenced palm species richness across Malesia. It has been demonstrated that palm diversity is strongly correlated with climatic factors, especially water (Bjorholm et al. 2005, Kissling et al. in press). Poor palm diversity can be observed directly in the Lesser Sunda Islands, most likely due to the relatively dry, seasonal climate of that region. Palm hotspots such as the Sunda region have experienced extended periods of climatic stability (Morley 2000), providing favourable conditions for lineage diversification in palms. However, seasonal climates that developed in the Late Neogene and Quaternary very likely affected palm diversity in the region and potentially contributed to the bimodal distribution (Dransfield 1981, Morley 1998).

The palm family has the potential to yield many novel insights into the evolution of plant diversity in space and time in Southeast Asia. We consider the construction of robust, well-sampled phylogenies for the purpose of divergence time estimation and formal biogeographic analysis to be a major priority to achieve this end. Macroecological analysis is also required to explore the spatial and environmental determinants of species richness patterns across the region, for example to test formally whether or not Wallacean palm diversity significantly deviates from expected species richness. Special attention should be paid to groups with distributions spanning Wallacea such as *Pigafetta*, perhaps employing phylogeographic methods. Our review demonstrates that through phylogenetic advances Malesian palms are beginning to give up their secrets. However, a multidisciplinary approach, such as we suggest here, would prove immeasurably more informative.

Acknowledgements

We are grateful to Christine Bacon, Finn Borchsenius, John Dransfield, Daniel Kissling, Bob Morley, Alexandra Muellner, James Richardson and Richard Saunders for providing many thoughtful comments on an early draft of this chapter. Anders Barfod, Saw Leng Guan and Tim Utteridge generously shared insights and unpublished information. We thank the organisers of the SAGE conference for the opportunity to share our ideas in this volume.

References

Ali, J. R. and Aitchison, J. C. (2008). Gondwana to Asia: plate tectonics, paleogeography and the biological connectivity of the Indian sub-continent from the Middle Jurassic through latest Eocene (166–35 Ma). *Earth-Science Reviews*, **88**, 145–66.

Asmussen, C. B. and Chase, M. W. (2001). Coding and noncoding plastid DNA in palm systematics. *American Journal of Botany*, **88**, 1103–17.

Asmussen, C. B., Baker, W. J. and Dransfield, J. (2000). Phylogeny of the palm family (Arecaceae) based on *rps*16 intron and *trn*L-*trn*F plastid DNA sequences. In *Monocots: Systematics and Evolution*, ed. K. L. Wilson and D. A. Morrison. Melbourne, Australia: CSIRO, pp. 525–35.

Asmussen, C. B., Dransfield, J., Deickmann, V. et al. (2006). A new subfamily classification of the palm family (Arecaceae): evidence from plastid DNA phylogeny. *Botanical Journal of the Linnean Society*, **151**, 15–38.

Bacon, C. D. and Baker, W. J. (2011). *Saribus* resurrected. *Palms*, **55**, 109–16.

Bacon, C. D., Baker, W. J. and Simmons, M. P. (in press). Miocene dispersal drives island radiations in the palm tribe Trachycarpeae (Arecaceae). *Systematic Biology*.

Baker, W. J. and Dransfield, J. (2000). Towards a biogeographic explanation of the calamoid palms. In *Monocots: Systematics and Evolution*, ed. K. L. Wilson and D. A. Morrison. Melbourne, Australia: CSIRO, pp. 545–53.

Baker, W. J., Coode, M. J. E., Dransfield, J. et al. (1998). Patterns of distribution of Malesian vascular plants. In *Biogeography and Geological Evolution of SE Asia*, ed. R. Hall and J. D. Holloway. Leiden, The Netherlands: Backhuys, pp. 243–58.

Baker, W. J., Asmussen, C. B., Barrow, S. C., Dransfield, J. and Hedderson, T. A. (1999). A phylogenetic study of the palm family (Palmae) based on chloroplast DNA sequences from the trnL-trnF region. *Plant Systematics and Evolution*, **219**, 111–26.

Baker, W. J., Hedderson, T. A. and Dransfield, J. (2000). Molecular phylogenetics of *Calamus* (Palmae) and related rattan genera based on 5S nrDNA spacer sequence data. *Molecular Phylogenetics and Evolution*, **14**, 218–31.

Baker, W. J., Savolainen, V., Asmussen-Lange, C. B. et al. (2009). Complete generic-level phylogenetic analyses of Palms (Arecaceae) with comparisons of supertree and supermatrix approaches. *Systematic Biology*, **58**, 240–56.

Bayton, R. P. (2005). *Borassus L. and the Borassoid palms: systematics and evolution*. PhD thesis, University of Reading.

Bayton, R. P. (2007). A revision of *Borassus* L. (Arecaceae). *Kew Bulletin*, **62**, 561–86.

Berry, E. W. (1914). The Upper Cretaceous and Eocene floras of South Carolina and Georgia. *U.S. Geological Survey, Professional Paper*, **84**.

Bjorholm, S., Svenning, J. C., Skov, F. and Balslev, H. (2005). Environmental and spatial controls of palm (Arecaceae) species richness across the Americas. *Global Ecology and Biogeography*, **14**, 423–9.

Bremer, K. (2000). Early Cretaceous lineages of monocot flowering plants. *Proceedings of the National Academy of Sciences of the United States of America*, **97**, 4707–11.

Brown, G. K., Nelson, G. and Ladiges, P. Y. (2006). Historical biogeography of *Rhododendron* section *Vireya* and the Malesian Archipelago. *Journal of Biogeography*, **33**, 1929–44.

Burrett, C., Duhig, N., Berry, R. and Varne, R. (1991). Asian and South-Western Pacific continental terranes derived from Gondwana, and their biogeographic significance. *Australian Systematic Botany*, **4**, 13–24.

Crié, L. (1892). Recherches sur les palmiers silicifiés des terrains Crétacés de l'Anjou. *Bulletin de la Société des Études Scentifiques d'Angers*, **21**, 97–103.

Crisp, M. D., Isagi, Y., Kato, Y., Cook, L. G. and Bowman, D. M. J. S. (2010). *Livistona* palms in Australia: ancient relics or opportunistic immigrants? *Molecular Phylogenetics and Evolution*, **54**, 512–23.

Couvreur, T. L. P., Forest, F. and Baker, W. J. (2011). Origin and global diversification patterns of tropical rain forests: inferences from a complete genus-level phylogeny of palms. *BMC Biology*, 9, 44.

Donoghue, M. J. and Smith, S. A. (2004). Patterns in the assembly of temperate forests around the Northern Hemisphere. *Philosophical*

Transactions of the Royal Society of London Series B, **359**, 1633–44.

Dowe, J. L. (2009). A taxonomic account of *Livistona* R.Br. (Arecaceae). *Gardens' Bulletin Singapore*, **60**, 185–344.

Dransfield, J. (1981). Palms and Wallace's line. In *Wallace's Line and Plate Tectonics*, ed. T. C. Whitmore. Oxford: Clarendon Press, pp. 43–56.

Dransfield, J. (1987). Bicentric distributions in Malesia as exemplified by palms. In *Biogeographical Evolution of the Malay Archipelago*, ed. T. C. Whitmore. Oxford: Clarendon Press, pp. 60–72.

Dransfield, J. (1988). The palms of Africa and their relationships. In *Proceedings of the Eleventh Plenary Meeting of the Association for the Study of the Flora of Tropical Africa*, Vol. 25, ed. P. Goldblatt and P. P. Lowry. St. Louis, MO: Missouri Botanical Garden, pp. 95–103.

Dransfield, J., Irvine, A. K. and Uhl, N. W. (1985). *Oraniopsis appendiculata*, a previously misunderstood Queensland palm. *Palms*, **29**, 56–63.

Dransfield, J., Uhl, N. W., Asmussen, C. B., Baker et al. (2005). A new phylogenetic classification of the palm family, Arecaceae. *Kew Bulletin*, **60**, 559–69.

Dransfield, J., Uhl, N. W., Asmussen, C. B. et al. (2008). *Genera Palmarum: The Evolution and Classification of Palms*. London: Royal Botanic Gardens, Kew.

Friis, E. M., Pedersen, K. R. and Crane, P. R. (2004). Araceae from the Early Cretaceous of Portugal: evidence on the emergence of monocotyledons. *Proceedings of the National Academy of Sciences of the United States of America*, **101**, 16565–70.

Frodin, D. (2001). *Guide to the Standard Floras of the World*, 2nd edn. Cambridge: Cambridge University Press.

Gee, C. T. (1990). On the fossil occurrence of the mangrove palm *Nypa*. In *Paleofloristics and Paleoclimatic Changes in the Cretaceous and Tertiary*, ed. E. Knobloch and Z. Kvacek. Prague: Geological Survey Press, pp. 315–319.

Govaerts, R. and Dransfield, J. (2005). *World Checklist of Palms*. London: Royal Botanic Gardens, Kew.

Govaerts, R., Dransfield, J., Zona, S. F, Hodel, D. R. and Henderson, A. (2010). *World Checklist of Arecaceae* (http://www.kew.org/wcsp/). London: Royal Botanic Gardens, Kew.

Gruezo, W. S. and Harries, H. C. (1984). Self-sown, wild-type coconuts in the Philippines. *Biotropica*, **16**, 140–7.

Hahn, W. J. (2002). A molecular phylogenetic study of the Palmae (Arecaceae) based on *atp*B, *rbc*L, and 18S nrDNA sequences. *Systematic Biology*, **51**, 92–112.

Hahn, W. J. and Sytsma, K. J. (1999). Molecular systematics and biogeography of the Southeast Asian genus Caryota (Palmae). *Systematic Botany*, **24**, 558–80.

Hall, R. (1998). The plate tectonics of Cenozoic SE Asia and the distribution of land and sea. In *Biogeography and Geological Evolution of SE Asia*, ed. R. Hall and J. D. Holloway. Leiden, The Netherlands: Backhuys, pp. 99–131.

Hall, R. (2001). Cenozoic reconstructions of SE Asia and the SW Pacific: changing patterns of land and sea. In *Faunal and Floral Migrations and Evolution in SE Asia–Australasia*, ed. I. Metcalfe, J. M. B. Smith, M. Morwood and I. D. Davidson). Lisse, The Netherlands: A. A. Balkema (Swets and Zeitlinger Publishers), pp. 35–56.

Hall, R. (2002). Cenozoic geological and plate tectonic evolution of SE

Asia and the SW Pacific: computer-based reconstructions, model and animations. *Journal of Asian Earth Sciences*, **20**, 353–431.

Hall, R. (2009). Southeast Asia's changing palaeogeography. *Blumea*, **54**, 148–161.

Harley, M. M. (2006). A summary of fossil records for Arecaceae. *Botanical Journal of the Linnean Society*, **151**, 39–67.

Harley, M. M. and Baker, W. J. (2001). Pollen aperture morphology in Arecaceae: application within phylogenetic analyses, and a summary of the fossil record of palm-like pollen. *Grana*, **40**, 45–77.

Harley, M. M. and Morley, R. J. (1995). Ultrastructural studies of some fossil and extant palm pollen, and the reconstruction of the biogeographical history of subtribes Iguanurinae and Calaminae. *Review of Palaeobotany and Palynology*, **85**, 153–82.

Harries, H. C. (1978). Evolution, dissemination and classification of *Cocos nucifera* L. *Botanical Review*, **44**, 265–319.

Hartwich, S. J., Conran, J. G., Bannister, J. M., Lindqvist, J. K. and Lee, D. E. (2010). Calamoid fossil palm leaves and fruits (Arecaceae: Calamoideae) from Late Eocene Southland, New Zealand. *Australian Systematic Botany*, **23**, 131–40.

Heads, M. (2001). Ericaceae in Malesia: vicariance biogeography, terrane tectonics and ecology. *Telopea*, **10**, 311–449.

Heatubun, C. D. (2009). *Systematics and evolution of the palm genus* Areca. PhD thesis, Bogor Agricultural University.

Heatubun, C. D., Baker, W. J., Mogea, J. P. et al. (2009). A monograph of *Cyrtostachys* (Arecaceae). *Kew Bulletin*, **64**, 67–94.

Janssen, T. and Bremer, K. (2004). The age of major monocot groups inferred from 800+ *rbc*L sequences. *Botanical Journal of the Linnean Society*, **146**, 385–98.

Keim, A. P. (2003). *A monograph of the genus Orania Zippelius (Arecaceae: Oraniinae)*. PhD thesis, University of Reading.

Keppel, G., Lowe, A. J. and Possingham, H. P. (2009). Changing perspectives on the biogeography of the tropical South Pacific: influences of dispersal, vicariance and extinction. *Journal of Biogeography*, **36**, 1035–54.

Kissling, W. D., Baker, W. J., Balslev, H. et al. (in press). Quaternary and pre-Quaternary historical legacies in the global distribution of a major tropical plant lineage. *Global Ecology and Biogeography*.

Kreft, H., Jetz, W., Mutke, J., Kier, G. and Barthlott, W. (2008). Global diversity of island floras from a macroecological perspective. *Ecology Letters*, **11**, 116–27.

Kvacek, J. and Herman, A. B. (2004). Monocotyledons from the Early Campanian (Cretaceous) of Grünbach, Lower Austria. *Review of Palaeobotany and Palynology*, **128**, 323–53.

Lewis, C. E. and Doyle, J. J. (2002). A phylogenetic analysis of tribe Areceae (Arecaceae) using two low-copy nuclear genes. *Plant Systematics and Evolution*, **236**, 1–17.

Loo, A. H. B., Dransfield, J., Chase, M. W. and Baker, W. J. (2006). Low-copy nuclear DNA, phylogeny and the evolution of dichogamy in the betel nut palms and their relatives (Arecinae; Arecaceae). *Molecular Phylogenetics and Evolution*, **39**, 598–618.

Mabberley, D. J. (2008). *Mabberley's Plant Book*. New York: Cambridge University Press.

Marsh, S. T., Brummitt, N. A., de Kok, R. P. J. and Utteridge, T. M. A. (2009). Large-scale patterns of plant diversity and conservation priorities in South East Asia. *Blumea*, **54**, 103–108.

McLoughlin, S. (2001). The breakup history of Gondwana and its impact on pre-Cenozoic floristic provincialism. *Australian Journal of Botany*, **49**, 271–300.

Meerow, A. W., Noblick, L., Borrone, J. W. et al. (2009). Phylogenetic analysis of seven WRKY genes across the palm subtribe Attaleinae (Arecaceae) identifies *Syagrus* as sister group of the coconut. *PLoS ONE*, **4**, e7353.

Mittermeier, R. A., Myers, N. and Mittermeier, C. G. (1999). *Hotspots: Earth's Biologically Richest and Most Endangered Terrestrial Ecoregions*. Mexico City and Washington: CEMEX and Conservation International.

Moore, H. E. (1973a). The major groups of palms and their distribution. *Gentes Herbarum*, **11**, 27–140.

Moore, H. E. (1973b). Palms in the tropical forest ecosystems of Africa and South America. In *Tropical Forest Ecosystems in African and South America: A Comparative Review*, ed. B. J. Meggers, E. S. Ayensu and W. D. Duckworth. Washington: Smithsonian Institution Press, pp. 63–88.

Morley, R. J. (1998). Palynological evidence for Tertiary plant dispersals in the SE Asian region in relation to plate tectonics and dispersal. In *Biogeography and Geological Evolution of SE Asia*, ed. R. Hall and J. D. Holloway. Leiden, The Netherlands: Backhuys, pp. 211–34.

Morley, R. J. (2000). *Origin and Evolution of Tropical Rainforests*. Chichester, UK: Wiley.

Morley, R. J. (2003). Interplate dispersal paths for megathermal angiosperms. *Perspectives in Plant Ecology Evolution and Systematics*, **6**, 5–20.

Muller, J. (1968). Palynology of the Pedawan and Plateau sandstone formations (Cretaceous–Eocene) in Sarawak, Malaysia. *Micropaleontology*, **14**, 1–37.

Muller, J. (1981). Fossil pollen records of extant angiosperms. *The Botanical Review*, **47**, 1–142.

Norup, M. V., Dransfield, J., Chase, M. W. et al. (2006). Homoplasious character combinations and generic delimitation: a case study from the Indo-Pacific arecoid palms (Arecaceae: Areceae). *American Journal of Botany*, **93**, 1065–80.

Pan, A. D., Jacobs, B. F., Dransfield, J. and Baker, W. J. (2006). The fossil history of palms (Arecaceae) in Africa and new records from the Late Oligocene (28–27 Mya) of north-western Ethiopia. *Botanical Journal of the Linnean Society*, **151**, 69–81.

Phadtare, N. R. and Kulkarni, A. R. (1984). Affinity of the genus *Quilonipollenites* with the Malaysian palm *Eugeissona* Griffith. *Pollen et Spores*, **26**, 217–25.

Raes, N. and van Welzen, P. C. (2009). The demarcation and internal division of Flora Malesiana: 1857–present. *Blumea*, **54**, 6–8.

Roos, M. C., Kessler, P. J. A., Gradstein, S. R. and Baas, P. (2004). Species diversity and endemism of five major Malesian islands: diversity–area relationships. *Journal of Biogeography*, **31**, 1893–908.

Savolainen, V., Anstett, M. C., Lexer, C. et al.(2006). Sympatric speciation in palms on an oceanic island. *Nature*, **441**, 210–13.

Schrank, E. (1994). Palynology of the Yesomma Formation in Northern Somalia: a study of pollen spores and associated phytoplankton from the late Cretaceous Palmae Province. *Palaeontographica Abteilung B*, **231**, 63–112.

Tiffney, B. H. (1985). Perspectives on the origin of the floristic similarity between eastern Asian and eastern North America. *Journal of the Arnold Arboretum*, **66**, 73–94.

Trenel, P., Gustafsson, M. H. G., Baker, W. J. et al. (2007). Mid-Tertiary dispersal, not Gondwanan vicariance explains distribution patterns in the wax palm subfamily (Ceroxyloideae : Arecaceae). *Molecular Phylogenetics and Evolution*, **45**, 272–88.

Truswell, E. M., Kershaw, P. A. and Sluiter, I. R. (1987). The Australian–South-east Asian connection: evidence from the palaeobotanical record. In *Biogeographical Evolution of the Malay Archipelago*, ed. T. C. Whitmore. Oxford: Clarendon Press, pp. 32–49.

Uhl, N. W. and Dransfield, J. (1987). *Genera Palmarum, A Classification of Palms Based on the Work of Harold E. Moore Jr.* Lawrence, KS: L. H. Bailey Hortorium and the International Palm Society.

Uhl, N. W., Dransfield, J., Davis, J. I. et al. (1995). Phylogenetic relationships among palms: cladistic analyses of morphological and chloroplast DNA restriction site variation. In *Monocotyledons: Systematics and Evolution*, Vol. 2, ed. P. J. Rudall, P. J. Cribb, D. F. Cutler and C. J. Humphries. London: Royal Botanic Gardens, Kew, pp. 623–61.

van Steenis, C. G. G. J. (1950). The delimitation of Malesia and its main plant geographical distributions. *Flora Malesiana, Series 1*, **1**, lxx–lxxv.

van Welzen, P. C. and Slik, J. W. F. (2009). Patterns in species richness and composition of plant families in the Malay Archipelago. *Blumea*, **54**, 166–71.

Wallace, A. R. (1860). On the zoological geography of the Malay Archipelago. *Journal of the Proceedings of the Linnean Society: Zoology*, **4**, 172–84.

Whitmore, T. C. (1998). *An Introduction to Tropical Rainforests*. Oxford: Oxford University Press.

Wikström, N., Savolainen, V. and Chase, M. W. (2001). Evolution of the angiosperms: calibrating the family tree. *Proceedings of the Royal Society of London Series B*, **268**, 2211–20.

8

Historical biogeography inference in Malesia

CAMPBELL O. WEBB AND RICHARD REE

8.1 Introduction

The Malesian region (the floristic region incorporating the nation states of Indonesia, Malaysia, Singapore, Brunei, the Philippines and Papua New Guinea) has long been recognised as a place of great biogeographic interest (Wallace 1869), and has stimulated a large literature (e.g. Croizat 1958, MacArthur and Wilson 1967, Michaux 1991). The ancestors of the Malesian biota have arrived by three major routes: (1) Laurasian (including Boreotropical) clades arriving from the west (e.g. for plants, *Trigonobalanus* and *Lithocarpus*, Morley 2000); (2) Gondwanan clades arriving from the west, via Africa or the Indian raft and Sundaland (e.g. Dipterocarpaceae; many Annonaceae, Richardson et al. 2004, Crypteroniaceae, Moyle 2004); and (3) Gondwanan clades arriving from the east, via the Australian raft (e.g. Proteaceae, Barker et al. 2007; Cunoniaceae; Monimiaceae; *Phyllocladus*, *Nothofagus*, Morley 2000; *Eucalyptus*, Ladiges et al. 2003). This biotic interchange has occurred on a complex, ever-changing landscape of continents and large and small islands, with barriers to species' movements not only of great over-ocean distances, but also of fluctuating climates, which sometimes caused large areas of unsuitable habitat. Given these multiple interacting factors, it is not surprising that we still poorly understand the detailed historical movements of most clades of organisms. Yet this very complexity makes Malesia a fascinating area to study, and tempts us with the promise that detecting shared patterns of movement and

Biotic Evolution and Environmental Change in Southeast Asia, eds D. J. Gower et al. Published by Cambridge University Press.

interaction in this region will help us better understand the historical assembly of local biota in general. Typical biogeographic questions in the region include:

- When and how did taxa cross the narrowing divide between Asia and Melanesia?
- Why does Wallace's line clearly demarcate faunas but not floras?
- What was the nature and effect of interaction when ecologically similar taxa from West Malesia and from East Malesia encountered each other?
- Why have some lineages diversified only in the east or the west of Malesia despite having been present throughout the region for long periods?

Answering such questions will require the application of methods of historical biogeographic analysis across many clades, with a synthetic comparison of results (Sanmartín et al. 2008). Methods of biogeographic inference employed in the region include comparison of taxon lists from different areas ('Q-mode analysis', e.g. Simberloff and Connor 1979, van Balgooy, 1987, van Welzen et al. 2005, and 'R-mode analysis', e.g. Holloway, 1998), super-imposition of range maps (e.g. Baker et al. 1998), panbiogeography (Heads, 2003) and studies of fossils (e.g. Truswell et al. 1984, Morley 2000). Phylogenies, both morphology-based (Turner 1996, Ridder-Numan 1998) and molecular (Scharaschkin and Doyle 2005, Brown et al. 2006b, Pfeil and Crisp 2008), have added hypotheses about the relationships among taxa, and more recently, estimates of the ages of ancestral taxa. Phylogenies have enabled the search for generalised, hierarchical patterns of connected areas (e.g. area cladograms inferred by Brooks Parsimony Analysis; Ruedi 1996, van Welzen et al. 2003), and the reconstruction of ancestral ranges by treating areas as pseudo-characters (e.g. Crisp et al. 2010). 'Event-based' analyses have added explicit consideration of the probability of the processes of dispersal, speciation and extinction. In particular, dispersal–vicariance analysis (DIVA; Ronquist 1997) is an accessible method and has been widely applied (e.g. Kreier and Schneider 2006, Jønsson et al. 2010). However, the Malesian region presents challenges to the assumptions of most existing analytical biogeography methods:

1. The terranes of the region are generally still approaching each other and accreting rather than fragmenting. Hence the area relationships themselves are better represented as a coalescing network rather than a branching tree (Fig 8.1). Approaches that attempt to reconcile trees (using cladistic or other methods) to find a common area tree may thus be imposing an incorrect underlying area hypothesis (Holloway 1998).
2. Because of the relatively small scale of the region, and possibly the existence of many 'island' species, with good dispersal ability, many taxa exhibit widespread (multi-terrane) distributions (e.g. van Welzen et al. 2005). This creates extensive polymorphism when treating areas as pseudo-characters, and most phylogenetic ancestral state reconstruction programs (the most widely

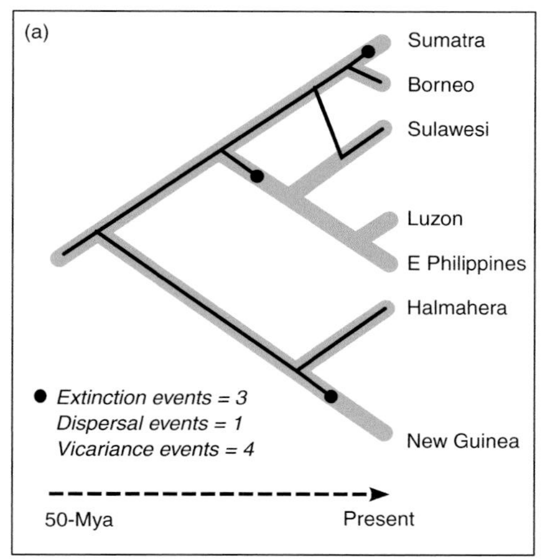

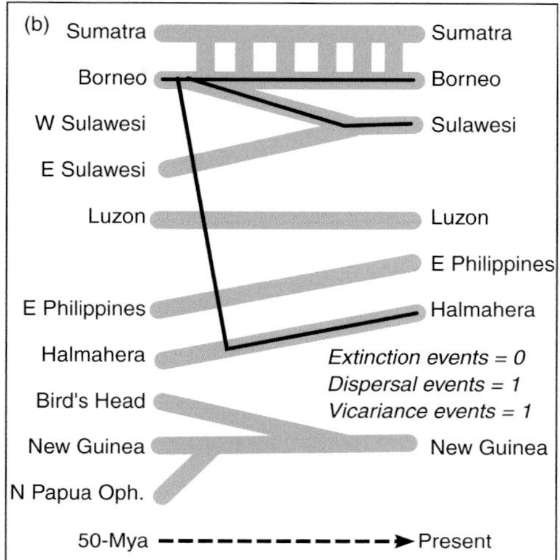

Figure 8.1 Schematic of hypotheses of land connectivity (grey) through time in Malesia, with associated organismal lineages. (a) A tree-like hypothesis of land vicariance that underlies tree-based analyses. (b) A connectivity-through-time network, based on terrane movement in Hall (2001). Clustering of the ends of the grey lines on the right side of (b) indicates spatial proximity; note similarity to Fig 9.26 in Morley (2000). The fine black lines represent the time–space trajectories of a currently widespread lineage (in Borneo, Sulawesi, Halmahera); solid circles indicate extinction events. For example, in (a), an initial vicariance event leaves descendants in the Halmahera–New Guinea area, and after the subsequent splitting of Halmahera and New Guinea, with associated lineage vicariance, the New Guinea lineage goes extinct. Note the extensive explanatory role for vicariance in (a), and its limited role in (b).

applied historical biogeography techniques; Sanmartín et al. 2008) are unable to accept polymorphisms (although this can be circumvented by recoding polymorphisms as combinations of independent present/absent characters, e.g. Hardy and Linder 2005).

3. The insular nature of Malesia means that dispersal, extinction and intra-island speciation will be the dominant processes structuring distributions (Keppel et al. 2009). Methods that assume vicariance to have a lower cost than dispersal (which include dispersal–vicariance analysis; Sanmartín et al. 2008, Kodandaramaiah, 2010) are likely to misinterpret the data.
4. Similarly, the changing distances between islands, and changing island areas, are not well represented by models with fixed costs (e.g. DIVA; Kodandaramaiah 2010).

Overall, reconstructing lineage movements in this complex area is unlikely to be successful without explicitly incorporating external information about temporal changes in area connectivity into our models. Recently, several stochastic modelling approaches have emerged that do incorporate and even test hypotheses about contemporary and historical land conformations. Sanmartín et al. (2008) applied Bayesian Markov chain Monte Carlo (MCMC) transition matrix methods to models of dispersal between island groups in the Canary Islands. Ree and Smith (2008) applied a maximum likelihood inference method (Ree et al. 2005) to reconstruct ancestral ranges for *Psychotria* in Hawaii. These methods have great promise in the Malesian region, for which there are well-developed hypotheses about terrane history (e.g., Hall 2001, Hall this volume, Chapter 3).

In this chapter, we apply the maximum likelihood method of Ree and Smith (2008), implemented in the program LAGRANGE, to an example case of reconstructing the ancestral distributions of *Rhododendron* section *Vireya*. We also introduce a new method for ancestral area reconstruction based on simulation of ranges on a given phylogeny (SHIBA). We compare both methods with the results of dispersal–vicariance analysis (DIVA) and discuss the benefits and drawbacks of the three methods. Finally, we highlight the critical need for more data for solving some of the most fascinating problems of Malesian historical biogeography.

8.1.1 Incorporating landscape history into historical biogeography inference

Some properties of phylogenetic methods that should improve our ability to correctly infer historical biogeography in complex regions like Southeast Asia are:

1. Incorporating empirical constraints on dispersal implied by the spatial configuration of areas, guided by the general principle that the dispersal rate between two areas should be inversely proportional to the distance or strength of the barrier between them, e.g. open water, mountain range or other inhospitable tract (e.g. MacArthur and Wilson 1967). This becomes increasingly important as the number of areas becomes large, because more areas means more complexity in spatial relationships (Ree and Sanmartín 2009).
2. Restricting inferences to ancestral ranges that are biologically plausible, for example those that form connected networks of possible gene flow. The rationale is that a species is unlikely to remain cohesive if its range is fragmented or disjunct, with sub-ranges that are disrupted by intervening areas in which the species is absent (Ree and Smith 2008). Ranges should therefore be 'contiguous', for example so that each area of a species' range should be connected to at least one other by a possible migration route. This spatial configuration is important when enumerating scenarios of range subdivision and inheritance at speciation events (see below), especially in the face of increasing geographic complexity as the number of areas becomes large. Species with disjunct ranges do occur in Malesia (van Welzen et al. 2005), but are rare.

3. Incorporating temporal information relevant to probabilities of range evolution through time. For example, the dynamics of area movement and accretion modelled by geologists could be linked to estimates of clade ages and lineage divergence times obtained from molecular clock analysis. To the extent that clades can be independently dated to absolute time, it is desirable to use this information in order to improve inferences of geographic ancestry (Ree et al. 2005), while acknowledging that geological models are also hypotheses, with varying degrees of error.
4. Along similar lines, including temporal information relevant to constraints on ancestral ranges. Examples of this include fossils constituting positive evidence that a lineage was present in a certain area at a particular time (Ree et al. 2005), or geological evidence of areas being uninhabitable (or underwater) at particular times (Ree and Smith 2008). The objective is to reject, or down-weight, historical inferences that are inconsistent with the evidence, for example, an ancestral range that includes an area before it exists, or does not include an area at a time when fossils place the lineage there.
5. Incorporating our understanding of how island area influences lineage survival. The positive relationship between island size and local species survival, part of the Theory of Island Biogeography (MacArthur and Wilson 1967), is a well-supported cornerstone of contemporary biogeography, and by applying it to historical hypotheses of lineage spread we are likely to greatly increase the accuracy of our reconstructions.

Thus, methods should ideally incorporate external hypotheses about land connectivity and size through time. At a minimum, they should include contemporary probabilities of dispersal events between different areas. Models can then be developed that allow the evolution of ranges at ancestral nodes of a branching phylogeny. Such models require an explicit statement of the spatial conformation of speciation events ('how do daughter species inherit spatial ranges?'). Finally an analytical engine can then optimise the fit of the model to the input data: phylogeny, current distribution, historical fossil locations, changing spatial conformations and even variation in the physical area of different spatial units. Uncertainty in phylogenetic, geological or event/process hypotheses can be assessed either through sensitivity analysis (i.e. re-running analyses with different combinations of parameters) or through incorporating informative prior probabilities in a Bayesian framework (e.g. Sanmartín et al. 2008). In the following examples we illustrate how desirable properties 1 and 2, above, can be addressed using LAGRANGE, and how 1 and 3–5 can be addressed using SHIBA.

LAGRANGE

This biogeographic inference method was inspired by perceived limitations of DIVA largely relating to its reliance on parsimony (Ree et al. 2005, Ree and Smith

2008). LAGRANGE retains a common conceptual framework with DIVA, in which geographic areas are discrete, species' ranges are coded as binary presence–absence data, and ranges evolve along phylogenetic branches by lineage movement (dispersal from one area to another, causing range expansion) and local extinction (extirpation within an area, causing range contraction). The primary objective also remains the same, namely inference of biogeographic history by fitting a model of range evolution, including geographic scenarios of speciation, to an observed phylogenetic tree with species ranges arrayed at its tips. With DIVA, this is achieved by finding the most parsimonious set of ancestral geographic ranges that minimises the number of dispersal and extinction events required to explain the data on the tree. With LAGRANGE, it involves optimising the likelihood of stochastic dispersal and local extinction along branches, and scenarios of range inheritance at nodes where lineages diverge. The primary differences of LAGRANGE from DIVA are: (1) probabilities of dispersal and local extinction are functions of time, and calculated analytically from a rate matrix; (2) geographic modes of speciation include a wider range of scenarios, including persistence of widespread ancestral ranges through cladogenesis events; and (3) flexible constraints reflecting spatial relationships, habitability, etc. of areas may be imposed on geographic ranges and rate parameters.

SHIBA

Another approach to estimating ancestral ranges is to simulate lineage movement in a discrete spatial and temporal model. During each time slice a species' range can expand via dispersal or contract via extinction, in an externally specified landscape. Where a phylogenetic branching event occurs within a time slice, the daughter taxa can be distributed to represent the products of either sympatric or allopatric speciation. This approach mirrors the 'experimental biogeography' (Posadas et al. 2006) of Colwell and Winkler (1984) and Haydon et al. (1994), but rather than generating simulated phylogenies, it takes a phylogeny as input and determines the probabilities of ancestral ranges and lineage movement. A necessary condition is that estimates for the ages of branching events in the phylogeny exist so that they can be anchored in the corresponding dynamic land-area model. To capture additional aspects of biogeographic reality, the extinction probability in an area can be modelled as a function of the physical area of that unit, and the dispersal probability a negative function of the distance to be crossed (Fig 8.2). This incorporates the fundamental components of the Island Biogeography Theory of MacArthur and Wilson (1967), and has led us to dub the software application that implements these methods SHIBA: 'simulated historical island biogeography analysis'. The software is open source, and available for download at http://phylodiversity.net/shiba/; it is a command-line C program, and can be compiled on any operating system. The input data are: (1) A distance matrix for spatial distance between x areas, for each of t time slices (i.e. of size $x \times x \times t$). A distance of zero between two areas implies they are directly connected. (2) A matrix of physical areas for each unit area, of size $x \times t$. (3) A list of n taxa, and a chronogram; for example a hypothesis for the phylogenetic relationships among the n taxa (with L edges), with node ages set to represent actual times of divergence. This chronogram is 'sliced' into t slices, in each of which a particular edge either branches (speciates into two or more new taxa) or continues as the same taxon. (4) An $L \times x$ presence/absence matrix of contemporary distribution for n taxa, and historical distribution of any fossils on the L edges. See Appendix B for an example of input data. The range evolution algorithm for each run is: (1) Choose a starting area or areas for the stem of the phylogeny, either at random or pre-determined. (2) Determine whether the lineage disperses to each of the other areas during the time slice, based on the distance between the existing location and the new one (Fig 8.2). A vagility parameter is 'tuned' during the simulation to match the total number of occurrences on islands with that observed (e.g. too low a dispersal rate will cause many islands to end up unoccupied). (3) Determine if the population in the original area goes extinct, based on the physical area of the site (Fig 8.2). While there has been some discussion about the actual form of the extinction rate versus area curve (Gilpin and Diamond 1976), we chose a model of *survival* = log *area* as the simplest to parameterise.

The speciation process model is similar to that employed in LAGRANGE: if a lineage is present in only one area, the speciation is sympatric (i.e. 'duplication'). If

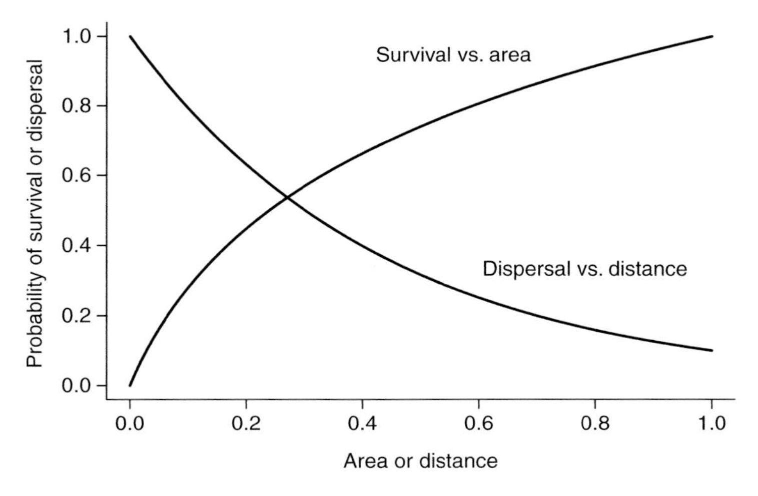

Figure 8.2 General form of the survival and dispersal curves in SHIBA: *survival* = $S \times$ log *area* and *dispersal* = $e^{-1 \times D \times distance}$, where S and D are constants (here set at 1.0). Areas and distances are expressed as proportions of maximum unit area and maximum interterrane distance.

the parent lineage is present in two or more areas, one is chosen randomly as the location of one daughter lineage, while populations in the other areas become the other daughter species (i.e. allopatric speciation). This is also sometimes described as a daughter species branching off of a continuing parent species.

The simulation is re-run many times. A run is terminated prematurely if (1) any lineage dies out in all areas before reaching the present, or (2) if a particular lineage does not pass though a specific fossil location in the specified time slice. If all lineages reach the present, the distribution of taxa is compared to the observed distribution. Only if the simulated distribution is the same as the observed distribution is a run kept as a 'success', and the ranges of its ancestral taxa recorded. After a pre-specified number of successes (r), the simulation terminates and histograms of locations of ancestral taxon locations are reported. These histograms can be described thus: 'A lineage L_p at time t_q occurred h_0 times out of r successful runs in area x_0, h_1 times in area x_1, h_3 times in area x_3, etc.' In this way, the probabilities of ancestral ranges and movements that can lead to the observed distribution are determined, given the model structure and parameters. Indeed, these simulations provide a probabilistic means to reject hypotheses; for example, if less than 925 runs placed the root taxon in an area A, we could with confidence say that the model does not support the historical scenario that the origin of the clade was in A.

8.2 An example: *Rhododendron* section *Vireya* in Malesia

8.2.1 The taxa

To demonstrate the application of these biogeographic methods in the Malesian region, we searched published studies for a clade of plants having species in both West and East Malesia, and a well-sampled molecular phylogeny with well-documented range data. The most suitable we could find was a study of Vireya rhododendrons by Brown et al. (2006b), which had detailed range maps, but included only 65 out of an estimated 300 species. *Vireya* is a primarily Malesian clade, with species easily recognised by their leaf and twig scales, and its species are much prized by gardeners. Brown et al. (2006b) mapped the 63 species (plus two outgroup species) onto 20 contemporary areas of endemism and constructed a molecular phylogeny from chloroplast markers (*trnT–trnL* and *psbA–trnH* intergenic spacer regions). They found that *Vireya* was divided into two major clades: one comprising species of Western and Central Malesia and one comprising species of New Guinea and further east. Using paralogy-free subtree analysis (Nelson and Ladiges 1996), Brown et al. explored the cladistic connections between areas, based on the *Vireya* phylogeny, and equivocally concluded that either (1) the clade is old, with *Vireya* having rafted north from Gondwanaland on the Australian continent, entering Malesia from the east, while sect. *Rhododendron* arrived from the west via India, or (2) the clade is much younger and only arrived from the west, via India, with the other *Rhododendron* clades (cf. similar hypothesis of Irving and Hebda 1993). Hence the outstanding questions in this example of *Vireya* in Malesia are: (1) How old is the clade? and (2) Did the clade diversify from west to east or from east to west?

8.2.2 The landscape

Increasingly detailed models of the complex tectonic history of the region (Hall 1998, 2001, 2009, this volume, Chapter 3) contribute much to our biogeographic understanding. Western Indonesia, on the Sunda shelf, has been part of the Eurasian mainland, and at equatorial latitudes, at least since the mid-Cretaceous (Morley 2000). The Australian plate broke away from Antarctica at *c.* 60 Ma and has approached the Asian plates from the South, only taking up its current conformation at *c.* 10 Ma. The islands of Wallacea are terranes that have variously been split from Sundaland (West Sulawesi), rafted in from the West Pacific (East Sulawesi, East Halmahera), arrived as outliers of the Australian plate (Seram, Buton), been extruded (West Halmahera, Sumbawa), or been uplifted (many small islands; Hall 2001). New Guinea is a complex assembly of Australian plate, recently accreted West Pacific terranes and a young, active centre.

For the SHIBA analysis presented here, which incorporates changing landscape conformations over time, we used the maps of Hall (2001, Figs 2–5), which present a hypothesis for terrane position and extent at 55, 45, 35, 25, 15 and 5 Ma. We digitised each image and computed (1) terrane area and (2) geographical distance among each pair of terranes, for each time period. While many separate terranes might be recognised from Hall's maps (the SHIBA software distribution includes data for 12), in this example we modelled four areas: Sundaland, Western Sulawesi, Luzon (North Philippines) and Central New Guinea, which are the current day locations of the taxa used in the SHIBA analysis. There is no theoretical limit to including more terranes, which might have acted as historical stepping-stones, but increasing the number of areas decreases the rate of finding solutions and slows the analyses. Values for distance and area were interpolated linearly for one time slice between each age of the landscape in the seven maps to give 12 time periods. Because we do not know the above-water land area for the terranes in the past, we made the assumption that terrane area is an appropriate substitute for land area.

8.2.3 LAGRANGE analysis

Rationale and methods

Our objective in revisiting *Vireya* with LAGRANGE was to explore the utility of its dispersal–extinction–cladogenesis (DEC) model in making historical biogeographic inferences in a region of substantial spatial complexity. A major challenge posed by

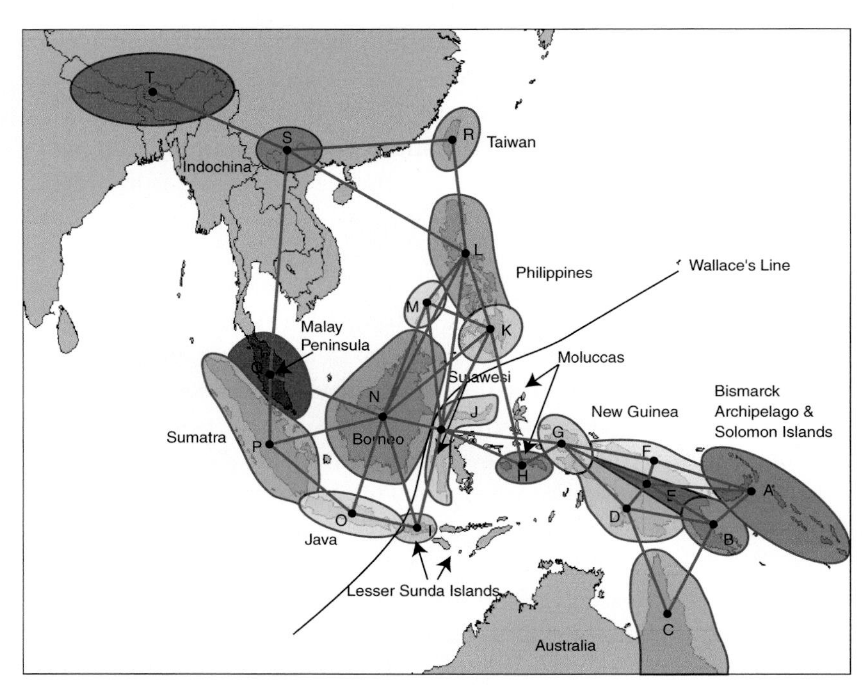

Figure 8.3 Areas of endemism for *Vireya* defined by Brown et al. (2006b), showing adjacency connections used to constrain dispersal routes in reanalysis using LAGRANGE. See plate section for colour version.

the $n = 20$ areas of endemism defined by Brown et al. (2006b) stems from the number of geographic ranges theoretically possible, $2^n = 1\ 048\ 576$, if all subsets of areas are considered valid ranges. Computing probabilities of ancestor-descendant range evolution for a phylogenetic branch requires integration of a $2^n \times 2^n$ rate matrix, meaning that for *Vireya*, the size of this matrix makes computing this integral impractical using standard software tools. Fortunately, however, this is neither necessary nor desirable if we use the spatial structure of areas to simplify the model.

Reasoning that dispersal and gene flow are most likely between spatially proximate areas, we created an 'adjacency graph' of nearest-neighbour relationships for the 20 areas of endemism based on their present-day positions (Fig 8.3). This graph allowed us to identify subsets of areas forming disconnected components, which we excluded from consideration on the grounds that they represented disjunct and therefore biologically unrealistic geographic ranges of species. In addition, we excluded ranges greater in extent than observed in extant species, setting the maximum number of areas to three (see below). This reduced the dimensions of the rate matrix considerably, to a manageable $n = 405$. We also used the adjacency graph to constrain dispersal, such that its rate parameter d was constant and symmetrical between connected areas and zero elsewhere.

Some caveats of this analysis must be noted. One is that we assumed static area positions, meaning that the constraints described above were constant across all branches of the *Vireya* phylogeny. This is, of course, a suboptimal approach. Ideally, we would have incorporated knowledge of terrane dynamics during the Cenozoic (Hall 2001, this volume, Chapter 3) by constructing a time series of adjacency graphs reflecting change in dispersal opportunities (as with the SHIBA analysis). Other caveats relate to the geographic range data from Brown et al. (2006b). In general we took these at face value, without considering issues such as incomplete sampling of extant species. However, in this particular case, we also reduced the size of the largest ranges, i.e. those exceeding three areas, in order for LAGRANGE to achieve convergence when optimizing dispersal and extinction rates. This meant modifying the given ranges of *Rhododendron malayanum*, *R. javanicum*, *R. zollingeri*, *R. culminicola* and *R. zoelleri*, by reducing each to a single representative area shared by its closest relative. The necessity of this step is perhaps clarified by considering the size of the phylogeny (63 leaf nodes) relative to the size of the transition matrix (405 elements square). We suspect that a larger phylogeny would be required for convergence in parameter estimation with a larger transition matrix. We acknowledge that these caveats reduce the empirical value of this analysis, and therefore emphasise it here as more of an illustration of how adjacency graphs can be used in spatially complex regions.

We inferred the phylogeny of the 65 *Vireya* species treated in Brown et al. (2006b) using sequences of *psbA–trnH* and *trnT–trnL* from Brown et al. (2006a). We aligned the sequences using MUSCLE (Edgar 2004) and estimated a maximum-likelihood tree given the GTRCAT nucleotide model implemented in RAxML 7.2.0 (Stamatakis 2006, Stamatakis and Ott 2008), with substitution parameters partitioned by marker. An identical topology and similar branch lengths were obtained from PHYML with the TVM models selected by Brown et al. (2006a) using AIC. We converted the tree to a relative-time chronogram by nonparametric rate smoothing (use of penalised likelihood was precluded by the presence of several terminal branches with effectively zero length), as implemented in APE (Paradis 2004). We then estimated maximum-likelihood values for rates of dispersal (d) and local extinction (e), using LAGRANGE with a DEC model constrained by the adjacency graph as described above. Finally, we used these parameter values to reconstruct maximum-likelihood ancestral ranges at each internal node on the tree.

Results

The phylogeny estimated using RAxML was congruent with the results of Brown et al. (2006a), in recovering a ‘Euvireya’ clade, within which was nested a pair of sister clades corresponding to those labelled ‘Eastern Malesia’ and ‘Western and Middle Malesia,’ respectively (Brown et al. 2006a, Fig 1). The relative-time chronogram of this tree, annotated with ancestral ranges inferred using LAGRANGE, is shown in Fig 8.4. At the root node of *Vireya*, a northern ancestral range was inferred

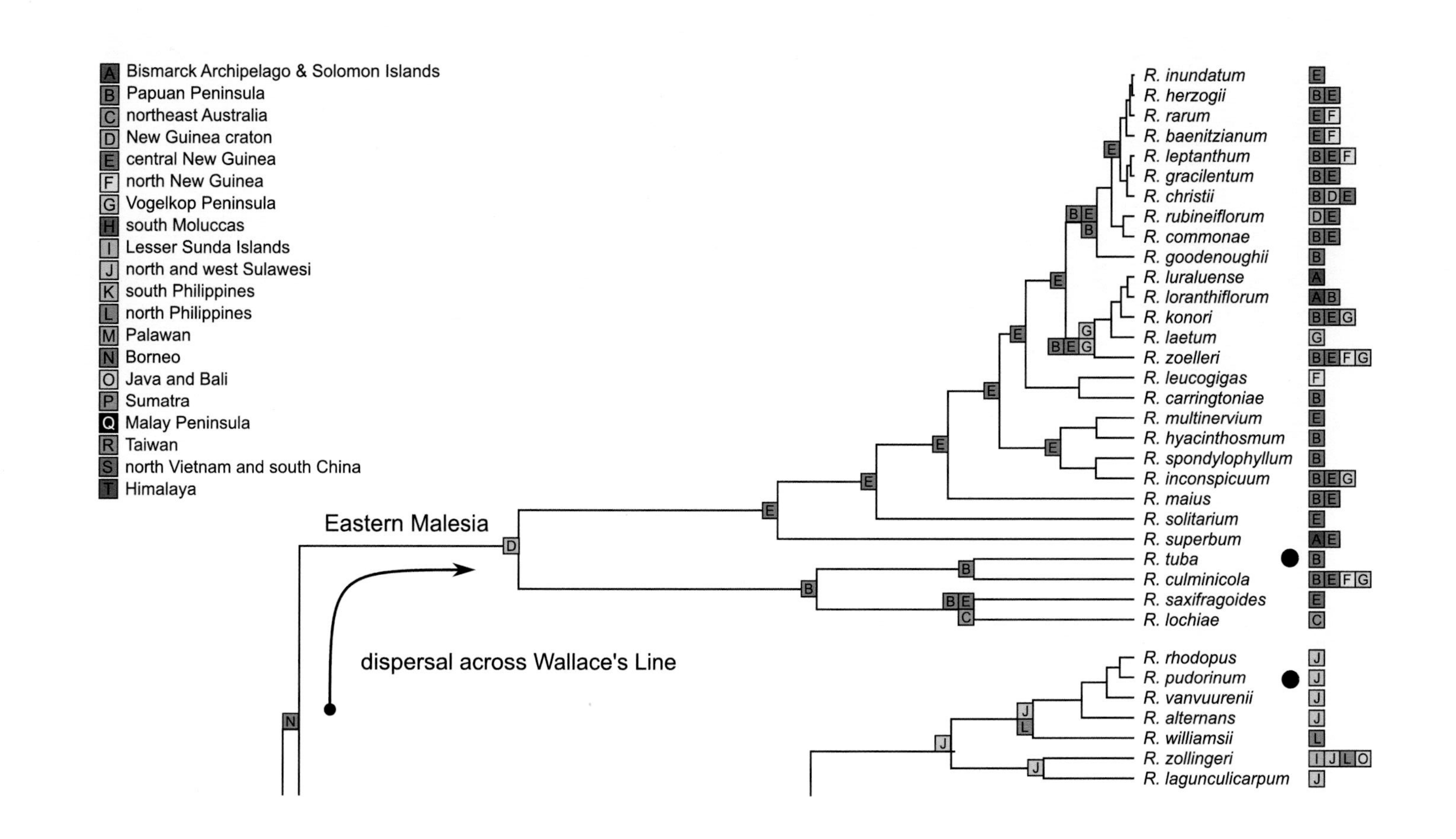

A Bismarck Archipelago & Solomon Islands
B Papuan Peninsula
C northeast Australia
D New Guinea craton
E central New Guinea
F north New Guinea
G Vogelkop Peninsula
H south Moluccas
I Lesser Sunda Islands
J north and west Sulawesi
K south Philippines
L north Philippines
M Palawan
N Borneo
O Java and Bali
P Sumatra
Q Malay Peninsula
R Taiwan
S north Vietnam and south China
T Himalaya
Eastern Malesia
dispersal across Wallace's Line
R. inundatum
R. herzogii
R. rarum
R. baenitzianum
R. leptanthum
R. gracilentum
R. christii
R. rubineiflorum
R. commonae
R. goodenoughii
R. luraluense
R. loranthiflorum
R. konori
R. laetum
R. zoelleri
R. leucogigas
R. carringtoniae
R. multinervium
R. hyacinthosmum
R. spondylophyllum
R. inconspicuum
R. maius
R. solitarium
R. superbum
R. tuba
R. culminicola
R. saxifragoides
R. lochiae
R. rhodopus
R. pudorinum
R. vanvuurenii
R. alternans
R. williamsii
R. zollingeri
R. lagunculicarpum

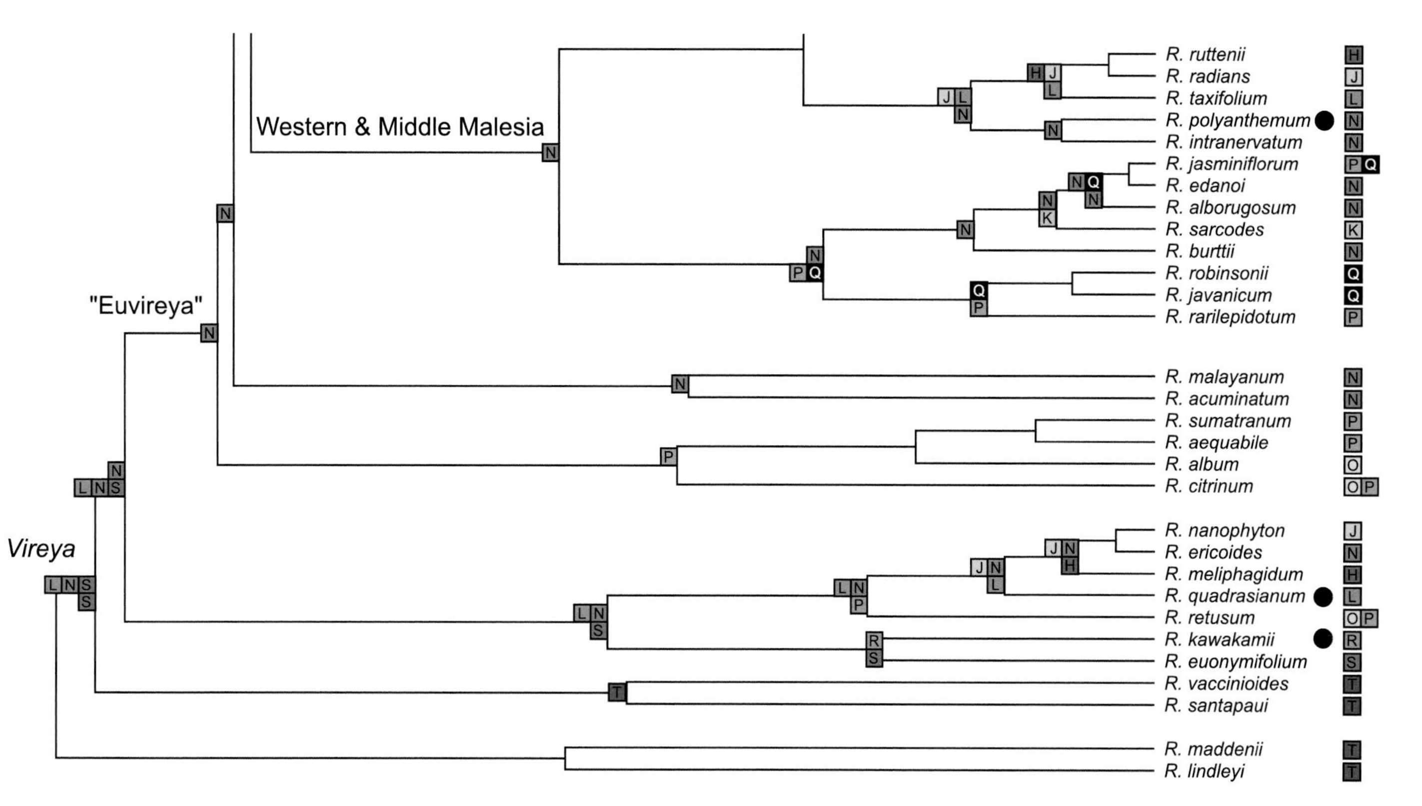

Figure 8.4 Results of LAGRANGE re-analysis of *Vireya* biogeography given 20 areas of endemism in southeast Asia, following Brown et al. (2006b). The relative-time chronogram was produced by applying nonparametric rate smoothing to a maximum likelihood (ML) molecular phylogeny. Nodes are annotated with ML estimates of ancestral geographic ranges. Rates of dispersal and local extinction were estimated subject to constraints implied by the adjacency graph shown in Figure 8.3 (see text for details). Colonisation of eastern Malesia is inferred as dispersal across Wallace's Line. Species used in the SHIBA runs are indicated with a dot. See plate section for colour version.

including southern China (S), Phillipines (L), and Borneo (N). From this origin, lineage movement was generally to the south and east, with Borneo serving as a hub. Migration across Wallace's Line was reconstructed along the stem branch of the 'Eastern Malesia' clade.

8.2.4 SHIBA analysis

Simulation details

Because SHIBA searches through the complete possibility space defined by a set of areas, times and lineage divisions, the rate of finding successful runs (i.e. giving observed taxa distributions) falls rapidly with increasing taxa and areas. Hence we chose a subset of *Vireya* taxa and Malesian areas to represent the biogeographic behaviour of the whole clade over the full set of terranes. We chose *R. kawakamii* and *R. quadrasianum* from the basal 'Pseudovireya' clade, *R. polyanthemum* and *R. pudorinum* from the 'western' clade and *R. tuba* from the 'eastern' clade (Fig 8.4). We chose areas: (1) Sundaland and mainland Asia, (2) West Sulawesi, (3) Luzon (North Philippines), and (4) Central New Guinea (Fig 8.5).

The utility of the SHIBA approach lies in its capacity to reconstruct lineage movements over shifting historical areas (also theoretically possible with LAGRANGE) and the intuitive interpretation of its output. We ran two analyses, one setting the root of the clade to 55 Ma, the other to 11 Ma, thus bracketing the likely range of ages for *Vireya* diversification in the region. We divided the chronogram of the five taxa (pruned from the chronogram used in the LAGRANGE run) into 12 'slices' (of 5 Ma and 1 Ma respectively) for which a land area and terrane distance model was prepared (see above). When they are known, fossils can be included in the model to constrain lineages to pass through particular places at certain times, but we were not aware of any *Vireya* fossils to use in this case. Survival parameter *S* was set to 1.0. Dispersal parameter *D* was initially set to 0.2, but is subsequently and repeatedly adjusted by the program to give simulated occurrence values (proportion of total possible areas × taxa occupied) that match the observed value (4 out of 20 = 0.2). A single area of the root taxon was chosen at random from the ancestral four areas. Random selections of multiple areas could also have been used as the root starting distribution, but in order to compare SHIBA results with other methods, and also to determine the most likely origin direction (east or west), we set the initial distribution to be a single area in these simulations. Simulations were run until 1000 successes were found.

Results

Both 'old root' and 'young root' simulations indicated that Sundaland and mainland Asia was most probably the ancestral area of root of *Vireya* (see bar chart at root node of Fig 8.5). In only 307 and 310 out of 1000 runs (in the 55 Ma and 11 Ma root-age analyses, respectively) was the ancestor of *R. tuba* already in New Guinea

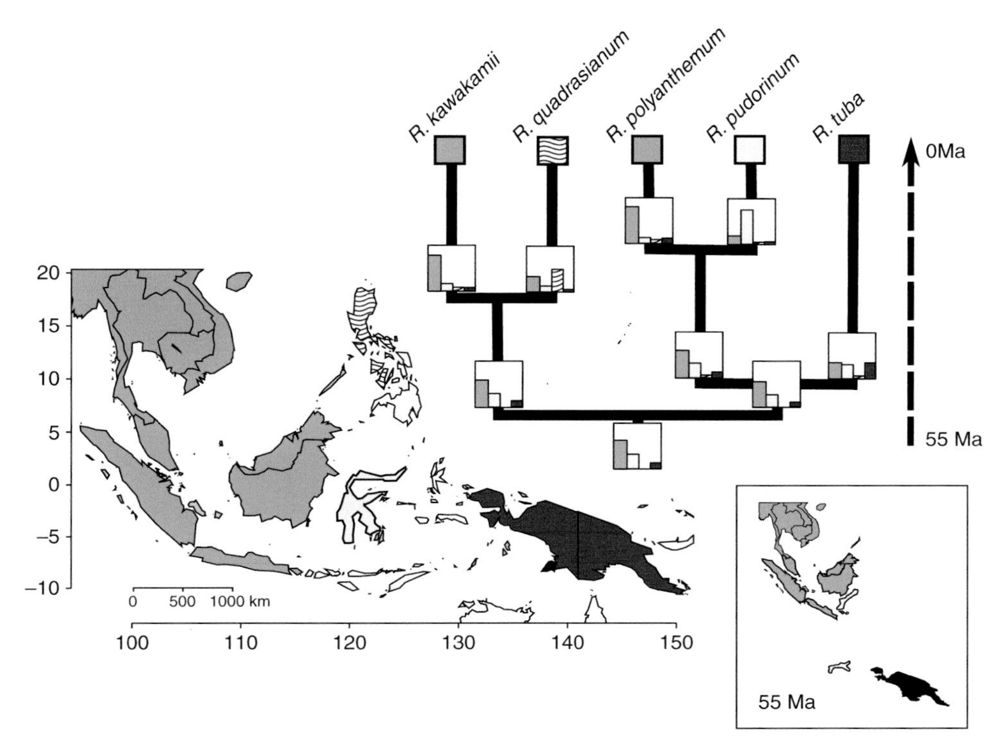

Figure 8.5 Results of SHIBA analysis of range evolution of five representative *Rhododendron* sect. *Vireya* species in Malesia, using four regions: mainland Southeast Asia and Sundaland (mid-grey), western Sulawesi (light grey), Northern Philippines (wavy lines), and mainland New Guinea (dark grey). Boxes on the branches of the phylogeny (pruned from Fig 8.4) indicate the proportion of runs (out of 1000) that the basal populations of a lineage were in a particular area. As an example, the direct ancestor of *R. quarasianum* was present in the Philippines in 478/1000 runs, and in Asia/Sundaland in 297/1000 runs. Inset figure shows approximate locations of areas at 55 Ma; the Luzon terrane was probably still submerged (Hall 2001).

at the time of its separation from the western lineage, indicating that it most likely dispersed into New Guinea soon after the origin of the eastern clade, followed by radiation within New Guinea (leading to *c.* 170 *Vireya* species). A different speciation scenario is indicated in the case of *R. pudorinum* and *R. polyanthemum*: the majority of runs show the ancestor of *R. pudorinum* to already be in Sulawesi and the ancestor of *R. polyanthemum* to be in Borneo (Sundaland) at the time of origination of these two lineages, indicating origination by allopatric speciation of their common ancestor.

One clear difference between the 'old root' and 'young root' simulations is the timing of arrival of the ancestors of *R. quadrasianum* in Luzon. Because the terrane was probably not above water at 55 Ma (Hall 2001), the Luzon physical area in our model was set to zero at 55 Ma, preventing this area being chosen as a starting area for the *Vireya* root. In the 11 Ma simulation, Luzon was near its present position and dry land, and we found that the *Vireya* root taxon occurred in Luzon in as many as 139 out of 1000 successes.

The success rate (number of runs giving the observed distribution divided by the total number of runs) is an indication of the likelihood of the solution, given the parameters chosen and the structure of the model. The success rate for a 55 Ma root was 6.71×10^{-7}, while that for a 11 Ma root was 6.49×10^{-7}. It is tempting to conclude that the older root age is slightly more probable, but these values are close and a more comprehensive sensitivity analysis would be needed to assess the confidence of any such conclusion.

8.2.5 DIVA analysis

Because DIVA is the most widely used phylogenetic biogeography tool, we conducted DIVA analyses that can be compared with the LAGRANGE and SHIBA analyses. For the LAGRANGE comparison, we used the distribution of the 63 taxa in 15 areas; collapsing areas N–T in Brown et al. (2006b) to a single 'Sundaland and mainland Asia' (N) area, because DIVA will only accept a maximum of 15 areas. We used the maximum likelihood phylogeny re-generated from the original sequence data, as described above. When the number of ancestral areas was not constrained, DIVA reconstructed a widespread *Vireya* root ancestor (most recent common ancestor of *R. santapaui* to *R. inundatum*; Fig 8.4), occurring from Mainland Asia to New Guinea (BCEFGHJLN or ABCEFGHJLN); these widespread solutions always have the lowest cost in DIVA in the absence of other constraints (Kodandaramaiah 2010). When we constrained the distribution to occur only in a maximum of three areas, the root taxon occurred only in 'Sundaland and mainland Asia' (area N). In this constrained model we also found that the distribution of the ancestor of the 'eastern' clade was only in Central New Guinea (area E), in agreement with both the LAGRANGE and SHIBA analyses. For the SHIBA comparison, with four areas and five taxa in DIVA, the *Vireya* root ancestor was again widespread in the unconstrained DIVA case, and limited to the 'Sundaland and Mainland Asia' area when the analysis was constrained to only two ancestral areas.

8.3 Discussion

8.3.1 *Vireya* example

All analyses, using the three different methods, inferred the ancestor of the *Vireya* clade to have occupied Western Malesia, and that the dispersal into New Guinea and Australia and subsequent radiation occurred relatively recently, thus supporting Hypothesis 2 ('Laurasian origin') of Brown et al. (2006b). Visual inspection of the phylogeny, showing the outgroup and the basally positioned taxa occurring in the west (Fig 8.4), plus the presence of fossil *Rhododendron* primarily in Laurasia (references in Brown et al. 2006b), intuitively suggest the same interpretation, so these results are not surprising. However, the extent to which this result is empirically robust depends on the ranges and phylogenetic positions of unsampled species. Because these represent a substantial proportion of the clade (only 63 out of *c.* 300 species were included), the ancestral ranges estimated here should be regarded as very preliminary in nature. If the unsampled taxa in a clade are missing at random, with respect to geographic area, we do not expect any systematic biases in our biogeographic reconstruction (i.e. the general issue of 'taxon sampling'; Ackerly 2000). It is, however, actually quite likely that there will be spatial structure in the unsampled taxa, caused by difficult access to parts of the total spatial range of the clade (e.g. poor sampling of Wallacea and West New Guinea).

Another question of interest relates to the age of *Vireya* and the eastern Malesian sub-clade. If the crown age of *Vireya* is old (e.g. Paleocene), and the relative branch lengths of our chronogram are correct (i.e. the eastern Malesian sub-clade is at least 50% of the age of the *Vireya* root), the most likely interpretation is a single long-distance dispersal event onto the New Guinea terranes at *c.* 35 Ma followed by a period of radiation in isolation. If the Vireya root age is young, dispersal into New Guinea would have involved shorter distances, and lineage diversification within New Guinea must then have occurred much faster. In both scenarios, it is notable that there has been no back-dispersal of New Guinea clade members into Western Malesia.

How might historical biogeographic analysis help resolve questions relating to clade age? Fossil-calibrated molecular clocks suggest that *Vireya* is relatively young: Milne (2004) estimated that subgenus *Rhododendron* is 46–32 Myr old, suggesting that the origin of Vireya is more recent. A more recent chronogram by Alex Twyford and James Richardson (pers. comm.), using the nuclear marker *RPB2*, and building on the work of Goetsch et al. (2005), dates the crown group age of *Vireya* at only *c.* 13 Ma. In the face of uncertainty about the statistical confidence and precision of these estimates, we suggest that biogeographic likelihoods are a potential source of relevant information. By this we mean comparing likelihoods of extant species ranges having evolved, given a history of area connections through time, and a set of alternative chronograms reflecting uncertainty in node ages. The rationale is that if the availability of dispersal routes has strongly influenced the movements of lineages over phylogenetic timescales, the maximum likelihood score of a LAGRANGE analysis and the success rate of a SHIBA run should be highest when the 'true' chronogram of a clade is overlaid on the land-connection history. For example, one could set the root age of a clade to be a free parameter (in LAGRANGE), or an axis along which to conduct sensitivity analyses (in SHIBA). This approach would lead to estimates of the root age (and ancestral range) that relate directly to biogeographic likelihoods. In this study, the difference between the success rate of the 55 Ma and 11 Ma root-age SHIBA runs, though small, demonstrates how, of the two levels, the former provided a slightly more likely solution; however, in this case we are not inclined to take this as strong evidence in favour of a Paleocene age for *Vireya*.

This idea in effect turns traditional biogeographic analysis on its head, and places weight on area history as a line of evidence relevant to making inferences about node ages and ancestral ranges. Naturally, this assumes that knowledge of area history can be obtained with greater confidence and precision than, say, divergence times. It also assumes that the 'true' area history has had a significant effect on the proliferation and movements of species in the clades of interest. The validity of both assumptions will be region- and clade-dependent. For Malesia, the generally accepted description of terrane movement has been relatively stable for more than a decade, indicating an increasing confidence in this hypothesis of area history. Similarly, while we still do not understand the details, we can be sure that the insular nature of Malesia has had a strong influence on species proliferation and movement.

A further possibility in the interaction of geological models with biogeographic models is that the latter could even be used to test the likelihood of the former. If confidence in a chronogram is high, and especially if there are fossils that can anchor particular ancestral lineages to particular areas, both LAGRANGE and SHIBA could be run over a range of geological models and the likelihood (or success rate) for each geological model used to indicate its validity.

8.3.2 Methodological approaches and future directions

While DIVA is an accessible and widely used tool, the uses of LAGRANGE and SHIBA described here demonstrate some important advantages, especially for the Malesian region. For example, neither method assumes vicariance to be the null model for widespread range inheritance, and both can model range evolution according to spatial and temporal constraints. They each in turn have particular strengths and weaknesses. LAGRANGE is more computationally efficient than SHIBA, and for that reason can handle larger numbers of areas and taxa. Neither method accounts for the phylogeny itself being the stochastic outcome of (possibly range-dependent) speciation and extinction processes (Ree and Sanmartín 2009). As a result, LAGRANGE analyses tend to infer artificially wide ancestral ranges and low local extinction rates, to condition on observed data (Lamm and Redelings 2009). SHIBA is slow relative to LAGRANGE because it attempts to search across all possible histories of range evolution. For example, a simulation of 1000 successes took *c.* 2.5 hours on a 3 Ghz 8-core Intel Xeon CPU with 16.2 GB

RAM; increasing simulation efficiency is a high priority for further development of SHIBA. This means that applying SHIBA may require the user to subsample taxa in the clade of interest, while being mindful of the issue of spatial bias in sampled taxa, as discussed above. On the other hand, SHIBA currently offers a more comprehensive framework for incorporating dynamic histories of land connection. Additionally, only SHIBA can currently incorporate variation in area size and its effect on extinction rates. These differences make LAGRANGE a better choice for reconstructing the biogeographic history of large clades in Malesia over the fairly recent past (when the islands' conformation was not greatly different than the present one), and SHIBA a better choice for smaller clades over a longer history of land movement.

Both methods use a similar algorithm for determining how daughter species are distributed spatially during vicariant speciation in a widespread species, invoking allopatric speciation with single areas for one daughter and remainder areas for the other (see Ree et al. 2005). These algorithms, while defensible as parsimonious solutions, do represent particular choices and as such may misrepresent the biology of speciation for some groups of organisms, and are open to criticism (e.g. Lamm and Redelings 2009). As is always the case with modelling complex systems, we would urge all users to be aware of the assumptions and simplifications involved. We intend in future to carry out a thorough assessment of the ability of SHIBA to infer biogeographic history in simulated test cases for which the history of lineage movement is fully known.

Finally, obtaining meaningful inferences from phylogenetic comparative methods will always depend on obtaining adequate numbers of representative taxa. To answer the most fundamental questions about biogeographic dynamics in Malesia, perhaps the greatest need is for more intensive sampling of both taxa and areas in molecular phylogenetic studies (Crisp et al. 1995, 2004). Despite the long history of botanical research in the Malesian region (Webb et al. 2010), few plant clades have well-sampled molecular phylogenies; the general requirement of fresh material to successfully extract DNA from plant species places a premium on new collections, especially from undersampled and remote regions. Several animal clades have been relatively well sampled (e.g. Evans et al. 2003, Honda et al. 2006), but we remain far from being able to recognise general patterns. More complete phylogenetic sampling will also allow analyses of historical biogeography to address key questions about disparities in species richness across the region. One kind of inference in particular need of methodological attention relates to links between range dynamics (the timing and direction of movement) and variation in regional rates of speciation and extinction (Ree and Sanmartín, 2009). In the interim, we suggest that process-based models for reconstructing historical biogeography in Malesia, tailored to the unique geological history of the region, will enable the most effective use of the data at hand.

Acknowledgements

We are grateful to A. Antonelli, L. Rüber and three anonymous reviewers for valuable comments, and to M. Donoghue and B. Moore for early discussions on modelling approaches. Webb was partially supported by a US NSF grant (DEB-1020868) during preparation of this chapter.

References

Ackerly, D. D. (2000). Taxon sampling, correlated evolution, and independent contrasts. *Evolution*, **54**, 1480–92.

Baker, W. J., Coode, M. J. E., Dransfield, J. et al. (1998). Patterns of distribution of Malesian vascular plants. In *Biogeography and Geological Evolution of SE Asia*, ed. R. Hall and J. D. Holloway. Leiden, The Netherlands: Backhuys Publishers, pp. 243–58.

Barker, N. P., Weston, P. H., Rutschmann, F. and Sauquet, H. (2007). Molecular dating of the 'Gondwanan' plant family Proteaceae is only partially congruent with the timing of the break-up of Gondwana. *Journal of Biogeography*, **34**, 2012–27.

Brown, G., Craven, L., Udovicic, F. and Ladiges, P. (2006a). Phylogeny of *Rhododendron* section *Vireya* (Ericaceae) based on two non-coding regions of cpDNA. *Plant Systematics and Evolution*, **257**, 57–93.

Brown, G. K., Nelson, G. and Ladiges, P. Y. (2006b). Historical biogeography of *Rhododendron* section *Vireya* and the Malesian Archipelago. *Journal of Biogeography*, **33**, 1929–44.

Colwell, R. K. and Winkler, D. W. (1984). A null model for null models in biogeography. In *Ecological Communities: Conceptual Issues and the Evidence*, ed. D. R. Strong, D. Simberloff, L. G. Abele and A. B. Thistle. Princeton, NJ: Princeton University Press, pp. 344–59.

Crisp, M. D., Linder, H. P. and Weston, P. H. (1995). Cladistic biogeography of plants in Australia and New Guinea: congruent pattern reveals two endemic tropical tracks. *Systematic Biology*, **44**, 457–73.

Crisp, M. D., Cook, L. and Steane, D. (2004). Radiation of the Australian flora: what can comparisons of molecular phylogenies across multiple taxa tell us about the evolution of diversity in present-day communities? *Proceedings of the Royal Society of London, Series B: Biological Sciences*, **359**, 1551–71.

Crisp, M. D., Isagi, Y., Kato, Y., Cook, L. G. and Bowman, D. M. J. S. (2010). *Livistona* palms in Australia: ancient relics or opportunistic immigrants? *Molecular Phylogenetics and Evolution*, **54**, 512–523.

Croizat, L. (1958). *Panbiogeography*. Caracas: Published by the author.

Edgar, R. C. (2004). MUSCLE: a multiple sequence alignment method with reduced time and space complexity. *BMC Bioinformatics*, **5**, 113.

Evans, B. J., Brown, R. M., McGuire, J. A. et al. (2003). Phylogenetics of fanged frogs: testing biogeographical hypotheses at the interface of the Asian and Australian faunal zones. *Systematic Biology*, **52**, 794–819.

Gilpin, M. E. and Diamond, J. M. (1976). Calculation of immigration and extinction curves from the species-area-distance relation. *Proceedings of the National Academy of Sciences of the USA*, **73**, 4130–4.

Goetsch, L., Eckert, A. J. and Hall, B. D. (2005). The molecular systematics of *Rhododendron* (Ericaceae): a phylogeny based upon RPB2 gene sequences. *Systematic Botany*, **30**, 616–26.

Hall, R. (1998). The plate tectonics of Cenozoic SE Asia and the distribution of land and sea. In *Biogeography and Geological Evolution of SE Asia*, ed. R. Hall, R. and J. D. Holloway. Leiden, The Netherlands: Backhuys Publishers, pp. 99–131.

Hall, R. (2001). Cenozoic reconstructions of SE Asia and the SW Pacific: changing patterns of land and sea. In *Faunal and Floral Migrations and Evolution in SE Asia- Australasia*, ed. I. Metcalfe, J. M. B. Smith, M. Morwood and I. D. Davidson. Lisse, The Netherlands: A. A. Balkerma (Swets and Zeitlinger Publishers), pp. 35–56.

Hall, R. (2009). Southeast Asia's changing palaeogeography. *Blumea*, **54**, 148–61.

Hardy, C. R. and Linder, H. P. (2005). Intraspecific variability and timing in ancestral ecology reconstruction: a test case from the Cape Flora. *Systematic Biology*, **54**, 299–316.

Haydon, D., Radtkey, R. R. and Pianka, E. R. (1994). Experimental biogeography: interactions between stochastic, historical, and ecological processes in a model archipelago. In *Species Diversity in Ecological Communities: Historical and Geographical Perspectives I*, ed. R. E. Ricklefs and D. Schluter. Chicago: University of Chicago Press, pp. 117–30.

Heads, M. (2003). Ericaceae in Malesia: vicariance biogeography, terrane tectonics and ecology. *Telopea*, **10**, 311–449.

Holloway, J. D. (1998). Geological signal and dispersal noise in two contrasting insect groups in the Indo-Australian tropics: R-mode analysis of pattern in Lepidoptera and cicadas. In *Biogeography and Geological Evolution of SE Asia*, ed. R. Hall and J. D. Holloway. Leiden, The Netherlands: Backhuys Publishers, pp. 291–314.

Honda, M., Ota, H., Murphy, R. W. and Hikida, T. (2006). Phylogeny and biogeography of water skinks of the genus *Tropidophorus* (Reptilia: Scincidae): a molecular approach. *Zoologica Scripta*, **35**, 85–95.

Irving, E. and Hebda, R. (1993). Concerning the origin and distribution of rhododendrons. *Journal of the American Rhododendron Society*, **47**, 139–62.

Jønsson, K. A., Bowie, R. C. K., Moyle, R. G. et al. (2010). Historical biogeography of an Indo-Pacific passerine bird family (Pachycephalidae): different colonization patterns in the Indonesian and Melanesian archipelagos. *Journal of Biogeography*, **37**, 245–57.

Keppel, G., Lowe, A. J. and Possingham, H. P. (2009). Changing perspectives on the biogeography of the tropical South Pacific: influences of dispersal, vicariance and extinction. *Journal of Biogeography*, **36**, 1035–1054.

Kodandaramaiah, U. (2010). Use of dispersal–vicariance analysis in biogeography: a critique. *Journal of Biogeography*, **37**, 3–11.

Kreier, H.-P. and Schneider, H. (2006). Phylogeny and biogeography of the staghorn fern genus *Platycerium* (Polypodiaceae, Polypodiidae).

American Journal of Botany, **93**, 217–25.

Ladiges, P. Y., Udovicic, F. and Nelson, G. (2003). Australian biogeographical connections and the phylogeny of large genera in the plant family Myrtaceae. *Journal of Biogeography*, **30**, 989–98.

Lamm, K. S. and Redelings, B. D. (2009). Reconstructing ancestral ranges in historical biogeography: properties and prospects. *Journal of Systematics and Evolution*, **47**, 369–82.

MacArthur, R. H., and Wilson, E. O. (1967). *The Theory of Island Biogeography*. Princeton, NJ: Princeton University Press.

Michaux, B. (1991). Distributional patterns and tectonic development in Indonesia: Wallace reinterpreted. *Australian Systematic Botany*, **4**, 25–36.

Milne, R. I. (2004). Phylogeny and biogeography of Rhododendron subsection *Pontica*, a group with a tertiary relict distribution. *Molecular Phylogenetics and Evolution*, **33**, 389–401.

Morley, R. J. (2000). *Origin and Evolution of Tropical Rainforests*. New York: Wiley.

Moyle, R. G. (2004). Calibration of molecular clocks and the biogeographic history of Crypteroniaceae. *Evolution*, **58**, 1871–3.

Nelson, G. and Ladiges, P. Y. (1996). Paralogy in cladistic biogeography and analysis of paralogy-free subtrees. *American Museum Novitates*, **3167**, 1–58.

Paradis, E. (2004). APE: analyses of phylogenetics and evolution in R language. *Bioinformatics*, **20**, 289–90.

Pfeil, B. E. and Crisp, M. D. (2008). The age and biogeography of citrus and the orange subfamily (Rutaceae: Aurantioideae) in Australasia and New Caledonia. *American Journal of Botany*, **95**, 1621–31.

Posadas, P., Crisci, J. V. and Katinas, L. (2006). Historical biogeography: a review of its basic concepts and critical issues. *Journal of Arid Environments*, **66**, 389–403.

Ree, R. H. and Sanmartín, I. (2009). Prospects and challenges for parametric models in historical biogeographical inference. *Journal of Biogeography*, **36**, 1211–20.

Ree, R. H. and Smith, S. A. (2008). Maximum-likelihood inference of geographic range evolution by dispersal, local extinction, and cladogenesis. *Systematic Biology*, **57**, 4–14.

Ree, R. H., Moore, B. R., Webb, C. O. and Donoghue, M. J. (2005). A likelihood framework for inferring the evolution of geographic range on phylogenetic trees. *Evolution*, **59**, 2299–311.

Richardson, J. E., Chatrou, L. W., Mols, J B., Erkens, R. H. J. and Pirie, M. D. (2004). Historical biogeography of two cosmopolitan families of flowering plants: Annonaceae and Rhamnaceae. *Philosophical Transactions of the Royal Society of London, Series B: Biological Sciences*, **359**, 1495–508.

Ridder-Numan, J. W. A. (1998). Historical biogeography of *Spatholobus* (Leguminosae- Papilionoideae) and allies in SE Asia. In *Biogeography and Geological Evolution of SE Asia*, ed. R. Hall and J. D. Holloway. Leiden, The Netherlands: Backhuys Publishers, pp. 259–77.

Ronquist, F. (1997). Dispersal–vicariance analysis: a new approach to the quantification of historical biogeography. *Systematic Biology*, **46**, 195–203.

Ruedi, M. (1996). Phylogenetic evolution and biogeography of southeast Asian shrews (genus *Crocidura*: Soricidae). *Biological Journal of the Linnean Society*, **58**, 197– 219.

Sanmartín, I., van der Mark, P. and Ronquist, F. (2008). Inferring dispersal: a Bayesian approach to phylogeny-based island biogeography, with special reference to the Canary Islands. *Journal of Biogeography*, **35**, 428–49.

Scharaschkin, T. and Doyle, J. A. (2005). Phylogeny and historical biogeography of *Anaxagorea* (Annonaceae) using morphology and non-coding chloroplast sequence data. *Systematic Botany*, **30**, 712–35.

Simberloff, D. and Connor, E. F. (1979). Q-mode and R-mode analyses of biogeographic distributions: null hypotheses based on random colonization. In *Contemporary Quantitative Ecology and Related Ecometrics*, ed. G. P. Patil and M. L. Rosenzweig. Fairland, MD: International Co-operative Publishing House, pp. 123–38.

Stamatakis, A. (2006). RAxML-VI-HPC: maximum likelihood-based phylogenetic analyses with thousands of taxa and mixed models. *Bioinformatics*, **22**, 2688–90.

Stamatakis, A. and Ott, M. (2008). Efficient computation of the phylogenetic likelihood function on multi-gene alignments and multi-core architectures. *Philosophical Transactions of the Royal Society B: Biological Sciences*, **363**, 3977–84.

Truswell, E. M., Kershaw, A. P. and Sluiter, I. R. (1984). The Australian-South-east Asian connection: evidence from the paleobotanical record. In *Biogeographical Evolution of the Malay Archipelago*, ed. T. C. Whitmore. Oxford: Clarendon Press, pp. 32–49.

Turner, H. (1996). Sapindaceae and the biogeography of Eastern Australia. *Australian Systematic Botany*, **8**, 133–67.

van Balgooy, M. M. J. (1987). A plant geographical analysis of Sulawesi. In *Biogeographical Evolution of the Malay Archipelago*, ed.T. C. Whitmore. Oxford: Clarendon Press, pp. 94–102

van Welzen, P. C., Turner, H. and Hovenkamp, P. (2003). Historical biogeography of Southeast Asia and the West Pacific, or the generality of unrooted area networks as historical biogeographic hypotheses. *Journal of Biogeography*, **30**, 181–92.

van Welzen, P. C., Slik, J. W. F. and Alahuhta, J. (2005). Plant distribution patterns and plate tectonics in Malesia. *Biologiske skrifter*, **55**, 199–217.

Wallace, A. R. (1869). *The Malay Archipelago*. London: Macmillan and Co.

Webb, C. O., Slik, J. W. F. and Triono, T. (2010). Biodiversity inventory and informatics in Southeast Asia. *Biodiversity and Conservation*, **19**, 955–972.

Appendix A

Input data for Lagrange

The following indicates the data taken as input for the LAGRANGE analysis. Full data can be found via the Lagrange page at http://www.reelab.net.

```
PHYLOGENY

   ((Rhododendron_santapaui:0.4837224249,Rhododendron_vaccinioides:0.4837224249):0.481
   376557,(((Rhododendron_euonymifolium:0.2488067527,Rhododendron_kawakamii:0.24880675
   27):0.2524890665,(Rhododendron_retusum:0.2629986131,(Rhododendron_quadrasianum:0.13
   70131223,(Rhododendron_meliphagidum:0.07023346485, ... (63 taxa total)

DISTRIBUTION

   63 20
   Rhododendron_santapaui             00000000000000000001
   Rhododendron_vaccinioides          00000000000000000001
   Rhododendron_euonymifolium         00000000000000000010
   Rhododendron_kawakamii             00000000000000000100
   Rhododendron_retusum               00000000000000110000
   Rhododendron_quadrasianum          00000000000100000000
   ... (63 rows total)

AREA NAMES

   A B C D E F G H I J K L M N O P Q R S T

RATE MATRIX

   1 1 0 0 1 1 0 0 0 0 0 0 0 0 0 0 0 0 0 0
   1 1 1 1 1 0 0 0 0 0 0 0 0 0 0 0 0 0 0 0
   0 1 1 1 0 0 0 0 0 0 0 0 0 0 0 0 0 0 0 0
   0 1 1 1 1 0 1 0 1 0 0 0 0 0 0 0 0 0 0 0
   1 1 0 1 1 1 1 0 0 0 0 0 0 0 0 0 0 0 0 0
   1 0 0 0 1 1 1 0 0 0 0 0 0 0 0 0 0 0 0 0
   ... (20 rows total)

INCLUDED DISTIBUTIONS

   10000000000000000000,11000000000000000000,11100000000000000000,
   11010000000000000000,11001000000000000000,11000100000000000000,
   10011000000000000000,10001000000000000000, ... (159 items total)
```

Appendix B

Input data for SHIBA

The following indicates the data taken as input for a set of SHIBA runs, though is not the input file itself. A '#' indicates a comment that is not involved in the computation. The data are: no. time slices, no. areas, matrix of area × time, distances matrices for each time, no. taxa, phylogeny with branch lengths, extant distribution and optional fossil constraints (e.g. fossil indicates that edge n_2 passed through area s_2), various parameters. A few other operational parameters (number of runs, stopping rules, etc.) are not shown.

```
TIME: 12  # t1: 60-55; t2: -50; t3: -45; t4: -40; t5: -35; t6: -30;
          # t7: -25; t8: -20; t9: -15; t10: -10; t11: -5; t12: 5-0 Ma


SPACE: 4  # s1: Borneo; s2: W. Sulawesi; s3: C. New Guinea; s4: Luzon


LAND AREA (arbitrary units):
           t1  t2  t3  t4  t5  t6  t7  t8  t9 t10 t11 t12
           ==  ==  ==  ==  ==  ==  ==  ==  ==  ==  ==  ==
     s1 :  10  10  10  10  10  10  10  10  10  10  10  10
     s2 :   3   3   3   3   3   3   3   3   3   3   3   3
     s3 :  12  12  12  12  12  12  12  12  12  12  12  12
     s4 :   0   0   3   3   3   3   3   3   3   3   3   3


DISTANCE BETWEEN LAND UNITS:
   TIME = t1:                     TIME = t2:
             s1   s2   s3   s4              s1   s2   s3   s4
           ==== ==== ==== ====            ==== ==== ==== ====
      s1 :    0  773 2690   na      s1 :     0  623 2546   na
      s2 :         0 1965   na      s2 :          0 1941   na
      s3 :              0   na      s3 :               0   na
      s4 :                  na      s4 :                   na

   TIME = t3...t11                TIME = t12:
                                            s1   s2   s3   s4
                                          ==== ==== ==== ====
           ...                      s1 :     0  591 1965 1173
                                    s2 :          0 1387 1245
                                    s3 :               0 1948
                                    s4 :                    0

TAXA: 5 # x1 = tuba, x2 = pudor., x3 = polya., x4 = quadr., x5 = kawak.


PHYLOGENY (branch lengths in units of time):
   ( ( x1:9, ( x2:5 , x3:5 )n3:4 )n2:2 , ( x4:6 , x4:6 )n4:5 )n1:1 ;


EXTANT DISTRIBUTION AND FOSSIL CONSTRAINTS:
                 n1 n2 x1 n3 x2 x3 n4 x4 x5
                 == == == == == == == == ==
           s1 :   .  .  0  .  0  1  .  0  1
           s2 :   .  .  0  .  1  0  .  0  0
           s3 :   .  .  1  .  0  0  .  0  0
           s4 :   .  .  0  .  0  0  .  1  0


NUMBER OF START AREAS: 1
POSSIBLE START AREAS: s1 OR s2 OR s3 OR s4
PROBABILITY OF SURVIVAL 1.0
PROBABILITY OF DISPERSAL 0.2
```

9

Biodiversity hotspots, evolution and coral reef biogeography: a review

David R. Bellwood, Willem Renema and Brian R. Rosen

9.1 Introduction

It is widely acknowledged that the Southeast Asian region is an area of outstanding biological and geological interest. It is an area of exceptional biodiversity, which contains both marine and terrestrial biodiversity hotspots (Myers et al. 2000, Roberts et al. 2002). The convoluted and still highly active tectonic history of the region has resulted in a complex archipelago characterised by highly localised terrestrial faunas with large numbers of endemic species (Hall 2002, Sodhi et al. 2004). Terrestrial hotspots are defined as areas of exceptional endemism (usually combined with some measure of environmental damage) (Myers et al. 2000). This focus on endemism has resulted in the identification of three small regional terrestrial hotspots within the Indo-Australian Archipelago (IAA). The marine system stands in marked contrast to the terrestrial pattern, with the IAA sitting in the middle of a single massive global marine hotspot. This feature dominates patterns of tropical marine biodiversity and spans over two thirds of the world's tropical equatorial oceans. This pattern is observed in most tropical shallow marine taxa (see review of Hoeksema 2007). Although the exact location of the most diverse area differs between taxa, in all cases, the IAA contains the hottest part of the world's largest biodiversity hotspot (Fig 9.1).

Biotic Evolution and Environmental Change in Southeast Asia, eds D. J. Gower et al. Published by Cambridge University Press.

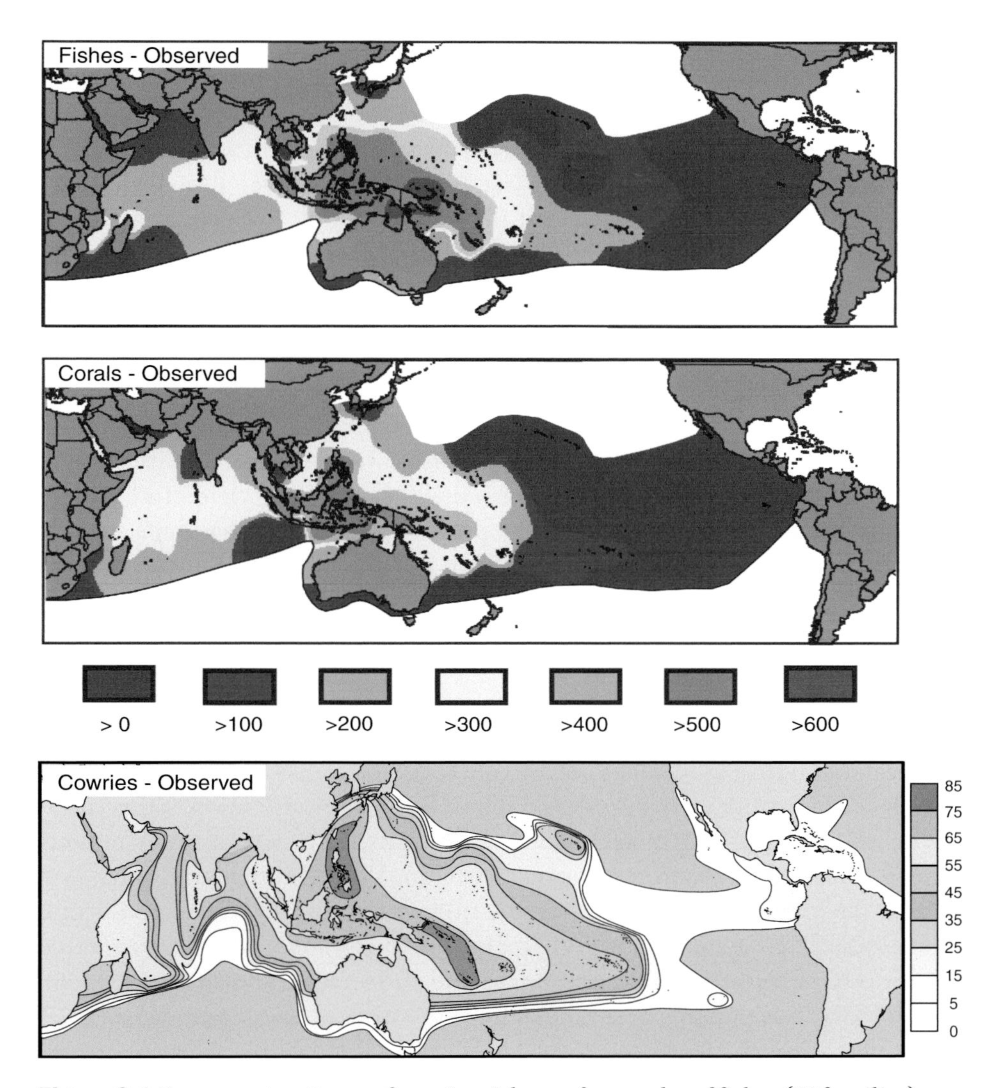

Figure 9.1 Congruent patterns of species richness for coral reef fishes (13 families), corals and cowries, showing the distinctive 'bulls-eye' biodiversity 'hotspot' in the Indo-Australian Archipelago (after Bellwood et al. 2005, Bellwood and Meyer 2009a). See plate section for colour version.

This area has been known by a wide variety of names reflecting political, geological, geographic and/or taxonomic boundaries. In an earlier review, Hoeksema (2007) lists 17 different terms for this exceptional region, each of them with specific advantages and limitations. The term Indo-Australian Archipelago was selected herein. It follows Weber and de Beaufort (1913). The term minimises colonial descriptors, has a broad geographic spread and emphasises the key characteristic of this region: it is the world's most spectacular archipelago. It is the latter term that is critical in our selection. The Southeast Asian Archipelago would be a better option but to avoid yet another term, IAA has been retained throughout to facilitate comparisons with previous studies (e.g. Renema et al. 2008). Broadly defined as the marine region between 15° N and 15° S and 100–150° E, it encompasses all of the previously recognised regions of high biodiversity. Regardless of the taxon, data or analytical approach, the IAA hotspot is a consistent feature of marine biogeographic studies.

The exceptional terrestrial richness of this area has been known since the days of Darwin and Wallace, while in the marine realm, Bleeker, Döderlein, Gerth, Rumphius, Valenciennes and Weber (among others) produced an ever-expanding catalogue of taxa from the seventeenth to the early twentieth century, attesting to the exceptional diversity of this region. Subsequent expeditions yielded additional new species, and continue to do so to the present. Only in a few groups such as fishes, corals and gastropods have there been any indications of an asymptote, and even here this is restricted to the most easily collected shallow-water families. Small gobies and deep-sea fishes are as poorly known as their invertebrate counterparts in these regions. However, the data in several groups have now reached a level where their biogeographic patterns are open to critical and quantitative evaluation.

Our understanding of tropical marine biogeography has developed relatively slowly. Based on earlier work by others, Ekman (1953) drew attention to the 'Indo-Malayan' region as a 'centre and focus' of diversity, from which other regions around it 'recruited the main contingent of its fauna'. This is the classic 'bull's-eye' pattern of the IAA hot spot, which was first fully quantified by Wells (1954) using 'isopangeneric' lines (contours of coral generic diversity). Within a few years this pattern was to be reported in numerous marine and terrestrial taxa (Stehli 1968), although today, it is perhaps best known from the work of Stehli and Wells (1971) based on coral genera.

The sheer geographic size and taxonomic extent of the pattern captures the imagination and demands an explanation. However, progress has been slow. One of the immediate problems is that the plethora of hypotheses that have been proposed to explain the high diversity of species in the tropics (e.g. Roy et al. 1998, Hawkins et al. 2003, Jablonski et al. 2006) are less easy to apply to this bull's-eye pattern where there is a need to explain both latitudinal and longitudinal gradients. As noted by Rosen (1981, 1984, 1988a), conventional hypotheses that were widely applied to, and appeared to account for, latitudinal gradients could not be applied to longitudinal gradients. The problem with the longitudinal gradients is that although they display comparable changes in biodiversity to those seen along a latitudinal axis, they do not have comparable correlated changes in temperature, solar irradiance, seasonality or productivity. This includes all of the usual correlates along a latitudinal gradient that have been used to explain the observed latitudinal changes in biodiversity (Bellwood et al. 2005). The most parsimonious explanation would enable us to explain both latitudinal and longitudinal gradients (Rosen 1984). Given the above, this would probably have to include both ecological and geological evidence.

9.2 Characterising the IAA hotspot: patterns of species richness

This review will focus on the evidence from fishes and corals. It is in these two groups that the data are most complete and in which the widest range of analytical approaches have been applied. A comprehensive review by Hoeksema (2007) has provided a detailed overview of the other taxa in the region, early works, and matters pertaining to the terminology of the region. A more idiosyncratic perspective is provided by Veron et al. (2009), while Heads (2005) provides a panbiogeographic view. Earlier reviews include Rosen (1984, 1988a), Potts (1985), Veron (1995), Palumbi (1997), Paulay (1997), Briggs (1999) and Bellwood and Wainwright (2002). In all cases, shallow-water corals and/or fishes were the primary descriptors of the IAA biodiversity hotspot.

Recent analyses of the composition of shallow-water fish and coral faunas throughout the Indo-Pacific reveal the exceptional similarity in species richness patterns between the two groups (even azooxanthellate corals exhibit a comparable pattern; Cairns 2007). Despite one being a predominantly clonal, long-lived, sedentary, symbiotically photosynthetic invertebrate and the other a group of mobile, short-lived vertebrates, the two groups show a striking degree of biogeographic similarity. Examination of the community composition of fishes and corals revealed that variation in the taxonomic composition of the two groups (in terms of the number of species in each family at a given location) was similar along both latitudinal and longitudinal gradients (Bellwood and Hughes 2001). For example, the number of wrasse species in the family Labridae was consistently about 20% of the total number of reef fish species, damselfishes 15%, acroporid corals 35% of all coral species and poritid corals 10%, and so on (Fig 9.2). The species composition, in terms of the proportion of species belonging to a given family, remained approximately the same regardless of the distance from the hotspot or the direction taken. Furthermore, the proportion of species in each family at a given location did not differ from that expected from a random allocation from the total species pool, for both fishes and corals (Bellwood and Hughes 2001). In effect, the Red Sea is a random but depauperate subset of the reef faunas found in Indonesia or on the Great Barrier Reef. The main difference among locations is in the total species

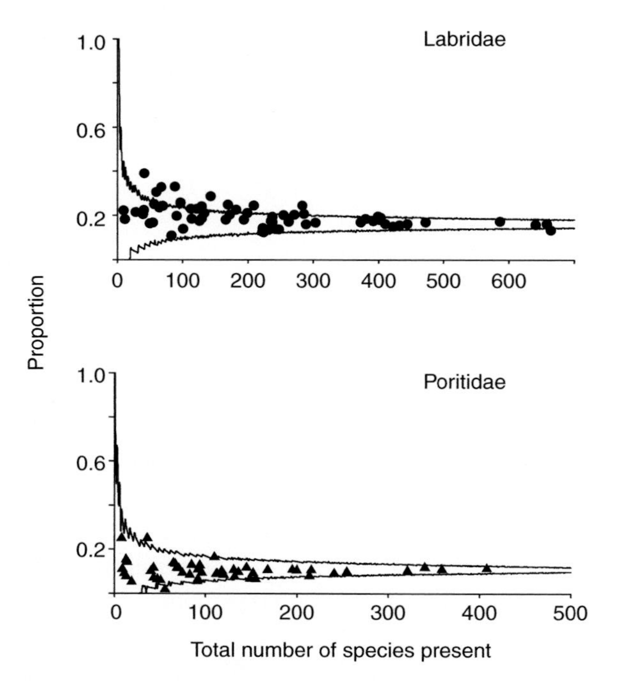

Figure 9.2 The predictability of coral reef assemblages. Each point shows the contribution of wrasses (family Labridae) or massive corals (family Poritidae) to the total species pool of fishes or corals, respectively, at a single geographic location. Whether one moves longitudinally or latitudinally from the IAA hotspot the proportions remain the same. In terms of its composition, the Red Sea fauna is just a depauperate subset of the Indo-Pacific fauna. The two converging lines indicate 95% confidence intervals based on a random allocation from the total Indo-Pacific species pool. For the majority of locations the number of species in a family does not differ from the numbers expected if the species were drawn at random from the available species pool (after Bellwood and Hughes 2001).

richness. In terms of biogeography, the key question is: what determines the total fish or coral species richness at any given location?

One of the first steps is to examine the taxonomic structure of the hotspot. Looking at the distribution of geographic ranges, it is clear that the majority of reef fish and coral species have very large ranges (Hughes et al. 2002). However, it is the spatial distribution of these various ranges that will characterise the size and nature of the hotspot. In terrestrial systems, as noted above, a great deal of the biodiversity in the IAA is represented by endemics. The marine IAA hotspot is, however, composed primarily of species with relatively large geographic ranges (Connolly et al. 2003). Marine endemics are relatively evenly spread across locations within the Indian and Pacific Oceans (but see Reaka et al. 2008 for a possible exception). It is the stacking of multiple medium- to large-range species over the

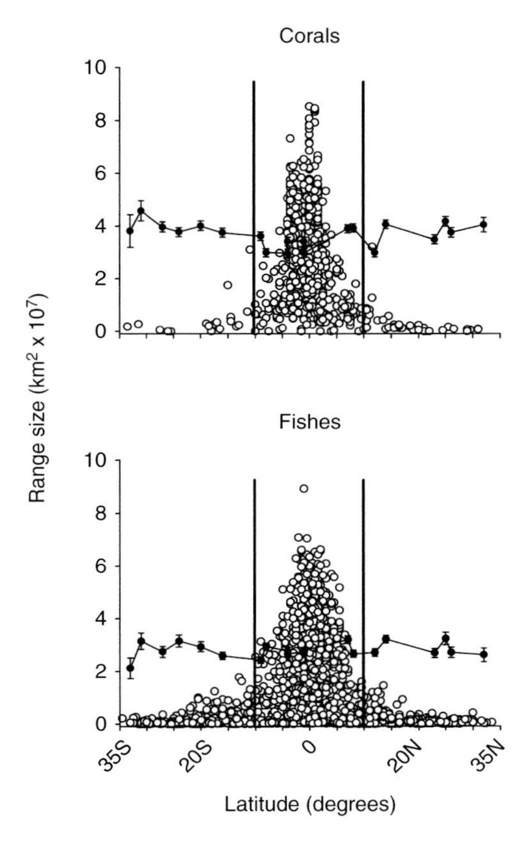

Figure 9.3 The latitudinal placement of the mid-range of individual coral and reef fish species as a function of range size. Each point represents a single species. The horizontal lines indicate the mean range size (± 2 SE). The vertical lines delimit the core of the hotspot at 10° N or 10° S (after Hughes et al. 2002).

hotspot that creates the bull's-eye pattern in fishes and corals (Fig 9.3). This has been termed the 'stack of pancakes' model (Connolly et al. 2003).

The 'stack of pancakes' model is based on current taxonomy. For some marine groups, the large ranges may represent cryptic species complexes with widespread allopatry (see Meyer et al. 2005, Malay and Paulay 2010). It remains to be determined how common cryptic species are and if overlap is sufficiently congruent to contribute significantly to the IAA biodiversity hotspot (Bellwood and Meyer 2009a). For corals and fishes, there is some evidence to suggest that the large geographic ranges are taxonomically sound and that the observed pattern of ranges will continue to be supported (Ayre and Hughes 2000, Klanten et al. 2004, Reece et al. 2010). The next step is to explain this pattern. Five main hypotheses have been presented to explain the observed variation in species richness.

9.3 Hypotheses that explain the hotspot

Explanations for the bull's-eye shaped hotspot have focused on a mechanism that would explain a region of diversity rather than a linear zone of diversity (e.g. the tropics). This has revolved around a number of 'centre-of' hypotheses. One of the first sets of alternate hypotheses was formulated by Potts (1985) who, building on earlier theories of Greenfield (1968) and Rosen (1984), proposed three models incorporating ecological and historical evidence to explain the bull's-eye pattern for corals. This included a Centre-of-Origin theory (Potts' Model I). By 1997, the various alternatives had been condensed into three or four 'centre-of' hypotheses (Palumbi 1997). These hypotheses have formed the cornerstones of studies of marine tropical biogeography and biodiversity for the last 30 years. In the following sections we will review these alternative hypotheses, the evidence for each, and the results of recent analytical examinations of the hotspot and its associated biodiversity gradients. In addition, we will examine the consequences of the IAA marine biodiversity hotspot for coral reef ecosystems and the people who depend on them.

9.3.1 Centre of Origin hypothesis (CoOr)

This hypothesis has its origins in the dispersalist theories of Darwin and Darlington, modified as a special case, with species radiating out from a single point of origin over a stable geological landscape. Following terrestrial ideas, it was readily applied to the marine realm where the vast majority of species have a planktonic larval phase and a demonstrable capacity to disperse, often over considerable distances. Applied to the coral reef realm in 1968 by Greenfield, it has subsequently been applied to a wide range of taxa. There are two basic versions.

The CoOr as a 'fountain' of new species (a spreading dye model)

This idea proposes that species originate in the IAA and each spreads by its own means as it expands its geographic range through geological time. This model makes no explicit assumptions about the mechanism driving range expansion and as such is presumably driven by chance (see Rosen 1988b for a discussion of dispersal versus range expansion).

The CoOr as a source of superior competitors

This formulation provides an explanation for the range expansion, with the suggestion that new species, with increasingly superior competitive abilities, displace older inferior competitors towards the periphery. This was the model proposed by Stehli and Wells (1971) in their early description of the bull's-eye pattern.

The CoOr hypothesis has been most consistently identified as the primary mechanism by Briggs who, in a series of publications, has provided evidence that may support this hypothesis (Briggs 1999). This hypothesis has been applied to many groups, including reef fishes (Allen 1975), with tentative routes of colonisation indicated. The CoOr hypotheses carry several specific predictions, namely that the youngest species will be found mostly in the centre (the IAA), that there will be a peripheral halo of predominantly older species and that older species will have larger geographic ranges. In addition, geological evidence should indicate that most of the stratigraphically earliest records of any given taxon will be concentrated within the CoOr (Wilson and Rosen 1998).

Interestingly, in the simplest conception of this model, and if competition were effective, there would be a dynamic long-term equilibrium between central origination rate, displacement rate by competition and peripheral extinction rate. This would generate a flat distribution pattern, not a bull's-eye pattern. Conversely, a bull's eye pattern therefore implies a lag in the displacement of older taxa from the centre, in relation to the central origination rate.

9.3.2 Centre of Overlap hypothesis (CoOl)

CoOl: vicariance

The Centre of Overlap (CoOl) hypothesis arose from an increasing realisation that there were distinct and often congruent boundaries between closely related species. The presence of such boundaries in land animals has been known since the days of Sclater, de Candolle and Wallace in the mid 1800s (see Humphries and Parenti 1999) with the well-known Wallace's Line passing through Indonesia and separating two major continental terrestrial faunas. Springer (1982) noted a similar faunal boundary in marine taxa approximately corresponding to the Andesite line in the West Pacific (which in turn more or less corresponds to the western limit of the Pacific Plate). However, in the marine realm, similar divisions were also seen in closely related taxa and the locations of these divisions were often congruent among species (locations are summarised in Blum 1989, Bellwood and Wainwright 2002). These biogeographic boundaries were well recognised, but any subsequent expansion or movement of ranges would lead to an overlap of geographic ranges and a localised increase in species richness in the overlap zone. Rosen (1988b, Table 2.1) listed at least seven possible modes of distributional change. The plethora of islands, currents and putative biogeographic boundaries in the IAA make it highly likely that this area exhibited many such divisions that were susceptible to frequent changes through time. In this respect, it may be considered to be a 'dynamic mosaic' (cf. 'overlapping mosaic' in Rosen 1984, and Rosen's 1992 discussion of Jablonski et al.'s (1985) 'dynamic palaeogeography'), of ever-changing ecological and biogeographic distributions, driven by continual tectonic, eustatic, climatic, oceanographic and geomorphological ('TECOG') processes and events. This CoOl hypothesis was proposed prior to the widespread application of cladistics and vicariance biogeography (Woodland 1986) but was readily applied to the burgeoning field of phylogenetics and phylogeography, with phylogenetic studies identifying additional putative vicariance events (e.g. Barber et al. 2000). Initially applied to reef fishes (Woodland 1986), it has found support in other groups including corals (Hoeksema 1989, Wallace et al. 2000, Wallace 2001), crustaceans (Fransen 2007), and gastropods (Williams and Reid 2004, Meyer et al. 2005, Reid et al. 2006). We consider that 'dynamic mosaics' would be an important factor in CoOl models, since many taxa appear to have subsequently expanded their ranges into mutually overlapping patterns, hence their diversity contribution to the hotspot.

Some mosaics may result in almost entirely allopatric distributions (Meyer et al. 2005). These non-overlapping mosaics do not stack (Bellwood and Meyer 2009a). However, they can produce a bull's-eye pattern, but only in terms of regional species totals. In terms of α-diversity, unlike ranges in a 'stack of pancakes' model, they make no contribution to any local species enrichment.

CoOl: peripheral oceanic islands

An early version of the centre of overlap in a slightly different form was proposed by Rosen (1984) who like Ladd (1960) suggested that species might have arisen on peripheral oceanic islands, regardless of their geological origin. Potts (1985) summarised this as his Model II: Vicariance-and-Refuge. Although originally based on coral genera, the model was readily applicable to other taxa. It differed from later CoOl models in that Rosen had suggested that the highest probability of speciation would be among the optimally spaced island clusters of the western Pacific. Subsequently, taxa moved unidirectionally through time towards the IAA as the islands moved westward on the Pacific Plate (see also 'Noah's Arks' and island integration, under 'CoAc by faunas' below).

CoOl models are less specific in their predictions with regard to the location of the youngest species, and thus also earliest records in the fossil record, although in CoOl-vicariance the youngest species should be either side of primary biogeographic boundaries, while in CoOl-peripheral oceanic islands they would be concentrated in peripheral islands and archipelagos.

9.3.3 Centre of Accumulation hypothesis (CoAc)

This hypothesis is the opposite to the centre of origin hypothesis. In the CoAc hypothesis it is proposed that species arise in peripheral locations, around, or at some distance from the margin of the bull's eye, and that they subsequently moved into the centre. This differs from CoOl-peripheral oceanic islands in that new species can be from anywhere outside the centre, and differs from CoOl-vicariance in that it does not require overlap with related taxa. There are two distinct CoAc formulations reflecting underlying biological or geological mechanisms.

CoAc by individuals

This hypothesis was initially proposed by Ladd (1960), and places the emphasis on isolation on peripheral oceanic islands in underpinning the speciation process. This is followed by changes in distribution through time with a net movement leading to a concentration in the centre. Species may subsequently die out in the peripheral locations.

This model, COAc by individuals, has a logical link with CoOl-peripheral oceanic islands. However, the CoAc model goes a step further in explaining the abundance in the IAA by presenting a 3-stage model with peripheral origins, net migration into the IAA, and enhanced survival in the IAA.

One specific explanation of the underlying mechanism is provided by the vortex model (Jokiel and Martinelli 1992) in which westerly currents carry the larvae of marine taxa predominantly to the western margins of ocean basins. In this scenario, the IAA hotspot represents the exceptional accumulation of Pacific Ocean taxa (a hotspot of flotsam), with a smaller analogous accumulation along the east coast of Africa (Connolly et al. 2003). In this specific case, there should be no Indian Ocean taxa in the bull's eye, but IAA species could 'leak' through the Indonesian archipelago into the Indian Ocean. It must be noted that the model of Jokiel and Martinelli (1992) is based largely on modern configurations of oceanographic currents. However, as the observed biogeographical patterns must have accumulated over considerable (i.e. geological) amounts of time, the model assumes that these current configurations must have remained more or less constant over millions of years. This is in spite of long-standing, well-established evidence for large-scale climatic and oceanographic change on this time scale. While it is likely that modern current configurations have influenced the observed distribution patterns, other longer-term TECOG factors (see 'dynamic mosaic' above) are just as likely to have played a part too. It is important to try and match the temporal scales of envisaged causes to their effects.

CoAc by faunas

This hypothesis reflects the accumulation of entire faunas on moving land masses. The assimilation of island arcs on the north coast of New Guinea and cumulative docking of numerous land fragments in the IAA from Asia, Gondwana and Australia, each with their respective faunas, may enhance the species richness in the IAA by the merging of entire faunas, cf. McKenna's (1973) biogeographic 'Noah's Arks'. This mechanism has many terrestrial examples, though the palaeogeographical reconstructions are undergoing revision (this volume, Chapter 3). The 'island integration' model of Rotondo et al. (1981), developed further by Rosen (1984), is effectively a case of multiple 'Noah's Arks'. The westward movement of oceanic islands on the Pacific plate potentially merges them and their biotas with land masses and biotas on the other side of the convergent boundary of the Pacific Plate, including the IAA. However, there are as yet no clear cases of marine 'Noah's Arks' that can be compared with terrestrial examples.

In the CoAc hypothesis, specific predictions about ages of taxa are hard for CoAc by faunas, but for CoAc by individuals they are clear. Young taxa will be on the periphery, older taxa near the centre, and the age of taxa should again increase with larger geographic ranges. Geological evidence for pure CoAc models should indicate first occurrences outside the IAA, in CoAc by individuals predominantly in the West Pacific, and in CoAc by faunas on particular terrains.

When comparing centre of overlap with centre of accumulation models, it is important to emphasise that the critical difference is in the relationships between species with overlapping ranges. In the CoOl model, overlap consists of numerous cases of overlapping sister species (= biogeographical sympatric distributions), whereas sister relationships are unnecessary in the CoAc models.

In many ways the distinctions between the various hypotheses, or subsets thereof, are not that well resolved. The essential distinction is whether origination of the taxa now found in the IAA hotspot has taken place predominantly within the centre or outside. The underlying mechanisms are all broadly similar with an interacting mélange of vicariance (through allopatric or parapatric speciation), extinction and changes in geographic distributions over time, each operating at different spatial or temporal scales. The key issues are in the location and timing of origination.

9.3.4 Centre of Survival hypothesis (CoS)

In many ways this is a composite hypothesis. Originally proposed by Heck and McCoy (1978) who identify the IAA as a potential refuge for taxa, it has in recent years been presented as a centre of survival. In recognising the likelihood of multiple sources of species (from the centre and elsewhere) this hypothesis seeks to explain why most remain within the centre, regardless of their geographic origins (Barber and Bellwood 2005). In this way, the emphasis is on persistence and survival in a refuge, rather than origination. In this form, the key aspect in this hypothesis is therefore regional variation in the relative rates of extinction (the centre is the region of least extinction) and it makes no assumptions about the rates or location of origination. Contrary to Heck and McCoy (1978), more recent developments of the model often also assume changes in geographic distributions over time. Survival models also propose that significantly lower rates of extinction in the IAA are due to more habitats, including deep areas (Hoeksema 2007), larger total reef area and larger population sizes (Bellwood et al. 2005). This is, in many ways, inversely complementary to Model III of Potts (1985), which explains the low diversity of species away from the centre as a consequence of high local extinction rates in isolated peripheral locations.

In the particular case of corals, a major difficulty of Heck and McCoy's model is their proposal that the modern pattern is essentially a residue of a Cretaceous Tethyan one, shaped into an IAA centre by peripheral extinctions. They did not consider there to have been any significant contribution by later originations or distributional changes, in spite of strong evidence from the fossil record (Wilson and Rosen 1998), and, more recently, from molecular systematics (Renema et al. 2008). One of the few exceptions to the general observation of widespread endemics and highest diversity in the IAA is the land hermit crab *Calcinus* (Malay and Paulay 2010). In this group, species appear to exhibit highest rates of origination and survival in West Pacific oceanic regions, rather than the IAA; a pattern that appears to reflect the terrestrial associations of this group. Although they have a pelagic larval stage, the biogeography of this group strongly reflects on the geographic distribution of suitable oceanic islands, the obligate adult habitat. As such, their distribution patterns may more strongly reflect terrestrial biogeographic patterns and processes than their wholly aquatic counterparts.

9.3.5 The Mid Domain Effect (MDE) and null models

One of the underlying, but often implicit, assumptions of all of the above hypotheses is that the hotspot represents a deviation from the expected distribution in which species richness is uniform across the two oceans. However, this may not be an appropriate null expectation, especially if most species have large geographic ranges. If the ranges are randomly placed within a bounded domain (such as the Indo-Pacific) the resultant pattern of species richness forms a peak in the middle of the domain: the Mid Domain Effect (MDE) (Colwell and Lees 2000). In some respects, it is a product of an edge effect in the placement of ranges. The size and shape of the central peak depends on the size distribution of the ranges but a peak in the middle, a bull's-eye hotspot, is the null expectation. One would therefore predict a hotspot based on nothing more than a random placement of species geographic ranges (Connolly et al. 2003, Bellwood et al. 2005). In the Indo-Pacific the mid domain lies close to the IAA (just northeast of New Guinea), and one would expect the hotspot to be centred here. Indeed, of all the geographic and environmental variables examined by Bellwood et al. (2005) the MDE explained more of the variation among locations in coral reef fish species richness than any other variable.

The strength of this relationship was even sufficient to overcome potential limitations in the placement of the mid domain. To be conservative, Bellwood et al. (2005) defined the Indo-Pacific domain as the waters bounded by the African and American coastlines (hard boundaries). However, for many reef species, the East Pacific Barrier (a 5000 km wide stretch of open ocean between the Marquesas and the Galápagos Islands) is an uninhabitable region, a 'soft' barrier. The effective eastern margin of the Indo-Pacific domain is marked by the Line Islands and

French Polynesia. If this is the case, the resultant mid domain is very close to the centre of the IAA hotspot for both fishes and corals (Bellwood et al. 2005).

Overall, the MDE predicts that there will be a peak in species richness in the centre of a domain. In our example, the mid domain lies close to the IAA.

9.4 Inferred mechanisms of species origination and accumulation

Regardless of the hypothesis, most (and especially CoOr, CoOl and CoAc hypotheses) have invoked glacioeustasy, especially through the later Cenozoic, as a common mechanism. Sea-level changes potentially result in the isolation of small marine basins, changes in currents, new land bridges and changing levels of population connectivity (Rosen 1984). Any consequent longer-term isolation of populations may lead to speciation or local extinction. Subsequent reconnection and intermixing will produce higher biodiversity where populations have survived and diverged, or lead to increasingly depauperate faunas where populations have declined to the point of local extinction. Successive cycles of high and low sea-level stands can magnify these dynamic mosaic effects. With numerous such cycles, especially the wide-amplitude fluctuations of the Pleistocene, this was regarded as a clear mechanism for driving variation in biodiversity and has been termed the Pleistocene species pump (Hubbs 1958). There are several lines of evidence to suggest that this was an important factor in terrestrial systems in the IAA and the same argument has been applied to numerous marine taxa in support of 'Centre of' hypotheses, for example CoOr (McManus 1985), CoOl (Woodland 1986) and CoAc (Potts 1985).

Four different models of the Pleistocene speciation pump have been applied to the IAA region. The first two models highlight speciation during sea-level low stands, either (I) as a result of isolation between the Indian and Pacific Oceans (e.g. Benzie 1999, Lessios et al. 2001, Crandall et al. 2008, Vogler et al. 2008) or (II) as a result of isolation in small basins inside the IAA (McManus 1985, Arcos and Fleminger 1986, Woodland 1986, Springer and Williams 1990, Barber et al. 2006, Landry et al. 2007). The third model (III) predicts that speciation will occur on isolated atolls (Potts 1985) or optimally clustered islands (Rosen 1984) during sea-level high stands, with subsequent range expansion to the IAA following sea-level falls. Finally, (IV) Verbruggen et al. (2005) hypothesised that geographic isolation of siblings of the calcareous green algae *Halimeda* may have taken place following Pleistocene or Pliocene periods of climatic cooling during which time subtropical species occupied larger distribution ranges. Regardless of the mechanisms, however, speciation should be accelerated during the past 2.5 Myr as a result of the increased amplitude of sea-level changes. Unfortunately, both fossil and phylogenetic evidence suggests that speciation rates were lower during this period (Renema et al. 2008, Rocha and Bowen 2008, Williams and Duda 2008). Some population studies have suggested that there is restricted population-level gene flow within the IAA (Timm and Kochzius 2008), with the locations of clades reflecting geographically isolated regions during sea-level low stands. However, these divisions are currently only recognised at a population level.

9.4.1 Tests of hypotheses: ecological correlates

Explanations for the location of the IAA hotspot often emphasise that the area is exceptional in terms of the abundance of habitats, reef area, potential for isolation, location and so on (summarised in Hoeksema 2007). Explicit tests of the explanatory power of these various ecological factors as correlates of the observed biodiversity patterns are, however, extremely limited. One of the first was by Fraser and Currie (1996) who examined the relationship between reef coral genera richness, in both the Indo-Pacific and Atlantic, and a range of environmental variables. They concluded that sea surface temperature (SST) is the main environmental variable that explains variation in coral generic richness. However, the lack of species level analyses and the merging of data from the two main biogeographic regions (the Atlantic and Indo-Pacific, with a respective generic richness ratio of approximately 1:4) limits comparisons with other studies.

Using Indo-Pacific species-level data, Bellwood and Hughes (2001) found reef area to explain the greatest proportion of variation among locations in both fish and coral species richness across the two oceans. Longitude and latitude explained a much smaller proportion of the variation in species richness.

The first explicit test of the MDE in coral reef ecosystems was by Connolly et al. (2003), who found that the MDE could account for a large proportion of the peak in species richness, especially along a latitudinal gradient. In a longitudinal direction, however, there was a positive anomaly in the IAA and along the east coast of Africa, and a negative anomaly in the east Pacific. Basically, the IAA was found to be 'hotter' than one would expect based on a MDE null model. A consideration of the ecological factors that may account for this deviation from the MDE suggested that reef area and current patterns may be significant drivers of reef fish and coral species richness. A comparable study looking at bryopsidale algae (Kerswell 2006) revealed a similar pattern and again, habitat area and ocean currents were identified as the most plausible drivers of observed diversity patterns.

Bellwood et al. (2005) subsequently examined two-dimensional species richness patterns in fishes and corals explicitly incorporating both MDE and ecological variables in combined models (using non-linear and linear models, which allowed for spatial autocorrelation). They found that 62% to 79% of the variation in species richness across the Indo-Pacific domain could be accounted for by just two variables: distance from the mid domain and present-day reef area. If the East

Pacific barrier is considered in defining the domain, the total variance explained is even greater.

These models are instructive but they are only correlative and incorporate no historical evidence. They are based on current day distributions and current day ecological factors. While they may provide an indication of the factors that led to the formation of the IAA hotspot and those that maintain it today, they do not allow us to begin to explore the relative contribution of the three or four main 'centre of' hypotheses, nor can they inform us of the history of the hotspot or the component species. A critical test of these hypotheses and a better understanding of the history of the hot spot and its component species require historical evidence. Fortunately, there are several lines of evidence available: endemics, phylogenies and fossils. These will be examined below.

9.4.2 Tests of hypotheses: endemics

Of all the lines of evidence that can be used to test the various hypotheses, endemics has been one of the most popular. Over the last 25 years, endemics have been used widely in the evaluation of the various models. It has been widely assumed, although often implicitly, that endemics mark the location where species arose. This assumption was most explicit in Mora et al. (2003) who stated that 'we assume that centres of endemism (areas with a high proportion of endemics) contain a preponderance of recently derived species that are yet to expand their ranges (neo-endemics) and thus provide insights into areas where species are most likely to originate'. And indeed, Mora et al. (2003) subsequently used endemics to identify the IAA as a centre of origin. However, there are two problems with this conclusion. First, most workers agree that although endemics are spread across the entire region, the majority are located in peripheral locations, outside the IAA (e.g. Hughes et al. 2002, Jones et al. 2002, Roberts et al. 2002, Connolly et al. 2003) (although this may change as molecular evidence reveals more cryptic species or questions the status of endemics; McCafferty et al. 2002). Second, there are inherent problems with the use of endemics in that: (1) theory suggests that most do not mark the geographic origin of a species (especially in widely dispersing species like reef fishes and corals), and (2) molecular data suggest that many endemics are not particularly young.

Given that species can arise without a period of relatively localised endemism, and range contractions can result in endemism as well (Fig 9.4), endemism may mark reliction as much as origination. Indeed, based on recent data on reef fishes and cowries, the ages of endemics suggest that in marine taxa with the potential to expand their geographic ranges, endemism is a poor predictor of their ages (Bellwood and Meyer 2009a, b). Endemism merely reflects a limited realisation of their range expanding potential; it indicates nothing about the relative age of the taxon. In a comparable vein, there was a poor relationship

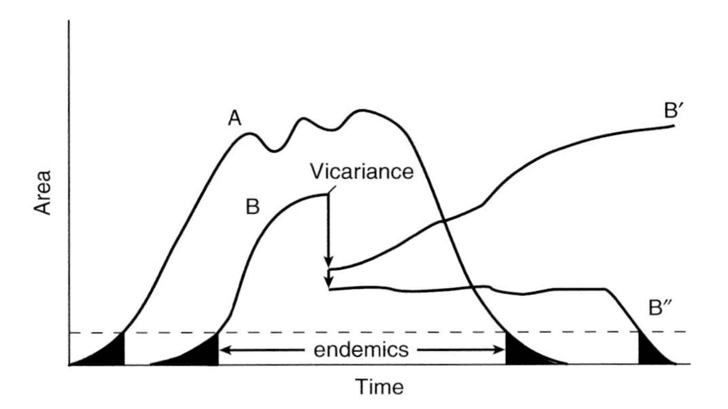

Figure 9.4 Changes in geographic ranges through time. Marine endemics are defined by their small geographic ranges (e.g. a restricted geographic region, <10% of all geographic range sizes or a limited area <0.5 million km^2). Using an areal definition of an endemic (marked by the dashed line), species can arise without a period of endemism. Species with spatially restricted origins will be endemics shortly after origination (e.g. founder effect, etc.) and prior to extinction (A). However, vicariance can result in species origination without a period of endemism (species B yields sister species B′ and B″). Only B″ becomes an endemic before extinction. Indeed, the only time that all species are endemics is shortly before extinction (in ecological if not geological timescales). In strongly dispersing or widespread species, endemism is likely to mark reliction and may be a poor indicator of the geographic origins of a species (after Bellwood and Meyer 2009a).

between age and geographic range, contrary to predictions in CoOr and CoAc hypotheses.

There is also a practical limitation with endemics in that at the point of first discovery of a species it is usually regarded as endemic (i.e. known from a single location or very small area). Thus, centres of endemism are likely to mark areas of recent taxonomic activity. As knowledge and sampling continue such localised endemism disappears (Razak and Hoeksema 2003, Hoeksema et al. 2010).

9.4.3 Tests of hypotheses: phylogenies and fossils

Although fishes and corals are the key taxa considered in descriptions of the IAA hotspot, their fossil record is currently of limited potential for hypothesis testing. Fossil records for the fishes are primarily restricted to an exceptional conservation Lagerstätte in the Eocene deposits of Monte Bolca, in northern Italy (Bellwood 1996, Bellwood and Wainwright 2002). Although corals have a much better fossil record (Rosen 2000, 2002), fossil species of the IAA are beset with taxonomic difficulties (K.G. Johnson and B.R. Rosen, unpublished data). The availability of phylogenetic hypotheses for the two groups contrasts with their respective fossil records. For corals, there is a limited number of morphological (e.g. Hoeksema 1989, Pandolfi 1992) and molecular phylogenies (e.g. Fukami et al. 2008), while there are numerous fish phylogenies, especially those based on molecular evidence (e.g. McCafferty et al. 2002, Bellwood et al. 2004a, 2010, Read et al. 2006, Smith and Wheeler 2006, Cowman et al. 2009). Despite the patchy nature of the fossil and phylogenetic evidence, the resultant patterns are remarkably congruent (Santini and Winterbottom 2002, Halas and Winterbottom 2009) with a similar, if incomplete, record for numerous groups of tropical marine fishes, corals, molluscs, algae, mangrove trees and large benthic foraminifera. Indeed, the evidence, both fossil and phylogenetic, was recently reviewed with a suggestion that the tropical oceans have been characterised by 'hopping hotspots' (Renema et al. 2008).

Renema et al. (2008) identified three separate marine biodiversity hotspots during the last 45 Myr. Sequentially, the hotspots were located in western Europe in the Eocene, then Arabia and the IAA in the Miocene, before leaving a single hotspot in the IAA today (Fig 9.5). In many cases the component taxa, in terms of families and genera, have moved with the hotspots. This is particularly clear for the eastward movement of higher taxa (genera, families) in corals (Wilson and Rosen 1998, Rosen 2002, Wallace and Rosen 2006), molluscs (Harzhauser et al. 2007), large benthic foraminifera (Renema 2007), and to a lesser extent, reef fishes (Bellwood 1996, Bellwood and Wainwright 2002). However, many species appear to have arisen while the hotspot was in its current position in the IAA (Cowman and Bellwood 2011). In effect, hotspots hop while taxa slide (from place to place over time).

It is interesting to speculate on why the hotspots moved and what was significant about those specific locations at those times (Renema et al. 2008). Individual hotspots arise, proliferate for tens of millions of years, senesce and then die. The geographic location and timing of the hotspots exhibits a remarkable correlation with the locations and timing of major tectonic events. Although climate was undoubtedly a factor, with cooling of the Mediterranean in the later Neogene being a clear example (Rosen 2002), tectonics appears to be a major driver. In each case the origination of a hotspot was associated with the collision and subsequent close apposition of major continental components of lithospheric plates. This process drives intense tectonic activity in the area leading to the formation of shallow enclosed continental seas and island arcs, localised uplift and microplate rotations. This geological complexity triggers a concomitant increase in ecologically important factors including new current patterns, increased areas of shallow habitats and new isolated islands. This, in turn, leads to complex patterns of population connectivity, with changing population sizes, and increased division or isolation leading to spatially variable rates of extinction and origination. It is believed that these ecological factors underpin the proliferation of species and the formation of hotspots in these specific locations. In each time slice the hotspots therefore correspond to areas that are operating as a 'dynamic mosaic' (explained above). As

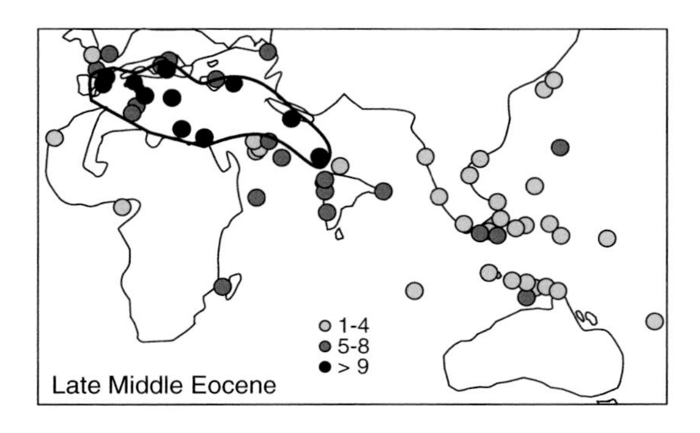

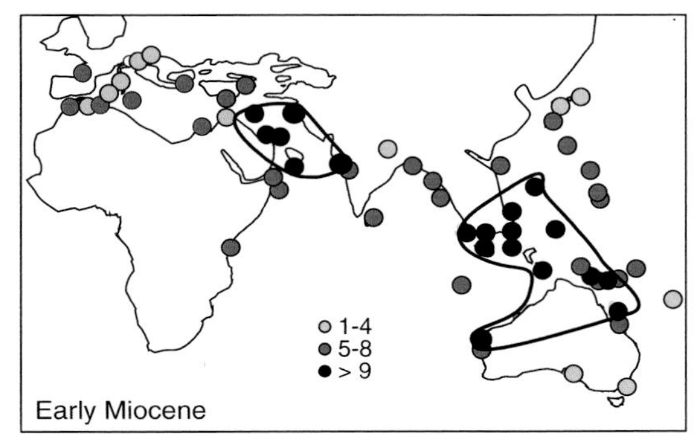

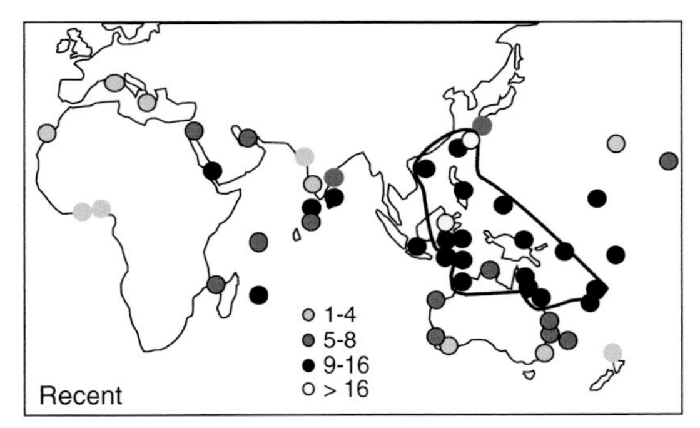

Figure 9.5 Hopping hotspots. Generic α-diversity of large benthic foraminifera in the late middle Eocene (42 to 39 Ma), the Early Miocene (23 to 16 Ma) and the Recent. The hotspots have moved over time from the West Tethys to Arabia and finally the IAA. Although the hotspots appear to hop the component taxa may move from place to place. Modified after Renema et al. (2008).

component land masses move closer together on their respective plates, however, the seas become shallower, seaways begin to close, oceanic basins disappear and eventually, through subsidence and collision, the marine realm is extinguished, and with it the remnants of the hotspot. During this process terrestrial influxes increase and thus change coastal habitats.

Prior to the Neogene, the combined Alpine, Mediterranean and Middle Eastern regions would have been dynamically and palaeogeographically analogous in this respect to Southeast Asia today. Today the European and Arabian hotspots only exist as fossil deposits in the hills and mountains of Europe, the Middle East and Pakistan. Moreover their erstwhile existence can only be inferred from the fossil record, not from analysis of modern biotas alone.

These patterns of origination, proliferation, senescence and death of hotspots are most clearly seen in the relative complete records of the large benthic foraminifera but the same pattern holds for many tropical groups (Renema et al. 2008; see also Williams 2007, Williams and Duda 2008). Thus, in many cases, tectonics appear to underpin patterns of biodiversity, mediated through ecological processes, in a wide range of taxa. It is likely that the ecological processes that operate today were important in the past and that increases in the area of carbonate platforms in the IAA, driven at least in part by tectonics, and an increase in the proportions of corals in the deposits (Wilson and Rosen 1998, Wilson 2008), was a significant factor in the proliferation of coral reef taxa in this area (Williams 2007, Cowman et al. 2009, 2011, Bellwood et al. 2010). Today the IAA stands as the only marine tropical hotspot and it is likely to persist for a long while (a minor biodiversity peak has been recorded in the west Indian Ocean, but this may reflect an uneven research effort (Hoeksema 1989, Connolly et al. 2003)). Tectonically the IAA is still in late middle age, although in eastern Indonesia we may be in early senescence, with the demise of former large carbonate platforms and the uplift of islands, resulting in fringing reefs that are under considerable continental influence (Wilson and Rosen 1998, Wilson 2008, 2011, Renema 2010).

In considering the role of drivers of large-scale diversity patterns it must be noted that the IAA cannot be regarded as a single location (discussed in detail by Hoeksema 2007). Patterns are a result of the interaction between ecological characteristics of the study taxa, the current environmental conditions in the area of interest and the (geological) history of the area (Fig 9.6). The ecological properties of a group, for example, environmental preferences and capacity for range expansion, can be assumed to vary only little within its distribution range. Environmental parameters and geological history, however, are rarely constant over geographic areas. Within a given area, therefore, different factors will dominate the biogeographic history. For example, in western Indonesia, late Cenozoic glacial cycles caused repeated emergence and submergence of the Sunda shelf, limiting the taxa in that region to relatively rapidly colonising species. The shallow depth, monsoonal conditions and the presence of a number of large neighbouring land masses also caused relatively poor conditions for reef-associated biota, but fairly good conditions for mangroves. In eastern Indonesia, however, the much deeper basins ensured that there was a continuity of shallow marine habitats during the glacial periods. As a consequence, faunas in western Indonesia are

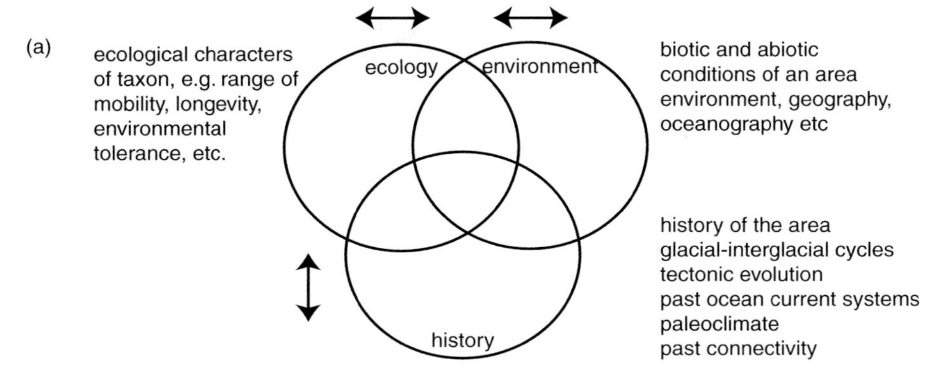

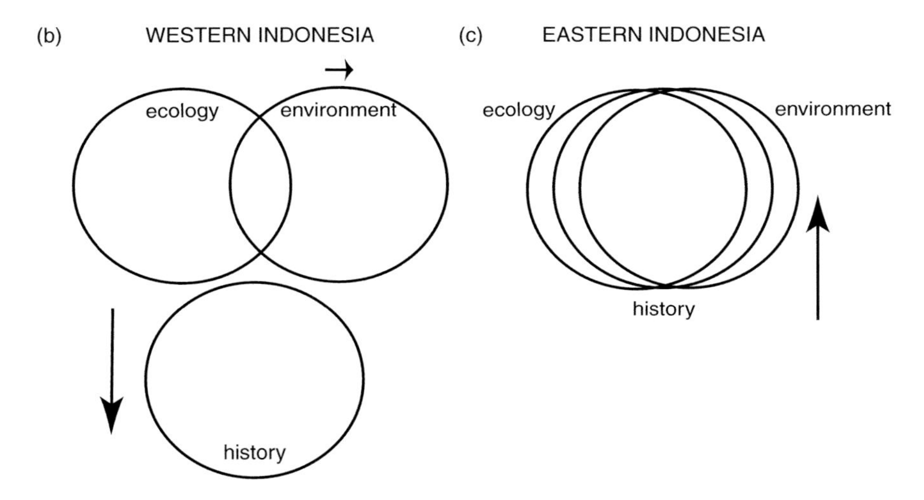

Figure 9.6 The relative contribution of ecological, environmental and historical factors in shaping regional biodiversity patterns. In western Indonesia the extensive Sunda Shelf was exposed during glacial lows resulting in extensive habitat loss. Patterns today are, therefore, primarily driven by the interaction between the ecology of taxa and current environmental factors. In the central and eastern regions, continuity of habitat has fostered a closer interaction between the three factors.

dominated by extirpation and immigration, whereas in central-eastern Indonesia origination and range expansion are likely to be more important.

Whether the current hotspot is predominantly a cradle or grave for species, a centre of origin or accumulation, is still to be determined. Nevertheless, the area has been a region of significantly higher biodiversity than elsewhere for over 20 Myr, and from recent evaluations of reef taxa it appears that the proliferation of modern trophic groups occurred while this area was the dominant hotspot. For reef fishes at least, there appears to have been two major periods of innovation. Firstly, a profound change was recorded shortly after the K/T boundary with the appearance of most modern groups including, most importantly, the fish herbivores (Bellwood 1996, 2003). This changed the nature of coral reefs and may have underpinned the success of scleractinian corals on reefs. On Recent coral reefs, herbivorous fishes maintain a regime of intense grazing that opens substrates for fast growing scleractinian corals (Done et al. 1996) and provides corals with a competitive advantage over tropical algae (Bellwood et al. 2006, Hughes et al. 2007). The second wave of innovation was in the Miocene. Many modern feeding types do not appear to have been represented in the Eocene (foraminifera feeders, cleaners, specialist detritus feeders). The first appearance of these trophic modes (at least in the Labridae) was not until the Miocene (Cowman et al. 2009), at a time when we had just a single hotspot in the IAA. A similar pattern is seen in several other groups with marked Miocene diversification (Renema et al. 2008, Williams and Duda 2008, Bellwood et al. 2010). In this respect, the IAA may mark the geographic location for the origin of modern coral reefs, hence also the cradle of functionally modern coral reef ecosystems.

9.5 Consequences of the biodiversity hotspot

The uneven biodiversity across the Indo-Pacific has resulted, to some extent, in an uneven degree of attention with high-diversity reefs being the focus for conservation. Hotspots are regarded as vulnerable and worthy targets of attention. However, this begs the question: Why are we protecting biodiversity and hotspots? For scientific interest in the high diversity locations, for the rare endemic species or habitats contained therein, for the people who depend on them for food, or for the inferred ecosystem benefits of biodiversity? These issues are difficult to resolve, but the evidence suggests that for marine systems there are not necessarily any more endemics in hotspots than elsewhere (Hughes et al. 2002), while intensive research into the ecosystem benefits of biodiversity on coral reefs has identified repeated cases where critical ecosystem functions, or processes, are supported by just a few species (Done et al. 1996, Bellwood et al. 2003, 2006, Fox and Bellwood 2007, Hoey and Bellwood 2008). The benefits of biodiversity, therefore, may rely on a few critical species rather than total richness or biodiversity per se.

There are approximately 5000 species of fishes on coral reefs around the world. In the IAA alone there are probably over 2500 fish species. Yet many reefs around the world, including the Red Sea, persist and appear to be comparatively robust with little more than 250 reef fish species. Clipperton Atoll survives with just 101 reef fish species (Robertson and Allen 1996). Similarly, the major framework building of coral reefs can be achieved with a handful of corals, as in the late Miocene reefs of the Mediterranean, which are built entirely of *Porites* (Esteban 1996), or Clipperton Atoll with just seven coral species (Glynn et al. 1996). It is not high biodiversity per se that is required to build a reef, but sufficiently rapid production and accumulation of *in situ* bioconstructors. This suggests that while many of the fishes seen on reefs are attractive, in terms of ecosystem function they represent nothing more than a colourful irrelevance; aquatic baubles on the coral reef tree. While biodiversity conservation may be a logical fist step in protecting reefs when we know little of ecosystem processes or function, more targeted protection of critical functional groups may be a more effective approach (Walker 1995, Bellwood et al. 2004b). Biodiversity (including species richness and phylogenetic diversity) alone may not be the best conservation metric and the apparent value of hotspots may be misleading. People still depend for their livelihoods on low diversity reefs (Hughes et al. 2003). So far, it appears that on coral reefs, when it comes to biodiversity, it is the quality not the quantity that counts.

9.6 Conclusion

Numerous authors have commented that there is unlikely to be a single explanation for the IAA marine biodiversity hotspot or that any one hypothesis will be correct (Rosen 1984, Hoeksema 1989, 2007, Palumbi 1997, Bellwood and Wainwright 2002, Bellwood and Meyer 2009a). It is most likely that it will be a combination of various hypotheses and that the relative contribution of each to the overall pattern will change as data, taxa and methodologies change. Indeed, the stomatopods have markedly different patterns of endemism than fishes and corals (Reaka et al. 2008) even though they display a similar hotspot. Likewise the bryopsidale algae (Kerswell 2006) respond more to temperature than reef area, perhaps reflecting a greater thermal sensitivity. Nevertheless, the strength of the consensus among taxa is striking. This includes:

1. Numerous families, in disparate taxa, that all show the same pattern: a single IAA hotspot with latitudinal and longitudinal gradients.
2. Species richness patterns that often differ little from random expectations (in terms of species compositions and in species richness patterns = the MDE).
3. Strong correlations between total species richness and a limited range of key environmental factors (primarily present-day habitat area, SST, ocean currents).
4. Perhaps most striking, a shared evolutionary and geological history. Most IAA taxa are pre-Pleistocene (primarily Miocene) with many taxa having previous representatives in the three Cenozoic hotspots, in Europe, Arabia and the IAA. Some IAA taxa have migrated half way around the globe moving through the

hotspots, their locations and fate underpinned by tectonic activity and other TECOG factors.

5. A shared and challenging future, with common responses to habitat loss as a result of climate change and direct human disturbance.

There always have been and always will be hotspots. Their history and location may be driven by tectonics but we have the capacity to shape their future. The two main environmental drivers of coral reefs, reef area and SST, have both been dramatically changed by human activity (Bellwood and Hughes 2001, Hughes et al. 2003). We are changing the future of reefs. Direct exploitation had degraded many reefs in the IAA before climate change became a recognisable problem. Now climate change challenges the future of reefs as an ecosystem. It is clear that we need to address human-induced greenhouse gas emissions as our first priority. In the meantime remedial measures are necessary, including the protection of critical functional groups and habitats, reduction of pollution and unsustainable fishing practices, and habitat protection in both exploited and marine protected areas. With luck, this will leave adequate areas of reef in a position to regenerate. Coral reefs have a long past and an uncertain future, but it is one that will be shaped substantially by our actions.

Acknowledgements

We thank R. Hall, S.T. Williams, K.G. Johnson and the SAGE organising committee for their kind invitation to attend the SAGE conference, where stimulating discussions with colleagues and friends laid the foundations for this review. In writing this review it was clear that our work was just part of a long continuum and that we owe a great deal to those who came before and those who travel this path with us. In particular, we would like to thank: G.R. Allen, P. Barber, G.A. Boxshall, S. Connolly, R. Hall, B.W. Hoeksema, T.P. Hughes, W.F. Humphreys, C.P. Meyer, J.M. Pandolfi, G. Paulay, D. Potts, J.E. Randall, D.G. Reid, V.G. Springer, P.C. Wainwright, C.C. Wallace, S.T. Williams, and M.E.J. Wilson. B.W Hoeksema, S.T. Williams, J. Tanner and an anonymous reviewer kindly provided insightful and helpful comments on an earlier draft. This work was supported by the Australian Research Council Centre of Excellence for Coral Reef Studies, and the Systematics Association.

References

Allen, G. R. (1975). *Damselfishes of the South Seas*. Neptune, NJ: TFH Publications.

Arcos, F. and Fleminger, A. (1986). Distribution of filter-feeding calanoid copepods in the eastern equatorial

Pacific. *California Oceanic Fisheries Investigations Reports*, **27**, 170–87.

Ayre, D. H. and Hughes, T. P. (2000). Genotypic diversity and gene flow in brooding and spawning corals along the Great Barrier Reef, Australia. *Evolution*, **54**, 1590–605.

Barber, P. H. and Bellwood, D. R. (2005). Biodiversity hotspots: evolutionary origins of biodiversity in wrasses (*Halichoeres*: Labridae) in the Indo-Pacific and New World tropics. *Molecular Phylogeny and Evolution*, **35**, 235–53.

Barber, P. H., Palumbi, S. R., Erdmann, M. V. and Moosa, M. V. (2000). Biogeography: a marine Wallace's line? *Nature*, **406**, 692–3.

Barber, P. H., Erdmann, M. V. and Palumbi, S. R. (2006). Comparative phylogeography of three codistributed stomatopods: origins and timing of regional lineage diversification in the coral triangle. *Evolution*, **60**, 1825–39.

Bellwood, D. R. (1996). The Eocene fishes of Monte Bolca: the earliest coral reef fish assemblage. *Coral Reefs*, **15**, 11–19.

Bellwood, D. R. (2003). Origins and escalation of herbivory in fishes: a functional perspective. *Paleobiology*, **29**, 71–83.

Bellwood, D. R. and Hughes, T. P. (2001). Regional-scale assembly rules and biodiversity of coral reefs. *Science*, **292**, 1532–4.

Bellwood, D. R. and Meyer, C. P. (2009a). Searching for heat in a marine biodiversity hotspot. *Journal of Biogeography*, **36**, 569–76.

Bellwood, D. R. and Meyer, C. P. (2009b). Endemism and evolution in the coral triangle: a call for clarity. *Journal of Biogeography*, **36**, 2010–12.

Bellwood, D. R. and Wainwright, P. C. (2002). The history and biogeography of fishes on coral reefs. In *Coral Reef Fishes: Dynamics and Diversity in a Complex Ecosystem*, ed. P. F. Sale. San Diego, CA: Academic Press, pp. 5–32.

Bellwood, D. R., Hoey, A. S. and Choat, J. H. (2003). Limited functional redundancy in high diversity systems: resilience and ecosystem function on coral reefs. *Ecology Letters*, **6**, 281–5.

Bellwood, D. R., van Herwerden, L. and Konow, N. (2004a). Evolution and biogeography of marine angelfishes (Pisces: Pomacanthidae) *Molecular Phylogeny and Evolution*, **33**, 140–55.

Bellwood, D. R., Hughes, T. P., Folke, C. and Nystrom, M. (2004b). Confronting the coral reef crisis. *Nature*, **429**, 827–33.

Bellwood, D. R., Hughes, T. P., Connolly, S. R., and Tanner, J. (2005). Environmental and geometric constraints on Indo-Pacific coral reef biodiversity. *Ecology Letters*, **8**, 643–51.

Bellwood, D. R., Hughes, T. P. and Hoey, A. S. (2006). Sleeping functional group drives coral reef recovery. *Current Biology*, **16**, 2434–9.

Bellwood, D. R., Klanten, S., Cowman, P. F. et al. (2010). Evolutionary history of the butterflyfishes (f: Chaetodontidae) and the rise of coral feeding fishes. *Journal of Evolutionary Biology*, **23**, 335–49.

Benzie, J. A. H. (1999). Genetic structure of coral reef organisms: ghosts of dispersal past. *American Zoologist*, **39**, 131–45.

Blum, S. D. (1989). Biogeography of the Chaetodontidae: an analysis of allopatry among closely related species. *Environmental Biology of Fishes*, **25**, 9–31.

Briggs, J. C. (1999). Coincident biogeographic patterns: Indo-West Pacific Ocean. *Evolution*, **53**, 326–35.

Cairns, S. D. (2007). Deep-water corals: an overview with special reference to diversity and distribution of deep-water Scleractinian corals. *Bulletin of Marine Science*, **81**, 311–22.

Colwell, R. K. and Lees, D. C. (2000). The mid-domain effect: geometric constraints on the geography of species richness. *Trends in Ecology and Evolution*, **15**, 70–6.

Connolly, S. R., Bellwood, D. R. and Hughes, T. P. (2003). Indo-Pacific biodiversity of coral reefs: deviations from a mid-domain model. *Ecology*, **84**, 2178–90.

Cowman, P. F. and Bellwood, D. R. (2011). Coral reefs as drivers of cladogenesis: expanding coral reefs, cryptic extinction events and the development of biodiversity hotspots. *Journal of Evolutionary Biology*, **24**, 2543–62.

Cowman, P. F., Bellwood, D. R. and van Herwerden, L. (2009). Dating the evolutionary origins of wrasse lineages (Labridae) and the rise of trophic novelty on coral reefs. *Molecular Phylogenetics and Evolution*, **52**, 621–31.

Crandall, E. D., Jones, M. E., Munoz, M. M. et al.(2008). Comparative phylogeography of two seastars and their ectosymbionts within the Coral Triangle. *Molecular Ecology*, **17**, 5276–90.

Done, T. J., Ogden, J. C., Wiebe, W. J. and Rosen, B. R. (with contributions from the BIOCORE Working Group) (1996). Biodiversity and ecosystem function of coral reefs. In *Functional Roles of Biodiversity: A Global Perspective, SCOPE Report*, ed. H. A. Mooney, J. H. Cushman, E. Medina, O. E. Sala and E. D. Schultz. Chichester, UK: John Wiley and Sons, pp. 393–429.

Ekman, S. (1953). *Zoogeography of the Sea*. London: Sidgwick and Jackson.

Esteban, M. (1996). An overview of Miocene reefs from Mediterranean areas: general trends and facies models. In *Models for Carbonate Stratigraphy from Miocene Reef Complexes of Mediterranean Regions*, ed. E. K. Franseen, M. Esteban, W. C. Ward and J.-M. Rouchy. Tulsa, OK: SEPM (Society for Sedimentary Geology). *Concepts in Sedimentology and Paleontology*, 5, 3–53.

Fox, R. J. and Bellwood, D. R. (2007). Quantifying herbivory across a coral reef depth gradient. *Marine Ecology Progress Series*, **339**, 49–59.

Fransen, C. H. J. M. (2007). The influence of land barriers on the evolution of Pontoniine shrimps (Crustacea, Decapoda) living in association with molluscs and solitary ascidians. In *Biogeography, Time and Place: Distributions, Barriers, and Islands*, ed. W. Renema. The Netherlands: Springer, pp. 103–115.

Fraser, R. H. and Currie, D. J. (1996). The species richness–energy hypothesis in a system where historical factors are thought to prevail: coral reefs. *American Naturalist*, **148**, 138–59.

Fukami, H., Chen, C. A., Budd, A. F. et al. (2008). Mitochondrial and nuclear genes suggest that stony corals are monophyletic but most families of stony corals are not (Order Scleractinia, Class Anthozoa, Phylum Cnidaria). *PLoS ONE*, **3**, e3222.

Glynn, P. W., Veron, J. E. N. and Wellington, G. M. (1996). Clipperton Atoll (eastern Pacifc): oceanography, geomorphology, reef-building coral ecology and biogeography. *Coral Reefs*, **15**, 71–99.

Greenfield, D. W. (1968). The zoogeography of *Myripristis* (Pisces: Holocentridae). *Systematic Zoology*, **17**, 76–87.

Halas, D. and Winterbottom, R. (2009). A phylogenetic test of multiple proposals for the origins of the East

Indies coral reef biota. *Journal of Biogeography*, **36**, 1847–60.

Hall, R. (2002). Cenozoic geological and plate tectonic evolution of SE Asia and the SW Pacific: computer based reconstructions, model and animations. *Journal of Asian Earth Sciences*, **20**, 353–431.

Harzhauser, M., Kroh, A., Mandic, O. et al. (2007). Biogeographic responses to geodynamics: a key study all around the Oligo–Miocene Tethyan Seaway. *Zoologischer Anzuge*, **246**, 241–56.

Hawkins, B. A., Field, R., Cornell, H. V. et al. (2003). Energy, water, and broad-scale geographic patterns of species richness. *Ecology*, **84**, 3105–17.

Heads, M. (2005). Towards a panbiogeography of the seas. *Biological Journal of the Linnean Society*, **84**, 675–723.

Heck, K. L. and McCoy, E. D. (1978). Long-distance dispersal and the reef-building corals of the Eastern Pacific. *Marine Biology*, **48**, 349–56.

Hoeksema, B. W. (1989). Taxonomy, phylogeny and biogeography of mushroom corals (Scleractinia: Fungiidae). *Zoologische Verhandelingen, Leiden*, **254**, 1–295.

Hoeksema, B. W. (2007). Delineation of the Indo-Malayan centre of maximum marine biodiversity: the Coral Triangle. In *Biogeography, Time and Place: Distributions, Barriers, and Islands*, ed. W. Renema. Dordrecht, The Netherlands: Springer, pp. 117–78.

Hoeksema, B. W., Dautova, T. N., Savinkin, V. S. et al. (2010). The westernmost record of the coral *Leptoseris kalayaanensis* in the South China Sea. *Zoological Studies*, **49**, 325.

Hoey, A. S. and Bellwood, D. R. (2008). Cross-shelf variation in the role of parrotfishes on the Great Barrier Reef. *Coral Reefs*, **27**, 37–47.

Hubbs, C. L. (1958). General conclusions. In *Zoogeography*, ed. C. L. Hubbs. Washington DC: American Association for the Advancement of Science, pp. 468–79.

Hughes, T. P., Bellwood, D. R. and Connolly, S. R. (2002). Biodiversity hotspots, centres of endemicity, and the conservation of coral reefs. *Ecology Letters*, **5**, 775–84.

Hughes, T. P., Baird, A. H., Bellwood, D. R. et al. (2003). Climate change, human impacts, and the resilience of coral reefs. *Science*, **301**, 929–33.

Hughes, T. P., Rodrigues, M. J., Bellwood, D. R. et al. (2007). Phase shifts, herbivory, and the resilience of coral reefs to climate change. *Current Biology*, **17**, 360–5.

Humphries, C. J. and Parenti, L. R. (1999). *Cladistic Biogeography. Interpreting Patterns of Plant and Animal Distributions*, 2nd edn. New York: Oxford University Press.

Jablonski, D., Flessa, K. W. and Valentine, J. W. (1985). Biogeography and paleobiology. *Paleobiology*, **11**, 75–90.

Jablonski, D., Roy, K. and Valentine, J. W. (2006). Out of the tropics: evolutionary dynamics of the latitudinal diversity gradient. *Science*, **314**, 102–106.

Jokiel, P. L. and Martinelli, F. J. (1992). The vortex model of coral-reef biogeography. *Journal of Biogeography*, **19**, 449–58.

Jones, K. M. M., Fitzgerald, D. G. and Sale, P. F. (2002). Comparative ecology of marine fish communities. In *Handbook of Fish Biology and Fisheries*, vol. 2, ed. P. J. B. Hart and J. D. Reynolds. Oxford: Blackwell Publishing, pp. 341–58.

Kerswell, A. P. (2006). Global biodiversity patterns of benthic marine algae. *Ecology*, **87**, 2479–88.

Klanten, S. O., van Herwerden, L., Choat, J. H. and Blair, D. (2004). Patterns of lineage diversification in the genus *Naso* (Acanthuridae). *Molecular Phylogenetics and Evolution*, **32**, 221–35.

Ladd, H. S. (1960). Origin of the Pacific island molluscan fauna. *American Journal of Science*, **258(A)**, 137–50.

Landry, C., Geyer, L. B., Arakaki, Y., Uehara, T. and Palumbi, S. R. (2007). Recent speciation in the Indo-West Pacific: rapid evolution of gamete recognition and sperm morphology in cryptic species of sea urchin. *Proceedings of the Royal Society B-Bioogical Sciences*, **270**, 1836–47.

Lessios, H. A., Kessing, B. D. and Pearse, J. S. (2001). Population structure and speciation in tropical seas: global phylogeography of the sea urchin *Diadema*. *Evolution*, **55**, 955–75.

Malay, M. C. D. and Paulay, G. (2010). Peripatric speciation drives diversification and distributional pattern of reef hermit crabs (Decapoda: Diogenidae: *Calcinus*). *Evolution*, **64**, 634–62.

McCafferty, S., Bermingham, E., Quenouille B. et al. (2002). Historical biogeography and molecular systematics of the Indo-Pacific genus *Dascyllus* (Teleostei: Pomacentridae). *Molecular Ecology*, **11**, 1377–92.

McKenna, M. C. 1973. Sweepstakes, filters, corridors, Noah's Arks and beached Viking funeral ships in palaeogeography. In *Implications of Continental Drift to the Earth Sciences*, ed. D. H. Tarling and S. K. Runcorn. London: Academic Press, 1, 295–308.

McManus, J. W. (1985). Marine speciation, tectonics and sea-level changes in Southeast Asia. *Proceedings of the 5th International Coral Reef Congress*, **4**, 133–8.

Meyer, C. P., Geller, J. B. and Paulay, G. (2005). Fine scale endemism on coral reefs: archipelagic differentiation in turbinid gastropods. *Evolution*, **59**, 113–25.

Mora, C., Chittaro, P. M., Sale, P. F., Kritzer, J. P. and Ludsin, S. A. (2003). Patterns and processes in reef fish diversity. *Nature*, **421**, 933–6.

Myers, N., Mittermeier, R. A., Mittermeier, C. G., da Fonseca, G. A. B. and Kent, J. (2000). Biodiversity hotspots for conservation priorities. *Nature*, **403**, 853–8.

Palumbi, S. R. (1997). Molecular biogeography of the Pacific. *Coral Reefs*, **16**, S47–S52.

Pandolfi, J. M. (1992). Successive isolation rather than evolutionary centres for the origination of Indo-Pacific reef corals. *Journal of Biogeography*, **19**, 593–609.

Paulay, G. (1997). Diversity and distribution of reef organisms. In *Life and Death of Coral Reefs*, ed. C. Birkeland. New York: Chapman Hall, pp. 298–353.

Potts, D. C. (1985). Sea-level fluctuations and speciation in Scleractinia. *Fifth International Coral Reef Congress*, **4**, 127–32.

Razak, T. B. and Hoeksema, B. W. (2003). The hydrocoral genus *Millepora* (Hydrozoa: Capitata: Milleporidae) in Indonesia. *Zoologische Verhandelingen, Leiden*, **345**, 313–36.

Read, C. I., Bellwood, D. R. and van Herwerden, L. (2006). Ancient origins of Indo-Pacific coral reef fish biodiversity: a case study of the leopard wrasses (Labridae: *Macropharyngodon*). *Molecular*

Phylogenetics and Evolution, **38**, 808–19.

Reaka, M. L., Rodgers, P. J. and Kudla, A. U. (2008). Patterns of biodiversity and endemism on Indo-West Pacific coral reefs. *Proceedings of the National Academy of Sciences*, **105**, 11474–81.

Reece, J. S., Bowen, B. W., Joshi, K., Goz, V. and Larson, A. (2010). Phylogeography of two moray eels indicates high dispersal throughout the Indo-Pacific. *Journal of Heredity*, **101**, 391–402.

Reid, D. G., Lal, K., Mackenzie-Dodds, J. et al. (2006). Comparative phylogeography and species boundaries in *Echinolittorina* snails in the central Indo-West Pacific. *Journal of Biogeography*, **33**, 990–1006.

Renema, W. (2007). Fauna development of larger benthic foraminifera in the Cenozoic of Southeast Asia. In *Biogeography, Time and Place: Distributions, Barriers, and Islands*, ed. W. Renema. Dordrecht, The Netherlands: Springer, pp.179–215.

Renema, W. (2010). Is increased calcarinid (foraminifera) abundance indicating a larger role for macro-algae in Indonesian Plio-Pleistocene coral reefs? *Coral Reefs*, **29**, 165–73.

Renema, W., Bellwood, D. R., Braga, J. C. et al. (2008). Hopping hotspots: global shifts in marine biodiversity. *Science*, **321**, 654–7.

Roberts, C. M., McClean, C. J., Veron, J. E. N. et al. (2002). Marine biodiversity hotspots and conservation priorities for tropical reefs. *Science*, **295**, 1280–4.

Robertson, D. R. and G. R. Allen (1996). Zoogeography of the shorefish fauna of Clipperton Atoll. *Coral Reefs*, **15**, 121–31.

Rocha, L. A. and Bowen, B. W. (2008). Speciation in coral-reef fishes. *Journal of Fish Biology*, **72**, 1101–21.

Rosen, B. R. (1981). The tropical high diversity enigma: the corals' eye view. In *Chance, Change and Challenge. The Evolving Biosphere*, ed. P. H. Greenwood and P. L. Forey. Cambridge: British Museum (Natural History) and Cambridge University Press, pp.103–29.

Rosen, B. R. (1984). Reef coral biogeography and climate through the late Cainozoic: just islands in the sun or a critical pattern of islands? In *Fossil and Climate*, ed. P. Brenchley. Chichester, UK: John Wiley and Sons, *Geological Journal Special Issue* 11, 201–62.

Rosen, B. R. (1988a). Progress, problems and patterns in the biogeography of reef corals and other tropical marine organisms. *Helgoländer Meerensunters*, **42**, 269–301.

Rosen, B. R. (1988b). Biogeographical patterns: a perceptual overview. In *Analytical Biogeography: An Integrated Approach to the Study of Animal and Plant Distributions*, ed. A. A. Myers and P. S. Giller. London: Chapman and Hall, pp. 23–55.

Rosen, B. R. (1992). Empiricism and the biogeographical black-box: concepts and methods in marine paleobiogeography. *Palaeogeography, Palaeoclimatology, Palaeoecology*, **92**, 171–205.

Rosen, B. R. (2000). Algal symbiosis, and the collapse and recovery of reef communities: Lazarus corals across the K–T boundary. In *Biotic Response to Global Change: The Last 145 Million Years*, ed. S. J. Culver and P. F. Rawson. Cambridge: Cambridge University Press, pp. 164–80.

Rosen, B. R. (2002). Biodiversity: old and new relevance for palaeontology. *Geoscientist*, **12**, 4–9.

Rotondo, G. M., Springer, V. G., Scott, G. A. J and Schlanger, S. O. (1981). Plate movement and island integration: a possible mechanism in the formation of endemic biotas, with special reference to the Hawaiian-islands. *Systematic Zoology*, **30**, 12–21.

Roy, K., Jablonski, D., Valentine, J. W. and Rosenberg, G. (1998). Marine latitudinal diversity gradients: tests of causal hypotheses. *Proceedings of the National Academy of Sciences*, **95**, 3699–702.

Santini, F. and Winterbottom, R. (2002). Historical biogeography of Indo-western Pacific coral reef biota: is the Indonesian region a centre of origin? *Journal of Biogeography*, **29**, 189–205.

Smith, W. L. and Wheeler, W. C. (2006). Venom evolution widespread in fishes: a phylogenetic road map for the bioprospecting of piscine venoms. *Journal of Heredity*, **97**, 206–17.

Sodhi, N. S., Koh, L. P., Brook, B. W. and Ng, P. K. L. (2004). Southeast Asian biodiversity: an impending disaster. *Trends in Ecology and Evolution*, **19**, 654–60.

Springer, V. G. (1982). Pacific plate biogeography with special reference to shorefishes. *Smithsonian Contributions to Zoology*, **367**, 1–182.

Springer, V. G. and Williams, J. T. (1990). Widely distributed Pacific plate endemics and lowered sea-level. *Bulletin of Marine Science*, **47**, 631–40.

Stehli, F. G. (1968). Diversity gradients in pole location. Pt.1. The Recent model. In *Evolution and Environment*, ed. E. T. Drake. New Haven, CT: Yale University Press.

Stehli, F. G. and Wells, J. W. (1971). Diversity and age patterns in hermatypic corals. *Systematic Zoology*, **20**, 115–26.

Timm, J. and Kochzius, M. (2008). Geological history and oceanography of the Indo-Malay Archipelago shape the genetic population structure in the false clown anemonefish (*Amphiprion ocellaris*). *Molecular Ecology*, **17**, 3999–4014.

Verbruggen, H., De Clerck, O., Schils, T., Kooistra, W. H. C. F. and Coppejans, E. (2005). Evolution and phylogeography of *Halimeda* section *Halimeda* (Bryopsidales, Chlorophyta). *Molecular Phylogenetics and Evolution*, **37**, 789–803.

Veron, J. E. N. (1995). *Corals in Space and Time: The Biogeography and Evolution of the Scleractinia*. Sydney, Australia: UNSW Press.

Veron, J. E. N, Devantier, L. M., Turak, E. et al. (2009). Delineating the Coral Triangle. *Galaxea, Journal of Coral Reef Studies*, **11**, 91–100.

Vogler, C., Benzie, J., Lessios, H., Barber, P. and Worheide, G. (2008). A threat to coral reefs multiplied? Four species of crown-of-thorns starfish. *Biology Letters*, **4**, 696–9.

Walker, B. (1995). Conserving biological diversity through ecosystem resilience. *Conservation Biology*, **9**, 747–52.

Wallace, C. C. (2001). Wallace's line and marine organisms: the distribution of staghorn corals (*Acropora*) in Indonesia. In *Faunal and Floral Migrations and Evolution in SE Asia-Australasia*, ed. I. Metcalfe, J. M. B. Smith, M. Morwood and I. Davidson. Lisse, The Netherlands: Swets and Zeitlinger, pp. 171–81.

Wallace, C. C. and Rosen, B. R. (2006). Diverse staghorn corals (*Acropora*) in high-latitude Eocene assemblages: implications for the evolution of modern diversity patterns of reef corals. *Proceedings of the Royal Society*

(Series B Biological Sciences) **273**, 975–82.

Wallace, C. C., Paulay, G., Hoeksema, B. W. et al. (2000). Nature and origins of unique high diversity reef faunas in the Bay of Tomini, Central Sulawesi: the ultimate 'centre of diversity'? *Proceedings of the 9th International Coral Reef Symposium, Bali, Indonesia*, **1**, 185–92.

Weber, M. and de Beaufort, L. F. (1913). *The Fishes of the Indo-Australian Archipelago. Malacopterygii, Myctophoidea, Ostariophysi: Siluroidea*. Leiden, The Netherlands: Brill.

Wells, J. W. (1954). Recent corals of the Marshall Islands, *United States Geological Survey Professional Paper* **260**, I, 385–459.

Williams, S. T. (2007) Origins and diversification of Indo-West Pacific marine fauna: evolutionary history and biogeography of turban shells (Gastropoda, Turbinidae). *Biological Journal of the Linnean Society*, **92**, 573–92.

Williams, S. T. and Duda, T. F. (2008). Did tectonic activity stimulate Oligo-Miocene speciation in the Indo-West Pacific? *Evolution*, **62**, 1618–34.

Williams, S. T. and Reid, D. G. (2004). Speciation and diversity on tropical rocky shores: a global phylogeny of snails of the genus *Echinolittorina*, **58**, 2227–51.

Wilson, M. E. J. (2008). Global and regional influences on equatorial shallow-marine carbonates during the Cenozoic. *Palaeogeography Palaeoclimatology Palaeoecology*, **265**, 262–74.

Wilson, M. E. J. (2011). SE Asian carbonates: tools for evaluating environmental and climatic change in equatorial tropics over the last 50 million years. In *The SE Asian Gateway: History and Tectonics of Australia–Asia Collision*, ed. R. Hall, M. A. Cottam and M. E. J. Wilson. *Geological Society of London Special Publication*, 355, 347–72.

Wilson, M. E. J. and Rosen, B. R. (1998). Implications of paucity of corals in the Paleogene of SE Asia: Plate tectonics or centre of origin? In *Biogeography and Geological Evolution of SE Asia*, ed. R. Hall and J. D. Holloway. Leiden, The Netherlands: Backhuys Publishers, pp.165–95.

Woodland, D. J. (1986). Wallace's line and the distribution of marine inshore fishes. In *Indo-Pacific Fish Biology, Second International Conference, Tokyo, Japan*, 29 July–3 Aug., Tokyo, Japan, ed. Uyeno, T. et al. The Ichthyological Society of Japan, pp. 453–60.

10

Tsunami impacts in the marine environment: review and results from studies in Thailand

GORDON L. J. PATERSON, MICHAEL A. KENDALL, CHITTIMA ARYUTHAKA, NICHOLAS J. EVANS, YAOWALUK MONTHUM, CHAWAPORN JITTANOON, PATRICK CAMPBELL, LESLIE R. NOBLE AND KEVIN M. O'NEILL

10.1 Introduction

On 26 December 2004 a tsunami, generated by tectonic movements off the coast of Sumatra, struck coastlines throughout Southeast Asia causing considerable destruction and loss of life. Over 270 000 people died and in excess of US$9.9 billion of damage was done (www.alnap.org/pool/files/tsunami-stats-facts.pdf). Although both the media and national agencies reacted to these events with considerable surprise, tsunamis are more frequent than is perhaps appreciated. The majority of tsunamis occur in areas of high tectonic activity such as the Pacific Rim. Recorded tsunami frequency in Southeast Asia and the Western Central Pacific since 1900 is shown in Fig 10.1. Between 1893 and 2005 there were 28 tsunami events greater than intensity six in the Indonesian and West Pacific area (Global Tsunami Database: http://tsun.sscc.ru/) and globally between 1983 and 2001 there were 157 events, most of which caused little damage (Lander et al. 2003). There have been only 12 large, transoceanic tsunamis since 1755 (Gusiakov 2009; plus the 2011 Japanese event).

Biotic Evolution and Environmental Change in Southeast Asia, eds D. J. Gower et al. Published by Cambridge University Press.

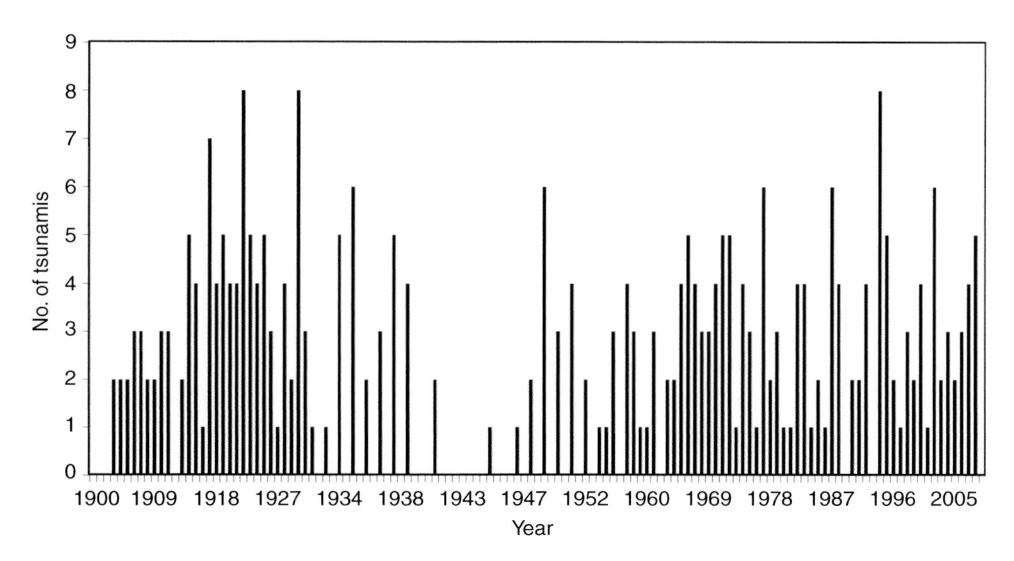

Figure 10.1 Frequency of tsunami events in the Indonesian archipelago and Western Central Pacific. Source: National Geophysical Data Centre (http://www.ngdc.noaa.gov/hazard/tsu_db.shtml).

The infrequency of damaging tsunamis led to their being given a low priority in disaster and relief planning prior to 2004. The Southeast Asian Tsunami has resulted in re-evaluation but effective planning must be based on sound knowledge of the consequences of tsunami impacts and the rate at which the environment recovers from damage. Research into the marine ecosystems of Southeast Asia is patchy both in terms of location and habitats. While there is a significant focus on coral reef science, little is known of the ecology of the sediments that form the basis for so many valuable fisheries and that underpin the livelihoods of many people.

In this chapter we review the limited literature available about tsunamis, the nature of the impacts and their effects on natural communities, both from previous events and the 2004 tsunami. We also present our recent research based on short time-series studies undertaken in Thailand after the 2004 tsunami on soft sediment and coral reef communities. These studies directly evaluate the changes that have occurred, and they allow the tsunami disturbance and recovery to be put into a local context. Finally, the ecological disturbance and recovery created by tsunamis are compared to and contrasted with other large-scale environmental disturbances in Thailand, especially storms and typhoons.

10.2 Review of tsunami environmental impacts

A range of tsunami impacts has been reported:

- Physical damage to the structure of coral reefs and mangrove forests caused by the passing of a succession of large, high-energy waves. While these effects can be severe on the open coast, they may be intensified by coastal morphology.
- The suspension of inshore and intertidal sediment leading to the displacement and loss of the biota and changes in sediment granulometry (Kendall et al. 2006, Grzelak et al. 2009). Associated effects are short-term increases in water column turbidity, which causes both impairment of feeding in some species and a loss of planktonic productivity. While the passage of the wave may release some nutrients, it also has the potential to release buried toxic material (Szczuciński et al. 2005, 2006, 2007).
- Damage cause by wave-driven debris. This can cause severe damage to the biota of hard substrata as well as impacting coastal infrastructure such as aquaculture facilities and town sewage systems.
- The introduction of saline waters to terrestrial soils and aquifers; such changes may have a long-lasting impact on agriculture as soils can no longer support traditional crops (Kume et al. 2009, MCleod et al. 2010).

In sediment ecosystems the changes brought about by the passage of the tsunami wave series are particularly severe. Vast quantities of sediment are moved, causing major changes to shoreline morphology. Sand is stripped away from some areas and deposited in others. As the waves pass through shallow water the seabed and the shallow-living animals within in it are lifted into suspension; not only is there substantial defaunation, but also the physical characteristics of the habitat are considerably changed. Tsunamis have been shown to be responsible for both long-term (Bryant et al. 1992) and short-term sediment compositional changes (Altaff et al. 2005, Kendall et al. 2006, Kotwicki and Szczuciński 2006). These, in turn, have strong implications for the ability of species with clear sediment preferences to recolonise and ecosystem services to recover.

10.2.1 Historical record

Prior to the 2004 event, records of tsunami effects make little mention of the marine environment; those records that do exist are summarised in Table 10.1.

It is clear that in the aftermath of these major destructive events, assessment of the marine environment has not been a major priority. Where records do occur they normally relate to the terrestrial deposition of marine material, which is a strong indicator of offshore disturbance. It is also evident that the impacts of a tsunami cannot be separated from those of associated seismic activity where uplift of the shore or shallow coral reefs has been reported.

10.2.2 The 2004 Asian Tsunami

This is the first tsunami for which we have large-scale information on the effects on natural marine assemblages (Wilkinson et al. 2006, Stoddart 2007). Surveys

Table 10.1 Summary of major tsunami events and their biological and environmental impacts

Location	Date	Impact	Reference
Krakatoa	1883	Waves up to 40 m deposited coral blocks up to hundreds of metres inland. Nevertheless, massive coral reefs survived.	Sokoliev et al. 1992
Chile and Peru	1960	No reports of direct tsunami impact but tectonic movement lifted the intertidal zone by around 1m causing the death of the biota and lasting changes in patterns of zonation. The waves reached Hawaii where damage to coral was reported.	Castilla 1998
Sulawasi, Indonesia	1969	Coral reefs fringing the region were damaged with coral being deposited on land.	Sokoliev et al. 1992
Flores, Indonesia	1992	Waves up to 25 m in height travelled up to 700 m inland. Impact of this tsunami on local coral reefs was minimal; the damage recorded on reef flats and shallow reefs in Maumere Bay was similar to that found after a typhoon. Fish, shell and coral debris was deposited up to 150 m inland, the slope of the beach was changed but no ecological changes in sediment habitats were reported.	Lander et al. 2003 Tomascik et al. 1997
Peru	1996	Although no reports were made about the impact on the natural environment, sand was deposited up to 100 m inland and the noise noted, as it dragged pebbles and shells down the beach slope, suggested major erosion of the near shore and intertidal environments.	Lander et al. 2003
Papua New Guinea	1998	Damage to mangroves up to 1300 m inland but some palms growing near the shore survived. Mangroves may have exacerbated the impact of the tsunami.	Lander et al. 2003
Sea of Marmora, Turkey	1999	Tons of fish washed ashore. Damage to infrastructure caused extensive pollution.	Lander et al. 2003

of marine habitats, particularly coral reefs, were undertaken within months of the event providing interesting insights into the effects of large-scale tsunamis. A useful summary of the geology and physical aspects of the tsunami is given by Spencer (2007).

Coral reefs

Coral reefs across the whole of Indonesia were assessed as having 30% damage to 97 250 ha of reefs (Gunawan et al. 2006). The damage was inconsistent between sites, with some reefs structurally damaged by the earthquake while nearby reefs were minimally affected. Many reefs showed moderate tsunami-related impacts and some reefs were entirely destroyed.

The reef systems off Sumatra, those closest to the epicentre, showed mixed levels of tsunami damage. Hagan et al. (2007) reported that, from a rapid survey of the region, 57% of the coral reefs appeared untouched, and that only 15% of the sites surveyed had severe damage. The main effect was in overturning large and tabular coral although fragmentation of upright colonies of taxa such as *Heliopora* was reported. Campbell et al. (2007) reported that the reef fish fauna appeared unaffected following the tsunami and that the event did not alter existing conservation priorities.

In Thailand there are long-term data on coral reefs, which greatly facilitated the assessment of the impact of the tsunami; only 13% of 174 coral systems surveyed along the Andaman Coast showed signs of severe damage, while 40% of sites surveyed showed no visible impact (Phongsuwan et al. 2006, Phongsuwan and Brown, 2007). Damage was patchy and perhaps specific to certain areas. Shallower fringing reefs near the shore were badly affected with large coral heads being toppled and other colonies smothered by sediment. Nevertheless, the authors predicted that most coral systems affected by the tsunami would recover fully in 10 years. A similar figure of 10 to 15 years for coral reef recovery is cited for sites throughout the Indian Ocean. Chavanich et al. (2008) suggested that the slope of the reef and the aspect of the coast partially determined the severity of the impact, with shallow sloping reefs suffering more damage than those with a steep drop off. Their results from the Similan Islands, off the Andaman coast of Thailand, also indicated that massive and tabular corals suffered greater damage than foliose and branching taxa.

In the Andaman and Nicobar Islands, Patterson Edwards et al. (2006) reported that damage to reefs ranged from severe to minor or unaffected; the most severe effects resulted from uplift of large sections of reef as a result of the earthquake. Other severe damage occurred in channels, which concentrated the force of the tsunami. On the Indian mainland, coral reef damage was not extensive with some branched or massive corals being damaged mostly by debris in the water or by being smothered by silt. As with other areas around the Indian Ocean, damage to coral reefs in Sri Lanka was patchy with extensive damage in some reefs. There are reports of large *Porites* heads being uprooted and deposited as far as 150 m inland, together with overturning or fragmentation of branching species such as *Acropora* (Rajasuriya et al. 2006). The tsunami waves also mobilised coral debris that had accumulated following a widespread coral-bleaching episode in 1998. Following the tsunami, the amount of coral rubble in some reefs areas doubled, burying live corals. The effects of water movement on large and tabular corals were noted from sites in the Maldives, particularly the channels between the islands (Gischler and Kikinger 2007). Here many of these large corals colonies were situated on unconsolidated coral rubble and so were susceptible to being knocked over.

Mangroves

The impact of a tsunami on a mangrove ecosystem is twofold, the most obvious being the destruction of the trees themselves. In areas of Thailand most heavily affected by the 2004 tsunami all trees were broken and killed within 50 m or more of the sea. The damage to the seaward trees protected the remaining forest, which might extend a considerable distance from the coast and it is also possible that the trees protected human settlements in some places. In other places the mangrove channels focused the tsunami wave, increasing its force and the damage caused to people and their property (S. Nimsantijaroen pers. comm. 2005). Less obvious was the impact of suspended sediment being deposited across the forest floor and completely changing the composition and ecological function of the invertebrate community living there (Kendall et al. 2009). Mangroves may be significant sources of nutrients for the inshore ecosystems and heavy sedimentation may impede this vital ecosystem service. In other parts of the Thai coast, reports indicate that some mangrove forests sustained little or no major damage (Gunawan et al. 2006, Phongsuwan et al. 2006, Tun et al. 2006); the differences between sites was generally related to the aspect of the coast relative to the wave front.

In the Andaman Islands tsunami damage ranged from negligible to almost total destruction, and was related to the position of the islands within the archipelago, with the middle islands being especially badly affected (Patterson Edwards et al. 2006). Erosion and changing sea levels were factors thought to be responsible for the death of mangrove trees. It appears from reports and personal observations that only the seaward edges of impacted forests were damaged.

Sediment systems

Many of the reports in Lander et al. (2003) and Wilkinson et al. (2006) describe changes to the topography of local beaches but do not mention any changes to the fauna. Seagrass beds were mostly undamaged by the tsunami because most occur in sheltered coastal areas with little wave exposure. The greatest seagrass damage was reported from the Andaman Islands but there was no detailed description of the changes (Patterson Edwards et al. 2006). Damage to seagrass was thought to have occurred when the waves withdrew scouring the seabed and tearing out plants.

In a study of the meiofauna of beaches on the Chennai coast, Altaff et al. (2005) reported that there were distinct changes to the slope of the beach but that the grain size composition of the sediment returned to pre-tsunami levels within days to weeks, indicating habitat resilience. The abundance of meiofauna decreased after the tsunami and there was a shift in the dominance patterns of the main groups, a possible response to changed grain size. However, meiofaunal abundance density increased after the third day, and coincided with changes in sediment and biogeochemical factors. A month after the tsunami the beach meiofauna was broadly similar to that before the passage of the wave. In Thailand, Kotwicki and Szczuciński (2006) studied a range of sandy beaches on Phuket and Kho Khao Islands with differing degrees of exposure and tsunami impact. They noted that, despite severe erosion in some of their study beaches, meiofauna assemblages appeared to be similar to those from non-impacted beaches, suggesting that either the meiofauna had not been so severely affected or had rapidly recovered. These findings point to a remarkably fast recovery both of the fauna but also of the environment and contrast with Kendall et al. (2006) who noted that four months after the tsunami the beach sediment at sites in Ranong Province further to the north had not returned to pre-disturbance condition. The difference in recovery between beaches in Chennai and in Thailand is probably related to the local physical environment.

Despite their importance to fisheries, there are no long-term data that relate to the offshore soft sediments and their biota. Nevertheless, Kendall et al. (2009) were able to compare pre- and post-tsunami patterns of sediment and recorded substantial shifts in the distribution of fine-grained material. However, given the period separating the two surveys, it was not possible to conclude unequivocally the changes were a direct consequence. Evidence supporting the disturbance of the bottom in water shallower than 16 m came from the size frequency of burrowing sea urchins; samples contained few animals that pre-dated the tsunami but were heavily dominated by individuals settling immediately after it. They assumed that rapid larval recruitment came from larvae supplied by animals living in deeper, undisturbed areas or from larvae already in the water column and which survived the waves and subsequent increases in suspended matter.

In Malaysia post-tsunami inshore fisheries landings were reported to have dropped by more than half (Tun et al. 2006). This is not necessarily a biological effect because, throughout Southeast Asia, the tsunami caused substantial destruction of boats and fishing gear and in the months that followed there was a great reluctance of people to eat seafood that they thought might have been contaminated by human remains.

In Ranong Province, Thailand the provision of new equipment allowed fishing to resume a year after the tsunami. Although there were anecdotal reports of decreases in stocks, these are difficult to attribute directly to the tsunami or to natural cycles. A year after the tsunami, village surveys suggested little concern about productivity of the fisheries (Kendall et al. 2009).

Rocky shores

Rocky habitats are poorly known and rarely studied in tropical areas. Nevertheless they do have interesting assemblages of plants and animals. The impact of tsunami on the organisms of the littoral fringe reflected the disturbance created. In a study of littorinid snails, Sanpanich et al. (2006) found that on rocky shores in Phuket and Kho Khao littorinid snail survival was patchy and that species were often missing from areas of tsunami impact. However, the authors noted that the species concerned had wide geographic distributions and planktonic larvae, facilitating recolonisation. Patterson Edwards et al. (2006) reported that barnacles disappeared from many rocky habitats and reefs on the Indian coast to be replaced with fine turf algae, rubble and sponges. In the rocky intertidal areas of southern Ranong, Kendall et al. (2006) also found that there had been considerable mortality of barnacles caused by scouring and settlement of fine sediment that dried into a hard layer. There were signs of recruitment in the following year. The authors noted that mussels had been eliminated from the lower tidal zone and had begun to recolonise only after 4 years (Kendall pers. obs.).

Littoral terrestrial habitats

In northern Sumatra there was little damage to the estuary systems and wetlands. There was also no reported impact on turtle nesting sites (although Gunawan et al. 2006 suggested that the erosion of many beaches would have had an impact) in contrast to the Andaman and Nicobar Islands and Thai shores where turtle nests were exposed. Terrestrial trees with low or no salt tolerance were killed off in many areas by inundation of freshwater aquifers. Kendall et al. (2006) noted mortality in terrestrial trees, such as cashew, in the near-shore areas in Ranong but salt-tolerant species such as *Cassuarina* survived well.

The run-up of the tsunami waves caused mortality in ground-nesting seabirds on Latham Island in Tanzania (Crawford et al. 2006). Sheppard (2007) reported that there were no immature individuals or chicks of the brown booby (*Sula leucogaster*) in North Brother of the Chagos Islands and suggested that this was due to the colony having been inundated. On islands as far away as the Crozet Archipelago breeding of king penguins (*Aptenodytes patagonicus*) was affected (Viera et al. 2006). Yet on other islands the nesting birds were unaffected despite signs of tsunami impact.

The conclusion that can be drawn from the various reports and surveys is that the impact of the tsunami on the natural environment was patchy, with geography playing a large role in whether an area was impacted or not. Damage to many areas, though apparently severe, was within the normal range of effects following other major environmental disturbances.

10.3 Impacts of 2004 tsunami in Ranong and Phang Nga Provinces, Thailand

One difficulty in trying to assess the effects of the tsunami was that few localities had been surveyed before the tsunami, or surveys that were carried out have not always been part of regular monitoring. It is difficult, therefore, to place the 2004 event in the context of natural cycles and periodic large-scale disturbance. The focus on coral reefs by post-tsunami surveys stems partly from the fact that many have been part of survey programmes and so have data collected before 2004. Time-series data are not common for other habitats, even economically important mangroves. In this section, new data on sediment habitats along the coast of Ranong and Phang Nga Provinces in Thailand taken as part of a short-term monitoring project from 2001 to 2004 and following the tsunami from 2006 to 2008 (Kendall et al. 2006, 2009) are presented. Offshore coral reefs were assessed over three consecutive years as part of UK Royal Air Force Sub-Aqua Club diving expeditions. These studies concentrated on the Similan Islands off Phang Nga province and supplement observations made by Phongsuwan et al. (2006) and Chavanich et al. (2008).

10.3.1 Intertidal sediment systems

Four sites along the coast of southern Ranong and northern Phang Nga Provinces were selected to assess the impact of the tsunami on intertidal habitats. All have been part of biodiversity surveys pre-tsunami and then part of a programme monitoring recovery. Three of the sites were impacted by the 2004 tsunami: Thalae Nok, an open sandy beach; Kampuan, a tidal estuary; Ao Khoei, a semi-open beach with mixed sediment lying over rock. The fourth site, Mae Hang, is a semi-sheltered seagrass bed thought to have been unaffected or at worst subject to only minor disturbance (Fig 10.2). The same quantitative collecting methodology was employed at all sites (see Kendall et al. 2006, Barrio Frojan et al. 2006 for details). The data presented here supplement the initial reports of Kendall et al. (2006, 2009).

The initial reports of Kendall et al. (2006) showed that the sediments of the beaches along this part of the coast were severely disrupted and mixed by the succession of tsunami waves. The first hypothesis considered was that the tsunami

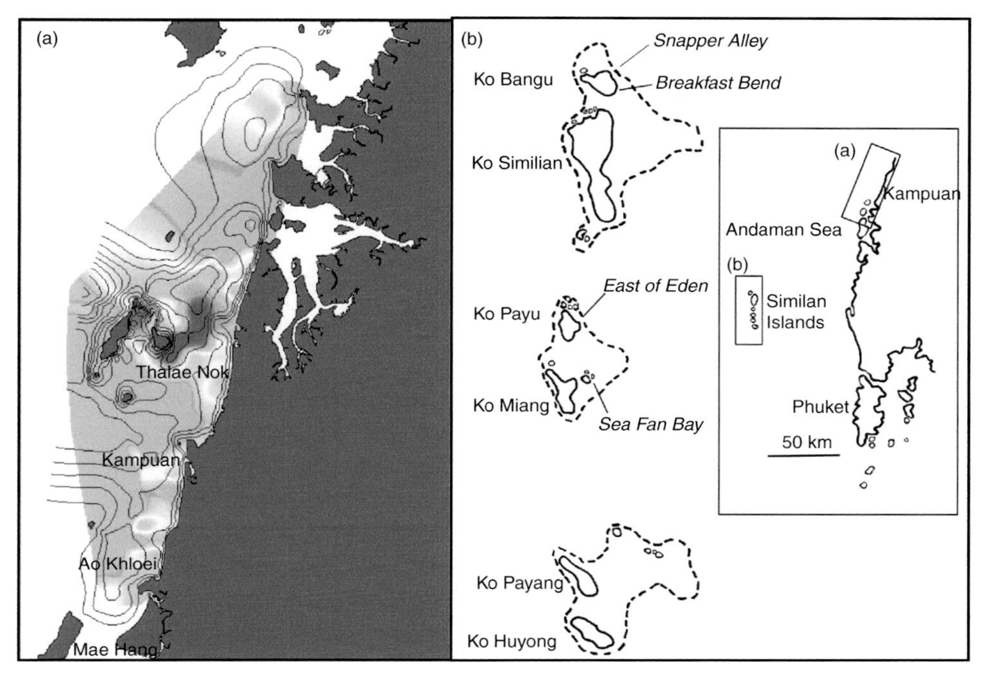

Figure 10.2 Localities for the studies of the impact of the 2004 Tsunami. (a) The four intertidal study areas on the coast of southern Ranong and northern Phang Nga provinces. (b) The dive localities in the Similan Islands off the west coast of Thailand.

would have decreased species abundances by removing existing fauna and, as tropical systems are less dynamic than boreal, recovery would take longer. The second hypothesis tested was that the tsunami would have changed the faunal assemblages. A major disturbance such as the tsunami would allow a different suite of species to colonise, resulting in a shift of dominance and composition. To test these hypotheses, changes in the polychaete fauna (identified to the family level) from pre- and post-tsunami sampling periods at each site were examined.

Polychaete abundance

In general there was no difference in polychaete abundance before and after the tsunami. Figure 10.3 indicates that interannual variation is a key feature of all these sites. The open beach site, Thalae Nok, had very low abundance and low numbers of families. There was a long period between successive sampling which precludes any major comparisons of pre- and post-tsunami but it is clear from the post-tsunami samples that the variation observed is high between time samples (Fig 10.3a). The Kampuan site is on a dynamic, small estuary and the sediment is subjected to deposition and erosion following monsoons. There was a suggestion

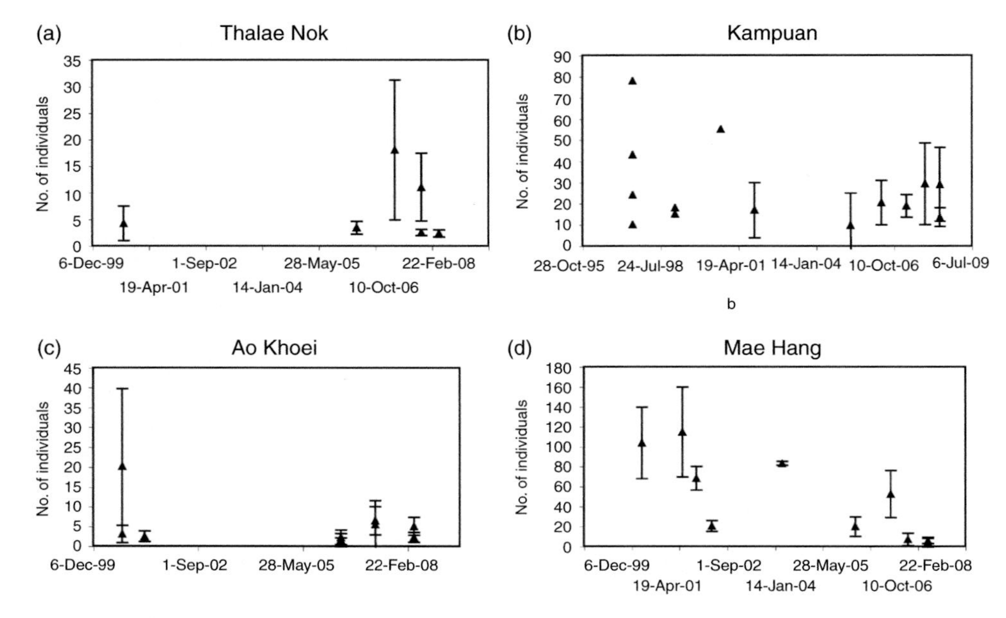

Figure 10.3 Comparison of polychaete abundance at four sites. (a) Thalae Nok, abundances were too low to perform ANOVA analyses; *t*-test of pooled pre- and post-tsunami samples was not significant $p > 0.05$. (b) Kampuan, ANOVA analyses showed significant differences between samples $p < 0.001$, $df = 11$, $F = 6.558$; Tukey–Kramer post hoc test indicated that August 1997, September 2000 and November 2001 had significantly higher abundance than pre- and post-tsunami samples. (c) Ao Khoei, ANOVA analyses indicated significant difference but post hoc testing showed that this was attributed to differences between post-tsunami samples and not those collected before and after impact. (d) Mae Hang, ANOVA and post hoc analyses point to the abundance of three pre-tsunami samples as being significantly higher than post-impact. There were not enough successful replicates to include the April 2004 samples but here the abundance is higher than those post-tsunami. Error bars are standard error of the mean.

that post-tsunami abundance was lower than pre-event but it was not statistically significant (see Fig 10.3b).

The same overall pattern can be seen in Ao Khoei, where there is a high degree of variance between sampling dates but with no obvious pre- and post-tsunami pattern (Fig 10.3c). Abundance at the seagrass bed site, Mae Hang, was the highest of the sites surveyed (Fig 10.3d). This site appeared to be the exception to the patterns noted above. While sample variance was often high, there appeared to be a trend of decreasing abundance through time. There is a significant difference between the abundance, with pooled pre-tsunami samples having a higher overall abundance compared with post-event samples (*t*-test, $p < 0.0001$). ANOVA results indicate a significant difference between samples; post-hoc testing indicates that three samples taken in May 2000, May 2001 and October 2001 had significantly higher mean density than those from the other sampling periods (both pre- and post-tsunami) indicative that the apparent decline in abundance with time may be due to particularly high densities recorded early in the monitoring programme. Mae Hang occasionally suffers from inundation of sand, which covers large areas of the seagrass. Such events are thought to be associated with strong onshore winds. The drop in abundance from May 2001 to October 2001 and subsequently may be the result of the sand inundation noted by Kendall et al. (2009). There is a rise in April 2004, which can be explained by some areas of seagrass escaping the sand inundation. However, the movement of sand by normal tidal processes may well have affected a greater area in subsequent years. It is possible the passage of the tsunami close to this site contributed to this movement of sediment.

Polychaete family richness

Changes in the mean numbers and total number of families recorded at the sites are given in Fig 10.4. In general, there is no obvious change in the family composition recorded pre- and post-tsunami.

The open beaches of Thalae Nok do not support a diverse fauna and number of families recorded was low. Changes that were observed cannot be ascribed to any particular event and probably represent normal interannual variation. Analyses of the composition of the samples using similarity analyses also fail to show any changes that can be linked to the tsunami, with pre-tsunami time samples clustering within post-tsunami ones (Fig 10.5a). The relatively high similarity of the samples from the different dates points to the same families being present. The differences between the dates were due to changes in abundance rather than changes to the families that occur (Capitellidae, Cirratulidae and Orbiniidae were the most abundant).

High variability in Kampuan reflects the dynamic nature of this estuary. The data indicate that there was variability in both the mean and total numbers of families reaching a low point between 2001 and 2006 (Fig 10.4b). Again this is not linked to the tsunami but may be related to other interannual events such as the monsoon. However, it is noted that the sampling regularity was not sufficient to capture interannual variation resulting from seasonal changes. The similarity analyses suggested that there is perhaps more structure relating to changes in family composition pre- and post-tsunami (Fig 10.5b). However, this was due mainly to two outliers both with low numbers of families and high abundance in families not recorded in other sample periods. For example, the samples from 8 November 2001 had only three families recorded and were dominated by small individuals belonging to the family Nereididae. This family was not present in other samples. The composition of the families changes slightly in that many families found at the beginning and at the end of the study were missing from the middle.

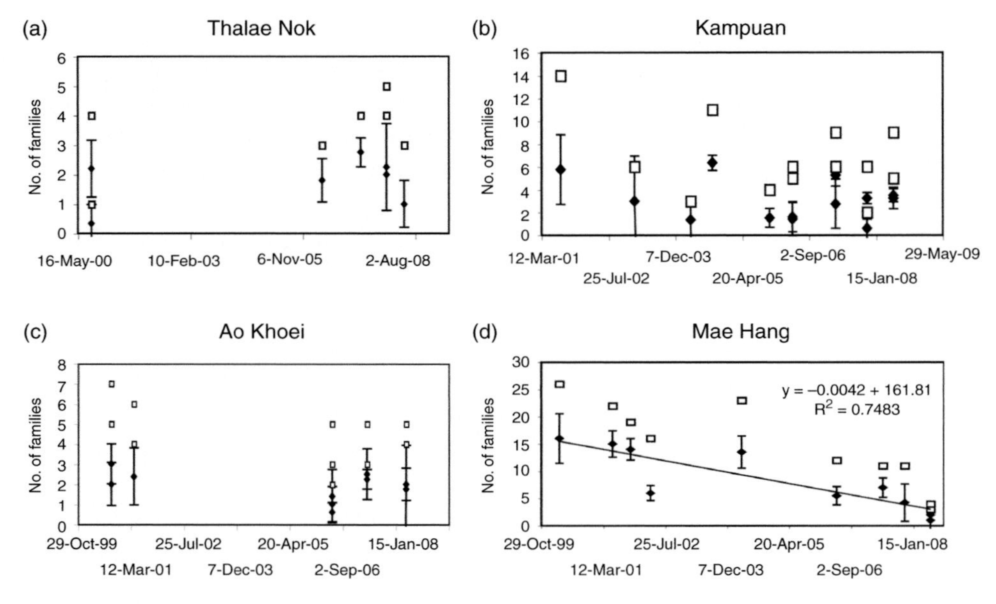

Figure 10.4 Comparison of the mean number of families at each sampling time (black diamonds) and total number of families recorded (open squares). (a) Thalae Nok, no obvious differences between pre- and post-samples. (b) Kampuan, again ANOVA analyses with post hoc testing failed to find significant differences that could be attributed to tsunami effects. (c) Ao Khoei, ANOVA analyses did not find any differences between sampling periods. (d) Mae Hang, ANOVA and post hoc analyses indicated that three of the four pre-tsunami samples had higher numbers of families than post-tsunami ones (ANOVA df = 8, $p < 0.001$, $F = 18.356$). Linear regression points to a significant decrease in mean numbers of families through time. Error bars are standard error of the mean.

Analyses of the family numbers at Ao Khoei showed no major shifts (Fig 10.4c). Mean numbers of families were similar pre- and post-tsunami. There appears to be a decrease in the abundance of families post-tsunami but this is not significant. Changes were observed in the families Capitellidae, Orbiniidae and Paraonidae, which were common at the start of the study but absent from samples taken after the tsunami. Again, although it is tempting to attribute this observation to changes in sediment characteristics associated with the tsunami, there are no data to support this and interannual variation cannot be ruled out. ANOSIM analyses indicated that there were no significant compositional differences between pre- and post-samples.

The decrease in abundance noted in Mae Hang is also reflected in the mean and total numbers of families recorded in the samples with time (Fig 10.4d). This decrease appears to be statistically significant. Post hoc test of the ANOVA analyses point to significant differences between samples at the start of the study and

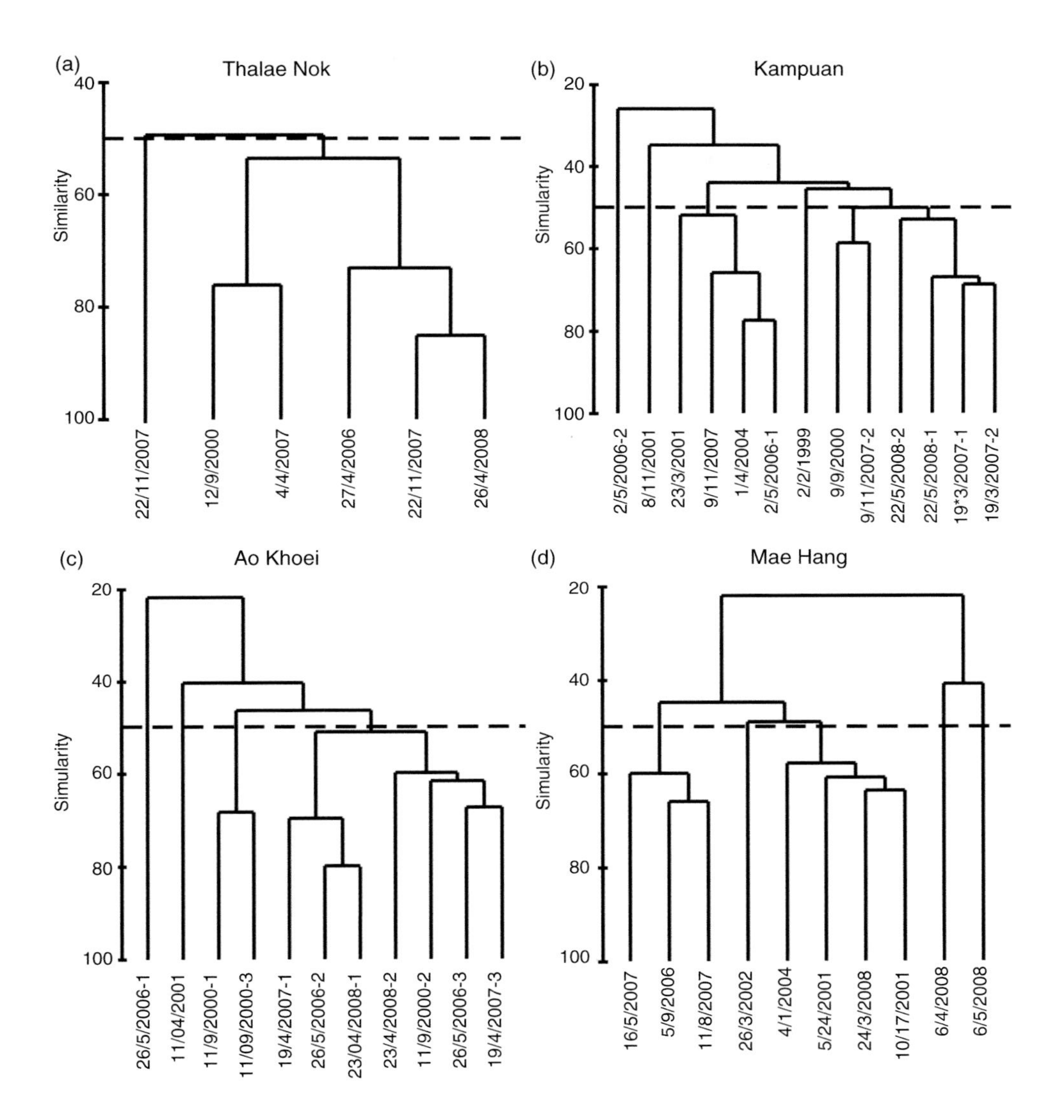

Figure 10.5 Analyses of species composition. PRIMER was used to analyse sample familial similarity between sampling dates. Bray–Curtis similarity was used after root transformation, the results was clustered using group averaging. ANOSIM analyses were also performed. (a) Thalae Nok. (b) Kampuan. (c) Ao Khoei. (d) Mae Hang. None of the analyses of the first three sites indicated separation of samples attributable to the tsunami but the exception was Mae Hang, where there is a distinct pre- and post-tsunami separation. However, the ANOSIM significance level of sample statistic was 0.8%, close, but not statistically significant.

those after the tsunami (see Fig 10.4 caption for details). A regression of the mean number of families with sampling date shows a negative correlation ($R^2 = 0.748$). The similarity analyses show a more distinct separation between early, pre-tsunami and later, post-tsunami samples. However, this is not due to a shift in the family composition but to the decrease in abundance and resulting absence of some families.

Discussion

All the sites showed a similar basic pattern with variation between sampling periods often more obvious than any pre- and post-tsunami effects, with the exception of the seagrass beds at Mae Hang. Such changes that did occur cannot be attributed unequivocally to the tsunami. The impacted beaches were not visited until 5 months after the event and it is entirely possible that recovery from the impacts was rapid. The studies from meiofauna indicate that recovery in dynamic open beaches was remarkably fast (Altaff et al. 2005) and this may apply to intertidal macrofauna (Kendall et al. 2006). It appears that the tsunami did not result in any lasting decrease in the abundance of polychaetes. Similarly, the tsunami did not seem to result in a shift in family composition of polychaetes. This suggests that the same basic groups recolonised after the disturbance, although perhaps in lower numbers at some sites. This observation is in agreement with the potential for a faster than predicted recovery. Also the fauna did not alter despite shifts in sediment, most becoming coarser after the tsunami (Kendall et al. 2006).

It has to be noted that we are dealing with a higher taxonomic level and that it is possible that different species were able to colonise the beaches and mudflats after the tsunami. However, the species to family ratio is quite low in our samples (G.L.J. Paterson and M.A. Kendall pers. obs.) so we would expect that these results based on the family level would be an accurate indicator of general trends at lower taxonomic levels. A general conclusion is that intertidal sediment faunas are as dynamic as those in the boreal systems and that considerably more research is needed to establish the ecological drivers in tropical sediment ecosystems.

10.3.2 Offshore assessment of the tsunami in Thailand: the Similan Islands

Introduction

A series of dive surveys were undertaken with RAF Sub-Aqua Association divers within the Similan Islands National Park in 2006, 2007 and 2008 to monitor the reef systems post-tsunami and to supplement the assessment by Phongwasan et al. (Phongsuwan et al. 2006, Phongsuwan and Brown, 2007). The sites surveyed range from those which experienced tsunami impacts to those that were reported as untouched. The Similans were selected because they have minimal anthropogenic impacts (Fig 10.2). Four localities were investigated: Ko Payu, 'East of Eden'; Ko Bangu, 'Snapper Alley'; Ko Bangu, 'Breakfast Bend'; Ko Miang, 'Sea Fan Bay'. At each locality two sites were investigated; one at a shallow depth (at about 6 m allowing for tidal variation) and at a deeper one (about 15 m) using a survey scheme based on Reef Check (Hodgson et al. 2006). Sites were located by GPS because permanent markers could not be used in a National Park. Although the precise location of the sites varied between the different surveys, they were all in close proximity to a particular site (within 50 m). The sites were selected as being representative for survey after an initial diver assessment of the reef system in the survey area. 'Snapper Alley' and 'Sea Fan Bay' were regarded as having been impacted by the tsunami, and 'East of Eden' and 'Breakfast Bend' as being relatively un-impacted on the basis of literature reports.

Survey methodology

At each site, surveying followed the basic scheme given by Reef Check with one main difference. Hard coral recording was expanded (based on the AIMS life forms system for recording coral cover, English et al. 1997) because some hard corals, for example branching and tabular forms of *Acropora* species, were reported to be impacted/more susceptible to tsunami impact, while species with a massive growth form, for example, *Porites*, were both less susceptible and also more resistant to bleaching and other stressors (Brown and Bythell 2005). In the 2007 and 2008 surveys five additional fish groups were added to give a wider coverage to different feeding guilds. Similarly, the range of invertebrates had increased.

Results: fish surveys

Overall, the results from the fish surveys indicate that fish stocks appear to be healthy with only rabbit fish (siganids) showing any significant differences in numbers between the study sites in 2008. There were no significant differences in the other fish species surveyed between severely impacted sites and unimpacted sites in 2008, including Reef Check indicator species from butterfly fish (chaetodontids) and parrotfish (scarine labrids).

In 2007 only, species of grouper (epinephelinids) showed significant differences between sites, but the pattern of difference was not correlated with impacted or unimpacted sites. The higher relative numbers of algal grazers including tangs (acanturids) and parrotfish (scarine labrids) might indicate the presence of increased algal cover (perhaps from mechanical damage or bleaching) but should also facilitate recolonisation by corals. The low numbers of snapper (lujanids) (also seen in the 2006 survey), sweetlips (haemulids) and jacks (carangids) might reflect merely the overall distribution and habits of these groups rather than a disturbance effect, particularly because overfishing and the aquarium trade do not appear to be important factors affecting the reefs surveyed.

Results: invertebrate belt transects

Overall, the results from the invertebrate survey do not show any evidence for changes that can be ascribed to tsunami damage, climatic changes or anthropogenic stressors. At the sites investigated, the Reef Check indicator species (Hodgson et al. 2006) were present and the reefs investigated appeared healthy with a diverse range of marine macroinvertebrates.

Results: transects and point data

The point survey method provides a reasonable assessment of the major components of the reefs (hard corals, rubble, sand, rock) and of the general diversity of the reefs. It allows semi-quantitative assessment of the relative proportions of the various type of substratum present together with the 'life-forms' of the hard corals found. Point data are not always accurate and may miss certain forms, for example a massive coral, because the point is on immediately adjacent bare rock. This shortcoming is particularly the case with encrusting organisms. So in 2006 and 2007 video transects were carried out to supplement the point data. Overall there was a reasonable concordance between the video and the point surveys, with the video survey providing a better resolution of dead and damaged coral and of encrusting organisms.

The effect of the tsunami on table corals was very marked and a high proportion were either overturned or broken. In 2008 almost half the colonies surveyed had been damaged, for example, overturned and/or broken. Of the broken colonies, only one was completely dead while the others showed a range of survival of the polyps from 20% (18.17% from digital image) to 98% (95.71% from digital image) – effectively complete survival. Overall the health of the table corals in 2008 appeared to be good.

The results from the sites showed that, overall, the reefs investigated in the Similan Islands were healthy. The damage observed in the Similans was in agreement with patterns observed elsewhere in Thailand (Phongsuwan and Brown 2007). The limited extent of damage together with its localised nature was evident. Contrasting results reflect the intrinsically patchy nature of the reefs together with the limited and localised nature of the damage of the 2004 tsunami. For example, 'Deep Six' had been reported as severely damaged by the 2004 tsunami but when visited as part of general diving in 2007 appeared to be in good condition with evidence of regeneration. The overall comparisons of the sites in the 2006, 2007 and 2008 surveys did not show any significant separation of the 'impacted' and 'unimpacted' sites (Fig 10.6). Although in Fig 10.6 there was a slight break point in the 'scree slope' indicative of a separation between the sites, this was not considered significant and was not seen in 2008.

The results of the fish survey and to a lesser extent the invertebrate survey showed a lower level of abundance and diversity in 2008 than in 2006 and 2007, but still indicated that the overall health of the coral reef systems in the Similan

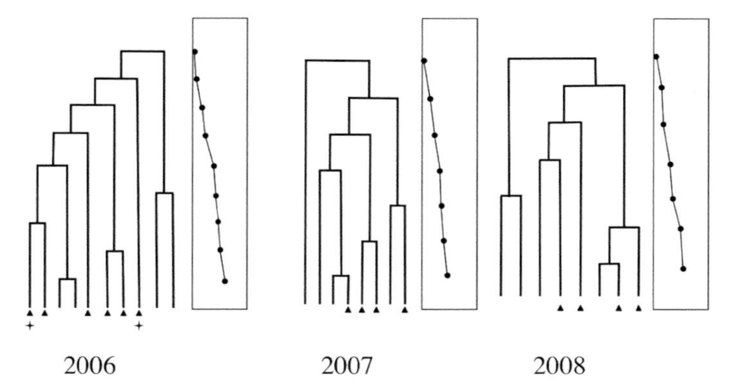

Figure 10.6 Dendrograms derived from two-way hierarchical clustering (Ward's Method) showing associations between various sites in 2006, 2007 and 2008. The black triangles indicate sites considered to have been severely impacted by the 2004 tsunami. The '+' in the 2006 survey dendrogram indicate two severely impacted sites at 'Bungalow Bay' Koh Racha Yai which were omitted from the 2007 and 2008 surveys because they were not within the Similan Islands. The 'scree slopes' graphs in the boxes adjacent to the dendrograms indicate any major discontinuities that would appear as an abrupt change in the gradient or a step change. There are no significant changes.

Islands was good with, in general, no marked or consistent differences between tsunami impacted and unimpacted sites. The Reef Check fish indicator species, baramundi cod (*Cromileptes altivelis*), bumphead parrotfish (*Bulbometopon muricatum*), humphead wrasse (*Chelilinus undulatus*) and moray eel (Muraenidae) were recorded in the 2006 and 2007 surveys while being absent from a significant proportion of Indo-Pacific reefs. Moray eels were also noted in 2008. High numbers of butterflyfish (Chaetodontidae) were indicative of healthy reefs and the low number of snapper (Lujanidae), sweetlips (Haemulidae) and jacks (Carangidae) may merely reflect normal variation in abundance. The high numbers of algal grazers should mitigate any phase shifts to algal dominated reef systems. The results of surveys of coralivorous fish made at four sites in 2007 and at one site in 2008 were indicative of a healthy reef.

Discussion

The results of these surveys of the Similan Islands are similar to other observations across the Indian Ocean showing that the impact of the tsunami was patchy. While many reefs were damaged, the vast majority escaped any serious disturbance. The surveys also showed that even in damaged areas, coral colonies were not totally killed and were recovering. Predictions made after the 2007 survey (Evans et al. 2007) of possible phase shifts from coral to algal dominated reefs in damaged areas did not appear to be fulfilled. In 'Snapper Alley' during 2007, algae were dominant at some sites but in the 2008 survey algae were rarely recorded.

Such transient changes show the importance of time series studies in establishing an appropriate baseline. Tsunami damaged areas are undergoing regeneration with evidence of coral recruitment, but the trajectory of this regeneration is more difficult to predict without further study. The surveys also showed that adjacent sites can show marked differences in species and lifeform structure. However, the lack of concordance between impacted and unimpacted sites within and between surveys shows the limited and very localised effect of damage from the 2004 tsunami. The results from the 2006, 2007 and 2008 surveys were in agreement with other published reports on the pattern of the effects of the 2004 tsunami on coral reefs in Thailand and suggest that the reefs of the Similan Islands were healthy and recovering.

10.4 Discussion

The general conclusion from this and previous studies is that, devastating though the tsunami was from a human perspective, it had limited effect on the natural environment. Even areas heavily disturbed by tsunami waves showed signs of biological recovery by 2009. There is no doubt that the tsunami was a major environmental disturbance causing considerable damage to reefs and intertidal areas but this was not consistent from place to place, even over short distances. Characteristics of the beaches and near coast were particularly important in determining wave damage and this and geographical barriers largely determined the severity of the impact (Yeh 2009). Nevertheless, it is important to recognise the resilience of the marine flora and fauna. Even damaged corals and reefs showed signs of recovery by 2009, and the data presented above indicate that the intertidal zone showed no lasting effects of this massive disturbance.

To understand why natural communities showed such resilience, it is important to place tsunami-generated disturbance in the context of other large-scale disturbances in the marine environment. Figure 10.7 shows the frequency of typhoons and storms recorded for the Andaman coast of Thailand. A total of 49 tropical storms and cyclones were recorded for this region between 1951 and 2004 (typhoons being much less common than storms). Storms have a similar impact on the near-shore marine environment to tsunamis because the majority of the disturbance is created by wave action although storms generate less debris-inflicted damage. The impact of a typhoon is extensive because there is a large area around the storm centre where wave height can be substantial (e.g. Typhoon Linda; Thai Marine Meteorological Centre, http://www.marine.tmd.go.th/) and a typhoon can track over hundreds of kilometres of sea. In some areas of Southeast Asia, cyclone-typhoons are more frequent (OCHA Regional Office for Asia Pacific; Tropical Storms worldwide, http://www.solar.ifa.hawaii.edu/Tropical/).

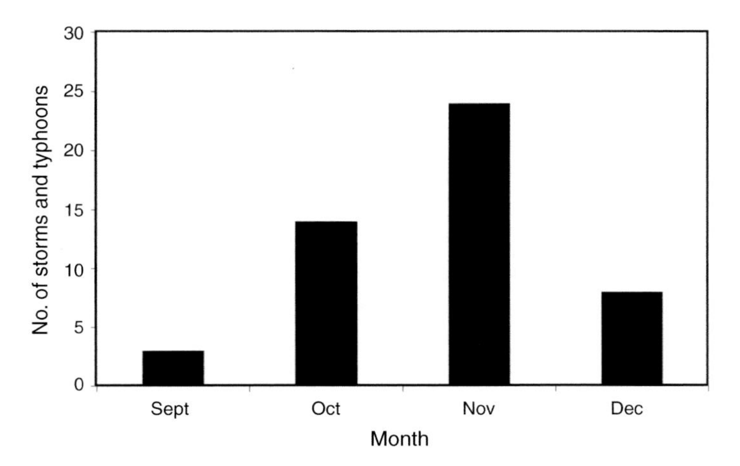

Figure 10.7 Frequency of typhoons in southern Thailand 1950–2003 (Source: http://www.marine.tmd.go.th).

The majority of these events occur post-monsoon, with a peak in November, and hence have a similar timing to the 2004 Tsunami. It is interesting to consider that had the tsunami occurred at a different time of the year (and potentially at a different time in the life cycle of dominant species) there might have been a different outcome and impact.

It is clear that large-scale, intense, punctuated disturbances are a feature of most tropical systems and that coastal flora and fauna are adapted to survive such disturbances. Similarly, as in all disturbance impacts, different elements of the ecosystem are not equally affected and recovery and recolonisation is usually possible from areas less impacted. However, recovery, particularly of reef systems, depends on a period of relative stability. A worrying aspect of recovery for Southeast Asian systems is the increasing anthropogenic pressure and associated disturbance. Future large-scale environmental disturbances may have greater effects because they may be impacting on already damaged ecosystems.

10.5 Conclusions

The 2004 tsunami mobilised the scientific community to assess the marine environment across a large area of the Indian Ocean. For the first time it was possible to survey and make assessments of the impact of a tsunami. This was important for two reasons: first it placed the disturbance and impact of such large-scale events into context, and second, it enabled the scientific community to gauge the resilience of the ecosystems upon which so many people depend. It has become clear that tsunami impacts are just one of a number of punctuated, large-scale, disturbance events experienced by natural marine assemblages. Recovery from them is part of the normal ecological cycle, with different groups recovering faster than others, depending on habitat and life cycle. Thus, long-lived corals will take longer to recover than short-lived meiofauna. The scale of the impact and subsequent recovery trajectories therefore will follow predictions from disturbance ecology.

In terms of the recovery of marine resources exploited by coastal stakeholders, many key offshore habitats were not as badly impacted as was feared immediately after the tsunami struck. Nursery areas such as mangroves and seagrass beds were, overall, not heavily impacted and although offshore sediment systems may have undergone substantial shifts they returned quickly to pre-tsunami conditions. The longer time taken by human communities to recover may have masked short-term changes.

The research undertaken following the tsunami highlighted a continuing problem in marine ecology, namely the lack of long-term time series. With a few notable exceptions for coral reefs and some mangroves, there are few long-term studies of key environments such as offshore sediment habitats, which underpin local fisheries efforts. In Southeast Asia such studies are particularly important given the increasing pressure on marine resources. The high species richness of marine assemblages poses challenges in that many of the key indicator species of crustaceans and polychaetes are new to science, and the regional taxonomic research base is under-resourced. It is vital that funding bodies and regional governments recognise that long-term monitoring and assessment are key tools in sustainable management of marine resources for the long-term benefit of their people.

Acknowledgements

We would like to acknowledge the resilience of the local people in Ranong and Phang Nga where we worked, and thank them for their hospitality and support. We thank the staff of the Ranong Marine Coastal Resources Station for their support in the field programme: Wisai Kongkaew, Decha Daugnamol, Predamon Kamwachirapitak and Phaothep Cherdsukjai; our colleagues at the NHM: Paul Clark, Mary Spencer Jones and Chris Allen for their help in the field. We also thank Kasetsart University's Research and Development Institute for its support and providing accommodation during the projects. The following colleagues helped with the polychaete sorting for which we are grateful: Chanokphon Jantharakhantee, Agnes Mittermeyer, Russell Doollan, Nick Higgs. Our thanks to Jim Chimonides for producing the map. This work was supported by contracts from EuropeAid 'Coastal Biodiversity in Ranong' (Contract: THA/B76200/1999/0342/DEV/ENV) and EU Asian Pro IIB 'Tsunami Impacts in Laem Son', which we gratefully acknowledge. Additional funding for fieldwork was granted to GLJP and NJE by the NHM's Department of Zoology and to MAK by PML. The participation of PC in the fieldwork was made possible by grants from the Percy Sladen Trust and the BSAC Jubilee Trust.

References

Altaff, K., Sugumaran, J. and Naveed, M. S. (2005). Impact of tsunami on meiofauna of Marina beach, Chennai, India. *Current Science* (Bangalore), **89**, 34–8.

Barrio Frojan, C., Kendall, M. A., Paterson, G. L. J. et al. (2006). Patterns of polychaete diversity in tropical intertidal habitats. *Scientia Marina*, **70S3**, 239–48.

Brown, B. E. and Bythell, J. C. (2005). Perspectives on mucus secretion in reef corals. *Marine Ecology Progress Series*, **296**, 291–309.

Bryant, E. A., Young, R. W. and Price, D. M. (1992). Evidence of tsunami sedimentation on the southeastern coast of Australia. *Journal of Geology*, **100(6)**, 753–65.

Campbell, S. J., Pratchett, M. S., Anggoro, A. W. et al. (2007). Disturbance to coral reefs in Aceh, northern Sumatra: impacts of the Sumatra-Andaman tsunami and pre-tsunami degradation. *Atoll Research Bulletin*, **544**, 55–78.

Castilla, J. C. (1998). Earthquake-caused coastal uplift and its effects on rocky intertidal kelp communities. *Science*, **42**, 440–5.

Chavanich, S., Viyakarn, V., Sojisuporn, P., Siripong, A. and Menasveta, P. (2008). Patterns of coral damage associated with the 2004 Indian Ocean tsunami at Mu Ko Similan Marine National Park, Thailand. *Journal of Natural History*, **42**,177–87.

Crawford, R. J. M., Asseid, B. S., Dyer, B. M. et al. (2006). The status of seabirds at Latham Island, Tanzania. *African Journal of Marine Science*, **28**, 99–108.

English, S., Wilkinson, C. and Baker, V. (1997). *Survey Manual for Tropical Marine Resources*. Townsville, Australia: Australian Institute of Marine Science.

Gischler, E. and Kikinger, R. (2007). Effects of the tsunami of 26 December 2004 on Rasdhoo and northern Ari atolls, Maldives. *Atoll Research Bulletin*, **544**, 93–103.

Grzelak, K., Kotwicki, L. and Szczuciński, W. (2009). Monitoring of sandy beach meiofaunal assemblages and sediments after the 2004 Tsunami in Thailand. *Polish Journal of Environmental Studies*, **18**, 43–51.

Gunawan, C. A., Allen, G., Bavestrello, G. et al. (2006). Status of coral reefs in Indonesia after the December 2004 Tsunami. In *Status of Coral Reefs in Tsunami Affected Countries: 2005*, ed. C. Wilkinson, D. Souter and J. Goldberg. Townsville, Australia: GCRMN, Australian Institute for Marine Studies, pp. 43–56.

Gusiakov, V. K. (2009). Tsunami history: recorded. *The Seas*, **15**, 23–53.

Hagan, A. B., Foster R., Perera, N. et al. (2007). Tsunami impacts in Aceh Province and north Sumatra, Indonesia. *Atoll Research Bulletin*, **544**, 37–54.

Hodgson, G., Hill, J., Kiene, W. et al. (2006). *Reef Check Instruction Manual: A Guide to Reef Check Coral Reef Monitoring.* Pacific Palisades, CA: Reef Check Foundation.

Kendall, M. A., Paterson, G. L. J., Aryuthaka, C. et al. (2006). Impact of the 2004 Tsunami on intertidal sediment and rocky shore assemblages in Ranong and Phangnga Provinces, Thailand. *Phuket Marine Biological Center Research Bulletin*, **67**, 63–75.

Kendall, M. A., Aryuthaka, C., Chimonides, J. et al. (2009). Post-tsunami recovery of shallow water biota and habitats on Thailand's Andaman coast. *Polish Journal of Environmental Studies*, **18**, 69–75.

Kotwicki, L. and Szczuciński, W. (2006). Meiofaunal assemblages and sediment characteristics of sandy beaches on the west coast of Thailand after the 2004 Tsunami event. *Phuket Marine Biological Center Research Bulletin*, **67**, 39–47.

Kume, T., Umetsu, C. and Palanisami, K. (2009). Impact of the December 2004 Tsunami on soil, groundwater and vegetation in the Nagapattinam district, India. *Journal of Environmental Management*, **90**, 3147–54.

Lander, J. F., Whiteside, L. S. and Lockridge, P. A. (2003). Two decades of global tsunamis: 1982–2002. *The International Journal of The Tsunami Society*, **21**(1), 1–88.

Mcleod, M. K., Slavich, P. G., Irhas, Y. et al. (2010). Soil salinity in Aceh after the December 2004 Indian Ocean tsunami. *Agricultural Water Management*, **97**, 605–13.

Patterson Edwards, J. K., Kulkanri, S., Jeyabaskaran, R. et al. (2006). The effects of the 2004 Tsunami on mainland India and the Andaman and Nicobar Islands. In *Status of Coral Reefs in Tsunami Affected Countries: 2005*, ed. C. Wilkinson, D. Souter and J. Goldberg. Townsville, Australia: GCRMN, Australian Institute for Marine Studies, pp. 85–97.

Phongsuwan, N. and Brown, B. E. (2007). The influence of the Indian Ocean tsunami on coral reefs of western Thailand, Andaman Sea, Indian Ocean. *Atoll Research Bulletin*, **544**, 79–91.

Phongsuwan, N., Yeemin, T., Worachananant, S. et al. (2006). Post-tsunami status of coral reefs and other coastal ecosystems on the Andaman Sea coast of Thailand. In *Status of Coral Reefs in Tsunami Affected Countries: 2005*, ed. C. Wilkinson, D. Souter and J. Goldberg. Townsville, Australia: GCRMN, Australian Institute for Marine Studies, pp. 63–77.

Rajasuriya, A., Perera, N., Karunarathna, C., Fernando, M. and Tamelander, J. (2006). Status of coral reefs in Sri Lanka after the tsunami. In *Status of Coral Reefs in Tsunami Affected Countries: 2005*, ed. C. Wilkinson, D. Souter and J. Goldberg. Townsville: GCRMN, Australian Institute for Marine Studies, pp. 99–110.

Sanpanich, K., Wells, F. E. and Chitramvong, Y. (2006). Effects of the 26 December 2004 Tsunami on littorinid molluscs near Phuket, Thailand. *Journal of Molluscan Studies*, **72**, 311–13.

Sheppard, C. R. C. (2007). Effects of the tsunami in the Chagos Archipelago. *Atoll Research Bulletin*, **544**, 135–48.

Sokoliev, S. L., Ngo, C. H. and Kim, K. S. (1992). *Calatog of Tsunamis in the Pacific 1962–1982.* Results of Researches on the International

Geophysical Projects. Moscow: Academy of Science of the USSR.

Spencer, T. (2007). Coral reefs and the tsunami of 26 December 2004: generating processes and ocean-wide patterns of impact. *Atoll Research Bulletin*, **544**, 1–36.

Stoddart, R. D. (ed.) (2007) Tsunamis and coral reefs. *Atoll Research Bulletin*, **544**, 1–164.

Szczuciński, W., Niedzielski, P., Rachlewicz, G. et al. (2005). Contamination of tsunami sediments in a coastal zone inundated by the 26 December 2004 tsunami in Thailand. *Environmental Geology*, **49**, 321–31.

Szczuciński, W., Chaimanee, N., Niedzielski, P. et al. (2006). Environmental and geological impacts of the 26 December 2004 Tsunami in coastal zone of Thailand: overview of short and long-term effects. *Polish Journal of Environmental Studies*, **15**, 793–810.

Szczuciński, W., Niedzielski, P. et al. (2007). Effects of rainy season on mobilization of contaminants from tsunami deposits left in a coastal zone of Thailand by the 26 December 2004 tsunami. *Environmental Geology*, **53**, 253–64.

Tomascik, T., Mah, A J., Nontji, A. and Moosa, M. K. (1997). *The Ecology of the Indonesian Seas. Part 1.* Oxford: Oxford University Press, pp. 1–642.

Tun, K., Yusuf, Y. and Amri, A. Y. (2006). Post-tsunami status of coral reefs in Malaysia. In *Status of Coral Reefs in Tsunami Affected Countries: 2005*, ed. C. Wilkinson, D. Souter and J. Goldberg. Townsville, Australia: GCRMN, Australian Institute for Marine Studies, pp. 57–62.

Viera, V.M., Le Bohec, C., Cote, S. D. and Groscolas, R. (2006). Massive breeding failures following a tsunami in a colonial seabird. *Polar Biology*, **29**, 713–16.

Wilkinson, C. (2006). Earthquakes, tsunamis and other stresses to coral reefs and coastal resources. In *Status of Coral Reefs in Tsunami Affected Countries: 2005*, ed. C. Wilkinson, D. Souter and J. Goldberg. Townsville, Australia: GCRMN, Australian Institute for Marine Studies, pp. 31–41.

Yeh, H. (2009). Tsunami impacts on coastlines. *The Seas*, **15**, 333–69.

11

Coalescent-based analysis of demography: applications to biogeography on Sulawesi

BEN J. EVANS

11.1 Introduction

If we were to travel back in time 18 000 years to Southeast Asia, we would encounter a radically different physical landscape. We would of course find a much more subtle impact of humans then compared to now, but perhaps equally striking would be a completely different extent of land and sea. During the last glacial period and the ones that preceded it, large regions that are now ocean were above sea level. It was possible during these intermittent but protracted periods to travel over land from mainland Asia to the islands of Borneo, Sumatra and Java, or from Australia to the island of New Guinea. These land connections account for faunal affinities between islands in the western part of Southeast Asia (the 'Sunda Region') and the eastern part of Southeast Asia (the 'Sahul Region'). Between these regions is Wallacea (Sulawesi, the Moluccan Archipelago and the Lesser Sunda Islands), which includes islands that were never connected by dry land to mainland Asia or to Australia. Abrupt transitions in flora and fauna in this region prompted biogeographers to demarcate the interface between the Asian and Australian faunal zones with boundaries such as Wallace's Line (Fig 11.1a; Wallace 1863, Huxley 1868, Lydekker 1896, George 1981).

Biotic Evolution and Environmental Change in Southeast Asia, eds D. J. Gower et al. Published by Cambridge University Press.

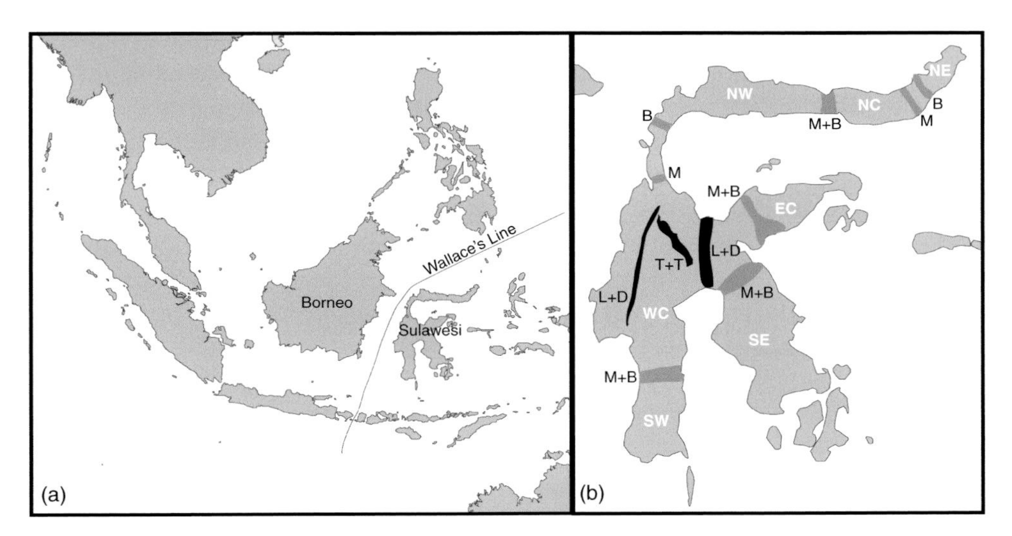

Figure 11.1 Geographical locations of major biogeographical boundaries in the Indonesian Archipelago and on Sulawesi. (a) Position of Wallace's Line relative to Borneo and Sulawesi. (b) Locations of geographically coincident boundaries of differentiated lineages in more than one taxon. In white font are the Southwest (SW), Southeast (SE), West-central (WC), East-central (EC), Northwest (NW), North central (NC) and Northeast (NE) areas of endemism defined by Sulawesi endemics from the genus *Macaca* (M) and *Bufo* (B) with boundaries in grey. Many boundaries of these lineages are shared (M + B). Other shared and probably more ancient boundaries (in black) are apparent in the genera *Limnonectes* and *Draco* (L + D) and the genera *Tarsius* and *Thoopterus* (T + T).

Wallacea is both a biogeographical transition zone and also a site of remarkable species endemism. Sulawesi Island, the largest in Wallacea, is arguably also the most complex in terms of geological and biogeographical history. Sulawesi is about 174 000 km^2, which is slightly bigger than the state of New York and about twice as large as Portugal. Forged from the accretion of at least four major palaeoislands (Hall 2001, this volume, Chapter 3), this island was subject to periodic marine inundation in multiple regions, such as the Tempe Depression at the base of the southwest peninsula and the Gorontalo Depression in the middle of the north peninsula. The ecology of Sulawesi is remarkably variable and different portions experience distinct rainfall cycles and have different climates and soil substrates (Whitten et al. 2002). Being an island composed largely of peninsulas, overland distances between different parts of Sulawesi can be quite large (up to ~1300 km), so geographic distance has the potential to contribute to genetic differentiation (Wright 1943). It is no surprise, therefore, that substantial population structure and/or high species diversity exists in many of Sulawesi's endemic terrestrial fauna (van der Vecht 1953, Musser 1987, Bridle et al. 2001, Evans et al. 2003a, b, c, 2008, Larson et al. 2005, McGuire et al. 2007, Brown et al. 2010, Musser et al. 2010).

What is surprising, however, is the extent to which biogeographical patterns coincide across evolutionary lineages on Sulawesi. While no two lineages can have identical evolutionary histories, physical aspects of a habitat such as the margins of land and water or altitudinal and climatic gradients can lead to biogeographical similarities among co-distributed (sympatric) species in terms of, for example, the geographical distribution of distinct populations, divergence time between populations, or the extent of migration between populations (Avise 2000). Recent studies of terrestrial vertebrates on Sulawesi have recovered similar patterns in multiple, distantly related groups that suggest a shared evolutionary impact of the ecological and geological factors.

11.1.1 Biogeographical patterns on Sulawesi

One shared pattern is exemplified by the Sulawesi macaques (genus *Macaca*) and the Celebes toad (*Bufo celebensis*). In both groups the geographic distributions of differentiated populations include three differentiated populations/species on the north peninsula, one on each of the other peninsulas (southwest, southeast, and east central), and one in the centre (Fig 11.1b; Evans et al. 1999, 2001, 2003b, c, 2008). The geographic location of the margins of differentiated areas of endemism is essentially identical in Sulawesi macaques and Celebes toads at the Gorontalo Depression between the northwest and north central area of endemism (AOE), at the Tempe Depression between the southwest and west-central AOE, and at the Bongka River between the northern margin between the west-central and east-central AOE (Fig 11.1b). In both groups, differentiation between the north-central and the northeast AOE is low compared to pairwise comparisons between other areas of endemism (Evans et al. 2003b, c, 2008). Exceptions to this general pattern of similar patterns of differentiation occur on the margin between the west-central and northwest AOE, which in toads is situated north of the margin in macaques, and also between the north-central and the northeast AOE, which in toads is east of the margin in macaques, at least for mitochondrial DNA.

Another shared pattern of differentiation is exemplified by flying lizards (genus *Draco*) and fanged frogs (genus *Limnonectes*; Fig 11.1b; Iskandar and Tjan 1996, Evans et al. 2003a, McGuire et al. 2007, Setiadi et al. 2010). Some of these species exhibit differentiation similar to some key features to the macaques and toads (for example, across the Gorontalo and Tempe Depression), but species ranges differ in key aspects that may reflect a longer residence on Sulawesi. Species of flying lizards and fanged frogs on the North peninsula both have ranges that extend down the west side of the West central AOE (Evans et al. 2003a, McGuire et al. 2007, Setiadi et al. 2011). Both groups also have a species or multiple species whose range spans the east central, southeast and east portion of the West central AOE (East of Lake Poso) (Evans et al. 2003a, McGuire et al. 2007, Setiadi et al. 2011).

A third shared pattern is substantial differentiation across the Palu–Koro fault that runs from the city of Poso at the base of the north peninsula to the Sea of Bone, which lies between the southwest and southeast peninsulas. The boundary between differentiated populations roughly corresponds to this fault in tarsiers (genus *Tarsius*) and in the bat *Thoopterus nigrescens* (Fig 11.1b; Campbell et al. 2007, Merker et al. 2009). This pattern is not found in Sulawesi macaques or Celebes toads and may reflect an underlying cause that pre-dates the arrival of these groups to Sulawesi. Analysis of tarsier vocalisations suggests that in addition to deep divergences across the Palu–Koro fault, areas of endemism exist in tarsiers that share some features with Sulawesi macaques and Celebes toads (Shekelle et al. 1997, Shekelle and Leksono 2004).

Perhaps not surprisingly, there is incomplete information about divergence, population structure and geographic ranges of most of Sulawesi's vertebrates, including its large vertebrates. Moreover, some species appear to share some features of these general biogeographical patterns but not others. For example, squirrels of the genus *Prosciurillus* have species distributions that bear similarities to the first and second distributional patterns described above (Musser et al. 2010). In particular, *P. leucomus* occurs throughout the north peninsula, *P. topapuensis* and *P. weberi* occur in the west central portion of Sulawesi, and a fossil record exists for another species in the southwest peninsula (Musser et al. 2010). *Prosciurillus alstoni* occurs in the southeast and west-central portion of Sulawesi but also in the northern region of West central Sulawesi (Musser et al. 2010) – this distribution is not shared with any of the species discussed above. Further sampling is needed to iron out the precise distributions of these squirrel species and their parasites, and the degree to which these distributions correspond with other Sulawesi endemics (Musser et al. 2010). Likewise, babirusas (genus *Babyrousa*) – an endemic pig of Sulawesi – from North Sulawesi appear to be morphologically divergent from populations on the Togian, Buru and Sula islands, but the margins of differentiation and the ranges of differentiated forms are not known (Meijaard and Groves 2002). Interestingly, a babirusa fossil of an extinct southwestern peninsula population appears distinct from the North Sulawesi population (Meijaard and Groves 2002). Another example is the anoa (genus *Bubalus*), the endemic dwarf buffalo of Sulawesi. There are reports of morphological and cytological variation, but limited sampling prevents the conclusive determination of whether multiple divergent lineages are co-distributed (sympatric) or not (Burton et al. 2005). In a survey by Burton and colleagues (2005) individuals that were morphologically distinguishable on the basis of horn shape, body size, pelage colour and tail length were sampled higher than and lower than 1000 m above sea level, suggesting that variation in Sulawesi anoas may not be associated with altitudinal gradients (that is, that there is not a 'highland' and 'lowland' anoa even though there may be multiple species of anoa). In the Celebes warty pig, *Sus celebensis*, mitochondrial DNA from the north peninsula is diverged from that of the rest of Sulawesi (Larson et al. 2005), but it is still unclear whether there exists further population subdivision on Sulawesi.

Some vertebrate groups, such as fanged frogs (Evans et al. 2003a, Setiadi et al. 2011) and shrews (genus *Crocidura*, Ruedi et al. 1998), have multiple sympatric forms in different parts of Sulawesi. This offers the opportunity to test whether sympatric forms are more or less closely related than non-sympatric forms. For example, after accounting for the effect of shared evolutionary history, variation in morphology and life history among sympatric fanged frogs on Sulawesi suggests the action of natural selection (Setiadi et al. 2011). This supports an adaptive radiation of these frogs on Sulawesi to occupy ecological niches that are not available on the Philippines – an archipelago that supports a higher diversity of frog families than Sulawesi (Setiadi et al. 2011). Undoubtedly, of course, many species have biogeographical histories that are not similar to these rough patterns (Bridle et al. 2004). One study of endangered Sulawesi tortoises (*Indotestudo forstenii*), for example, did not find substantial genetic differences among captive individuals, although the exact geographic provenance of these samples is unknown (Ives et al. 2008). However, if we ignore patterns that are not universally shared by all species, conservation decisions will have little biological basis (Evans et al. 2004). Patterns of differentiation shared by multiple (but not all) species therefore provide a useful guide for conservation prioritisation on Sulawesi. Arguably, we should target multiple areas of genetic endemism on different portions of this island, especially regions that currently do not have large national parks (Evans et al. 2003c, 2008). According to a census in 2000, about 15 million people live on Sulawesi. Habitat alteration is occurring at an alarming pace, including within protected areas (Bickford et al. 2007), and there is urgent need for further conservation measures.

11.1.2 Coalescence, divergence population genetics and isolation-migration models

Molecular polymorphism is influenced by demography, ancestry, mutation, genetic drift and natural selection. The evolutionary history of molecular polymorphism can be represented by a gene tree (a genealogy) that traces the evolutionary history of different alleles back in time to a single ancestral allele. This 'reverse evolution' backwards in time from descendant to ancestor is known as coalescence. Theory has been developed that decouples the mutational process that generates molecular polymorphism, and the coalescent genealogical process that accounts for its evolutionary history (Hudson 1991). This theory has been applied to a number of problems in evolutionary genetics including inferences concerning natural selection, recombination, phylogenetics and migration (Nielsen and Wakeley 2001, Nordborg 2001, Edwards et al. 2007).

Speciation occurs when an ancestral lineage diverges into two or more descendant lineages. Speciation could result from a physical barrier to migration that emerges within the ancestral lineage, in which case all parts of the genome begin diverging at the same time except those portions that are polymorphic in the ancestral lineage. Divergence of ancestral polymorphisms by definition begins before the time of speciation. Alternatively, speciation could occur in a stepwise fashion wherein incompatibilities between descendant lineages (genetic, morphological, behavioural) encoded by different parts of the genome accumulate over time. In the second scenario, reproductive isolation of many different parts of the genome occurs at different times (de Queiroz 1998, Wu 2001). Study of these alternative scenarios using multilocus polymorphism data, is facilitated by coalescent methods that permit comparison of demographic models that have and that do not have migration after divergence, an approach called divergence population genetics (Kliman et al. 2000). A useful feature of these methods is that realistic evolutionary phenomena can be incorporated into demographic models including intra- and interlocus recombination, biparental or uniparental inheritance, variation among loci in the rate of mutation, changes in population size, gene flow among populations and speciation without gene flow (Hey and Nielsen 2004, Becquet and Przeworski 2007).

Multiple approaches have been developed to study divergence population genetics (Wakeley and Hey 1997, Kliman et al. 2000, Becquet and Przeworski 2007, Hey and Nielsen 2004, Leman et al. 2005, Putnam et al. 2007). This study uses one of them, implemented by the program MIMAR (Becquet and Przeworski 2007), to explore how violations of model assumptions could affect estimation of parameters of demographic models (Fig 11.2a). Typically the null model (hereafter the isolation model) is one in which a single ancestral population diverges into two descendant populations with no subsequent gene flow. The alternative model (hereafter the isolation–migration model) permits ongoing gene flow (symmetrical or asymmetrical) between the descendant populations. Estimated parameters include the mutation parameters for the ancestral and both descendant populations ($\theta = 4N_e\mu$ for diploid biparentally inherited loci), the divergence time (τ), and (for the alternative model only) the magnitude of gene flow in terms of effective number of migrants per generation in each direction (m_{12} and m_{21}, respectively). Instead of directly analysing sequence data, MIMAR uses four summary statistics (s_1, s_2, ss and sf) that are affected by the parameters of the model (that is, θ_1, θ_2, θ_A, τ, m_{12} and m_{21}). The summary statistics concern the distribution among populations of derived mutations, categorising them as being either fixed in one population or the other (sf), polymorphic in one population but not the other (s_1) or vice versa (s_2), or as polymorphisms shared by both populations (ss) (Wakeley and Hey 1997). MIMAR uses a Markov chain Monte Carlo approach based on genealogical coalescence to estimate the posterior probabilities of the model parameters (detailed in

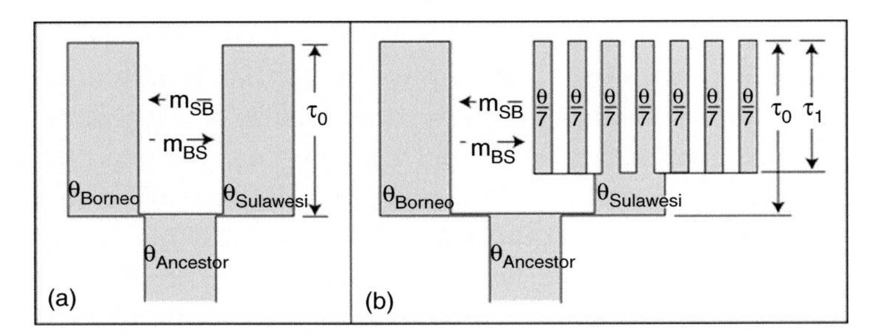

Figure 11.2 Demographic parameters used for simulations and estimated by MIMAR analysis. All simulations were performed without migration between Sulawesi and Borneo ($m_{SB} = m_{BS} = 0$) and the simulations were analysed with models without or with asymmetric migration between Sulawesi and Borneo. (a) Control simulations were performed with essentially no population structure on Borneo or on Sulawesi ($\delta = 0.001$, see text). (b) Simulations were also performed with a model with asymmetric population structure ($\delta = 0.5$) in which the Sulawesi lineage split into 7 demes at time $\tau_1 = 0.5^*\tau_0$ after divergence with Borneo.

Becquet and Przeworski 2007). The aim of this analysis is essentially to assess how frequently coalescent simulations with alternative parameter values generate data with summary statistics similar to those of the observed data.

11.1.3 Violations of models used in divergence population genetics

For computational reasons, models used to study divergence population genetics make simplifying assumptions concerning the evolutionary history of the species being studied – assumptions such as constant population size of each of the three lineages (the ancestral and both descendant lineages), constant rates of migration between the descendant lineages and no population structure in the ancestral or descendant lineages. It is therefore of interest to understand the impact of making these assumptions in situations when there is reason to suspect that they are not true.

Recently, two studies explored the impact of violations of assumptions of the standard models used in divergence population genetics. Becquet and Przeworski (2009) tested the impact of violating assumptions due to ancestral population structure, early gene flow immediately after speciation but not later on, and secondary contact – no gene flow immediately after speciation followed by gene flow after a period of reproductive isolation. Using two approaches in divergence population genetics – MIMAR (Becquet and Przeworski 2007) and IM (Hey and Nielsen 2004), Becquet and Przeworski identified conditions in which biologically plausible violations of model assumptions could lead to biased estimation of model parameters. For example, if the ancestral population is not panmictic, IM and MIMAR both tended to overestimate the ancestral effective population size (Becquet and

Przeworski 2009). Becquet and Przeworski (2009) also found that violation of the assumption of constant rate of migration led to an inappropriate inference of no migration if migration after divergence occurred only in an early phase. Strasburg and Rieseberg (2010) examined other violations including symmetrical population structure in both descendant populations, linkage among loci and gene flow from an unsampled species, using the programs IM and IM_A (Hey and Nielsen 2004, 2007). Strasburg and Rieseberg concluded that estimates of model parameter were generally robust to the violation they considered. For example, analysis of simulations with symmetrical population structure generated 90% highest posterior densities for model parameters that included the true parameter value about as frequently as the control (Strasburg and Rieseberg 2010).

Here I have explored another violation of the standard model used in divergence population genetic analysis that is related to studies of dispersal to the Indonesian island of Sulawesi. I explored how 'asymmetric population structure' – when one of the descendant populations develops substantial population structure while the other remains essentially panmictic – affects conclusions drawn from divergence population genetics analysis using the program MIMAR. This is relevant to studies of dispersal for Sulawesi biogeography because many groups appear to have more substantial population structure on Sulawesi compared to similarly sized portions of Borneo.

11.2 Methods

In an attempt to make the results of these simulations relevant to at least one dataset, I performed simulations to mimic roughly a recent analysis of macaque monkeys from Borneo and Sulawesi by Evans et al. (2010). This includes using the following features: (1) a single dispersal of a macaque ancestor to Sulawesi τ_0 generations ago counting back in time from the present followed by (2) the Sulawesi population being subdivided into seven (equally sized) demes which subsequently exchange no migration from the present going back in time for τ_1 generations, where $\tau_1 < \tau_0$ (Fig 11.2). Simulations matched the dataset of Evans et al. (2010) in terms of the number of loci (29), number of silent sites per locus (for an average of 459 silent sites per locus), the number of autosomal, X-chromosome and Y-chromosome loci (14 autosomal, 14 X-chromosome, 1 Y-chromosome), inheritance and mutation rate scalars of each locus, and the number of individuals sequenced for each locus for each *Macaca* species in the study, which included *M. nemestrina*, *M. nigra*, *M. nigrescens*, *M. hecki*, *M. tonkeana*, *M. maura* and *M. ochreata* (Table 11.1). Mitochondrial DNA sequences were not included by Evans et al. (2010) to avoid polarisation of polymorphisms at this rapidly evolving locus. Additionally, parameter values of the simulations roughly approximate estimated

Table 11.1 Data and sampling from Evans et al. (2010) including gene acronym (Name), number of silent sites sequenced (bp), mutation rate scalar (m scalar), number of samples from each of eight species or populations, and counts of 4 types of segregating sites (s_1, s_2, s_S, s_F) used in MIMAR analysis (see MIMAR documentation for details).

Gene	bp	μ scalar	*M. nigra*	*M. nigrescens*	*M. hecki*	*M. tonkeana* (West)	*M. tonkeana* (East)	*M. maura*	*M. ochreata*	*M. nemestrina* (Borneo)	S_1	S_2	S_s	S_f
ASIP	439	1	2	0	4	6	2	0	2	18	6	3	0	0
ATXN10	423	1.7	2	2	2	6	2	2	2	12	6	4	0	0
GPR15	138	0.9	2	2	4	6	2	2	2	16	2	2	0	0
IRBP	452	0.8	2	2	4	6	0	2	2	8	7	4	0	1
KFL10	133	1.3	2	2	4	6	2	0	2	16	3	2	1	0
PDYN	113	0.8	2	2	4	6	2	0	2	18	0	2	0	0
TRIM22	179	1	2	0	2	0	2	0	2	8	3	0	0	0
AFP	751	1	2	0	2	4	0	0	2	16	7	8	1	1
APOE	583	1.2	0	0	2	4	2	0	2	18	7	2	0	6
B2M	644	0.7	2	2	2	4	2	0	2	18	1	9	0	6
beta	376	0.9	2	2	4	4	2	2	2	24	3	5	0	0
CCL2	862	1.2	2	2	4	4	2	0	2	16	15	21	1	0
NRAMP	421	0.6	2	2	4	6	0	2	2	18	7	3	0	0
TTR	858	1.2	0	0	2	2	2	0	2	8	9	13	0	2
AMLEX	454	0.2	1	1	2	3	1	1	1	15	1	0	0	0

CXORF15	506	0.8	1	1	2	2	1	1	1	11	1	1	0	1
DBX	462	1	1	1	2	3	1	1	1	15	4	1	1	1
EFiAX	457	0.9	1	1	0	1	0	0	1	12	2	0	0	1
NLGN4X	212	2	1	1	2	3	1	1	1	15	0	4	1	0
PRX	314	2	1	1	2	3	1	1	1	15	3	9	1	0
RPMX	495	0.5	1	1	2	3	1	1	1	11	3	0	0	0
SMCX	385	0.7	1	1	2	2	1	1	1	12	0	1	0	0
SOX3	209	1.5	1	0	1	3	1	1	0	11	0	0	0	0
TBL1X	1357	1.2	1	1	2	3	1	1	1	12	12	15	1	0
TMSB4X	418	0.6	1	1	2	3	1	1	1	11	0	1	0	0
USP9X	187	0.9	1	1	2	3	1	1	1	14	1	1	0	1
UTX	186	0.3	1	1	2	3	1	1	1	15	0	0	0	0
ZFX	188	0.6	1	1	2	3	1	1	1	13	0	0	0	0
Y	1104	1.6	1	1	2	3	1	1	1	8	15	0	0	4

values for macaques from Sulawesi and Borneo, including divergence time, θ_{Borneo}, $\theta_{Sulawesi}$ and $\theta_{Ancestor}$ (see below).

Coalescent simulations were performed using ms (Hudson 2002). Custom PERL scripts were used to concatenate these simulations and convert them into input files for MIMAR analysis. Following Becquet and Przeworski (2009), simulations and MIMAR analysis were performed without intralocus recombination. The model I considered had five parameters including three mutation parameters $\theta_{Ancestor}$, θ_{Borneo}, $\theta_{Sulawesi}$, the divergence time between the descendant lineages on Borneo and Sulawesi (τ_0) and the time after colonization that the Sulawesi population diverged simultaneously into seven lineages that did not subsequently exchange migrants until the present (τ_1). For all simulations, $\theta_{Ancestor} = \theta_{Borneo} = \theta_{Sulawesi} = 0.002$ (per site) which is similar to results from macaque monkeys on Borneo (Evans et al. 2010). τ_0 was set to $4.4N_e$ generations, a value chosen because it falls within the 95% confidence interval estimated by Evans et al. (2010).

To incorporate mutation rate heterogeneity and differences in effective population size of autosomal DNA, the X-chromosome and the Y-chromosome, θ for each locus was multiplied by a mutation rate scalar (Table 11.1) and by an inheritance scalar. Mutation rate scalars were estimated by Evans et al. (2010) using an outgroup (baboons). Inheritance scalars for each locus were estimated by maximum likelihood (Evans et al. 2010) using a demographic model with unequal proportion of males and females due to sex-specific variation in reproductive success. The maximum likelihood scalars for aDNA, xDNA and yDNA were 1.00, 0.87 and 0.18, respectively, instead of the ideal expectation of 1.00, 0.75 and 0.25. These maximum likelihood estimates are consistent with field observations of macaque societies that suggest that there is higher variation in reproductive success in males compared to females (Dittus 1975, de Ruiter et al. 1992, Keane et al. 1997, Van Noordwijk and Van Schaik 2002, Widdig et al. 2004).

In order to simulate population structure on Sulawesi, the simulations included division of one descendant lineage into seven subpopulations at time τ_1, where $\tau_1 = \delta\tau_0$ and δ is equal to 0.001 or 0.5. When $\delta = 0.001$, the MIMAR model is essentially not violated so this was treated as a control simulation. Twenty simulations were performed for each value of δ, and all of them were analysed under models with and without asymmetric migration parameters between Borneo and Sulawesi. Each model was analysed twice using a different random seed for MIMAR using the sharcnet computer cluster (www.sharcnet.ca).

For each simulation, model comparison was performed using a goodness of fit test (Becquet and Przeworski 2007). This involves comparison of the summary statistics used in MIMAR analysis and also summary statistics not used in the analysis, including F_{ST} (Hudson et al. 1992), the average pairwise number of nucleotide differences for Sulawesi and Borneo, and Tajima's D (Tajima 1989) for Sulawesi and Borneo. Summary statistics were calculated from the simulations using PERL scripts and the isolation-migration model was considered better if one or more of the test statistics fit the data significantly worse in the isolation model (Fig 11.2; Becquet and Przeworski 2007, 2009).

11.3 Results

The data analysed in this study were generated using simulations with no migration between any of the descendant populations. At least three sources of variation could contribute to differences between the actual and inferred model parameter values. First, even though each simulation was performed with fixed parameter values ($\theta_{Ancestor}$, θ_{Borneo}, $\theta_{Sulawesi}$, τ_0, τ_1), stochastic differences among simulations in the coalescent will impact the posterior distribution of the parameter values recovered by MIMAR analysis. Second, the simulations with population structure on Sulawesi (that is, where $\delta = 0.5$) are expected to have a larger value for $\theta_{Sulawesi}$ because population subdivision with fixed deme size increases effective population size compared to a panmictic population with no subdivision (Nei and Takahata 1993). Third, variation among loci in mutation rates (encapsulated with a mutation rate scalar) could affect parameter estimates, although this effect should be similar in the control ($\delta = 0.001$) and test ($\delta = 0.5$) simulations. And fourth, any biases in the estimation of the posterior distribution of parameter values by MIMAR could also contribute differences in parameter estimates recovered from simulations.

11.3.1 No violations of the isolation model

As a first step, simulations were analysed with essentially panmictic populations in both of the descendant lineages that arose from the ancestral lineage (where $\delta = 0.001$). These simulations were analysed with an isolation model (no migration following divergence of the descendant lineages) and with the isolation-migration model (asymmetric migration after divergence). Goodness of fit tests indicated that in no case was the isolation-migration model preferred over the isolation model (data not shown). The mode of the posterior distributions of $\theta_{Sulawesi}$ and θ_{Borneo} was similar in both models and was close to the actual value of 0.002 in most simulations (Fig 11.3a). In analysis of the control simulations ($\delta = 0.001$) using the isolation model, 90% confidence intervals did not include the actual parameter value twice for $\theta_{Sulawesi}$, twice for θ_{Borneo}, five times for τ_0, and three times for $\theta_{Ancestor}$. In all cases the 90% confidence intervals derived from the same simulation included the actual value for more than one of these parameters. In analysis of the control simulations ($\delta = 0.001$) using the isolation-migration model, 90% confidence intervals did not include the actual parameter value twice for $\theta_{Sulawesi}$, twice for θ_{Borneo}, three times for τ_0, and four times for $\theta_{Ancestor}$. In analyses with the isolation model

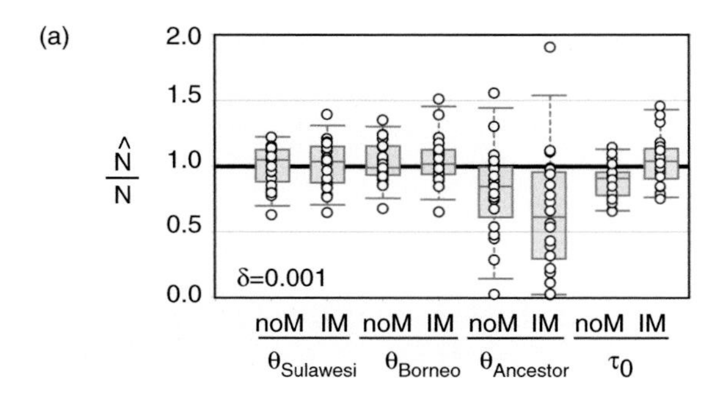

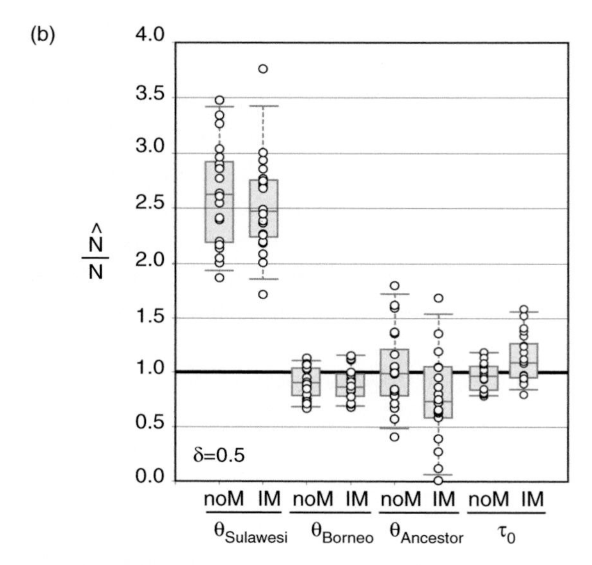

Figure 11.3 Parameter estimates recovered from MIMAR analysis of simulations (a) without ($\delta = 0.001$) and (b) with ($\delta = 0.5$) population structure on Sulawesi using a model without (noM) or with (IM) asymmetric migration between Sulawesi and Borneo. The *y*-axis is the ratio of the mode of the posterior distribution compared to the actual value used in the simulation ($\hat{N}/N$). Box and whiskers plots indicate the first and third quartiles, the medians, and the 95 percentiles of the modes of the posterior distributions.

and with the isolation–migration model, the dispersion of the estimates of θ_{Ancestor} was greater than the dispersion of estimates of θ_{Sulawesi} and θ_{Borneo} (Fig 11.3a).

11.3.2 Asymmetric population structure: a violation of the isolation model

When simulations generated with asymmetric population subdivision in one descendant lineage ($\delta = 0.5$) were analysed using the isolation model, parameter estimates were fairly robust to violations of model assumptions. Similar to the control simulations, in no instance did goodness-of-fit tests prefer the isolation–migration model over the isolation model (data not shown). When these simulations were analysed using the isolation model, 90% confidence intervals did not include the actual parameter value three times for θ_{Borneo}, three times for τ_0, and once for $\theta_{Ancestor}$. The 90% confidence interval did not include the actual value for θ_{Borneo} and for τ_0 for analysis of one simulation. When these simulations were analysed using the isolation–migration model, 90% confidence intervals did not include the actual parameter value three times for θ_{Borneo}, once for τ_0, and twice for $\theta_{Ancestor}$, and these inaccuracies were all from analyses of different simulations. As predicted by theory, estimates of $\theta_{Sulawesi}$ were higher when $\delta = 0.5$ (Fig 11.3). In the simulations where $\delta = 0.5$, the actual parameter value for $\theta_{Sulawesi}$ was lower than the 90% confidence interval in all of the analyses with the isolation model and also in all analyses with the isolation–migration model (Fig 11.3b).

11.4 Discussion

A population at mutation–drift equilibrium has a roughly constant level of polymorphism over time because variation is introduced by mutation at the same rate that it is removed by genetic drift. Divergence population genetics explores systems that are not in mutation–drift equilibrium because population size has changed, and because the system as a whole is no longer panmictic (even though each descendant population is panmictic under model assumptions). This study used simulations and a coalescent-based analysis – MIMAR (Becquet and Przeworski 2007) – to explore how violation of a demographic model used in divergence population genetics could impact inferences concerning non-equilibrium systems based on multi-locus DNA sequence data. Using the biogeographical history of macaque monkeys on the Southeast Asian islands of Sulawesi and Borneo as an example, data were simulated in which one descendant population on Sulawesi acquired population structure, whereas another on Borneo remained panmictic. Analysis of these simulations is of interest because multiple species on Sulawesi suggest that population structure on this island is quite common. In particular, multiple distantly related lineages such as toads and monkeys, tarsiers and bats, and flying lizards and fanged frogs, have similar geographic distributions of substantially differentiated populations on Sulawesi. The degree to which patterns are shared across species is the culmination of multiple variables including the timing of initial colonisation, dynamic abiotic factors such as marine inundation, ecological transitions and heterogeneity, and species-specific phenomena such as adaptation, ecological tolerance, genetic drift and demography.

Overall MIMAR analyses of the simulations performed here suggest that this model-based approach to studying biogeography is robust to extreme model violation by asymmetrical population structure in descendant populations. While dispersion of the parameter estimate for the ancestral polymorphism parameter θ_{Ancestor} tended to be larger than that of the polymorphism parameters for the extant populations (θ_{Sulawesi} and θ_{Borneo}), this difference was not substantially affected by asymmetric population structure in the descendant populations (Fig 11.3). Asymmetric population structure did not lead to an inappropriate inference of ongoing migration between Borneo and Sulawesi, and estimates of the timing of divergence of populations on Sulawesi and Borneo appeared to not be substantially biased. This finding is similar to those reported for symmetrical population structure by Strasburg and Rieseberg (2010). An unsurprising caveat to these conclusions is that asymmetric population structure did inflate estimates of the polymorphism parameter of the structured lineage (θ_{Sulawesi}), a result that is predicted by population genetic theory. Becquet and Przeworski (2009) recovered a similar result for θ_{Ancestor} in their study of a structured ancestral population. Similar to the findings of Becquet and Przeworski (2007), control simulations (without population structure on Sulawesi) produced broader posterior distributions for θ_{Ancestor} than for θ_{Sulawesi} or θ_{Borneo} (Fig 11.3).

In MIMAR, inferences about migration and ancestral population sizes are based on variance in coalescence times across loci (Becquet and Przeworski 2007, 2009). Larger ancestral populations are expected to have more variation among loci in coalescence times because of ancestral polymorphism at the moment of speciation. Migration between descendant lineages is also expected to increase variance in coalescence times across loci compared to a model with no migration. In an isolation model, all loci must coalesce at or before the time of speciation. Theory predicts that after ~9 to $12N_e$ generations, where N_e is the effective population size of each descendant population, greater than 95% of the loci should be reciprocally monophyletic in two diverged populations (Hudson and Coyne 2002). In the simulations analysed here, divergence time was $4.4N_e$ generations, so presumably many of the loci are not reciprocally monophyletic in each of the simulated descendant populations. This is also suggested by the observed data from Evans et al. (2010), on which the simulations were based, which have relatively few fixed differences between Sulawesi and Borneo (Table 11.1). These data are instead characterised by a preponderance of derived segregating sites that are found in only one or the other island, although relatively few polymorphisms are shared by populations on both islands.

In these simulations, population subdivision increases the overall effective population size (Nei and Takahata 1993). Consequently, alleles shared with the other descendant population (that is, the simulated panmictic population representing individuals from Borneo) due to ancestral polymorphism are less likely to be lost because of drift on Sulawesi. The net effect of population subdivision, therefore, could be to increase the variance in coalescence times across loci. This in turn could contribute to an incorrect signature of migration or errors in estimates of $\theta_{Ancestor}$. However, this does not appear to be the case for the parameter values considered here, and the recent analysis of macaque migration between Borneo and Sulawesi also did not recover significant support for migration (Evans et al. 2010). Instead of accounting for this variance with migration or a larger $\theta_{Ancestor}$, MIMAR correctly inflated the estimate of $\theta_{Sulawesi}$. This is undoubtedly a consequence of the large number of Sulawesi-specific polymorphisms compared to the number of polymorphisms shared between Borneo and Sulawesi in the simulations. The impact of asymmetric population structure could be more substantial if θ_{Borneo} were larger, if $(\tau_0 - \tau_1)$ were smaller, and/or if τ_0 were smaller, and this is an interesting avenue for further work. Also relevant to the conclusions of Evans et al. (2010) but not considered here is an assessment of what level of migration could be missed by coalescent-based analysis using isolation and isolation-migration models – that is, whether there could be an impact of asymmetric population structure on the incidence of false negatives for ongoing migration.

11.4.1 Future directions

When and from where did ancestors of Sulawesi endemics colonise this island? To what degree do patterns of differentiation and timings of population subdivision correspond among species? Did closely related Sulawesi endemics evolve *in situ*? We have only begun to address these questions and many exciting studies lie ahead. Model-based approaches offer a promising strategy to help us identify biogeographical trends and exceptions using molecular data. An obvious application of this research is for conservation planning on Sulawesi. While not all species will match areas of endemism identified in species discussed here, the available data argue strongly for a 'distributed' approach to conservation on Sulawesi wherein different portions of this island are independently targeted for management, as opposed to targeting a single ecosystem or region.

Acknowledgements

I am particularly indebted to Jim McGuire, Rafe Brown, Mohammad Iqbal Setiadi, Noviar Andayani and Jatna Supriatna for collaborating with me on studies of Southeast Asian biogeography. I also thank Celine Becquet and Jonathan Dushoff for statistical advice.

References

Avise, J. C. (2000). *Phylogeography. The History and Formation of Species.* Cambridge, MA: Harvard University Press.

Becquet, C. and Przeworski, M. (2007). A new approach to estimate parameters of speciation models with application to apes. *Genome Research*, **17**, 1505–19.

Becquet, C. and Przeworski, M. (2009). Learning about modes of speciation by computational approaches. *Evolution*, **63**, 2547–62.

Bickford, D. P., Supriatna, J., Andayani, N. et al. (2007). *Indonesia's Protected Areas Need More Protection: Suggestions from Island Examples*. Cambridge: Cambridge University Press.

Bridle, J. R., Garn, A., Monk, K. A. and Butlin, R. K. (2001). Speciation in *Chitaura* grasshoppers (Acrididae: Oxyinae) on the island of Sulawesi: colour patterns, morphology and contact zones. *Biological Journal of the Linnean Society*, **72**, 373–90.

Bridle, J. R., Pedro, P. M. and Butlin, R. K. (2004). Habitat fragmentation and biodiversity: testing for the evolutionary effects of refugia. *Evolution*, **58**, 1394–6.

Brown, R. M., Linkem, C. W., Siler, C. D. et al. (2010). Phylogeography and historical demography of *Polypedates leucomystax* in the islands of Indonesia and the Philippines: Evidence for recent human-mediated range expansion? *Molecular Phylogenetics and Evolution*, **57**, 598–619.

Burton, J. A., Hedges, S. and Mustari, A. H. (2005). The taxonomic status, distribution and conservation of the lowland anoa *Bubalus depressicornis* and mountain anoa *Bubalus quarlesi*. *Mammal Review*, **35**, 25–50.

Campbell, P., Putnam, A. S., Bonney, C. et al. (2007). Contrasting patterns of genetic differentiation between endemic and widespread species of fruit bats (Chiroptera: Pteropodidae) in Sulawesi, Indonesia. *Molecular Phylogenetics and Evolution*, **44**, 474–82.

de Queiroz, K. (1998). The general lineage concept of species, species criteria, and the process of speciation. In *Endless Forms: Species and Speciation*, ed. D. Howard and S. Berlocher. New York: Oxford Press, pp. 57–75.

de Ruiter, J. R., Scheffrahn, W., Trommelen, G. J. J. M. et al. (1992). Male social rank and reproductive success in wild long-tailed macaques. In *Paternity in primates: genetic tests and theories*. ed. R. D. Martin, A. F. Dixon and E. J. Wickings. Basel, Switzerland: Karger, pp. 175–191.

Dittus, W. (1975). Population dynamics of the toque macaque, *Macaca sinica*. In *Socioecology and Psychology of Primates*. ed. R. H. Tuttle. The Hague: Mounton, pp. 125–52.

Edwards, S. V., Lie, L. and Pearl, D. K. (2007). High-resolution species trees without concatenation. *Proceedings of the National Academy of Sciences*, **104**, 5936–41.

Evans, B. J., Morales, J. C., Supriatna, J. and Melnick, D. J. (1999). Origin of the Sulawesi macaques (Cercopithecidae, *Macaca*) as inferred from a mitochondrial DNA phylogeny. *Biological Journal of the Linnean Society*, **66**, 539–60.

Evans, B. J., Supriatna, J. and Melnick, D. J. (2001). Hybridization and population genetics of two macaque species in Sulawesi, Indonesia. *Evolution*, **55**, 1685–702.

Evans, B. J., Brown, R. M., McGuire, J. A. et al. (2003a). Phylogenetics of fanged frogs (Anura; Ranidae; *Limnonectes*): testing biogeographical hypotheses at the Asian–Australian faunal zone interface. *Systematic Biology*, **52**, 794–819.

Evans, B. J., Supriatna, J., Andayani, N. and Melnick, D. J. (2003b). Diversification of Sulawesi macaque monkeys: decoupled evolution of mitochondrial and autosomal DNA. *Evolution*, **57**, 1931–46.

Evans, B. J., Supriatna, J., Andayani, N. et al. (2003c). Monkeys and toads define areas of endemism on Sulawesi. *Evolution*, **57**, 1436–43.

Evans, B. J., Cannatella, D. C. and Melnick, D. J. (2004). Understanding the origins of areas of endemism in phylogeographic analyses: a reply to Bridle et al. *Evolution*, **58**, 1397–1400.

Evans, B. J., McGuire, J. A., Brown, R. M., Andayani, N. and Supriatna, J. (2008). A coalescent framework for comparing alternative models of population structure with genetic data: evolution of Celebes toads. *Biology Letters*, **4**, 430–3.

Evans, B. J., Pin, L., Melnick, D. J. and Wright, S. I. (2010). Sex-linked inheritance in macaque monkeys: implications for effective population size and dispersal to Sulawesi. *Genetics*, **185**, 923–37.

George, W. (1981). Wallace and his line. In *Wallace's Line and Plate Tectonics*, ed. T. C. Whitmore. Oxford: Clarendon Press, pp. 3–8.

Hall, R. (2001). Cenozoic reconstructions of SE Asia and the SW Pacific: changing patterns of land and sea. In *Faunal and Floral Migrations and Evolution in SE Asia–Australia*, ed. I. Metcalfe, J. Smith, M. Morwood and I. Davidson. Lisse, The Netherlands: Swets and Zeitlinger Publishers, pp. 35–56.

Hey, J. and Nielsen, R. (2004). Multilocus methods for estimating population sizes, migration rates and divergence time, with applications to the divergence of *Drosophila pseudoobscura* and *D. persimilis*. *Genetics*, **167**, 747–60.

Hey, J. and Nielsen, R. (2007). Integration within the Felsenstein equation for improved Markov chain Monte Carlo methods in population genetics. *Proceedings of the National Academy of Sciences*, **104**, 2785–90.

Hudson, R. R. (1991). Gene genealogies and the coalescent process. *Oxford Surveys in Evolutionary Biology*. **7**, 1–44.

Hudson, R. R. (2002). Generating samples under a Wright–Fisher neutral model of genetic variation. *Bioinformatics*, **18**, 337–8.

Hudson, R. R. and Coyne, J. A. (2002). Mathematical consequences of the genealogical species concept. *Evolution*, **56**, 1557–65.

Hudson, R. R., Slatkin, M. and Maddison, W. P. (1992). Estimation of levels of gene flow from DNA sequence data. *Genetics*, **132**, 583–9.

Huxley, T. H. (1868). On the classification and distribution of the Alectoromorphae and Heteromorphae. *Proceedings of the Zoological Society of London*, **1868**, 294–319.

Iskandar, D. T. and Tjan, K. N. (1996). The amphibians and reptiles of Sulawesi, with notes of the distribution and chromosomal number of frogs. In *Proceedings of the First International Conference on Eastern Indonesian-Australian Vertebrate Fauna. Manado, Indonesia, Nov 22–26*, ed. D. J. Kitchner and A. Suyanto, pp. 39–46.

Ives, I., Spinks, P. Q. and Shaffer, H. B. (2008). Morphological and genetic variation in the endangered Sulawesi tortoise *Indotestudo forstenii*; evidence of distinct lineages? *Conservation Genetics*, **9**, 709–713.

Keane, B., Dittus, W. P. and Melnick, D. J. (1997). Paternity assessment in wild groups of toque macaques *Macaca sinica* at Polonnaruwa, Sri Lanka using molecular markers. *Molecular Ecology*, **6**, 267–82.

Kliman, R. M., Andolfatto, P., Coyne, J. A. et al. (2000). The population genetics of the origin and divergence of the *Drosophila simulans* complex species. *Genetics*, **156**, 1913–31.

Larson, G., Dobney, K., Albarella, U. et al. (2005). Worldwide phylogeography of wild boar reveals multiple centers of pig domestication. *Science*, **3007**, 1618–21.

Leman, S. C., Chen, Y., Stajich, J. E., Noor, M. A. F. and Uyenoyama, M. (2005). Likelihoods from summary statistics: recent divergence between species. *Genetics*, **171**, 1419–36.

Lydekker, R. (1896). *A Geographical History of Mammals*. Cambridge: Cambridge University Press.

McGuire, J. A., Brown, R. M., Mumpuni, Riyanto, A. and Andayani, N. (2007). The flying lizards of the *Draco lineatus* group (Squamata: Iguania: Agamidae): a taxonomic revision with descriptions of two new species. *Herpetological Monographs*, **21**, 179–212.

Meijaard, E. and Groves, C. (2002). Upgrading three subspecies of babirusa (*Babyrousa* sp.) to full species level. *Asian Wild Pig News*, **2**, 33–9.

Merker, S., Driller, C., Pertwitasari-Farajallah, D., Pamungkas, J. and Zischler, H. (2009). Elucidating geological and biological processes underlying the diversification of Sulawesi tarsiers. *Proceedings of the National Academy of Sciences*, **106**, 8459–64.

Musser, G. G. (1987). The mammals of Sulawesi. In *Biogeographical Evolution of the Malay Archipelago*, ed. T. C. Whitmore. Oxford: Clarendon Press, pp. 73–93.

Musser, G. G., Durden, L. A., Holden, M. E. and Light, J. E. (2010). Systematic review of endemic Sulawesi squirrels (Rodentia, Sciuridae), with descriptions of new species of associated sucking lice (Insecta, Anoplura), and phylogenetic and zoogeographic assessments of sciurid lice. *Bulletin of the American Museum of Natural History*, **339**, 1–160.

Nei, M. and Takahata, N. (1993). Effective population size, genetic diversity and coalescence time in subdivided populations. *Journal of Molecular Evolution*, **37**, 240–44.

Nielsen, R. and Wakeley, J. (2001). Distinguishing migration from isolation: a Markov chain Monte Carlo approach. *Genetics*, **158**, 885–96.

Nordborg, M. (2001). Coalescent theory. In *Handbook of Statistical Genetics*, ed. D. Balding, M. Bishop and C. Cannings. Chichester, UK: Wiley, pp. 1–37.

Putnam, A. S., Scriber, J. M. and Andolfatto, P. (2007). Discordant divergence times among Z-chromosome regions between two ecologically distinct swallowtail butterfly species. *Evolution*, **61**, 912–27.

Ruedi, M., Auberson, M. and Savolainen, V. (1998). Biogeography of Sulawesian shrews: testing for their origin with a parametric bootstrap on molecular data. *Molecular Phylogenetics and Evolution*, **9**, 567–71.

Setiadi, M. I., McGuire, J. A., Brown, R. M. et al. (2011). Adaptive radiation and ecological opportunity in Sulawesi and Philippine fanged frogs (Limnonectes). *American Naturalist*, **178**(2), 221–40.

Shekelle, M. and Leksono, S. M. (2004). Rencana konservasi di Pulau Sulawesi:

dengan menggunakan tarsius sebagai 'flagship taxon'. *Biota*, **9**, 1–10.

Shekelle, M., Leksono, S. M., Ichwan, L. L. S. and Masala, Y. (1997). The natural history of the tarsiers of North and Central Sulawesi. *Sulawesi Primate Newsletter*, **4**, 3–8.

Strasburg, J. L. and Rieseberg, L. H. (2010). How robust are 'isolation with migration' analyses to violations of the IM model? A simulation study. *Molecular Biology and Evolution*, **27**, 297–310.

Tajima, F. (1989). Statistical method for testing the neutral mutation hypothesis by DNA polymorphism. *Genetics*, **123**, 585–95.

Van Noordwijk, M. A. and Van Schaik, C. P. (2002). Career moves: transfer and rank challenge decisions by male long-tailed macaques. *Behavior*, **138**, 359–95.

van der Vecht, J. (1953). The carpenter bees (*Xylocopa* Latr.) of Celebes. *Idea*, **9**, 57–9.

Wakeley, J. and Hey, J. (1997). Estimating ancestral population parameters. *Genetics*, **145**, 847–55.

Wallace, A. R. (1863). On the physical geography of the Malay Archipelago. *Journal of the Royal Geographical Society*, **1863**, 217–34.

Whitten, T., Henderson, G. S. and Mustafa, M. (2002). *The Ecology of Sulawesi*. Hong Kong: Periplus Editions Ltd.

Widdig, A., Bercovich, F. B., Streich, W. J., Sauermann, U., Nürnberg, P. and Krawczak, M. (2004). A longitudinal analysis of reproductive skew in male rhesus macaques. *Proceedings of the Royal Society of London B*, **271**, 819–26.

Wright, S. (1943). Isolation by distance. *Genetics*, **28**, 114–38.

Wu, C.-I. (2001). The genic view of the process of speciation. *Journal of Evolutionary Biology*, **14**, 851–65.

12

Aquatic biodiversity hotspots in Wallacea: the species flocks in the ancient lakes of Sulawesi, Indonesia

THOMAS VON RINTELEN, KRISTINA VON RINTELEN, MATTHIAS GLAUBRECHT, CHRISTOPH D. SCHUBART AND FABIAN HERDER

12.1 Introduction

Some of the world's most spectacular species radiations or species flocks are found in so-called 'ancient lakes'. These are long-lived lakes that have existed for 100 000 years (Gorthner et al. 1994, but see also Albrecht and Wilke 2008) or more (e.g. Lake Tanganyika and Lake Baikal). Ancient lakes are justifiably regarded as hotspots of diversification (e.g. Martens 1997, Rossiter and Kawanabe 2000), even if not all ancient lake species flocks match the diversity of the super-flock of East African cichlids (e.g. Kornfield and Smith 2000, Kocher 2004). Studies on the evolution of ancient lake organisms have continuously resulted in important insights into general patterns of speciation and radiation (e.g. Streelman and Danley 2003) ever since the seminal review of Brooks (1950).

During the last decade, smaller ancient lakes (*c*. <1 000 km^2), which are generally less well investigated, have attracted increasing attention. In Southeast Asia, four lakes or lake systems are regarded as putative ancient lakes (Martens 1997): Lake Inlé in Burma, Lake Lanao on Mindanao, and Lake Poso and the Malili lakes on

---

Biotic Evolution and Environmental Change in Southeast Asia, eds D. J. Gower et al. Published by Cambridge University Press.

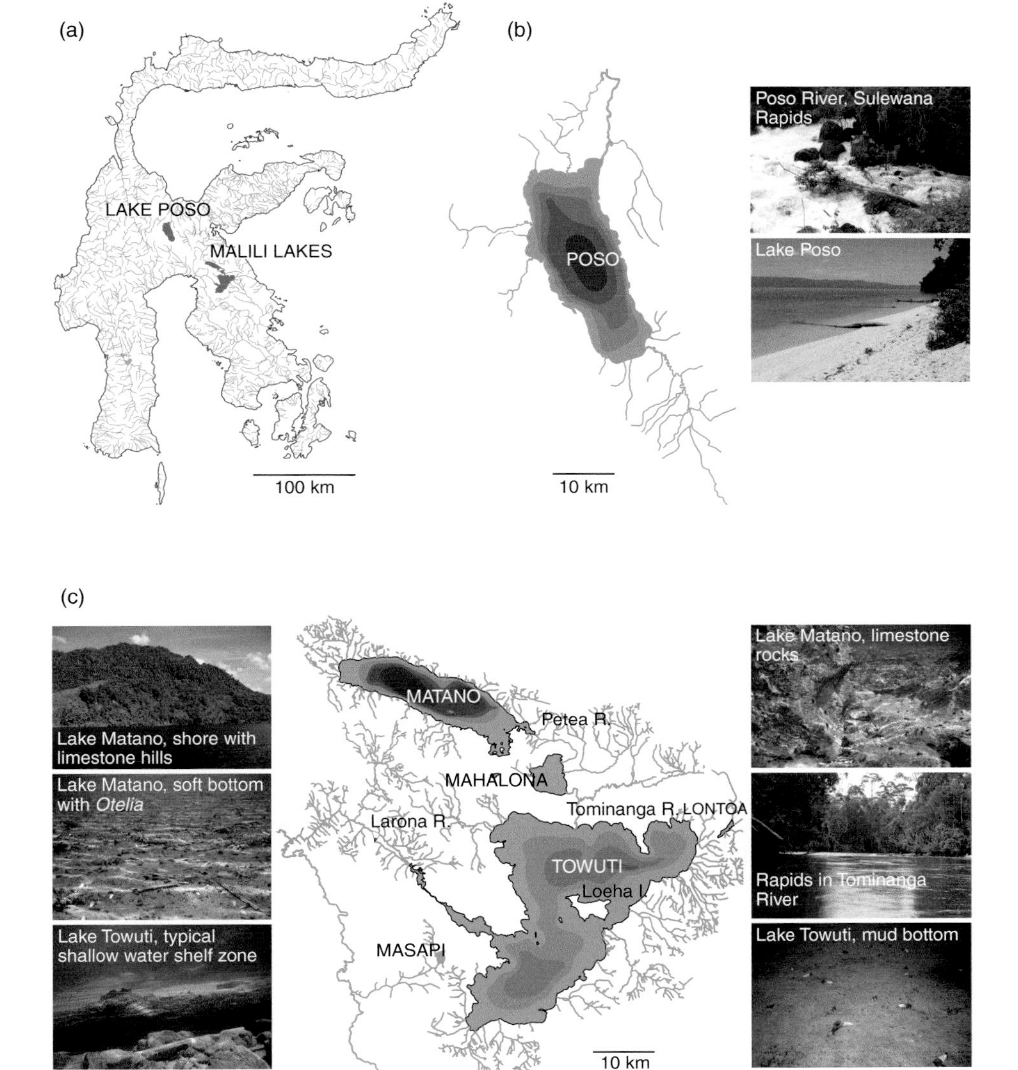

Figure 12.1 Indonesia, Sulawesi and the ancient lakes with characteristic habitats. (a) Sulawesi; the two ancient lake systems are highlighted in red. (b) Lake Poso. (c) Malili lake system. Lake names are printed in capital letters. Abbreviations: R., River; I., Island. See plate section for colour version.

Sulawesi (Fig 12.1). Among these lakes, only the Sulawesi lakes are clearly ancient lakes, both in terms of their estimated age and their fauna (see below for details), while the general lack of data prevents any firm conclusion about the other two systems.

Table 12.1 The ancient lakes of Sulawesi. Based on data from Abendanon (1915a, 1915b), Giesen et al. (1991), Giesen (1994) and Haffner et al. (2001).

Lake	Area (km^2)	Max. depth (m)	Transparency (Secchi disk) (m)
Poso	323.2	450	11
Matano	164.0	590	20
Mahalona	24.4	73	20
Towuti	561.1	203	22
Lontoa	1.6	3	<3
Masapi	2.2	4	<3

12.1.1 The ancient lakes of Sulawesi

Hydrology, geology and limnology

Both ancient lake systems on Sulawesi are located in the central mountains of the island (Fig 12.1a). Lake Poso (Fig 12.1b; Table 12.1) is a deep solitary lake, while the Malili system comprises five lakes sharing a common drainage (Fig 12.1c; Table 12.1). The three larger lakes of the Malili system are directly connected: Lake Matano flows into Lake Mahalona via the Petea River. In turn, Lake Mahalona spills into Lake Towuti via the Tominanga River. Lake Towuti is drained by the Larona River into the Gulf of Bone (Teluk Bone). Two smaller satellite lakes, Lake Lontoa (also known as Wawontoa or Lantoa) and Lake Masapi, are less directly connected to the system (see Fig 12.1c). The latter drains independently into the Malili River.

Lake Poso and Lake Matano are of tectonic origin, which accounts for their extraordinary depth. Lake Matano is situated in a strike-slip fault, the Matano fault, which was formed in the final juxtaposition process of South-, Southeast- and East Sulawesi since the Pliocene (*c.* 4 Ma) to the present day (Wilson and Moss 1999). The age of Lake Towuti has been estimated at 700 000 years (J. Russell pers. comm. 2011). Age estimates for the other lakes are lacking. The major ancient lakes of Sulawesi are oligotrophic, with a very low nutrient and organic content and a high transparency of up to 22 m in Lake Towuti (Giesen et al. 1991, Giesen 1994, Haffner et al. 2001, Crowe et al. 2008a, 2008b).

The lakes offer a wide range of habitats ranging from soft-bottom with sand and mud to steep rocky drop-offs (Fig 12.1). Typically, a shallow shelf zone (2–5 m) is followed by a steep slope with depth quickly increasing. Frequently, hard substrates predominate at the shoreline (0–2 m) and soft substrate in deeper water. Steep rocky areas and habitats dominated by gravel frequently interrupt these soft-bottom aggregations and dominate along some stretches of the larger lakes. Habitat heterogeneity is high in all lakes, although there are general differences between the lakes for instance in the amount of hard substrate available in deep water (>10 m), which is largely lacking in Lake Mahalona and Lake Towuti. Extensive sand beaches are characteristic for large areas of Lake Poso, but almost entirely lacking in the Malili lakes. In the connecting and draining rivers of both lakes systems, currents of varying strength form an additional factor determining the limnic environment.

Exploration, fauna and species flocks

The Sulawesi lakes were first explored (Lake Poso, 1895; Malili lakes, 1896) by the Swiss naturalists Paul and Fritz Sarasin (Sarasin and Sarasin 1905). The Sarasins, focusing on the lakes' snails, did not fail to recognise the vastly different morphology of the lake species and consequently regarded them as ancient elements (Sarasin and Sarasin 1898). Roughly a decade later, the Dutch geologist E.C. Abendanon investigated the geology of both lake systems and sampled the lakes in 1909/1910 (Abendanon 1915b). In 1932, Woltereck conducted the last biological expedition with a universal approach to the lakes when he visited the Malili lakes during his 'Wallacea expedition' (Woltereck 1941). Charles Bonne sampled Lake Poso in 1941 for molluscs (Haase and Bouchet 2006). For several decades no further research on the lakes was conducted. In the late 1980s and early 1990s new collections were made in the lakes by M. Kottelat (fishes; Kottelat 1990, 1991, Larson and Kottelat 1992), and P. Bouchet in 1993 (molluscs; Bouchet et al. 1995). New species were described as an outcome of every expedition. For a summary of prior taxonomic work on the major lake taxa see Herder et al. (2006a), von Rintelen et al. (2007a), Schubart and Ng (2008) and von Rintelen and Cai (2009).

This taxonomic activity revealed that almost every taxon studied had radiated within Sulawesi lakes, and the lake fauna was discussed in the context of ancient lake radiations (Brooks 1950).

Since the late 1990s, several research groups have been working on the evolution of the major taxa in the lakes and the results of these studies are reviewed in this chapter. The speciose and morphologically diverse endemic species flocks in the ancient lakes of Sulawesi provide an excellent opportunity to study mechanisms of speciation and diversification in a comparative way across several groups of organisms with very different intrinsic properties. In addition to the groups discussed here, a number of other taxa have likely radiated in the lakes, such as hydrobiid gastropods (Haase and Bouchet 2006) and gobiid fishes (Kottelat et al. 1993). However, the general lack of data apart from species descriptions prevents a further discussion of these taxa here.

12.2 The species flocks in the ancient lakes of Sulawesi

12.2.1 Snails

Both ancient lake systems host species flocks of the viviparous freshwater gastropod *Tylomelania* (Mollusca: Caenogastropoda: Pachychilidae; Fig 12.2a). *Tylomelania* is endemic to Sulawesi (von Rintelen and Glaubrecht 2005) where it is widely distributed with approximately 75 species currently recognised (von Rintelen et al. 2010a), the majority of which are endemic to the ancient lakes. Here we summarise current knowledge on the ecological and phylogenetic patterns that have emerged so far from this model system.

Species diversity and distribution

At least 53 species of *Tylomelania* occur in both ancient lake systems of Sulawesi (von Rintelen et al. 2010a), making it the most speciose of all Sulawesi lake radiations. All species are endemic to their respective lake system. A total of 28 species have been described from the Malili system (von Rintelen and Glaubrecht 2003, 2008, von Rintelen et al. 2007a) and 25 morphospecies are estimated to occur in Lake Poso and Poso River (von Rintelen et al. 2010a), albeit only seven taxa have so far been formally described from that system.

In the Malili system, 24 species (86%) are endemic to a single lake (Glaubrecht and von Rintelen 2008, von Rintelen and Glaubrecht 2008) and there are pronounced differences in species diversity in the different lakes (Table 12.2). The high level of endemism in each lake or river is a striking feature of the Malili system, which suggests, in contrast to the situation in Lake Poso, a strong influence of geographic factors in species divergence, resulting in allopatric speciation, which is perhaps not entirely unexpected given the spatial structure of the system (Fig 12.1c).

Lake colonisation, adaptive radiation and introgression

Molecular data (1535 bp of mitochondrial DNA from the 16S rRNA (16S) and the cytochrome oxidase subunit I (*COI*) genes) suggest four independent colonisation events in the lakes, three of these in the Malili lakes alone (Fig 12.3a; von Rintelen et al. 2004, von Rintelen et al. 2010a): four strongly supported clades are found within the lakes, one in Lake Poso and three in the Malili system, and riverine taxa are identified as sister groups to three of the four lacustrine clades.

Each colonisation event was followed by diversification into an array of morphologically distinct and ecologically specialised species (see below for details), the hallmark of an adaptive radiation (Schluter 2000). The occurrence of four independent adaptive radiations under identical (Malili system) or very similar (Poso) conditions offers the rare opportunity to study patterns of parallel evolution. These

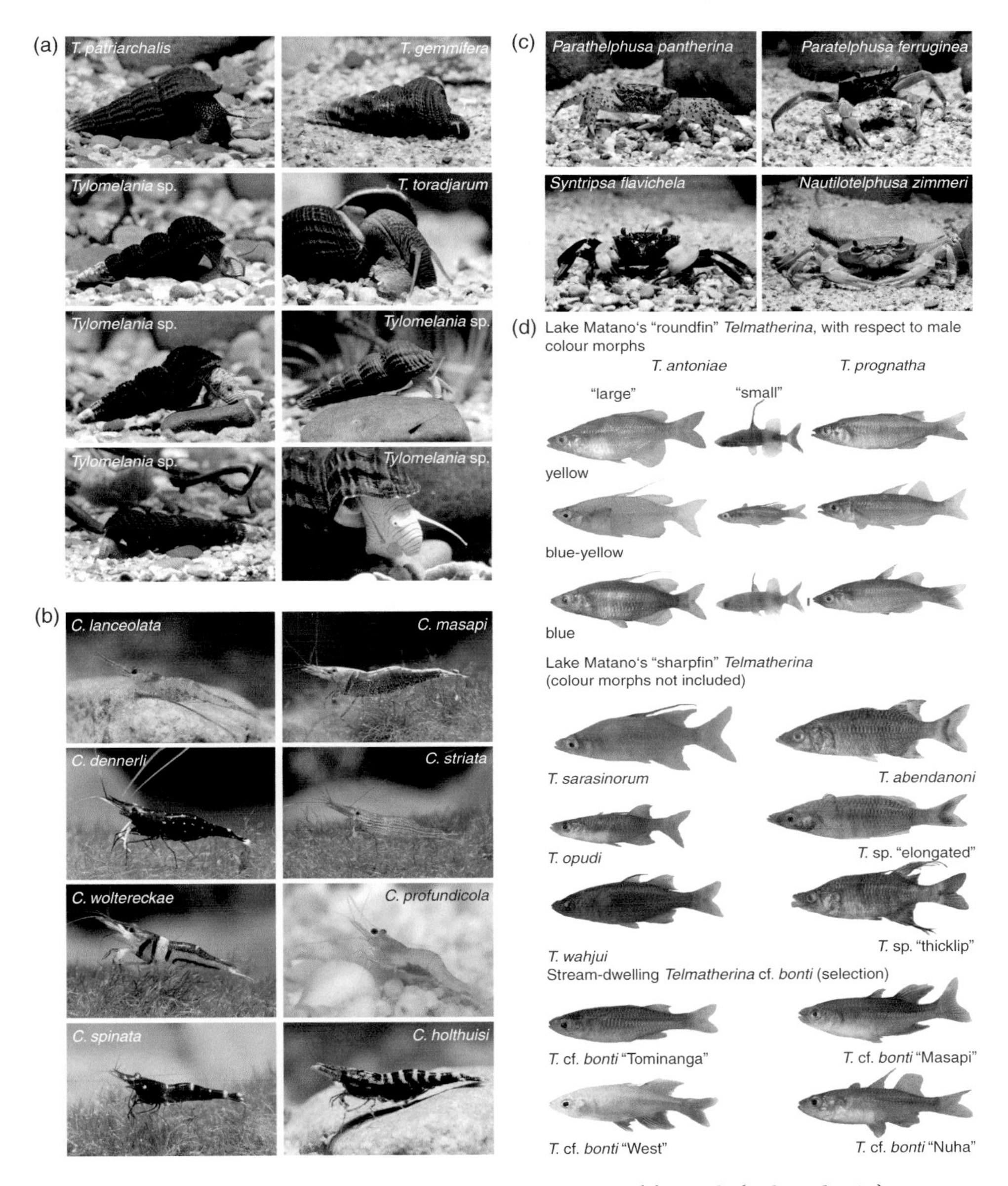

Figure 12.2 Characteristic species of the Sulawesi lakes. (a) Snails (*Tylomelania)* (b) Shrimps (*Caridina*). (c) Crabs (Gecarcinucidae). (d) Fishes (Telmatherinidae). Photographs of living animals (a–c) courtesy Chris Lukhaup. See plate section for colour version.

Table 12.2 Species diversity (morphospecies) in the ancient lakes of Sulawesi. The total count for each lake system includes also species endemic to rivers Petea, Tominanga, Larona and Poso.

	Tylomelania	Gecarcinucidae	*Caridina*	Telmatherinidae
Total	53	8	18	31
Malili lakes	28	5	14	31
Lake Matano	6	3	6	10
Lake Mahalona	9	3	7	7
Lake Towuti	10	3	13	12
Lake Masapi	1	1	1	–
Lake Lontoa	2	1	1	1–2
Lake Poso	25	3	4	–

adaptive radiations differ considerably in the extent of their diversification, with species numbers and respective morphological 'types' ranging from 3 to 25 (Malili 1: 13 spp.; Malili 2: 12 spp.; Malili 3: 3 spp.; Poso: 25 spp.).

The high level of support for the four major lake clades contrasts with virtually no resolution at the species level for *Tylomelania*. All lacustrine morphospecies for which more than one specimen or population has been sequenced appear polyphyletic in the molecular phylogeny (von Rintelen et al. 2004, 2007a, Glaubrecht and von Rintelen 2008). This lack of resolution is even more remarkable because there is no lack of genetic structure per se in the data, as there are several well-supported subclades within three of the four major lake clades. While this pattern may be caused by several factors (see e.g. Funk and Omland 2003), a pivotal role for introgressive hybridisation is indicated by genotype–phenotype mismatches and nuclear amplified fragment length polymorphism (AFLP) data (Glaubrecht and von Rintelen 2008, von Rintelen et al. 2010a).

Adaptive radiation through trophic specialisation

The morphology of the gastropod radula (rasping tongue) is highly diverse in all lacustrine lineages (Fig 12.4c) with three to six phenotypes found in each clade (von Rintelen et al. 2010a). The radula is a pivotal part of the alimentary system in gastropods, and radular morphological differences have been demonstrated to be indicative of food and substrate preferences (Hawkins et al. 1989). Indeed, radular morphology and substrate are highly correlated in all clades (von Rintelen et al. 2004, 2010). All species in the major lakes of the Malili system and Lake Poso are specialised on either soft (mud, sand) or hard (rock, sunken wood) substrates,

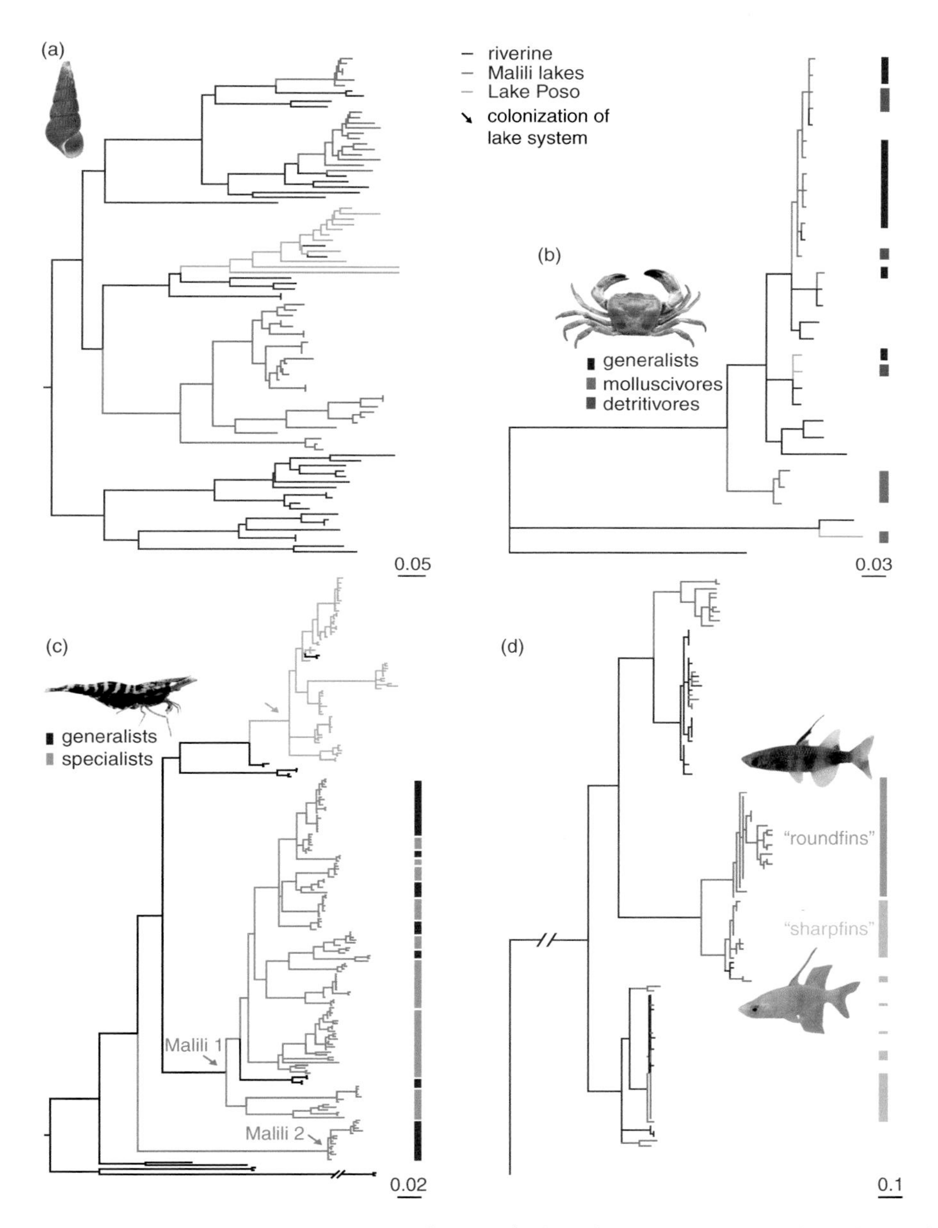

Figure 12.3 Molecular phylogenies (mtDNA) of the four major taxa in the Sulawesi lakes. (a) Snails (*Tylomelania)* (b) Crabs (Gecarcinucidae). (c) Shrimps (*Caridina*). (d) Fishes (Telmatherinidae). Bayesian inference (BI) based on mtDNA sequences. Lacustrine taxa are highlighted in red (Malili lakes) and green (Lake Poso), riverine species in black. The bars on the right highlight selected morphological or ecological groups. Modified or expanded from von Rintelen et al. 2004 (snails), von Rintelen et al. 2010b (shrimps), Schubart and Ng 2008 (crabs) and Herder et al. 2006 (fishes). Refer to these papers for methodological details. See plate section for colour version.

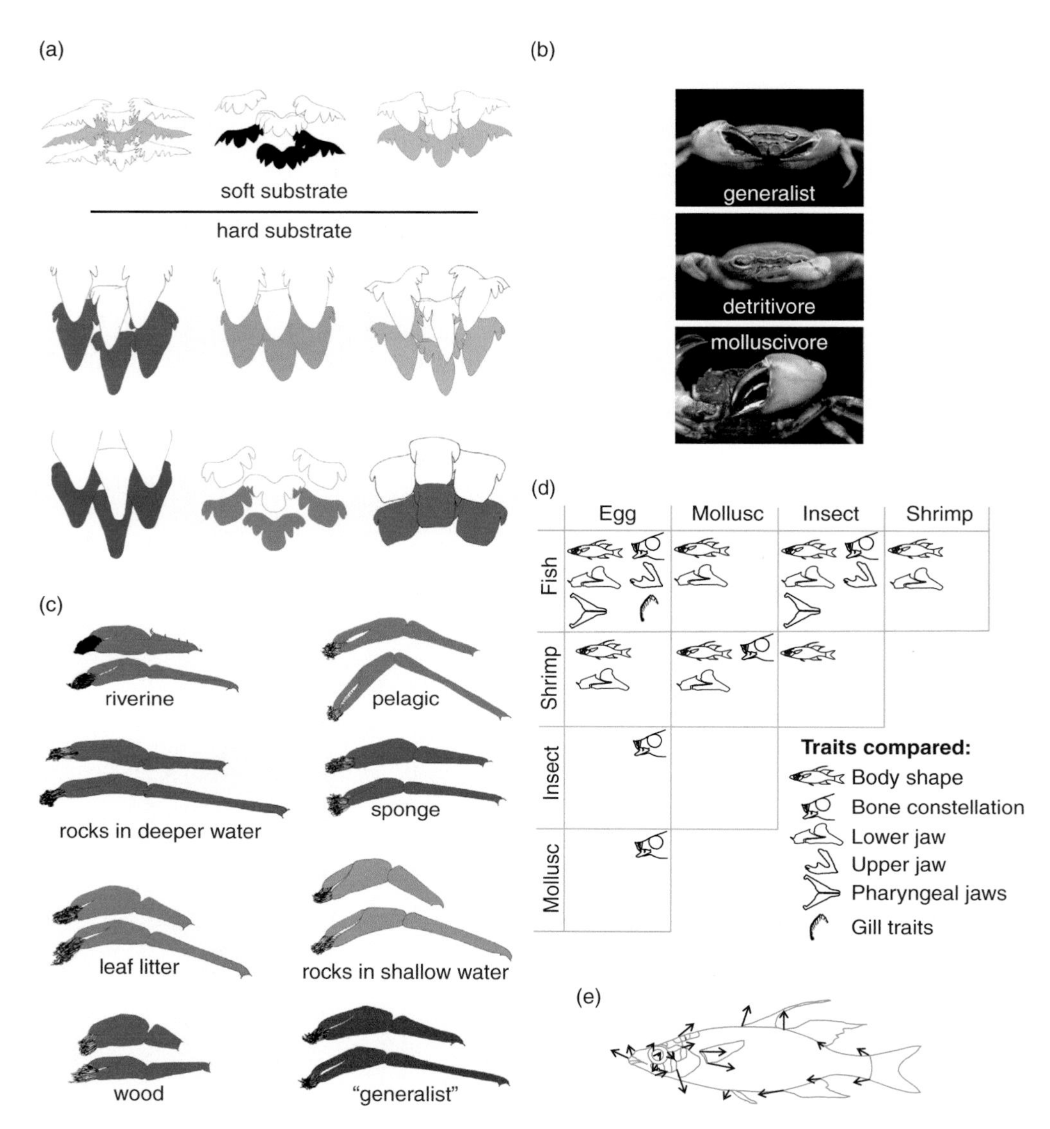

Figure 12.4 Habitat and trophic specialization in the Malili lakes. (a) Snails (*Tylomelania*) – major substrate-specific radula types highlighted by colour. (b) Crabs (Gecarcinucidae) – chela diversification in ecogroups defined by food preference. (c) Shrimps (*Caridina*) – cheliped diversification in substrate specific ecogroups highlighted by colour; modified from von Rintelen et al. (2010). (d) Fishes (Telmatherinidae): significant pairwise differentiation in candidate traits among major trophic groups as defined by stomach contents. (e) Body shape variation along a multivariate axis distinguishing shrimp-feeders from fish-feeders and other trophic specialists. Vector displacements indicate the direction of variation in shape, and show substantial differences in body depth, head shape, fin position and caudal peduncle length (d and e modified from Pfaender et al. 2010). See plate section for colour version.

with about 50% of species occurring on either substrate category (von Rintelen et al. 2007a); some taxa are specialised on an even finer scale, for example occurring on wood only (Glaubrecht and von Rintelen 2008, von Rintelen et al. 2010a). Hard substrate taxa in particular show a wide range of substrate and species-specific radular forms, and there is generally a tight correlation between enlargement of radular denticles and hard substrate (von Rintelen et al. 2004). The parallel substrate-specific occurrence of most radular forms in both ancient lake systems on Sulawesi supports a tight link between radular morphology and substrate. These observations suggest a functional role for the differences found, although a detailed understanding of the underlying mechanisms requires further investigation.

Habitat specialisation (substrate and to a lesser degree depth preferences) and radular differentiation in ancient lake *Tylomelania* enable at least five, possibly up to seven species to coexist at localities with sufficiently structured habitats (Glaubrecht and von Rintelen 2008, von Rintelen et al. 2010a). These data suggest a strong role for ecological factors in the diversification and possibly even speciation of *Tylomelania* (von Rintelen et al. 2010a).

Coevolution with crabs

Dramatic changes in snail shell morphology are associated with lake colonisation in both ancient lake systems. The species in each lacustrine clade share characteristic shell features, such as the presence of spiral and/or axial ribs (von Rintelen et al. 2004). Species can be distinguished by their shells, although intraspecific variability is rather high (von Rintelen and Glaubrecht 2003, von Rintelen et al. 2007a). In each lacustrine clade, convergent evolution of thicker shells relative to riverine species has occurred in almost all cases (von Rintelen et al. 2004). Shell thickness can be regarded as indicative of shell strength and thus serves as an estimator of resistance to crab predation. These findings coincide with the occurrence of different species of molluscivorous crabs of the Gecarcinucidae (see 12.2.2), which possess pronounced dentition on their chelae enabling them to crack shells, in each of the lakes. These data on lacustrine gastropod shell strength, structure and also the frequent occurrence of shell repair in lacustrine *Tylomelania*, in combination with the occurrence of large molluscivorous crabs, suggests that evolution in the face of crab predation is a driving factor in initial shell divergence upon colonisation of the lakes (von Rintelen et al. 2004). The presence of massive and dentulate chelae exclusively in the molluscivorous crab species makes it very likely that this is an example of true coevolution because the development of stronger gastropod shells is likely to provoke an evolutionary response in the crabs.

12.2.2 Crabs

Freshwater crabs of the family Gecarcinucidae (see Klaus et al. 2006) constitute another group of conspicuous macroinvertebrates that are abundant throughout the benthos of all Sulawesi lakes. Most of the species currently known from freshwater streams and lakes of Sulawesi belong to the genus *Parathelphusa*. Others were recently removed from this genus and placed into separate genera, because of their morphological distinctiveness (Chia and Ng 2006). This distinctiveness is especially common in crabs from lakes, and is most likely the result of morphological differentiation as a consequence of ecological specialisation. Establishment of new genera for specialists derived from a lineage of morphologically more conserved generalists often results in paraphyletic units, which is also true in the case of Sulawesi freshwater crabs (Schubart and Ng 2008; Fig 12.3b). The genus *Sundathelphusa* represents a second lineage of freshwater crabs on Sulawesi, which is distributed throughout North Sulawesi, but also occurs in Central Sulawesi as far south as the Poso catchment.

Compared with gastropods, shrimps and fishes, lacustrine species diversity is lower among Sulawesi freshwater crabs. However, in this group there has also been an increase of species recognised within the last decade. Until the year 2000, only one crab species was known from Lake Poso (*P. sarasinorum*) and three from the Malili lakes (*Parathelphusa matannensis*, *P. pantherina*, *Nautilothelphusa zimmeri*). An expedition to both lake systems in January 2000 resulted in the discovery of two new crab species from Lake Poso and two from the Malili lakes (Chia and Ng 2006, Schubart and Ng 2008). Currently, we are investigating whether small-sized crabs from the small satellite lakes Masapi and Lontoa also deserve distinct species status.

Ecology and morphology

There are almost no data on the ecology of freshwater crabs from Sulawesi and most of the available information is based on occasional opportunistic observations. Each of the four larger lakes (Poso, Matano, Mahalona and Towuti) is inhabited by three crab species, with only Mahalona and Towuti sharing the same species. However, the recent findings of new species in the larger lakes revealed a pattern suggesting the existence of three ecological niches that are consistently occupied by distinct ecotypes occurring within each lake. We have termed these ecotypes (1) the 'undifferentiated or unspecialised form', (2) the 'detritivore and burrowing form', and (3) the 'molluscivore' form (Schubart and Ng 2008). Most of the lake crabs are considerably larger than their close relatives from the rivers. Nevertheless, crabs of ecotype 1 seem to retain similar proportions of their chelae and ambulatory legs. Their feeding habits and behaviour must therefore be similar to crabs from freshwater streams, which are typically opportunistic generalists. It is noteworthy that crabs occupying this niche have consequently maintained their generic classification and are *Parathelphusa sarasinorum* in Poso, *P. pantherina* in Matano and *P. ferruginea* in Mahalona and Towuti. In addition to these generalists, there are two types of specialists, easily recognisable by the shape of the chelae. In one case, the size of the chelae is reduced to smaller tweezer-like appendages, as in the case of *Nautilothelphusa zimmeri*, which is found in the major lakes of the Malili system. This species mostly occurs on muddy or sandy substrate and is relatively difficult to find due its mimetic colouration and because it often remains buried in the soft sediment during daytime. The last pair of walking legs is modified to facilitate backward burrowing. We hypothesise that this species specialises on detritus as a food source and refer to it as the 'detritivore' eco- and morphotype (*Migmathelphusa olivacea* in Poso). There are also crabs with proportionally larger chelae that have strong molariform dentition on the cutting edges. This characteristic is otherwise only found in marine crabs that are known to feed on molluscs (and in molluscivorous crabs from Lake Tanganyika) (Vermeij 1994, Marijnissen et al. 2006). We assume and have observed that these crabs are specialised mollusc predators and thus constitute a third feeding guild, the 'molluscivores'. Two species of molluscivorous crabs of the Malili lakes have been placed in the new genus *Syntripsa* (Chia and Ng 2006). In Lake Poso the only species of *Sundathelphusa* of the ancient lakes acts as the resident molluscivore crab species, based on its chelar morphology, and was consequently named *S. molluscivora* (Schubart and Ng 2008).

Phylogeny and radiation

Figure 12.3b depicts the most recent mitochondrial phylogeny of Sulawesi freshwater crabs based on approximately 560 bp of 16S. River species not very closely related to one of the lake-colonising lineages are excluded. Molecular analyses reveal that there is a high likelihood of cryptic endemism in the rivers of Sulawesi: crabs from different watersheds can easily be separated genetically and many of them may represent undescribed species. *Sundathelphusa molluscivora* from Lake Poso gives the clearest evidence of independent colonisation of the lakes, because it belongs to a divergent lineage of freshwater crabs. However, the tree also reveals that the molluscivore crabs from the Malili lakes seem to have colonised the lakes independently, and probably earlier than all other taxa. This suggests that the best strategy for new lake colonisers was to specialise on the protein-rich gastropods as prey, possibly leaving vacant those niches exploiting other food sources. Currently the lakes do not harbour species that are closely related to these molluscivore forms. This argues against adaptive radiation and sympatric speciation during this first step of colonisation.

A second sweep of lake colonisation is much younger according to genetic distances among the species and compared with the closest river species. Most surprising is that the colonisation patterns continued to occur almost in parallel in the two independent lake systems. In both cases, a generalist and a detritivore lineage evolved more or less simultaneously, and probably in the presence of a molluscivore species. Moreover, in both cases, closely related river species are extant and cluster together in monophyletic lineages and generalist and detritivore forms are genetically very similar, to the point where some cannot be distinguished with mtDNA, as in the case of *Nautilothelphusa zimmeri* and *Parathelphusa ferruginea* in Towuti-Mahalona. During these speciation processes, adaptive radiation and possibly sympatric speciation (especially in the single lake system Poso) are likely scenarios.

Overall, we can summarise that currently eight endemic species of freshwater crabs are known from the ancient lakes in Sulawesi. This number will be increased to ten, if genetically separable crab populations from Masapi and Lontoa are recognised as distinct species. This species diversity is not profuse, but the parallel ecological diversification, niche segregation and speciation patterns make these crab species an intriguing case study in evolutionary biology. Ongoing studies with nuclear DNA show that hybridisation and introgression play an important role in their evolution and diversification (P. Koller and F. Kolbinger unpublished data).

12.2.3 Shrimps

Among the four major groups of organisms that have radiated in the central lakes of Sulawesi are atyid freshwater shrimps (Crustacea: Decapoda: Caridea). Both lake systems harbour endemic species flocks of *Caridina*, a genus that is generally widely distributed throughout the Indo-Pacific region. With over 200 species, it is the most speciose group within the family (De Grave et al. 2008).

Species of *Caridina* from the Malili lake system were first described by Woltereck (1937a, b). Only 70 years later, new species descriptions followed (Zitzler and Cai 2006, Cai et al. 2009). Today, 21 species are known from the ancient lakes of Sulawesi (including species from the lake drainages), six from Lake Poso and 15 from the Malili lake system (Table 12.2; von Rintelen and Cai 2009). Regarding reproductive biology, all lake species produce relatively large eggs (von Rintelen and Cai 2009). This egg size is typical of direct developers that lack planktonic larval stages and live in landlocked freshwater areas for their whole life (Lai and Shy 2009).

The Malili flock represents the largest radiation within the genus and even within the family. A similar radiation is thus far known only from Lake Tanganyika. However, the Tanganyikan species flock comprises three different genera of atyid freshwater shrimps with 11 species altogether (Fryer 2006). Fryer (2006: 136) stated that 'Tanganyikan atyids have undergone adaptive radiation in a manner analogous to that of the jaws and oral teeth of cichlid fishes'. The case for adaptive radiation in atyid shrimps has so far only been tested in the Poso (von Rintelen et al. 2007c) and Malili species flocks of *Caridina* (von Rintelen et al. 2010a). The latter will be explained in detail below. Although a direct comparison of Sulawesi and Tanganyika radiations is difficult because they belong to different genera, some of the hallmarks of adaptive radiation are apparently analogous, especially with respect to morphological differences in feeding appendages (Fryer 2006, von Rintelen et al. 2010a). These differences mainly refer to the chelipeds or the first and second pair of claw-bearing legs (Fig 12.4c) that are used to pick up food particles from the ground (Fryer 1960).

Lake colonisation, hybridisation and cryptic species

A molecular phylogeny based on two mitochondrial gene fragments of 16S and *COI* revealed three ancient lake clades in Sulawesi (Fig 12.3c). One clade comprises species from Lake Poso and two clades include all Malili species (von Rintelen et al. 2010a). Thus, at least two independent colonisations of the Malili lake system can be inferred. However, because one of the two clades consists of only a single species, von Rintelen et al. (2010a) suggested an independent colonisation event without subsequent radiation, while the ancestor of the other clade gave rise to a larger radiation with 14 species.

Another result of the molecular phylogeny was the occasional mismatch between morphological and molecular data (von Rintelen and Cai 2009, von Rintelen et al. 2010a). The majority of species are monophyletic in the molecular phylogeny, but seven are not. In some cases von Rintelen et al. (2011) suggested introgressive hybridisation or incomplete lineage sorting. However, there are also likely cases of cryptic species, especially when allopatrically distributed populations of a single morphospecies appear in separate clades that are not even sister to each other (for details compare von Rintelen and Cai 2009, von Rintelen et al. 2010a).

Ecological diversification

For the Malili shrimp species flock it has been questioned whether this radiation was adaptive because of the lack of apparent correlation between phenotypic and environmental traits (Woltereck 1937a). However, von Rintelen et al. (2010) distinguished eight different ecogroups defined by substrate differences and morphological characters (mainly the chelipeds; Fig 12.4c). Six ecogroups corresponded to a single habitat each, such as rocks or wood, and the remaining two to several substrates each (Fig 12.4c: a lacustrine 'generalist' and 'riverine', a species whose occurrence is limited to the Malili rivers only). Species within each ecogroup had similar chelipeds, while differences in the morphology of these feeding appendages between ecogroups could be considerable (von Rintelen et al. 2010a). Based on these results, the authors divided the Malili species flock into habitat generalists and specialists (Fig 12.3c). In summary, they suggest a phenotype-environment correlation as a crucial criterion for adaptive radiation (Schluter 2000).

A special case of ecological diversification within the Malili lakes is the first known instance of a sponge-dwelling shrimp in freshwater (von Rintelen et al. 2007b). The species *Caridina spongicola* forms an exclusive and probably commensal association with an as yet undescribed freshwater sponge in the outlet bay of Lake Towuti, the largest of the Malili lakes. Phylogenetic and ecological data suggest a comparatively recent origin of both taxa (von Rintelen et al. 2007b).

The evolution of colour patterns

Last but not least, the Sulawesi lake shrimp species show a variety of species-specific colour patterns (Fig 12.2c shows several examples from the Malili lakes) that are equally pronounced in both sexes and occasionally in juveniles (von Rintelen and Cai 2009). With such colouration, some lake species strongly resemble marine rather than typical freshwater shrimp. Although colourful species occur rarely at other localities (e.g. Yam and Cai 2003), this abundance of different body colourations within such a confined area is unique within the genus. However, it remains largely speculative whether the colour patterns of the Malili lake species have evolved under ecological (natural) selection and whether assortative mating plays a role in maintaining these patterns (von Rintelen et al. 2010a).

12.2.4 Fishes

Fish species flocks restricted to freshwater lakes are one of the prime examples of adaptive radiation, and have become increasingly popular models for investigating the mechanisms driving speciation processes (Schliewen et al. 1994, Schluter 2000, Kocher et al. 2004, Seehausen 2009, Vonlanthen et al. 2009). The Malili lakes system harbours endemic species or small species flocks of several fish families, the most conspicuous being the endemic radiation of small, sexually dimorphic sailfin silverside fish (Telmatherinidae: Atheriniformes) (Kottelat 1990, 1991). Closely related to rainbow fishes (Melanotaeniidae) from Australia and New Guinea, most sailfin silverside species of the Malili lakes flock are characterised by spectacular male colour polychromatisms, with typically either yellow, blue or blue-yellow courtship colouration (Herder et al. 2006a; Fig 12.2d). Except for two species restricted to small islands off Western New Guinea or Southwestern Sulawesi, all sailfin silverside species known so far occur only in the Malili lakes drainage or surrounding river systems. Most of the species are lake dwelling (lacustrine) and restricted either to the extraordinary deep and isolated hydrological head of the lakes system, Lake Matano, or to the less isolated lakes Mahalona and Towuti (Kottelat 1990, 1991, Herder et al. 2006a). Three genera with roughly 31 morphospecies are currently recognised: *Paratherina* (4 spp.), *Tominanga* (2 spp.), and *Telmatherina* (~23 spp.) (note that the species number for stream-dwelling *Telmatherina* is still tentative; see Herder et al. 2006a for a brief review on exploration history and taxonomy). *Paratherina* and *Tominanga* inhabit in- and offshore habitats of lakes Towuti and Mahalona, *Tominanga* occur in lakes Towuti and Mahalona, and also frequently in the river connecting these lakes (River Tominanga). Only two formally described species of *Telmatherina* are present in these 'lower lakes' and the streams surrounding them: the abundant lake-dwelling *Telmatherina celebensis*, and *Telmatherina bonti* which form a morphologically highly diverse group inhabiting nearly all permanent rivers and streams of the area, with strong indications for local differentiation (Herder et al. 2006a, b). All remaining lacustrine *Telmatherina* (morpho)species recognised so far are restricted to Lake Matano (Herder et al. 2006a).

The *Telmatherina* species flock of ancient graben-lake Lake Matano

Lake Matano is characterised by the endemism of all lacustrine sailfin silversides inhabiting it, its long-term stability as ancient and extraordinary deep graben-lake (Crowe et al. 2008a, b), and perfect settings for observational studies in its clear waters. These features make it the system of choice for several studies focusing on speciation, adaptive radiation and the maintenance of colour polymorphisms (see Herder and Schliewen 2010 for a comprehensive review). Molecular phylogenetic studies identified two sister clades of lacustrine *Telmatherina* radiating in Lake Matano, which can be recognised by the shape of their second dorsal- and anal fin. 'Sharpfins', characterised by conspicuously pointed, elongated fins in males of most morphospecies, are diverse in terms of morphospecies and body shapes (see Fig 12.2d). In contrast to the three morphospecies of 'roundfins', with rounded male fins, lake populations of sharpfins are genetically introgressed by stream populations (Herder et al. 2006b, Schwarzer et al. 2008; Fig 12.3d). Together, roundfins and sharpfins of Lake Matano were identified as an ancient monophyletic group, which is however not easily recognisable as such in analyses of nuclear and mitochondrial DNA data because of introgression from riverine invaders (Herder et al. 2006b). Morphological, ecological, behavioural and genetic analyses suggest that both radiations are adaptive with respect to habitat use and trophic ecology, but also indicate that individuals intermediate to distinct 'morphospecies' occur (Herder et al. 2006a, 2008, Cerwenka et al. (in press), Pfaender et al. 2010, 2011).

Sympatric speciation in Lake Matano's roundfin *Telmatherina*

A study on the three roundfin sailfin silverside morphospecies, integrating data on morphology, ecology, mating behaviour and gene flow, provided support for the hypothesis that these lake fish evolved in full sympatry in response to ecological selection pressure within Lake Matano (Herder et al. 2008). The three roundfin morphospecies are distinguished mainly by body depth, size, and snout shape. Males are polymorphic in courtship colouration, being yellow, blue, or blue-yellow (Fig 12.2d). The smallest and most abundant species is an offshore feeder, whereas the larger two species inhabit inshore areas where they segregate according to feeding modes and microhabitat use. Population structure and mating observations support substantial but incomplete reproductive isolation of the three morphospecies, but do not indicate any restrictions in gene flow between the conspicuous colour morphs (Herder et al. 2008). Genomic signatures suggest that the actions of selection are restricted to only small parts of the genome, a result fitting recent ideas about initial stages of ecological speciation (Nosil et al. 2009). Absence of population structure among populations within the lake and of differentiation among male colour morphs were recently confirmed also based on alternative molecular markers (Walter et al. 2009a, 2009b). Taken together, these data are consistent with incipient sympatric speciation in Lake Matano's roundfins, based on the criteria suggested by Coyne and Orr (2004). Accordingly, the small radiation of roundfins endemic to ancient Lake Matano adds one more case to the rare examples in nature where speciation is clearly not explained by separating effects of strict geographic isolation. The mechanism most likely driving roundfin speciation is adaptation to alternative modes of resource exploitation, leading to the alternative patterns of habitat use and trophic specialisation. The comparatively young age of the Matano species flock (Herder et al. 2006b, Roy et al. 2007), its compact spatial settings and the incipient character of its roundfin radiation renders this group a highly promising model group for further studies on the genomic consequences of ecological speciation.

Adaptive radiation in Lake Matano's sharpfins

Schluter's (2000) criteria provide a framework for evaluating the adaptive character of radiations. Common ancestry, rapid speciation and correlation of phenotypes with their environment and utility of the traits evolved are prerequisites for stating that a radiation is truly 'adaptive'. Both Lake Matano's endemic sharpfin *Telmatherina* and its sister clade of roundfins show some conspicuous trophic specialisations that suggest that traits related to feeding modes are adaptive. One sharpfin species is characterised by enlarged, puffy lips that are most likely an adaptation to shrimp feeding (Herder et al. 2006a, Pfaender et al. 2010). Another species specialised in eating the eggs of other sailfin silversides and is especially abundant at spawning sites where it uses alternative behavioural tactics for obtaining con- and heterospecific eggs (Gray et al. 2007, 2008a, Cerwenka et al. in press). To test the hypothesis that the sharpfin radiation is adaptive, a recent morphological study compared potential key traits such as shape and configuration of jaw bones, shape of the 'second set of jaws' in the fishes' throat (pharyngeal jaws), body shape, number of gill rakers and body size with stomach contents. Surprisingly it revealed fine-scale patterns of morphological differentiation among groups of sharpfins defined by stomach contents (Pfaender et al. 2010; see Fig 12.4d). The most distinct adaptations are in fish-, shrimp- and egg-feeding groups, with trait expression being widely consistent with other adaptive fish radiations. The strong correlation between morphological characters and feeding specialisation is detectable in the whole set of traits analysed, and serves as evidence for adaptation as a result of ecological selection pressure. Accordingly, this strongly supports the hypothesis that the sharpfin species flock fulfils the criterion of 'trait utility' (Schluter 2000) – however, the biomechanical function of the specialised traits remains to be demonstrated in detail. Although introgressed by stream populations, lake populations of sharpfins clearly also fulfil the criterion of common ancestry. In line with the absence of mitochondrial lineage sorting within sharpfins (Herder et al. 2006b), molecular (Roy et al. 2007) and geological (von Rintelen et al. 2004) age estimates clearly support rapid evolution. Phenotype-specific habitat use, as evident in roundfins, remains the last of Schluter's (2000) four criteria to be critically evaluated in sharpfins.

No evidence of speciation by sexual selection: colour polymorphic sailfin silversides

The spectacular male colour morphs are, at least in roundfins, clearly not linked to population structure (Herder et al. 2008, Walter et al. 2009b). This contrasts with findings in African cichlid fish radiations where the evolution of alternative male colour morphs is explained by alternative female mating preferences, finally leading to speciation (Seehausen et al. 2008). Male colour morphs in Lake Matano's *Telmatherina* are probably best explained by a polymorphism maintained by heterogeneous visual environments. Gray et al. (2008b) reported differential conspicuousness of alternative male colour morphs in two contrasting courting habitats (shady root versus open beach sites), corresponding to differential reproductive fitness. Given the theoretical plausibility that environmental heterogeneity may maintain colour polymorphisms in nature (Chunco et al. 2007, Gray and McKinnon 2007), patchy distribution of opposing visual environments or other environmental factors affecting light conditions (e.g. seasonal or diurnal effects) may provide an explanation for this phenomenon. Such hypotheses await further investigation. Interestingly, colour polymorphisms occur in all major groups of the radiation, including populations inhabiting alternative habitats such as shady, fast-flowing streams or calm and open beaches of the lakes (Herder et al. 2006a), supporting the idea of common mechanisms maintaining this conspicuous phenomenon.

12.3 Evolution of the Sulawesi lake species flocks

Parallel radiations of gastropods (von Rintelen et al. 2004, von Rintelen and Glaubrecht 2005, Glaubrecht and von Rintelen 2008, von Rintelen et al. 2010b), parathelphusid crabs (Schubart and Ng 2008, Schubart et al. 2008), atyid shrimps (von Rintelen et al. 2007c, 2010b) and telmatherinid fishes (Herder et al. 2006b, Herder and Schliewen 2010) in the lakes, particularly the Malili system, offer an outstanding opportunity to compare diversification patterns in organisms differing in fundamental biological properties within the same environmental setting.

Detailed comparative analyses, especially on the impact of direct interactions between these animal groups for speciation, are under way. While joint in-depth analyses are not yet available, the published results (see citations above) already suggest a number of commonalities between some or all groups.

Ecological speciation in response to alternative modes of resource use has clearly been demonstrated or appears at least highly likely in all of the Malili and Poso radiations analysed so far. Specialised radular morphology in gastropods, alternative forms of chelipeds or chelae in shrimps and crabs, and an array of traits including jaw function and body shape or the number of gill rakers in sailfin silversides clearly show signatures of ecological selection with respect to trophic resource use. In line with alternative modes of habitat use and utility of the adapted traits for fine-scaled trophic adaptations, these findings clearly support the idea that speciation following disruptive ecological selection pressure is the main force explaining local diversification observed in all these morphologically and ecologically highly distinct organisms. Accordingly, all these radiations are likely to constitute cases of adaptive radiation.

Although both Poso and the Malili lakes are among the oldest freshwater lakes on earth, they are young in terms of geological timeframes. Hence, it is not surprising that their endemic radiations are apparently in the early stages of species-flock formation. This includes significant levels of hybridisation or introgression among species of lake-dwelling lineages evolving in sympatry or after multiple colonisation events (von Rintelen et al. 2004, 2007c, Herder et al. 2008). Likewise, hybridisation occurs among lake- and stream species (von Rintelen et al. 2004, Schwarzer et al. 2008), resulting in complex patterns of mitochondrial diversity in sailfin silversides and gastropods that are not consistent with nuclear DNA data and morphology (von Rintelen et al. 2004, Herder et al. 2006b, Roy et al. 2007). Theory and a few animal case studies (e.g. Schliewen and Klee 2004, Seehausen 2004, Bell and Travis 2005, Stelkens and Seehausen 2009) suggest that reticulate evolution by hybridisation might promote adaptive divergence in species-flocks, a striking hypothesis that however remains to be tested critically in the case of radiations in the Malili lakes.

In contrast to adaptations affecting trophic morphology of all groups analysed so far, the conspicuous colour patterns occurring in sailfin silversides (body and fins), crustaceans (body and legs) and gastropods (body) most likely result from very different mechanisms. Colour most likely plays a role in species recognition and speciation in cichlid fishes, one of the prime model systems in speciation research (Streelman and Danley 2003, Seehausen et al. 2008). In Sulawesi's lacustrine atyid shrimp radiations, conspicuous colour patterns widely coincide with morphological species concepts, suggesting that colouration might indeed play a role in processes of divergence (von Rintelen et al. 2007c, 2010). Colour-driven diversification might also contribute to diversification in *Tylomelania* gastropods, because substantial numbers of morphospecies especially in Lake

Poso are characterised by bright pigmentation of the soft body, which appears difficult to explain solely in light of ecological selective pressures. However, visual systems and factors influencing patterns of mate choice are not yet explored in *Tylomelania* snails and atyid shrimps, such that further studies are required to test critically the hypothesis of colour-driven diversification in both groups. In the Malili lakes sailfin silverside radiation, conspicuous male colour polymorphisms correlate with mating success in different habitats, but are not associated to population divergence (Herder et al. 2008, Gray et al. 2008). Sailfin silversides have a pentachromatic visual system, ranging from UV to yellow including the spectra of *Telmatherina* male courtship colouration (Gray et al. 2008). Stable colour polymorphisms are the most likely explanation for this conspicuous phenomenon – there are so far no indications for a link between colour polymorphisms and speciation in sailfin silversides. Taken together, additional efforts will be required to provide a further elucidation of evolutionary responses common to all groups of organisms analysed in Sulawesi's ancient lakes so far, including potential universal patterns of radiation in the light of crucial differences explained by the organisms' particular biology.

12.4 Threats and conservation

The Sulawesi lakes are a 'hotspot' of Southeast Asian biodiversity and at the same time a scientific treasure trove for evolutionary biology research. Unfortunately, the lakes' environment is threatened by several factors. At both lake systems, the growth of local human communities is a general point of concern, because it seems inevitably coupled with pollution and habitat destruction. The continued activities of illegal loggers also threaten to damage lake habitats by erosion of surface soil. This effect is more severe at the Malili lakes, where the shores have been less densely populated and less accessible for a longer time than those of Lake Poso. Although all of these issues are notoriously difficult to tackle, efforts should at least be made to increase community awareness about the problems because they also have the potential to threaten the future of the local people.

More specific risks for the lake ecosystems stem from the activities of the nickel mining company P.T. Vale Indonesia at the Malili lakes, which, in addition to their role in opening up the area, directly impact the sensitive environment of the lakes. Although appreciable efforts have been made by this company to preserve the water quality in the lakes, less attention is paid to the importance of protecting the environment of the lakes and particularly the rivers connecting and draining them. The recently finished construction of a third hydroelectric dam at Larona River is just one example. A dam has also been constructed at Poso River just above the Sulewana Rapids, which are a hotspot of endemic species adapted to fast-flowing water. New risks have emerged only during the last years, when the collecting of live snails, crabs and shrimps for export started to resemble a 'gold rush' among pet traders fuelled by the demand from aquarium enthusiasts, and an increasing number of introduced alien freshwater species appeared in the lakes (Herder et al., in prep). This enumeration of actual and potential risks to the lake biota indicates that their conservation should have a high priority. The risk from pet traders may also have beneficial side effects, though, as it has significantly increased public awareness about the lakes' unique fauna in several countries.

Acknowledgements

We thank the editors for inviting us to contribute to this volume. We are especially grateful to Ristiyanti Marwoto, Renny K. Hadiaty and Daisy Wowor (Zoological Museum Bogor) for strongly supporting our research in Indonesia. We thank LIPI for permits to conduct research in Indonesia. We are most indebted to INCO for invaluable logistic support at the lakes during our many visits. A. Cerwenka, S. Chapuis, C. Dames, D. Franz, J. Frommen, B. Gregor, J. Herder, J. Jang, P. Koller, T. Miethe, A. Munandar, P. K. L. Ng, A. Nolte, J. Pfaender, T. Santl, J. Schwarzer, J. Simonis, and B. Stelbrink significantly contributed to the projects in field and/ or lab. We thank C. Lukhaup for allowing us to use his photographs. F. H. thanks U.K. Schliewen for the opportunity to study sailfin silversides under his supervision, and for his great support in many aspects. Our work was funded by research grants GL 297/1, GL 297/7 awarded to M.G., SCHU 1460/6 to C.D.S. and SCHL 567/2-1, 2, 3 to U.K. Schliewen from the Deutsche Forschungsgemeinschaft.

References

Abendanon, E. C. (1915a). *Midden-Celebes-Expeditie. Geologische en geographische doorkruisingen van Midden-Celebes (1909–1910)*, Atlas. Leiden, The Netherlands: E. J. Brill.

Abendanon, E. C. (1915b). *Midden-Celebes-Expeditie. Geologische en geographische doorkruisingen van Midden-Celebes (1909–1910)*, Vol. 2. Leiden, The Netherlands: E. J. Brill.

Albrecht, C. and Wilke, T. (2008). Ancient Lake Ohrid: biodiversity and evolution. *Hydrobiologia*, **615**, 103–40.

Bell, M. A. and M. P. Travis. (2005). Hybridization, transgressive segregation, genetic covariation, and adaptive radiation. *Trends in Ecology and Evolution*, **20**, 358–361.

Bouchet, P., Guerra, A., Rolán, E. and Rocha, F. (1995). A major new mollusc radiation discovered in the ancient lakes of Sulawesi. In *Abstracts of the 12th International Malacological Congress*, Vigo, pp. 14–15.

Brooks, J. L. (1950). Speciation in ancient lakes. *Quarterly Review of Biology*, **25**, 30–60, 131–76.

Cai, Y., Wowor, D. and Choy, S. (2009). Partial revision of freshwater shrimps from Central Sulawesi, Indonesia, with descriptions of two new species. *Zootaxa*, **2045**, 15–32.

Cerwenka, A. F., Schliewen, U. K. and Herder, F. (in press). Alternative egg-feeding tactics in *Telmatherina sarasinorum*, a trophic specialist of Lake Matano's evolving sailfin silversides fish radiation. *Hydrobiologia*.

Chia, O. C. K. and Ng, P. K. L. (2006). The freshwater crabs of Sulawesi, with descriptions of two new genera and four new species (Crustacea: Decapoda: Brachyura: Parathelphusidae). *Raffles Bulletin of Zoology Singapore*, **54**, 381–428.

Chunco, A. J., McKinnon, J. S. and Servedio, M. R. (2007). Microhabitat variation and sexual selection can maintain male color polymorphisms. *Evolution*, **61**, 2504–15.

Coyne, J. A. and Orr, H. A. (2004). *Speciation*. Sunderland, MA: Sinauer Associates.

Crowe, S. A., O'Neill, A. H., Katsev, S. et al. (2008a). The biogeochemistry of tropical lakes: a case study from Lake Matano, Indonesia. *Limnology and Oceanography*, **53**, 319–31.

Crowe, S. A., Jones, C. A., Katsev, S. et al. (2008b). Photoferrotrophs thrive in an Archean Ocean analogue. *Proceedings of the National Academy of Sciences USA*, **105**, 15938–43.

De Grave, S., Cai, Y. and Anker, A. (2008). Global diversity of shrimps (Crustacea: Decapoda: Caridea) in freshwater. *Hydrobiologia*, **595**, 287–93.

Fryer, G. (1960). The feeding mechanism of some atyid prawns of the genus *Caridina*. *Transactions of the Royal Society of Edinburgh*, **64**, 217–44.

Fryer, G. (2006). Evolution in ancient lakes: radiation of Tanganyikan atyid prawns and speciation of pelagic cichlid fishes in Lake Malawi. *Hydrobiologia*, **568**, 131–42.

Funk, D. J. and Omland, K. E. (2003). Species-level paraphyly and polyphyly: frequency, causes, and consequences, with Insights from animal mitochondrial DNA. *Annual Reviews of Ecology and Systematics*, **34**, 397–423.

Giesen, W. (1994). Indonesia's major freshwater lakes: a review of current knowledge, development processes and threats. *Mitteilungen des Internationalen Vereins für Theoretische und Angewandte Limnologie*, **24**, 115–28.

Giesen, W., Baltzer, M. and Baruadi, R. (1991). *Integrating Conservation with Land-use Development in Wetlands of South Sulawesi*. Bogor, Indonesia: Directorate General of Forest Protection and Nature Conservation.

Glaubrecht, M. and von Rintelen, T. (2008). The species flocks of lacustrine gastropods: *Tylomelania* on Sulawesi as models in speciation and adaptive radiation. *Hydrobiologia*, **615**, 181–99

Gorthner, A., Martens, K., Goddeeris, B. and Coulter, G. (1994). What is an ancient lake? In *Speciation in Ancient Lakes*, ed. K. Martens, B. Goddeeris and G. Coulter. Stuttgart, Germany: E. Schweizerbart'sche Verlagsbuchhandlung, pp. 97–100.

Gray, S. M. and McKinnon, J. S. (2007). Linking color polymorphism maintenance and speciation. *Trends in Ecology and Evolution*, **22**, 71–9.

Gray, S. M., Dill, L. M. and McKinnon, J. S. (2007). Cuckoldry incites cannibalism: male fish turn to cannibalism when perceived certainty of paternity decreases. *American Naturalist*, **169**, 258–63.

Gray, S. M., McKinnon, J. S., Tantu, F. Y. and Dill, L. M. (2008a). Sneaky egg-eating in *Telmatherina sarasinorum*, an endemic fish from Sulawesi. *Journal of Fish Biology*, **73**, 728–31.

Gray, S. M., Dill, L. M., Tantu, F. Y. et al.(2008b). Environment-contingent sexual selection in a colour polymorphic fish. *Proceedings of the Royal Society London B*, **275**, 1785–91.

Haase, M. and Bouchet, P. (2006). The species flock of hydrobioid gastropods (Caenogastropoda, Rissooidea) in ancient Lake Poso, Sulawesi. *Hydrobiologia*, **556**, 17–46.

Haffner, G. D., Hehanussa, P. E. and Hartoto, D. (2001). The biology and physical processes of large lakes of Indonesia: Lakes Matano and Towuti. In *The Great Lakes of the World (GLOW): Food-web, Health and Integrity*, ed. M. Munawar and R. E. Heck. Leiden, The Netherlands: Backhuys Publishers, pp. 183–92.

Hawkins, S. J., Watson, D. C., Hill, A. S. et al. (1989). A comparison of feeding mechanisms in microphagous, herbivorous, intertidal prosobranchs in relation to resource partitioning. *Journal of Molluscan Studies*, **55**, 151–65.

Herder, F. and Schliewen, U. K. (2010). Beyond sympatric speciation: Radiation of sailfin silverside fishes in the Malili lakes (Sulawesi). In *Evolution in Action – Adaptive Radiations and the Origins of Biodiversity*, ed. M. Glaubrecht. Heidelberg, Germany: Springer Verlag. pp. 465–83.

Herder, F., Schwarzer, J., Pfaender, J., Hadiaty, R. K. and Schliewen, U. K. (2006a). Preliminary checklist of sailfin silversides (Telostei: Telmatherinidae) in the Malili lakes of Sulawesi (Indonesia), with a synopsis of systematics and threats. *Verhandlungen der Gesellschaft für Ichthyologie*, **5**, 139–63.

Herder, F., Nolte, A. W., Pfaender, J. et al. (2006b). Adaptive radiation and hybridization in Wallace's dreamponds: evidence from sailfin silversides in the Malili lakes of Sulawesi. *Proceedings of the Royal Society London B*, **273**, 2209–17.

Herder, F., Pfaender, J. and Schliewen, U. K. (2008). Adaptive sympatric speciation of polychromatic 'roundfin' sailfin silverside fish in Lake Matano (Sulawesi). *Evolution*, **62**, 2178–95.

Klaus, S., Schubart, C. D. and Brandis, D. (2006). Phylogeny, biogeography and a new taxonomy for the Gecarcinucoidea Rathbun, 1904 (Decapoda: Brachyura). *Organisms, Diversity and Evolution*, **6**, 199–217.

Kocher, T. D. (2004). Adaptive evolution and explosive speciation: the cichlid fish model. *Nature Reviews Genetics*, **5**, 288–98.

Kornfield, I. and Smith, P. F. (2000). African cichlid fishes: model systems for evolutionary biology. *Annual Reviews of Ecology and Systematics*, **31**, 163–96.

Kottelat, M. (1990). Sailfin silversides (Pisces: Telmatherinidae) of Lakes Towuti, Mahalona and Wawontoa (Sulawesi, Indonesia) with descriptions of two new genera and two new species. *Ichthyological Exploration of Freshwaters*, **1**, 227–46.

Kottelat, M. (1991). Sailfin silversides (Pisces: Telmatherinidae) of Lake Matano, Sulawesi, Indonesia, with descriptions of six new species. *Ichthyological Exploration of Freshwaters*, **1**, 321–44.

Kottelat, M., Whitten, A. J., Kartikasari, S. N. and Wirjoatmodjo, S. (1993). *Freshwater Fishes of Western Indonesia and Sulawesi*. Jakarta, Indonesia: Periplus.

Lai, H.-T. and Shy, J.-Y. (2009). The larval development of *Caridina pseudodenticulata* (Crustacea: Decapoda: Atyidae) reared in the laboratory, with a discussion of larval metamorphosis types. *Raffles Bulletin of Zoology*, **20**, 97–107.

Larson, H. K. and Kottelat, M. (1992). A new species of *Mugilogobius* (Pisces: Gobiidae) from Lake Matano, central Sulawesi, Indonesia. *Ichthyological Exploration of Freshwaters*, **3**, 225–34.

Marijnissen, S. A. E., Michel, E., Daniels, S. R. et al. (2006). Molecular evidence for recent divergence of Lake Tanganyika endemic crabs (Decapoda: Platythelphusidae). *Molecular Phylogenetics and Evolution*, **40**, 628–34.

Martens, K. (1997). Speciation in ancient lakes. *Trends in Ecology and Evolution*, **12**, 177–82.

Nosil, P., Funk, D. J. and Ortiz-Barrientos, D. (2009). Divergent selection and heterogeneous genomic divergence. *Molecular Ecology*, **18**, 375–402.

Pfaender, J., Schliewen, U. K. and Herder, F. (2010) Phenotypic traits meet patterns of resource use in the radiation of 'sharpfin' sailfin silverside fish in Lake Matano. *Evolutionary Ecology*, **24**, 957–74.

Pfaender, J., Miesen, F. W., Hadiaty, R. K. and Herder, F. (2011). Adaptive speciation and sexual dimorphism contribute to form and functional diversity in Lake Mataho's roundfin sailfin silversides adaptive radiation. *Journal of Evolutionary Biology*, **24**, 2329–2345.

Rossiter, A. and Kawanabe, H. (eds) (2000). *Ancient Lakes: Biodiversity, Ecology and Evolution*. San Diego, CA: Academic Press.

Roy, D., Paterson, G., Hamilton, P. B., Heath, D. D. and Haffner, G. D. (2007). Resource-based adaptive divergence in the freshwater fish *Telmatherina* from Lake Matano, Indonesia. *Molecular Ecology*, **16**, 35–48.

Sarasin, P. and Sarasin, F. (1898). *Die Süßwassermollusken von Celebes*. Wiesbaden, Germany: Kreidel.

Sarasin, P. and Sarasin, F. (1905). *Reisen in Celebes ausgeführt in den Jahren 1893–1896 und 1902–1903*. Wiesbaden, Germany: Kreidel.

Schliewen, U. K., Tautz, D. and Pääbo, S. (1994). Sympatric speciation suggested by monophyly of crater lake cichlids. *Nature*, **368**, 629–32.

Schluter, D. (2000). *The Ecology of Adaptive Radiation*. Oxford: Oxford University Press.

Schubart, C. D. and Ng, P. K. L. (2008). A new molluscivore crab from Lake Poso confirms multiple colonisation of ancient lakes in Sulawesi by freshwater crabs (Decapoda: Brachyura). *Zoological Journal of the Linnean Society*, **154**, 211–21.

Schubart, C. D., Santl, T. and Koller, P. (2008). Mitochondrial patterns of intra- and interspecific differentiation among endemic freshwater crabs of ancient lakes in Sulawesi. *Contributions to Zoology*, **77**, 83–90.

Schwarzer, J., Herder, F., Misof, B., Hadiaty, R. K. and Schliewen U. K. (2008). Gene flow at the margin of Lake Matano's adaptive sailfin silverside radiation: Telmatherinidae of River Petea in Sulawesi. *Hydrobiologia*, **615**, 201–13.

Seehausen, O. (2004). Hybridization and adaptive radiation. *Trends in Ecology and Evolution*, **19**, 198–207.

Seehausen, O. (2009). Progressive levels of trait divergence along a 'speciation transect' in the Lake Victoria cichlid fish *Pundamilia*. In *Ecological Reviews: Speciation and Patterns of Diversity*, ed. R. Butlin, J. Bridle and D. Schluter. Cambridge: Cambridge University Press, pp. 155–76.

Seehausen, O., Terai, Y., Magalhaes, I. S. et al. (2008). Speciation through sensory drive in cichlid fish. *Nature*, **455**, 620–26.

Streelman, J. T. and Danley, P. D. (2003). The stages of vertebrate evolutionary radiation. *Trends in Ecology and Evolution*, **18**, 126–31.

Stelkens, R. and Seehausen, O. (2009). Genetic distance between species predicts novel trait expression in their hybrids. *Evolution*, **63**, 884–97.

Vermeij, G. J. (1994). The evolutionary interaction among species: selection, escalation and coevolution. *Annual Review of Ecology and Systematics*, **25**, 219–36.

Vonlanthen, P., Roy, D., Hudson, A. et al. (2009). Divergence along a steep ecological gradient in Lake whitefish (*Coregonus* sp.). *Journal of Evolutionary Biology*, **22**, 498–514.

von Rintelen, K. (2011). Intraspecific geographic differentiation and patterns of endemism in freshwater shrimp species flocks in ancient lakes of Sulawesi. In *Phylogeography and Population Genetics in Crustacea*, ed. C. Held, S. Koenemann and C. D. Schubart. Berlin: Springer, pp. 257–71.

von Rintelen, K. and Cai, Y. (2009). Radiation of endemic species flocks in ancient lakes: Systematic revision of the freshwater shrimp Caridina H. Milne Edwards, 1837 (Crustacea: Decapoda: Atyidae) from the ancient lakes of Sulawesi, Indonesia, with the description of eight new species. *Raffles Bulletin of Zoology*, **57**, 343–452.

von Rintelen, K., von Rintelen, T., Meixner, M. et al. (2007b). Freshwater shrimp-sponge association from an ancient lake. *Biology Letters*, **3**, 262–4.

von Rintelen, K., von Rintelen, T. von and Glaubrecht, M. (2007c). Molecular phylogeny and diversification of freshwater shrimps (Decapoda, Atyidae, *Caridina*) from ancient Lake Poso (Sulawesi, Indonesia) – the importance of being colourful. *Molecular Phylogenetics and Evolution*, **45**, 1033–41.

von Rintelen, K., Glaubrecht, M., Schubart, C. D., Wessel, A. and von Rintelen, T. (2010a). Adaptive radiation and ecological diversification of Sulawesi's ancient lake shrimps. *Evolution*, **64**, 3287–99.

von Rintelen, T. and Glaubrecht, M. (2003). New discoveries in old lakes: three new species of *Tylomelania* Sarasin and Sarasin, 1897 (Gastropoda: Cerithioidea: Pachychilidae) from the Malili lake system on Sulawesi, Indonesia. *Journal of Molluscan Studies*, **69**, 3–17.

von Rintelen, T. and Glaubrecht, M. (2005). Anatomy of an adaptive radiation: a unique reproductive strategy in the endemic freshwater gastropod *Tylomelania* (Cerithioidea: Pachychilidae) on Sulawesi, Indonesia, and its biogeographic implications. *Biological Journal of the Linnean Society*, **85**, 513–42.

von Rintelen, T. and Glaubrecht, M. (2008). Three new species of the freshwater snail genus *Tylomelania* (Caenogastropoda: Pachychilidae) from the Malili lake system, Sulawesi, Indonesia. *Zootaxa*, **1852**, 37–49.

von Rintelen, T., Wilson, A. B., Meyer, A. and Glaubrecht, M. (2004). Escalation and trophic specialization drive adaptive radiation of viviparous freshwater gastropods in the ancient lakes on Sulawesi, Indonesia. *Proceedings of the Royal Society London B*, **271**, 2541–9.

von Rintelen, T., Bouchet, P. and Glaubrecht, M. (2007a). Ancient lakes as hotspots of diversity: a morphological review of an endemic species flock of *Tylomelania* (Gastropoda: Cerithioidea: Pachychilidae) in the Malili lake system on Sulawesi, Indonesia. *Hydrobiologia*, **592**, 1–94.

von Rintelen, T., von Rintelen, K., and Glaubrecht, M. (2010b). The species flock of the viviparous freshwater gastropod *Tylomelania* (Mollusca: Cerithioidea: Pachychilidae) in the ancient lakes of Sulawesi, Indonesia: the role of geography, trophic morphology and colour as driving forces in adaptive radiation. In *Evolution in Action: Adaptive Radiations and the Origins of Biodiversity*, ed. M. Glaubrecht and H. Schneider. Heidelberg, Germany: Springer Verlag, pp. 485–512.

Walter, R. P., Haffner, G. D. and Heath, D. D. (2009a). Dispersal and population genetic structure of *Telmatherina antoniae*, an endemic freshwater sailfin silverside from Sulawesi, Indonesia. *Journal of Evolutionary Biology*, **22**, 314–23.

Walter, R. P., Haffner, G. D. and Heath, D. D. (2009b). No barriers to gene flow among sympatric polychromatic 'small' *Telmatherina antoniae* from Lake Matano, Indonesia. *Journal of Fish Biology*, **74**, 1804–15.

Wilson, M. E. J. and Moss, S. J. (1999). Cenozoic palaeogeographic evolution of Sulawesi and Borneo. *Palaeogeography, Palaeoclimatology, Palaeoecology*, **145**, 303–37.

Woltereck, E. (1937a). Systematisch-variationsanalytische Untersuchungen über die Rassen- und Artbildung bei Süßwassergarnelen aus der Gattung *Caridina* (Decapoda, Atyidae). *Internationale Revue der Hydrobiologie und Hydrographie*, **34**, 208–62.

Woltereck, E. (1937b). Zur Systematik und geographischen Verbreitung der Caridinen. *Internationale Revue der gesamten Hydrobiologie und Hydrographie*, **34**, 294–330.

Woltereck, R. (1941). Die Seen und Inseln der 'Wallacea'-Zwischenregion und ihre endemische Tierwelt. Erster Teil: Vorgeschichte und Aufgabe der Forschungsreise. *Internationale Revue der Hydrobiologie und Hydrographie*, **41**, 1–36.

Yam, R. S. W. and Cai, Y. (2003). *Caridina trifasciata*, a new species of freshwater shrimp (Decapoda, Atyidae) from Hong Kong. *Raffles Bulletin of Zoology*, **51**, 277–82.

Zitzler, K. and Cai, Y. (2006). *Caridina spongicola*, new species, a freshwater shrimp (Crustacea: Decapoda: Atyidae) from the ancient Malili lake system of Sulawesi, Indonesia. *Raffles Bulletin of Zoology*, **54**, 271–6.

13

Molecular biogeography and phylogeography of the freshwater fauna of the Indo-Australian Archipelago

Mark de Bruyn, Thomas von Rintelen, Kristina von Rintelen, Peter B. Mather and Gary R. Carvalho

13.1 Introduction

13.1.1 The Indo-Australian Archipelago and its freshwater fauna

The Indo-Australian Archipelago (IAA) houses one of the highest levels of species richness and endemism in the world, encompassing four of the world's 25 biodiversity 'hotspots', which coalesce within this region in Southeast Asia (Myers et al. 2000). Southeast Asia covers only around 4% of the earth's surface, but houses some 20–25% of global fauna and flora (Mittermeier et al. 1999, 2005, Myers et al. 2000). Freshwater biodiversity levels are also extremely high in Southeast Asia, although very understudied and undervalued (Dudgeon et al. 2006, and references therein). Freshwater biodiversity within this species-rich region is often overlooked by conservationists; for example, only 0.6% of papers published in the conservation literature from 1992–2001 dealt with Asian freshwater biodiversity (Dudgeon 2003). Around 15% of global freshwater fauna are found in Indonesia

Biotic Evolution and Environmental Change in Southeast Asia, eds D. J. Gower et al. Published by Cambridge University Press.

alone (Braatz et al. 1992, Dudgeon et al. 2006). It is likely that the Southeast Asian region houses the second richest freshwater fauna globally, second only to the Amazon in terms of species-richness and endemicity (Kottelat 2002, Dudgeon et al. 2006).

Several theories have been proposed to explain such high levels of diversification in the region. Major hypotheses include: (1) *centre of accumulation hypotheses* (Wallace 1869, Archbold et al. 1982, Audley-Charles 1983), whereby Southeast Asia is a region of admixture between predominantly Oriental and Australian or earlier (Laurasian and Gondwanan) biotas brought into contact through plate tectonic coalescence; and (2) *vicariance* (*eustasy* (Heaney 1986, Schmitt et al. 1995) and *refugium hypotheses* (Brandon-Jones 1998, Hewitt 2000, Gathorne-Hardy et al. 2002)), in which cyclical Pleistocene sea-level fluctuations (eustasy) acted as a speciation 'pump', as populations were isolated repeatedly, sometimes into discrete refugia. It is apparent from the literature that many studies on the evolution of the biota conducted in this region, however, focus on organisms that may be able to disperse widely across 'continuous' habitat (land/air), which may somewhat confound their true biogeographical history (such as for rodents and bats, Heaney 1986; bats, Schmitt et al. 1995; various taxa reviewed in Hewitt 2000; marine fishes, Briggs 2000). In contrast, freshwater-restricted taxa should be better suited for examining the relative effects of earth and climatic history, because they reflect well the underlying biogeographical history of a given region due to limited dispersal abilities, as their requirement for freshwater should normally restrict them (Bermingham and Avise 1986). It is highly likely that combinations of the above two primary hypotheses have shaped the region's biota, and thus freshwater-based studies focused at both fine and coarse taxonomic, temporal and geographical scales, are likely to prove most illuminating in our understanding of processes driving evolutionary diversification of the biota in the IAA.

Few molecular systematic studies have been carried out on freshwater taxa from the IAA (but see von Rintelen et al., this volume, Chapter 12), although an increasing body of work is accumulating on freshwater crustacea (de Bruyn et al. 2004a, 2004b, 2005, Murphy and Austin 2005, de Bruyn and Mather 2007, Page et al. 2007, von Rintelen et al. 2007). Phylogenetic studies on freshwater shrimp (*Caridina*, Page et al. 2007, *Macrobrachium*, de Bruyn et al. 2004a, Murphy and Austin 2005) from the IAA provide strong support for the centre of accumulation hypothesis, which predicts an exchange or accumulation of components of the Oriental and Australian freshwater fauna during the Miocene, when the Sunda (Oriental) and Sahul (Australian) continental shelves collided (Hall 2002). Interestingly, these and other freshwater crustacean studies (Baker et al. 2008) also suggest a more recent (probably of Pleistocene origin) exchange of freshwater taxa, which, in effect, is a consequence of the 'vicariance hypotheses' outlined above.

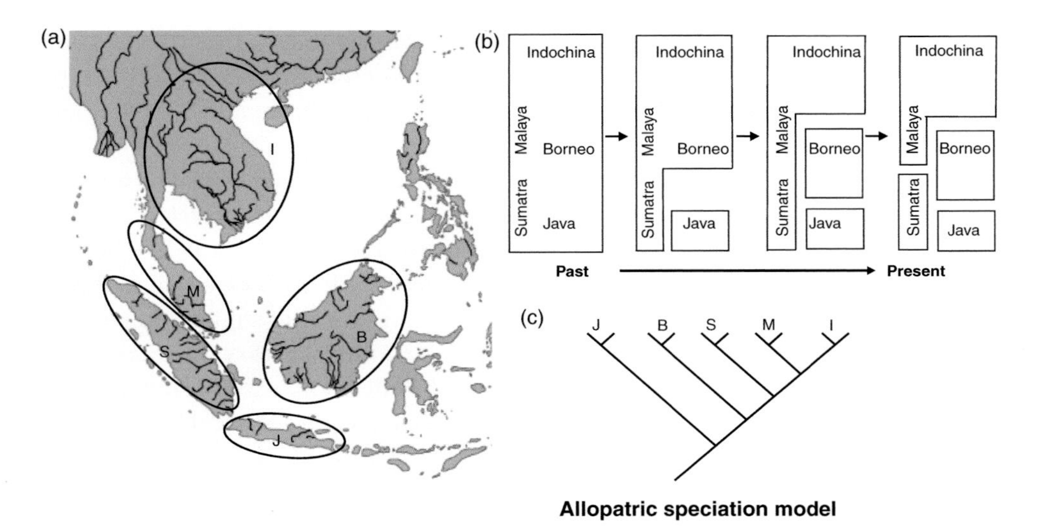

Figure 13.1 (a) Hypothetical sampling locations of a monophyletic group of freshwater aquatic taxa from SE Asia, I = Indochina, M = Malaya, S = Sumatra, B = Borneo, J = Java. (b) One cycle of sea-level rise leading to present geography (i.e. succession of island separation). (c). A priori prediction of phylogenetic relationships among the aquatic taxa after allopatric speciation resulting from successive vicariance of a widespread ancestor (redrawn after Ruedi and Fumagalli 1996: Fig 1).

Pleistocene lowering of sea levels associated with glacial cycling resulted in either a substantial reduction in geographic distances between fresh watersheds, or even the coalescence of drainage basins, which were formerly, and are presently, isolated from each other (Figs 13.1, 13.2). These processes may have facilitated the movement of freshwater and saltwater tolerant taxa (some shrimp of the genus *Macrobrachium*) among what are today geographically isolated freshwater systems, and resulted in a period of increased connectivity and range evolution of populations and/or sister taxa. Molecular data, although preliminary, are enabling researchers to tease apart and identify the relative contributions of various processes driving the evolutionary history of the freshwater fauna from this geologically complex and biologically rich region of our earth. Specifically, the ability to reconstruct past distributions within a temporal and spatial framework based on levels and patterns of genetic diversity and relatedness, allows direct testing of competing scenarios to explain current trends (see Figs 13.1, 13.2; Avise 1994, 2000). Moreover, the sequence of vicariant events and diversification may also be estimated from genetic data and compared with documented dates of geological events (see Figs 13.1, 13.2; Bermingham and Avise 1986, Page and Charleston 1998).

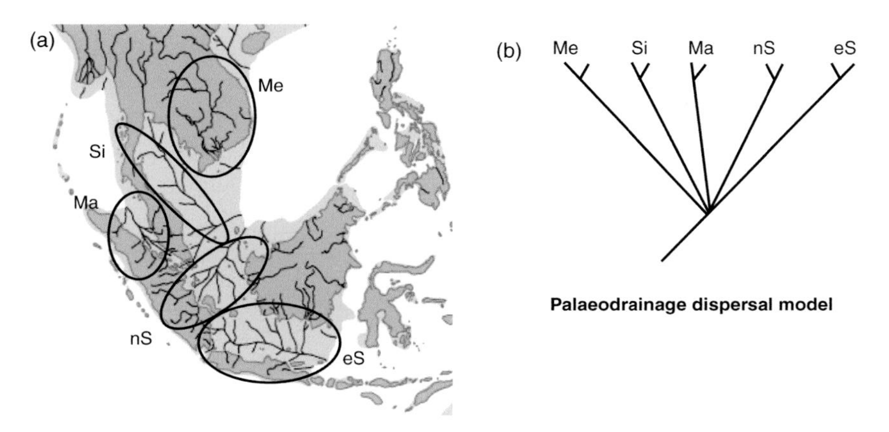

Figure 13.2 (a) Hypothetical sampling locations of a group of freshwater aquatic taxa from SE Asia; palaeodrainages from Voris (2000; reproduced with kind permission The Field Museum Chicago): Me = Mekong, Si = Siam, Ma = Malacca, nS = north Sunda, eS = east Sunda. (b). A priori prediction of phylogenetic relationships of the aquatic taxa after palaeodrainages facilitated connectivity, followed by allopatric speciation resulting from simultaneous vicariance, due to rising sea levels. Note that, in reality, multiple palaeodrainage systems are unlikely to have experienced simultaneous vicariance; sea-level rise was presumably not uniform throughout the region, but would have depended on multiple factors, including terrain elevation. Phylogenetic relationships in such a study might inform our understanding of the chronology of palaeodrainage system drowning across Sundaland.

13.1.2 Molecular biogeography and the origins and purview of phylogeography

Historical biogeography as a discipline is concerned with documenting the relationship of past events and processes on the geographical distributions of taxa. Species are therefore the fundamental units of analyses in historical biogeographic studies, and a phylogenetic 'tree' can be used to describe the observed genealogical pattern among related taxa. Alternatively, a general taxon area cladogram can be generated based on the distributional limits of multiple taxa, which may illustrate a shared environmental history. These approaches, while providing considerable insights into the historical effects of earth history events on the distribution of biological diversity, have an obvious limitation. Biological diversity is structured hierarchically at all levels, from the community level to the intraspecific level and below, that is, variation within a single species/individual. Intraspecific variation can also be strongly influenced by earth history and/or environmental and ecological processes within a given region, and thus can provide information on the historical biogeography of that region; however, this history may remain undiscovered if research is focused only at or above the species level. This will be most problematic when cryptic evolutionary lineages are present, which are not reflected in the current taxonomy of the organisms in question (Gómez et al. 2007). In such a situation, considerable information about the role of earth history in structuring the biota (including a vicariant history), which may be recorded at the intraspecific level, may be lost. Thus, a combination of molecular-based research programs aimed both above and below the species level may provide a more complete understanding of a region's historical biogeography.

To address this issue, some biogeographical questions have in recent times been examined using analyses of intraspecific variation, a discipline known as *phylogeography*, and defined as 'the study of the principles and processes governing the geographical distributions of genealogical lineages, including those at the intraspecific level' (Avise 1994: 233). The origins of this discipline can be traced back to the advent of genetic techniques that enabled rapid and fairly inexpensive screening of variation within and among species, thus largely eliminating a reliance on morphological traits as character states for analyses. With the advent of polymerase chain reaction (PCR) in the early 1980s, DNA sequence variation could be used to determine genealogical relations within and among species. For the first time, DNA-based studies incorporated both a spatial and temporal perspective, because mutational sequence changes accumulate over time (Arbogast et al. 2002). This was important, because earth-history events leave two distinct imprints on biological diversity – that of geography and time. A robust phylogeographic analysis of a species' biogeographical history would thus incorporate inferences about the geographical relationships among terminal taxa (i.e. individuals), and the chronology of causal events driving such pattern.

When traditional historical biogeographical and/or phylogeographical patterns are congruent in indicating disjunctions across multiple taxa (*comparative phylogeography*; Bermingham and Avise 1986), the distinction between the two methods is minimised. The explanation for the observed pattern can be relatively straightforward, that is, a widespread ancestral biota was fragmented by some vicariant event. However, the two approaches differ widely in their ability to explain incongruent patterns. Incongruent historical biogeographical patterns (i.e. an unresolved area cladogram or phylogenetic tree) are difficult to reconcile because of the age of the events under investigation; the true biogeographical history may be obscured by dispersal, extinctions of taxa, and/or overlapping earth history events (Cracraft 1988) – in other words 'the trace grows colder with time' (Zink 2002). Nonetheless, traditional historical biogeographical studies or phylogenetic studies aimed at the species level or above, are more appropriate for investigating relatively ancient earth history events, particularly when a good resolution of genealogical relationships and/or area cladograms is achieved. In contrast, phylogeographical studies often deal with relatively recent events, and genealogical relationships may reveal species' histories whether the phylogeny is resolved or not.

A resolved or structured gene tree will exhibit a pattern of reciprocal monophyly among geographical lineages. Reciprocal monophyly has been described as the 'currency' of phylogeography (Zink 2002), and permits the rejection of the hypothesis that (reciprocally monophyletic) groups are exchanging genes. Moreover, it is possible to determine how long the groups have been isolated from each other, either in a relative sense, or by applying a molecular clock to the data (Arbogast et al. 2002; but see Marko 2002). In contrast to the view put forward by some cladistic biogeographers (Nelson and Platnick 1981; Ebach and Humphries 2002), many recent studies (Donoghue and Moore 2003; de Queiroz 2005) report that the history of a species is equally likely to be shaped by gene flow resulting from dispersal events, as they are by a history of isolation resulting from vicariance. Such processes can result in an unstructured, or even a star-like gene tree (Slatkin and Hudson 1991), illustrating a *dynamic history of non-isolation*.

To circumvent this lack of resolution, new population genetic methods were developed, based on coalescent theory (Kingman 1982a, 1982b), that do not rely on a traditional structured phylogenetic tree to describe relationships among genotypes (Excoffier et al. 1992, Crandall and Templeton 1993, Excoffier and Smouse 1994). Today these methods are collectively known as minimum-spanning trees/networks (see Smouse 1998 for discussion, and review by Posada and Crandall 2001). An unstructured phylogeographical tree still allows for further inferences to be made about biogeographical history. Population genetic analyses based on coalescent theory can be applied to the spatial distribution of genotypes to provide information about the relative roles of gene flow (effective dispersal) (Slatkin 1989, Hudson et al. 1992) and vicariance. Similarly, genetic variation within and among populations can provide information about the demographic history of a species that is relevant to the biogeographical history of a region. For example, these data can be used to estimate effective population size (Fu 1994), and to determine historical changes in population size (past bottlenecks, range expansions related to habitat availability, etc.; Rogers and Harpending 1992), among many other applications (see Emerson et al. 2001 for review).

These developments have been accompanied by a move away from earlier descriptive phylogeographical studies that simply overlaid genealogical relationships upon geography, to a more formal framework that has allowed the testing of specific biogeographical hypotheses. When the geological history of a region is reasonably well-documented, biogeographical hypotheses can be erected a priori (Figs 13.1, 13.2), and tested using phylogeographical methodology. Moreover, a phylogeographical approach may allow one to distinguish between competing hypotheses when such hypotheses exist (Wallis and Trewick 2001). Alternatively, when the geological history of a region is poorly understood, phylogeographical analyses may reveal unexpected patterns (da Silva and Patton 1998) that can be used as testable hypotheses by earth scientists. 'Statistical phylogeography' (*sensu*

Knowles and Maddison 2002, Knowles 2009, Templeton 2009) provides a framework that renders it possible to distinguish among competing hypotheses, such as the use of a simulation approach to measure the discord between alternative tree topologies based on specific a priori hypotheses regarding population history. Coalescent models have been instrumental in driving the field of statistical phylogeography forward (see Evans this volume, Chapter 11, for an example of the use of coalescent-based models). Recently developed methods such as Approximate Bayesian Computation (ABC), and the hierarchical (comparative) version HABC, hold great promise for testing alternative biogeographic hypotheses (Beaumont et al. 2002, Hickerson et al. 2007).

Although most of the early phylogeographic studies focused on small mammals and marine fishes (reviewed in Avise 1994, 2000), Avise and co-workers recognised that phylogeographic patterns in freshwater aquatic taxa could provide good resolution for many biogeographic questions. One of their early influential papers found that intraspecific genetic breaks were congruent across four freshwater fish species, and were concordant with previously described historical biogeographical boundaries; that is, the distributional limits of species (Bermingham and Avise 1986). This good resolution results from the fact that historical connections among discrete drainages (and therefore gene flow) relies directly on the underlying earth and climatic history of the region (Figs 13.1, 13.2), which is not always the case in more mobile terrestrial or volant taxa. Thus, patterns observed in freshwater fauna permit strong inferences to be made about the biogeographical history of a given region (Lundberg 1993). Nonetheless, this fact is often overlooked, and phylogeographic studies of freshwater aquatic taxa remain limited compared with those, for example, on terrestrial mammals, reptiles, insects and birds.

13.2 Earth and climatic history of the Indo-Australian Archipelago

13.2.1 Earth history of the Indo-Australian Archipelago

Contrary to past river system connections in the IAA that may have facilitated connectivity among freshwater taxa, resulting from Pleistocene eustasy, long-standing barriers to connectivity such as the deep-sea trench of the Makassar Strait would presumably long have impeded the movement of these same taxa. The Makassar Strait, which houses the Indonesian Throughflow, acts as the western boundary for the Australian and Asian biotic transition zone, a region known as Wallacea (Dickerson 1928). Indeed, of all vertebrate groups, the distributions of primary freshwater fish fauna most clearly demarcate this boundary (Moss and Wilson 1998). This ancient deep-water barrier was formed in the early Tertiary (Eocene, *c.* 44 Ma; Moss and Wilson 1998, Hall 2009), and was first recognised by Wallace in the nineteenth century (Wallace 1859). Today, this well-known biogeographical boundary is known as 'Wallace's Line'. Subsequently, Huxley (1868) suggested an addition to this line, based on zoological data, and extended it north into the Philippines. Huxley placed his proposed demarcation to the east of Palawan (the most westerly of the Philippine islands), effectively linking Palawan biogeographically to Borneo and mainland Asia, and separating it from the rest of the Philippine Archipelago. Other zoogeographers have recognised various other 'lines' based on their study organism(s) of choice (see Mayr 1944 and Simpson 1977 for details).

Major zoogeographic boundaries such as these often result from far more ancient earth history events, rather than recent climatic fluctuations and associated eustatic change – namely plate-tectonic movements. Plate tectonics have played a fundamental role in the evolutionary history of many, if not all but the very youngest, of IAA taxa. The tectonic history of the region is extremely complex (see Hall, this volume, Chapter 3), the Philippine Archipelago being a prime example. Reconstructions by Hall (1996) suggest that the main land mass of the Philippines originated as a series of island arcs extending into the Pacific Ocean more than 50 Ma. However, the Philippine Archipelago may have only taken on its current shape over the last 5–10 million years (Myr), but the geological history of the region is still poorly understood. Similarly, Sulawesi has a complex history, and is believed to be a composite land mass of different geological origins. Sulawesi probably took on its present shape between the Pliocene and the present. For Wallacea as a whole, Hall (2001, 2009) postulated that most of the smaller islands emerged only within the last 5 Myr, and therefore the biota can only have populated much of Wallacea during this period (also see Halmahera's apparent emergence since 5 Ma; Cane and Molnar 2001). Thus, phylogenetic and phylogeographic data on species that occur in the IAA region may be useful not only for determining mechanisms that may have shaped biodiversity in the region, but may also contribute to our understanding of the region's geological history. This is particularly relevant for land masses as geologically complex as Sulawesi (this volume, Chapters 3, 11 and 12) and the Philippine Archipelago. Readers are directed to Hall (this volume, Chapter 3) for an extensive and updated review of the geological history of the IAA.

13.2.2 Pleistocene and earlier eustasy and the IAA

The impact that Pleistocene (~2 million to 10 000 years before the present) climate change had on driving evolutionary diversification has been debated for more than a century (Darwin 1859, Wallace 1862, Hofreiter et al. 2003, Hofreiter and Stewart 2009). Accumulating evidence from Europe and North America indicates that glacial events during the Pleistocene epoch resulted in major shifts in species distributions (Avise 2000, Hewitt 2000, Davis and Shaw 2001), and may have been the principal factor that contributed to the decline and eventual extinction of some species (Guthrie 2003, Shapiro et al. 2004). As climate oscillated through the Pleistocene, fossil (Coope 1994, Bennett 1997) and other evidence (reviewed in Hewitt 2000) shows that the geographical distributions of some species responded by expanding and contracting their ranges during times of glacial cycling. Furthermore, these responses may have been swift, and may have occurred repeatedly (Hewitt 2000). Understanding the causal mechanisms that generate or extinguish biodiversity is critical for safeguarding our natural resources, and is of particular urgency in regions of high species richness, such as the IAA.

Employing phylogeographic approaches (see 13.3.2 below) to the study of the evolutionary history of freshwater taxa in the IAA may prove particularly insightful, owing to this region's dynamic recent history. Miocene and Pleistocene sea-level changes are believed to have played an important role in the distribution and range evolution of both aquatic and terrestrial taxa within this region (Dodson et al. 1995, Voris 2000), which may have somewhat obscured historical biogeographical relationships that are based on species-level phylogenies or taxon area cladograms. Considering the dynamic history of this region, it is probable that significant levels of cryptic diversity exist, which will not be reflected in current taxonomy, further confounding historical biogeographic patterns. It is thus essential to understand how these recent eustatic processes have shaped the distribution and connectivity of species and populations in this region.

East of Wallace's Line, the Torres Strait land-bridge on the Sahul Shelf connected Australia and New Guinea periodically during the Pleistocene (and most probably prior to this), and is one such example of a major eustatic influence on the Australian/New Guinean biota. This land bridge was exposed for much of the Pleistocene, due to lowered sea levels resulting from climatic fluctuations and associated glacial maxima (Voris 2000). The Torres Strait land bridge played not only a significant role in the vicariance of marine taxa restricted to either side of this land bridge (Chenoweth et al. 1998), but also allowed an interchange of elements of the terrestrial (snakes, Rawlings and Donnellan 2003, Wüster et al. 2005) and freshwater biota (rainbowfishes, McGuigan et al. 2000; freshwater shrimp, de Bruyn et al. 2004b, de Bruyn and Mather 2007; freshwater crayfish, Baker et al. 2008) between Australia and New Guinea. For example, drainage basins that are today restricted to Australia or New Guinea, respectively, drained into palaeo-Lake Carpentaria during the Pleistocene (Fig 13.3; Torgersen et al. 1985, Voris 2000). This vast palaeo-lake may have provided ample opportunity for trans-Torresian interchange of species, and connectivity and gene flow among populations of freshwater organisms which are today isolated from each other due to a marine barrier (i.e. vicariance). Indeed, the inundation of Lake Carpentaria by rising sea levels is believed to have occurred sometime around 10–8.5 ka (Chivas et al. 2001). There is some debate on the duration of full freshwater conditions, with Lake age spanning some 80–8.5 ka (Torgersen et al. 1983, 1985, Chivas et al. 2001, Reeves et al. 2007, 2008), though many aquatic invertebrates found in this

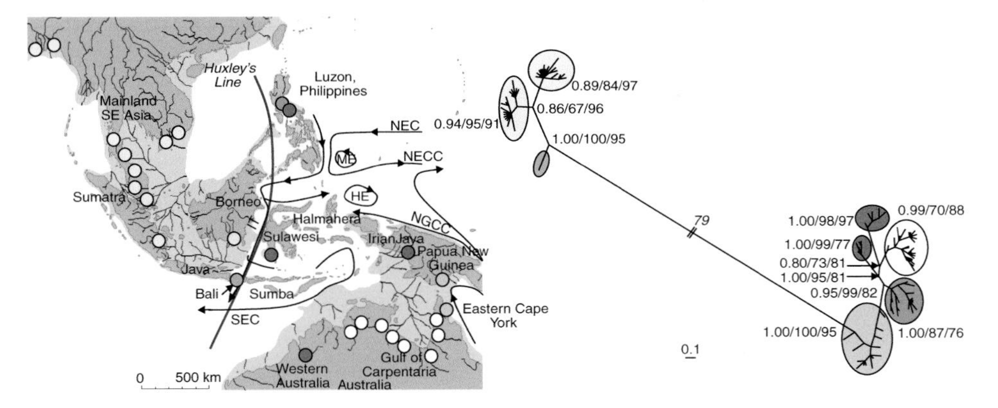

Figure 13.3 Map of sampling locations of *Macrobrachium rosenbergii* showing the distribution of lineages, and Bayesian consensus tree for 93 unique mitochondrial *COI* haplotypes obtained from sampling 861 giant freshwater prawns from 26 locations east and west of Huxley's Line (reproduced from de Bruyn and Mather 2007, with permission John Wiley and Sons). Major surface currents of interest are shown. Dashed line indicates seasonally reversing currents. North Equatorial Current (NEC); North Equatorial Counter Current (NECC); New Guinea Coastal Current (NGCC); Indonesian Throughflow (IT; location of Makassar Strait); Mindanao Eddy (ME); Halmahera Eddy (HE). Pale grey shading on map indicates –120 m sea-level contour, and major fresh watersheds at this time are shown (Voris 2000; reproduced with kind permission The Field Museum Chicago). Values at tree nodes indicate (in order), Bayesian posterior probabilities, neighbour joining bootstrap values, maximum likelihood bootstrap values. See plate section for colour version.

region also tolerate brackish water to varying degrees (some freshwater shrimp of the genera *Macrobrachium* and *Caridina*).

West of Wallace's Line, the Southeast Asian mainland (Indochina) was connected above sea level in recent geological time (*c.* 1 Ma–10 ka) to a number of islands, including Sumatra, Borneo, Java and parts of the Philippine Archipelago (Palawan) (Voris 2000). Sundaland existed for much of the Pleistocene, as sea levels dropped by around 120 m or more, and probably at various times in the more distant past (Hanebuth et al. 2009). Several vast palaeo-river systems are known to have existed (see for example Posamentier 2001) across Sundaland during the Pleistocene (Fig 13.2), and most probably at earlier times of low sea level (Plio- and Miocene; Lisiecki and Raymo 2005), and readers are directed to Voris (2000) and references therein (the excellent synthesis by Rainboth 1996 is also of considerable interest here), for an authoritative synthesis and review of the available data. In brief, four major palaeodrainage systems are noteworthy here: (1) *Malacca Straits River System*: drained northwest into the Andaman Sea, uniting present day rivers on the northeastern coast of Sumatra and the western Malay Peninsula;

(2) *Siam River System*: ran north across the Sunda Shelf, uniting rivers from central Sumatra, the Singapore Straits region, and most likely also the Gulf of Thailand. Some authors suggest alternative arrangements (see Voris 2000 for details); (3) *North Sunda River System*: the largest of the Sunda Shelf river systems ran north from the northeast coast of Sumatra, uniting rivers from this region with the large Kapuas River from Borneo, and drained into the ocean northeast of Natuna Island; (4) *East Sunda River System*: drained into the Java Sea near Bali, uniting rivers of northern Java, southern Borneo and southern Sumatra (Fig 13.2). Voris (2000) recognised that height ('depth') of maximal sea-level regression is only one factor to consider. Of equal or greater importance for our understanding of the biogeography of the aquatic fauna is the *duration* that sea levels were at any particular height, and thus for what *duration* various palaeo-river systems were in existence (Voris 2000: Fig 2). Phylogeographic data from freshwater aquatic taxa can contribute significantly to our understanding of past drainage history in the region, by testing alternative a priori hypotheses regarding the arrangement and evolution of these drainage systems (Figs 13.1, 13.2; de Bruyn et al. 2004b).

Recently, Sathiamurthy and Voris (2006) extended the analyses presented in Voris (2000) to explore the effects of recent sea-level change on this region, from 21–1 ka. Of particular note is the mid Holocene high-stand at ~4.2 ka, when sea levels were approximately 5 m higher than at present (Sathiamurthy and Voris 2006: Fig 26). Such a marine introgression would have inundated large parts of eastern Sumatra, Indochina (for example, the Chao Phraya and Mekong deltas) and Borneo, thereby displacing their freshwater faunas further upstream. Sathiamurthy and Voris also present evidence for the existence of 32 probable submerged palaeo-lakes on the Sunda Shelf, which were connected to traceable palaeo-river systems, and call for geological drilling programs to evaluate these hypotheses (Sathiamurthy and Voris 2006: Fig 27). A future challenge will be to formulate phylogeographic research programmes to determine the influence of these inundated lakes on the evolutionary history of freshwater aquatic taxa. Did they act as 'refugia' during these periods of immense and rapid environmental upheaval? Newly developed statistical tools for biogeographical analyses may enable us to tackle such intriguing questions (Lemmon and Lemmon 2008, Ree and Sanmartín 2009; Webb and Ree this volume, Chapter 8).

Recent palaeo-vegetation models by Cannon et al. (2009) provide additional evidence for the extraordinarily rapid emergence and evolution of Sundaland and its vegetation, and present novel hypotheses regarding the recent Pleistocene history of this region. Thus, the phylogeographic structure of freshwater taxa within the IAA region is likely to reflect a dynamic history influenced both by ancient earth-history events, and more recent sea-level fluctuations; however, studies (particularly phylogeographic) on the freshwater biota of the IAA remain rare (but see below).

13.3 The freshwater fauna of the Indo-Australian Archipelago as model biogeographic taxa

13.3.1 Phylogenetic biogeography of the freshwater fauna of the Indo-Australian Archipelago

Distribution patterns of freshwater organisms in the IAA have been used to infer biogeographic relationships since the late nineteenth century (see for example von Martens 1897 for a discussion of the biogeography of freshwater molluscs), even though the basic concepts on faunal affinities and boundaries in the region were almost exclusively based on terrestrial taxa (Wallace 1860, Mayr 1944). Irrespective of the habitat of the organisms studied, the affinities of the Wallacean fauna, especially that of Sulawesi, one of the 'anomalous islands' of Wallace (1880), received most attention. One early expedition, the exploration of Celebes (Sulawesi) 1893–96 and 1902–03 by the Sarasins (Sarasin and Sarasin 1898, 1905) was largely motivated by the desire to test Wallace's ideas on the faunistic relationships of the region (Sarasin 1896). A significant part of the biogeographic evidence from that expedition was based on freshwater taxa (Sarasin and Sarasin 1901).

The 'centre of accumulation' hypothesis predicts the exchange or admixture of 'Asian' and 'Australian' taxa in the IAA. The understanding of the origin of the fauna of two IAA island groups, the Philippines and the islands of Wallacea, are crucial in this context. Most of the Philippines and the Wallacean islands are generally regarded as oceanic islands, thus without a land connection to any surrounding area since their emergence (van Oosterzee 1997). Consequently, their fauna must owe its origins to dispersal. Sulawesi is a special case, because its complicated geological history (see above and Chapter 3) allows for the theoretical possibility of a vicariant origin of parts of its fauna either from the west (Sundaland) or the east (Australia–New Guinea) (Moss and Wilson 1998, Wilson and Moss 1999; but see Hall 2001, 2009 for a contrary view). Given that some Sulawesi freshwater taxa such as molluscs have long been assumed to represent 'ancient' elements (Sarasin and Sarasin 1898), the biogeographical relationships of this island's aquatic biota deserve some attention. There has historically been little doubt that the bulk of Sulawesi's fauna is of Asian origin (Weber 1902, de Beaufort 1926; for an overview see also van Oosterzee 1997); recently, land bridges have been replaced by dispersal as the favourite model for explaining how taxa arrived at the island in the first place (for an overview, see van Oosterzee 1997).

The early biogeographic studies of freshwater taxa in the region were hampered by the lack of two important prerequisites for biogeographic research, namely sufficient sampling and knowledge of the phylogeny of the studied organisms, both of which are necessary for avoiding distributional artefacts. Most of the older literature (before the rise of molecular systematics) on the distribution of

IAA freshwater taxa, particularly in invertebrates, should be viewed with extreme caution, as exemplified by the gastropod examples given below. In this review we focus on the results of molecular phylogenetic studies involving the IAA conducted within the last one or two decades.

Invertebrates

Molecular phylogenetic studies of IAA freshwater invertebrates are generally scarce, with the exception of work on the ancient lakes of Sulawesi (von Rintelen et al. this volume, Chapter 12). Only three groups have been studied in sufficient detail for biogeographic purposes, albeit with highly varying degrees of completeness regarding taxonomic and distributional coverage: molluscs (Glaubrecht et al. 2003, Glaubrecht and von Rintelen 2003, Köhler and Glaubrecht 2003, Park and Kim 2003, Köhler et al. 2004, von Rintelen and Glaubrecht 2005, 2006, Albrecht et al. 2007), decapod crustaceans (Page et al. 2007, Yeo et al. 2007, von Rintelen et al. 2008, Shih et al. 2009), and water beetles (Balke et al. 2004, 2009).

Corbiculid clams are widespread throughout the IAA, ranging from most of Eurasia into Australia, but with a gap in distribution on the Moluccas and the Lesser Sunda Isles. East and Southeast Asian populations of the freshwater lineage including among other species the widespread *Corbicula fluminea* share mtDNA haplotypes (*COI*) on a large geographic scale (Fig 13.4), such as between China and Java (Park and Kim 2003). In contrast, distinct lineages with several endemic species have been identified on Sumatra and Sulawesi (Glaubrecht et al. 2003, 2006, von Rintelen and Glaubrecht 2006). The Sumatran lineage is sister group to all other Asian corbiculids studied so far (Fig 13.4). On Sulawesi, two endemic lineages in the ancient lakes of the island may represent independent colonisations of the island, while at the same time haplotypes are shared between populations of the riverine and supposedly endemic Sulawesi species *C. subplanata* and the Javanese *C. javanica* (Fig 13.4), which in turn is possibly conspecific with the widespread *C. fluminea* (von Rintelen and Glaubrecht 2006). While the lack of reliable dates for the splits between these clades prevents any temporal correlation of their origin with geologic events, it seems plausible to assume that the present-day distribution of *Corbicula* in the IAA reflects both older (presumably pre-Pleistocene, the endemic clades on Sumatra and Sulawesi) and more recent (Pleistocene, the *C. fluminea* complex including Sulawesi's *C. subplanata*) dispersal processes. Given Pleistocene sea-level changes and the associated repeated land connections to the rest of Sundaland, it seems remarkable that the Sumatran lineage remained distinct. However, the direction of dispersal in *Corbicula* has clearly been from west to east – Sulawesi as well as Australia appears terminally within Asian clades.

In the groups of freshwater gastropods studied so far, the pattern is very different. Several viviparous genera of Pachychilidae (Caenogastropoda: Cerithioidea) occur in the IAA including the Philippines and Sulawesi, and on the Australian

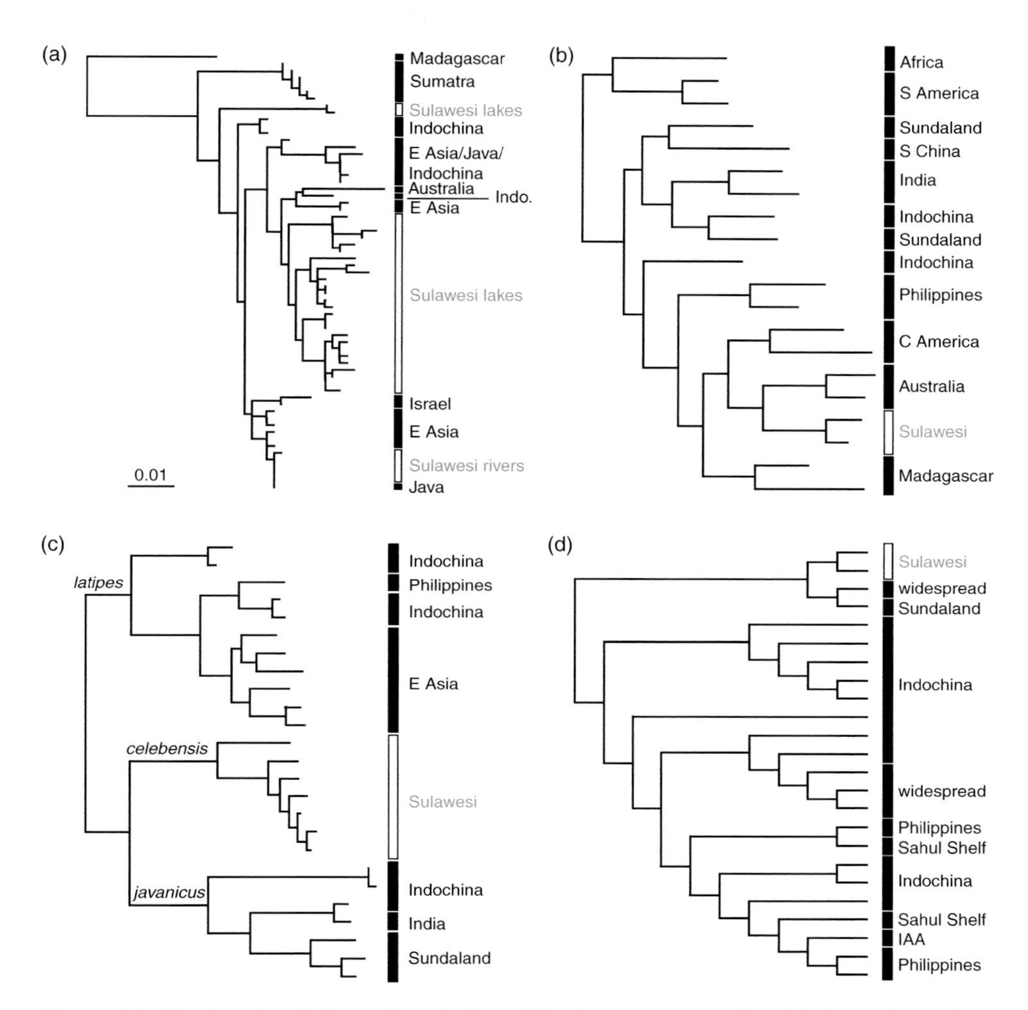

Figure 13.4 Molecular phylogenies of selected freshwater taxa in the IAA. Sample origins are indicated by bars and labels to the right of each tree. (a) Corbiculid clams (Maximum Likelihood (ML); modified from Glaubrecht et al. 2006). (b) Pachychilid snails (Maximum Parsimony (MP); modified from Köhler and Glaubrecht 2010). (c) Ricefishes (MP; modified from Takehana et al. 2005); (d) Homalopsid water snakes (ML; modified from Alfaro et al. 2008). Outgroups have been omitted. For details on methods and trees see cited references. IAA, taxon occurs in all subregions of the Indoaustralian Archipelago; widespread = taxon occurs in more than one subregion of the IAA = exact sample origin could not be derived from source publication.

Torres Strait Islands. Subsequent to their description, the Australian and Philippine species and the majority of the Sulawesi taxa have been assigned to a widespread Southeast Asian genus, *Brotia* H. Adams, 1866 (Thiele 1929). Three species from Sulawesi were described in a distinct genus, *Tylomelania* Sarasin and Sarasin 1897, subsequently regarded as a subgenus of another Asian genus, *Sulcospira* Troschel, 1858 (Thiele 1929). The distribution inferred from this taxonomy suggests a widespread Asian taxon crossing Wallace's Line possibly even into Australia. Recent molecular and anatomical work has challenged this view though (Fig 13.4; Glaubrecht and von Rintelen 2003, Köhler and Glaubrecht 2003, 2010, Köhler et al. 2004, von Rintelen and Glaubrecht 2005). The Philippine species have been shown to form a distinct lineage within pachychilids that resulted from a relatively basal divergence (Köhler and Glaubrecht 2003, Köhler et al. 2004). The Sulawesi species belong to one endemic lineage, *Tylomelania* (von Rintelen and Glaubrecht 2005), which is sister group to the equally endemic Torres Strait Islands taxon, *Pseudopotamis* (Glaubrecht and von Rintelen 2003, Köhler et al. 2004). The sister-group relationship between the Sulawesi and Torres Strait Islands taxa is supported by morphological data, which also suggest that (ovo-)viviparity is the ancestral condition of their common ancestor (Glaubrecht and von Rintelen 2003, von Rintelen and Glaubrecht 2005). Consequently, the probability of passive dispersal across sea water is extremely low for this group, prompting a vicariance hypothesis involving the drift of terranes, forming present day Sulawesi from the New Guinea margin in the Miocene, to explain the disjunct distribution on Sulawesi and the Torres Strait Islands (Glaubrecht and von Rintelen 2003, von Rintelen and Glaubrecht 2005). Irrespective of this particular hypothesis, the pachychilid example illustrates nicely the fallacies of using distribution data from taxonomically poorly studied groups in the absence of a phylogeny for biogeographic purposes, because the apparent patterns can be extremely misleading.

An Australia–Sulawesi link has also been tentatively assumed from phylogenetic work on another gastropod family, the Planorbidae, a diverse group of freshwater pulmonates. Using both mitochondrial and nuclear markers, Albrecht et al. (2007) showed a sister-group relationship for a highly derived Sulawesi endemic, *Protancylus*, and two genera endemic to Australia. All these genera were in turn sister to another genus from Australia, suggesting an Australian origin of the Sulawesi *Protancylus* (Albrecht et al. 2007). Other endemic genera from Sulawesi, New Guinea, and Australia could not be included in this analysis, nor was the geographic coverage within each region comprehensive, making the results somewhat preliminary. A relationship between Australian and Sulawesi taxa inferred for a third freshwater gastropod group, the hydrobioids, based on morphological data (Ponder and Haase 2005) should be regarded with even more caution until molecular data become available. The gastropod studies discussed here all suggest an Australian origin of Sulawesi taxa.

Two studies on potamid freshwater crabs (Yeo et al. 2007, Shih et al. 2009) have focused on the biogeography of the western part of the IAA. Yeo et al. (2007) found three deep lineages of pre-Pleistocene origin (Miocene to Pliocene) within the genus *Johora* on the Malay Peninsula and Tioman Island. They suggested that the origin of the entire group might be linked to the submergence of the Isthmus of Kra, subsequent diversification to the orogeny of the main range for the mainland clades, and sea-level changes for the Tioman clade/species, respectively. A more extensive study on the entire family (Shih et al. 2009) revealed the existence of two major clades within the largest subfamily with a wide distribution in the IAA, the Potamiscinae, an East Asian/Southeast Asian mainland clade, and a Sunda shelf Islands (Java, Borneo, Sumatra) clade. Shih et al. (2009) suggested that this major split occurred when sea-level highstands in the early Miocene isolated the Sunda Islands from mainland Southeast Asia. They also attributed the further subdivision of this group into two Bornean and a Java/Sumatra clade, to the severance of land connections by later Miocene sea-level fluctuations. Interestingly, the Philippines have been colonised twice. The north Philippine taxa originate from East Asia, while *Isolapotamon* from Mindanao must have come from Borneo possibly via the Sulu Islands. The diversification of potamids thus seems to be largely driven by pre-Pleistocene vicariant events, with possibly just a single instance of late Miocene dispersal into the Philippines.

Dispersal from the west was suggested for gecarcinucoid freshwater crabs that entered Southeast Asia from India most probably during the Miocene (Klaus et al. 2006). In a more comprehensive study Klaus et al. (2009) suggested a dispersal of this group via the Malayan Peninsula into Sundaland and the rest of the Archipelago, with four major lineages evolving on the Sunda shelf, possibly on Borneo. One of these IAA clades probably diversified on Sumatra and the Malay Peninsula, and subsequently dispersed back into continental East Asia. Given the distribution of the group with derived clades in the Philippines, several Wallacean islands and New Guinea–Australia, the authors suggested that dispersal by rafting is a probable explanation for the present distribution range of gecarcinucoids. Overall the pattern in gecarcinucoid crabs seems to be dominated by dispersal.

The available data from atyid freshwater shrimps (Page et al. 2007, von Rintelen et al. 2008) are rather more inconclusive. In a study on the origin of Australian *Caridina*, Page et al. (2007) showed a multiple origin of endemic Australian species or species groups with an abbreviated larval development. These taxa, restricted to freshwater (land-locked), lie within a clade of widespread Indo-Pacific species with free-swimming larvae that can tolerate brackish or even salt water, and are thus likely able to disperse widely. Similarly, von Rintelen et al. (2008) found that an endemic land-locked species of *Caridina* on Peleng Island east of Sulawesi is derived from species with free-swimming larvae, independently from the land-locked species known from parts of Sulawesi. While both studies clearly suggest the origin of the respective endemic shrimp fauna by dispersal, the origin of the speciose groups of land-locked species of *Caridina* from Sundaland and Sulawesi is still not known.

Two of the rare biogeographic studies on aquatic insects from the IAA used diving beetles as model organisms (Balke et al. 2004, 2009). They highlight the importance of long-range dispersal over water gaps and among continental land masses in this group. For example, in the genus *Rhantus*, Balke et al. (2009) suggest multiple transgressions of Wallace's Line. According to these authors, one of the most extreme cases of an island emigrant founding new continental lineages is the 'supertramp' species *Rhantus suturalis* that, originating from the remote New Guinea highlands, also colonised Wallacea on its way across major biogeographical boundaries (Balke et al. 2009).

Vertebrates

Several biogeographic studies of IAA freshwater vertebrates, the majority from fishes, have been carried out. Ricefishes (or medaka) of the genera *Oryzias* and *Xenopoecilus* were studied by Takehana et al. (2005) based on nuclear and mitochondrial DNA sequences (Fig 13.4). Phylogenetic analyses not only revealed that *Xenopoecilus* is a polyphyletic genus nested within *Oryzias*, but also, more relevant for this chapter, that members of both genera can be divided into three well-supported species groups that show a clear geographical structure (Fig 13.4; Takehana et al. 2005). All members of one of these clades, the *celebensis* species group, are endemic to Sulawesi. Its sister group, the *javanicus* species group, comprises species from the Sunda Shelf and from the Asian mainland. The third group, the *latipes* species group, and sister group to the other two clades, comprises species from the Asian mainland and from North Luzon. Takehana et al. (2005) suggested that the formation of the Makassar Strait, separating Sulawesi and Borneo, appeared to have been a barrier for dispersal in ricefishes and other freshwater fishes. Thus, Takehana et al. adapt the idea of historical vicariance between Sulawesi and the continental shelf as proposed by Whitten et al. (1987). By applying a molecular clock, these authors estimated that the *celebensis* species group has been isolated from the other groups for more than 29 Myr. However, this split considerably postdates the opening of the Makassar Strait 44 Ma, rendering a vicariant explanation unlikely.

To the best of our knowledge, the only molecular phylogenetic study of freshwater vertebrates from the IAA not obtained from fishes was published by Alfaro et al. (2008). They investigated the phylogeny and biogeography of aquatic snakes inferred from one nuclear and three mitochondrial genes (Fig 13.4). Within the family Homalopsidae, or rear-fanged water snakes, the authors suggested two dispersal events. Two species of the genus *Enhydris*, both endemic to Sulawesi, are sister taxa to the widespread species *E. plumbea* and are possibly derived from a single dispersal event of *E. plumbea* from Borneo across the Makassar Strait to Sulawesi, roughly timed between the early and late Pliocene. The second dispersal event was suggested for the only freshwater snake within the genus *Cerberus*,

C. microlepis. This species is probably derived from a Sunda Shelf population of a non-freshwater species that crossed the Philippine islands of Palawan and Mindoro to Luzon (Alfaro et al. 2008).

Semi-aquatic vertebrates from the IAA were studied by Evans et al. (2003a, b, 2008), and dispersal across the Makassar Strait was, for example, suggested for fanged frogs (Evans et al. 2003a).

13.3.2 Phylogeography and population structure of the freshwater fauna of the Indo-Australian Archipelago

Phylogeographic studies of freshwater taxa from the IAA are extremely limited. Here we review the few available datasets for invertebrates and vertebrates.

Invertebrates

The decapod crustacean *Macrobrachium rosenbergii* (giant freshwater prawn) is an ideal candidate species to investigate the biogeography of the IAA region using phylogeographic approaches because it: (1) occurs in freshwater, (2) is widespread, and (3) is locally abundant across its natural range. *Macrobrachium rosenbergii* is distributed from Pakistan in the west to southern Vietnam in the east, and south across Southeast Asia to New Guinea and northern Australia. *Macrobrachium rosenbergii* is most often associated with coastal river systems, because it is freshwater dependent as an adult, but requires brackish water for breeding and larval development (New and Singholka 1985). Our earlier mtDNA work (de Bruyn et al. 2004a) built on previous allozyme and morphological variation (Malecha 1977, 1987, Hedgecock et al. 1979, Lindenfelser 1984) that recognised two major forms of this species, restricted to either side of Huxley's Line (Fig 13.3). Taxonomists have since designated separate species status for these two forms (Wowor et al. 2009). The estimated timing of the split between these two lineages was consistent with Hall's (2001) interpretation of the most probable time of connectivity between the two regions, resulting from collision of the Asian and Australian plates, some 5–10 Ma. The phylogeographic structure of the eastern and western species revealed three major genetic lineages within the western species, and five major lineages within the eastern species (Fig 13.3; de Bruyn et al. 2004a, 2004b, 2005, de Bruyn and Mather 2007). Using coalescent population genetic analyses methods, results from these studies indicated that the major lineages within each species probably diverged during the Pleistocene, most probably at times of glacial maxima when sea levels were low and distances among discrete drainages were minimised, while the timing of divergence for populations within each of these lineages was most likely to have occurred during the Last Glacial Maximum. Interestingly, patterns and timing of species and lineage diversification identified in giant freshwater prawns (de Bruyn and Mather 2007) are closely mirrored by aquatic beetles (Balke et al. 2009), providing preliminary indications that mechanisms driving 'freshwater' diversification in the IAA region may be general ones, and may operate across diverse taxonomic groups. Thus, three hierarchical levels of information regarding the biogeographical history of the IAA have been identified in giant freshwater prawns at different scales (eastern and western species, lineages within each species, populations within lineages within each species), embedded within what was considered previously as a single species. It should be noted that the timings of the various events discussed above were reliant on the application of a molecular clock, the use of which has attracted heated debate from some members of the biogeographic community (Heads 2005).

Within the eastern giant freshwater prawn species, population genetic estimators of population expansion and migration events indicated a probable role for Lake Carpentaria as a conduit for gene flow among populations sampled from the Gulf of Carpentaria (Australia) and New Guinea (Fig 13.3; de Bruyn et al. 2004b). Similar broadly overlapping signatures were identified in a second freshwater decapod crustacean, the redclaw crayfish (*Cherax quadricarinatus*) (Baker et al. 2008), and from phylogenetic studies on melanotaeniid rainbowfishes (McGuigan et al. 2000). Recent geological evidence suggests that, just prior to the incursion of marine water around 10 ka, Lake Carpentaria was fresh and almost full, approaching a surface area of 600 × 300 km and a depth of up to 15 m (Chivas et al. 2001, Reeves et al. 2007). Because redclaw crayfish are an obligate freshwater species, a more recent marine connection between the Australian and New Guinean populations can be ruled out with some confidence. Giant freshwater prawns, on the other hand, can tolerate brackish water to some extent as larvae, thus marine dispersal, while unlikely, is one potentially confounding factor here. Interestingly, phylogeographic patterns characterised in freshwater prawns (and crayfish, although not discussed) were concordant with both plant and animal (terrestrial and aquatic vertebrates and invertebrates) taxa distributed across northern Australia, thus providing strong support for a common biogeographical history influencing the biota of this region (Bowman et al. 2010).

Within the western giant freshwater prawn species (now designated *Macrobrachium rosenbergii*), a major phylogeographic break was evident among populations north and south of the Isthmus of Kra on the Malay Peninsula. Initial work by Woodruff (2003) indicated that this biogeographic break, found in various terrestrial fauna, might have been the result of marine transgressions, splitting the Peninsula in two. A recent re-analysis (Woodruff and Turner 2009) suggests the observed patterns may result from faunal and floral 'compression', that is range contraction both north and south of the Isthmus, due to rapid and repeated sea-level fluctuations over the past 5 Myr. It was unclear from the work on freshwater prawns whether Sundaland palaeodrainages had facilitated connectivity, although the timing of both lineage and population diversification within this region indicated that events during the Pleistocene were the likely cause.

Further comparative phylogeographic analyses of obligate freshwater taxa from Sundaland and from the IAA in general, are required to clarify factors driving evolutionary diversification of the IAA's freshwater fauna.

Vertebrates

One of the earliest freshwater-based phylogeographic studies from the IAA examined mitochondrial variation (restriction fragment length polymorphisms, RFLPs) in a catfish species, *Hemibagrus nemurus*, sampled from Sundaland (Dodson et al. 1995). Results from this study indicated that the three main lineages identified (Indochinese, Sundaic and Sarawak groups) probably arose quite early in the Pleistocene, and the authors found limited evidence for connectivity among mainland and island lineages through the late Pleistocene. In contrast, a study on the same species using the more recently developed marker system of nuclear microsatellites identified a close link between peninsular Malaysian and Sarawak fishes, although some of the samples included in this study were from cultured stock, which may potentially have confounded results (Usmani et al. 2003). Similar conclusions were drawn from a more limited geographic coverage of the cyprinid fish *Barbodes gonionotus*, sampled from four sites (Mekong Delta, Malay peninsula, Southern Sumatra and East Java), utilising variation at the mitochondrial control region and a single microsatellite locus (McConnell 2004). The data provided additional evidence, however, for more recent Pleistocene connectivity among sites, although the coarse geographic sampling reduces the ability to determine whether patterns are congruent across both taxa mentioned above, and whether palaeodrainage systems facilitated connectivity through the Pleistocene.

Two genera of cyprinid fishes endemic to Asia, referred to as mahseer, have been studied by Nguyen et al. (2008) using three mitochondrial gene regions (16S, cytochrome *b*, ATPase). Interestingly, a molecular phylogeny obtained from sequences of the genus *Tor* revealed two contrasting phylogeographical patterns in two species (or possible species groups, because the authors discussed the species status of geographically separated lineages within a single species). In *T. douronensis*, two distinct lineages correspond to three distinct geographic regions, Borneo, the Mekong river system and Sumatra. In contrast, *T. tambroides*, occurring sympatrically with *T. douronensis* in mainland Asia and Sundaland, showed little genetic variation over its distribution range. As a possible scenario, Nguyen et al. (2008) suggested vicariance fragmentation for *T. douronensis* because their data indicate that the three geographically isolated lineages separated in the late Miocene, and the Sunda Shelf obviously limited the dispersal abilities of these lineages between Borneo, Sumatra and mainland Asia. According to Nguyen et al. (2008), the low levels of divergence between haplotypes of *T. tambroides* from different islands indicate some level of migration between populations during periods of low Pleistocene sea levels across Sundaland. The authors thus suggest dispersal as a likely scenario in one case and vicariance in the other.

Several intraspecific studies on various fish taxa from the Mekong Basin, and from proximate drainages in Thailand and Laos have indicated either: (1) low levels of genetic structure, suggesting high levels of gene flow within the Mekong drainage basin (catfishes, So et al. 2006a), or more commonly, (2) significant population structuring, indicating limited gene flow resulting from a probable combination of life-history characteristics and complex historical palaeodrainage rearrangements (catfishes, So et al. 2006b; cyprinids, Hara et al. 1998, Hurwood et al. 2008, Adamson et al. 2009).

13.4 Synthesis and conclusion

Although molecular systematic studies of the freshwater fauna of the IAA are in their infancy, they do hold great promise for recovering biogeographical signal. As the number and resolution of studies aimed at differing taxonomic, spatial and temporal scales increases, this should provide excellent resolution for improving our understanding of processes driving evolutionary diversification in this region. Findings discussed above are sometimes contradictory or inconclusive, which probably reflects to some extent the complex and dynamic geological and climatic history of the IAA.

Several common patterns are emerging, however: west of Wallace's Line, a probable Miocene dispersal into Southeast Asia from India (in turn possibly reflecting an ancestral connection to Gondwana) is suggested for freshwater crabs (Klaus et al. 2006, 2009) and shrimp (*Macrobrachium rosenbergii* species group, Johnson 1973, de Bruyn et al. 2004a), followed by a radiation across Sundaland. Miocene sea-level highstands have been invoked to have driven diversification in crabs (Shih et al. 2009) and shrimp (de Bruyn et al. 2005) in this region. The Isthmus of Kra barrier (Woodruff 2003, Woodruff and Turner 2009) has impacted on both these groups, above (crabs) and below (shrimp) the species level. Across Sundaland, Pleistocene palaeodrainage connections (Fig 13.2) facilitated connectivity to varying degrees among populations of some freshwater taxa (Dodson et al. 1995, McConnell 2004), although this pattern awaits a formal multi-species comparative analysis. Several studies have also revealed high levels of cryptic diversity, particularly within Sumatra, the Philippines, Sulawesi and Borneo (Evans et al. 2003a, b, Glaubrecht et al. 2003, 2006, Köhler and Glaubrecht 2003, Köhler et al. 2004, Takehana et al. 2005, von Rintelen and Glaubrecht 2006, de Bruyn and Mather 2007, Alfaro et al. 2008), and several uncovered patterns consistent with multiple colonisations of the Philippines from Borneo and/or Sulawesi.

Wallace's Line has been crossed from west to east during the Miocene by beetles (Balke et al. 2009) and shrimp (de Bruyn and Mather 2007). Subsequent radiation within Wallacea took place and, in the case of the beetles, a re-crossing of Wallace's Line and a subsequent Eurasian radiation (Balke et al. 2009). The same west to east crossing of Wallace's Line, this time to Sulawesi, was documented in clams (Glaubrecht et al. 2003, 2006, von Rintelen and Glaubrecht 2006), fish (Takehana et al. 2005), and frogs (Evans et al. 2003a, this volume, Chapter 11), while other workers report an eastern origin of the Sulawesi fauna, possibly by terrane drifting and accretion (gastropods – Glaubrecht and von Rintelen 2003, von Rintelen and Glaubrecht 2005, Albrecht et al. 2007, von Rintelen et al. this volume, Chapter 12), with a similar pattern observed in other taxa both above (fish, Sparks and Smith 2004) and below (shrimp, de Bruyn and Mather 2007) the species level. Sulawesi is occupied by both a diverse endemic and waif fauna, and holds great promise for understanding the geological history of the IAA region (this volume, Chapters 11 and 12). On the Sahul Shelf, Plio- and Pleistocene eustatic changes, land bridges, and the existence of palaeo-Lake Carpentaria facilitated the movement of taxa and populations between New Guinea and Australia, probably in both directions (fish, McGuigan et al. 2000, but see Cook and Hughes 2010, Huey et al. 2010; shrimp, de Bruyn et al. 2004b, de Bruyn and Mather 2007; crayfish, Baker et al. 2008; and snakes, Rawlings and Donnellan 2003, Wüster et al. 2005; see review on the biogeography of this region by Bowman et al. 2010).

The two main theories, set out at the beginning of this chapter for the exceptionally high levels of diversity found within this region, are likely operating over different geological, geographical, and temporal scales to drive diversification in this region. The IAA region, often thought of as an immense natural evolutionary laboratory, now provides the opportunity – resulting from recent enlightening palaeo-modelling studies and scientific syntheses (Sathiamurthy and Voris 2006, Cannon et al. 2009, Woodruff and Turner 2009, Woodruff 2010) – to examine the effects of rapid climate and habitat change on the biodiversity of this region. Ongoing phylogenetic and phylogeographic studies of freshwater taxa from the IAA should provide exceptionally detailed case studies, which may effectively be considered analogues for understanding future scenarios of rapid climate change.

The freshwater fauna of the IAA, representing millions of years of unique evolutionary history, are under extreme and sustained threat (Bickford et al. this volume, Chapter 17, Page et al. this volume, Chapter 16), and consolidated research programmes are required. We urge biologists to communicate and integrate the outcomes of their research programmes into existing and future conservation policies, public outreach and education, and the sociopolitical process. The Wallacea Research Group (initiated recently by the second author of this chapter), currently comprising 31 researchers and/or research groups worldwide, is a very positive first step along this path (see: http://www.wallacea.info/).

Acknowledgements

MdB acknowledges funding from a European Community Marie Curie Incoming International Fellowship (FP6 MIIF-CT-2006-39798), and a National Geographic Society Research Grant (8541-08). We thank the editor of this chapter, Lukas Rüber, for the invitation to contribute towards this volume, and Lukas Rüber, Harold Voris and two anonymous reviewers for helpful suggestions and comments that improved the manuscript.

References

Adamson, E. A. S., Hurwood, D. A., Baker, A. M. and Mather, P. B. (2009). Population subdivision in Siamese mud carp *Henicorhynchus simaensis* in the Mekong River basin: implications for management. *Journal of Fish Biology*, **75**, 1371–92.

Albrecht, C., Kuhn, K. and Streit, B. (2007). A molecular phylogeny of Planorboidea (Gastropoda, Pulmonata): insights from enhanced taxon sampling. *Zoologica Scripta*, **36**, 27–39.

Alfaro, M. E., Karns, D. R., Voris, H. K., Brock, C. D. and Stuart, B. L. (2008). Phylogeny, evolutionary history, and biogeography of Oriental-Australian rear-fanged water snakes (Colubroidea, Homalopsidae) inferred from mitochondrial and nuclear DNA sequences. *Molecular Phylogenetics and Evolution*, **2**, 576–93.

Arbogast, B. S., Edwards, S. V., Wakeley, J., Beerli, P. and Slowinski, J. B. (2002). Estimating divergence times from molecular data on phylogenetic and population genetic timescales. *Annual Review of Ecology and Systematics*, **33**, 707–40.

Archbold, N. W., Pigram, C. J., Ratman, N. and Hakim, S. (1982). Indonesian Permian Brachiopod fauna and Gondwana–South-East Asia relationships. *Nature*, **296**, 556–558.

Audley-Charles, M. G. (1983). Reconstruction of eastern Gondwanaland. *Nature*, **306**, 48–50.

Avise, J. C. (1994). *Molecular Markers, Natural History and Evolution*. New York: Chapman and Hall.

Avise, J. C. (2000). *Phylogeography: The History and Formation of Species*. Cambridge, MA: Harvard University Press.

Baker, N., de Bruyn, M. and Mather, P. B. (2008). Patterns of molecular diversity in wild stocks of the redclaw crayfish (*Cherax quadricarinatus*) from northern Australia and Papua New Guinea: impacts of Plio-Pleistocene landscape evolution. *Freshwater Biology*, **53**, 1592–605.

Balke, M., Ribera, I. and Vogler, A. P. (2004). MtDNA phylogeny and biogeography of Copelatinae, a highly diverse group of tropical diving beetles (Dytiscidae). *Molecular Phylogenetics and Evolution*, **32** 866–80.

Balke, M., Ribera, I., Hendrich, L. et al. (2009). New Guinea highland origin of a widespread arthropod supertramp. *Proceedings of the Royal Society B*, **279**, 2359–67.

Beaumont, M. A., Zhang, W. and Balding, D. J. (2002). Approximate Bayesian computation in population genetics. *Genetics*, **162**, 2025–35.

Bennett, K. J. (1997). *Evolution and Ecology, the Pace of Life*. Cambridge: Cambridge University Press.

Bermingham, E. and Avise, J. C. (1986). Molecular zoogeography of freshwater fishes in the Southeastern United States. *Genetics*, **113**, 939–65.

Bowman, D. M. J. S., Brown, G. K., Braby, M. F. et al.(2010). Biogeography of the Australian monsoon tropics. *Journal of Biogeography* **37**, 201–16.

Braatz, S. (1992). Conserving biological diversity: a strategy for protected areas in the Asia–Pacific Region. *World Bank Technical Paper*, **193**, 1–66.

Brandon-Jones, D. (1998). Pre-glacial Bornean primate impoverishment and Wallace's Line. In *Biogeography and Geological Evolution of SE Asia*, ed. J. D. Holloway and R. Hall. Leiden, The Netherlands: Backhuys, pp. 393–404.

Briggs, J. C. (2000). Centrifugal speciation and centres of origin. *Journal of Biogeography*, **27**, 1183–8.

Cane, M. A. and Molnar, P. (2001). Closing of the Indonesian seaway as a precursor to east African aridification around 3–4 million years ago. *Nature*, **411**, 157–62.

Cannon, C. H., Morley, R. J. and Bush, A. B. G. (2009). The current refugial rainforests of Sundaland are unrepresentative of their biogeographic past and highly vulnerable to disturbance. *Proceedings of the National Academy of Sciences of the USA*, **106**,11188–93.

Chenoweth, S. F., Hughes, J. M., Keenan, C. P. and Lavery, S. (1998). When oceans meet: a teleost shows secondary intergradation at an Indian Pacific interface. *Proceedings of the Royal Society Series B*, **265**, 415–20.

Chivas, A. R., García, A., der Kaars, S. et al. (2001). Sea-level and environmental changes since the last interglacial in the Gulf of Carpentaria, Australia: an overview. *Quaternary International*, **83**–85, 19–46.

Cook, B. D. and Hughes, J. M. (2010). Historical population connectivity and fragmentation in a tropical freshwater fish with a disjunct distribution (pennyfish, *Denariusa bandata*). *Journal of the North American Benthological Society*, **29**(3), 1119–31.

Coope, G. R. (1994). The response of insect faunas to glacial-interglacial climatic fluctuations. *Philosophical Transactions of the Royal Society of London B*, **344**, 19–26.

Cracraft, J. (1988). Deep-history biogeography: retrieving the historical pattern of evolving continental biotas. *Systematic Zoology*, **37**, 221–36.

Crandall, K. A. and Templeton, A. R. (1993). Empirical tests and some predictions from coalescent theory with applications to intraspecific phylogeny reconstruction. *Genetics*, **134**, 959–69.

Darwin, C. (1859). *The Origin of Species*. London: John Murray.

Davis, M. B. and Shaw, R. G. (2001). Range shifts and adaptive responses to Quaternary climate change. *Science*, **292**, 673–9.

da Silva, M. N. F. and Patton, J. L. (1998). Molecular phylogeography and the evolution and conservation of Amazonian mammals. *Molecular Ecology*, **7**, 475–86.

de Beaufort, L. F. (1926). *Zoögeographie van den Indischen Archipel*. Haarlem, The Netherlands: Bohn.

de Bruyn, M. and Mather, P. B. (2007). Molecular signatures of Pleistocene sea-level changes that affected connectivity among freshwater shrimp in Indo-Australian waters. *Molecular Ecology*, **16**, 4295–307.

de Bruyn, M., Wilson, J. C. and Mather, P. B. (2004a). Huxley's line demarcates extensive genetic divergence between eastern and western forms of the giant freshwater prawn, *Macrobrachium rosenbergii*. *Molecular Phylogenetics and Evolution*, **30**, 251–7.

de Bruyn M., Wilson J. C. and Mather P. B. (2004b). Reconciling geography and genealogy: phylogeography of giant freshwater prawns from the Lake Carpentaria region. *Molecular Ecology*, **13**, 3515–26.

de Bruyn, M., Nugroho, E., Hossain, Md. M., Wilson, J. C. and Mather P. B. (2005). Phylogeographic evidence for the existence of an ancient biogeographic barrier: the Isthmus of Kra Seaway. *Heredity*, **94**, 370–8.

de Queiroz, A. (2005). The resurrection of oceanic dispersal in historical biogeography. *Trends in Ecology and Evolution*, **20**, 68–73.

Dickerson, R. E. (1928). *Distribution of Life in the Philippines*. Manila: Bureau of Printing.

Dodson, J. J., Colombani, F. and Ng, P. K. L. (1995). Phylogeographic structure in mitochondrial DNA of a South-east Asian freshwater fish, *Hemibagrus nemurus* (Siluroidei; Bagridae) and Pleistocene sea-level changes on the Sunda shelf. *Molecular Ecology*, **4**, 331–46.

Donoghue, M. J. and Moore, B. R. (2003). Toward an integrative historical biogeography. *Integrative and Comparative Biology*, **43**, 261–70.

Dudgeon, D. (2003). The contribution of scientific information to the conservation and management of freshwater biodiversity in tropical Asia. *Hydrobiologia*, **500**, 295–314.

Dudgeon, D., Arthington, A. H., Gessner, M. O. et al. (2006). Freshwater biodiversity: importance, threats, status and conservation challenges. *Biological Reviews* **81**, 163–82.

Ebach, M. C. and Humphries, C. J. (2002). Cladistic biogeography and the art of discovery. *Journal of Biogeography*, **29**, 427–44.

Emerson, B. C., Paradis, E. and Thébaud, C. (2001). Revealing the demographic histories of species using DNA sequences. *Trends in Ecology and Evolution*, **16**, 707–16.

Evans, B. J., Brown, R. M., McGuire, J. A. et al. (2003a). Phylogenetics of fanged frogs: testing biogeographical hypotheses at the interface of the Asian and Australian faunal zones. *Systematic Biology*, **52**, 794–819.

Evans, B. J., Supriatna, J., Andayani, N. et al. (2003b). Monkeys and toads define areas of endemism on Sulawesi. *Evolution*, **57**, 1436–43.

Evans, B. J., McGuire, J. A., Brown, R. M., Andayani, N. and Supriatna, J. (2008). A coalescent framework for comparing alternative models of population structure with genetic data: evolution of Celebes toads. *Biology Letters*, **4**, 430–3.

Excoffier, L. and Smouse, P. E. (1994). Using allele frequencies and geographic subdivision to reconstruct gene trees within a species: molecular variance parsimony. *Genetics*, **136**, 343–59.

Excoffier, L., Smouse, P. E. and Quattro, J. M. (1992). Analysis of molecular variance inferred from metric distances among DNA haplotypes: application to human mitochondrial DNA restriction sites. *Genetics*, **131**, 479–91.

Fu, Y-X. (1994). Estimating effective population size or mutation rate using the frequencies of mutations of various classes in a sample of DNA sequences. *Genetics*, **138**, 1375–86.

Gathorne-Hardy, F. J., Syaukani, Davies, R. G., Eggleton, P. and Jones, D. T. (2002). Quaternary rainforest refugia in south-east Asia: using termites (Isoptera) as indicators. *Biological Journal of the Linnean Society*, **75**, 453–66.

Glaubrecht, M. and von Rintelen, T. (2003). Systematics and zoogeography of the pachychilid gastropod *Pseudopotamis* Martens, 1894 (Mollusca, Gastropoda, Cerithioidea): a limnic relict on the Torres Strait Islands, Australia? *Zoologica Scripta*, **32**, 415–35.

Glaubrecht, M., von Rintelen, T. and Korniushin, A. V. (2003). Toward a systematic revision of brooding freshwater Corbiculidae in Southeast Asia (Bivalvia, Veneroida): on shell morphology, anatomy and molecular phylogenetics of endemic taxa from islands in Indonesia. *Malacologia*, **45**, 1–40.

Glaubrecht, M., Fehér, Z. and von Rintelen, T. (2006). Brooding in *Corbicula madagascariensis* (Bivalvia, Corbiculidae) and the repeated evolution of viviparity in corbiculids. *Zoologica Scripta*, **35**, 641–54.

Gómez, A., Wright, P. J., Lunt, D. H. et al. (2007). Mating trials validate the use of DNA barcoding to reveal cryptic speciation of a marine bryozoan taxon. *Proceedings of the Royal Society London Series B*, **274**, 199–207.

Guthrie, R. D. (2003). Rapid body size decline in Alaskan Pleistocene horses before extinction. *Nature*, **426**, 169–71.

Haase, M. and Bouchet, P. (2006). The radiation of hydrobioid gastropods (Caenogastropoda, Rissooidea) in ancient Lake Poso, Sulawesi. *Hydrobiologia*, **556**, 17–46.

Hall, R. (1996). Reconstructing Cenozoic Asia. In *Tectonic Evolution of Southeast Asia*, ed. R. Hall and D. J. Blundell. Bath, UK: The Geological Society Publishing House, pp. 153–84.

Hall, R. (1998). The plate tectonics of Cenozoic SE Asia and the distribution of land and sea. In *Biogeography and Geological Evolution of SE Asia*, ed. J. D. Holloway and R. Hall. Leiden, The Netherlands: Backhuys, pp. 99–131.

Hall, R. (2001). Cenozoic reconstructions of SE Asia and the SW Pacific, changing patterns of land and sea. In *Faunal and Floral migrations and Evolution in SE Asia-Australalasia*, ed. I. Metcalfe, J. M. B. Smith, M. Morwood and I. Davidson. Lisse, The Netherlands: A. A. Balkema Publishers, pp. 35–56.

Hall, R. (2002). Cenozoic geological and plate tectonic evolution of SE Asia and the SW Pacific: computer-based reconstructions, model and animations. *Journal of Asian Earth Sciences*, **20**, 353–431.

Hall, R. (2009). Southeast Asia's changing palaeogeography. *Blumea*, **54**, 148–61.

Hanebuth, T. J. J., Stattegger, K. and Bojanowski, A. (2009). Termination of the last glacial maximum sea-level lowstand: the Sunda-Shelf data revisited. *Global and Planetary Change*, **66**, 76–84.

Hara, M., Sekino, M. and Na-Nakorn, U. (1998). Genetic differentiation of natural populations of the snake-head fish, *Channa striatus* in Thailand. *Fisheries Science*, **64**, 882–5.

Heads, M. (2005). Dating nodes on molecular phylogenies: a critique of molecular biogeography. *Cladistics*, **21**, 62–78.

Heaney, L. R. (1986). Biogeography of mammals in SE Asia: estimates of rates of colonization, extinction and speciation. *Biological Journal of the Linnean Society*, **28**, 127–65.

Hedgecock, D., Stelmach, D. J., Nelson, K., Lindenfelser, M. E. and Malecha, S. R. (1979). Genetic divergence and biogeography of natural populations of *Macrobrachium rosenbergii*. *Proceedings of the World Mariculture Society*, **10**, 873–9.

Hewitt, G. (2000). The genetic legacy of the Quaternary ice ages. *Nature*, **405**, 907–13.

Hickerson, M. J., Stahl, E. and Takebayashi, N. (2007). msBayes: pipeline for testing comparative phylogeographic histories using hierarchical approximate Bayesian computation. *BMC Bioinformatics*, **8**, 268.

Hofreiter, M. and Stewart, J. (2009). Ecological change, range fluctuations and population dynamics during the Pleistocene. *Current Biology*, **19**, R584–R594.

Hofreiter, M., Serre, D., Rohland, N. et al. (2003). Lack of phylogeography in European mammals before the last glaciation. *Proceedings of the National Academy of Sciences of the USA*, **101**, 12963–8.

Hudson, R. R., Slatkin, M. and Maddison, W. P. (1992). Estimation of levels of gene flow from DNA sequence data. *Genetics*, **132**, 583–9.

Huey, J. A., Baker, A. M. and Hughes, J. M. (2010). High levels of genetic structure in the Australian freshwater fish, *Ambassis macleayi*. *Journal of the North American Benthological Society*, **29**(3), 1148–60.

Hurwood, D. A., Adamson, E. A. S. and Mather, P. B. (2008). Evidence for strong genetic structure in a regionally important, highly vagile cyprinid (*Henicorhynchus lobatus*) in the Mekong River Basin. *Ecology of Freshwater Fish*, **17**, 273–83.

Huxley, T. H. (1868). On the classification and distribution of the Alectoromorphae and Heteromorphae. *Proceedings of the Zoological Society of London*, 294–319.

Johnson, D. S. (1973). Notes on some species of the genus *Macrobrachium* (Crustacea: Decapoda: Caridea: Palaemonidae). *Journal of the Singapore National Academy of Sciences*, **3**(3), 273–91.

Kingman, J. (1982a). The coalescent. *Stochastic Processes and their Applications*, **13**, 235–48.

Kingman, J. (1982b). On the genealogy of large populations. In *Essays in Statistical Science*, ed. J. Gani and E. Hannan. London: Applied Probability Trust, pp. 27–43.

Klaus, S., Schubart, C. D. and Brandis, D. (2006). Phylogeny, biogeography and a new taxonomy for the Gecarcinucoidea Rathbun, 1904 (Decapoda, Brachyura). *Organisms, Diversity and Evolution*, **6**, 199–217.

Klaus, S., Brandis, D., Ng, P. K. L., Yeo, D. C. J. and Schubart, C. D. (2009). Phylogeny and biogeography of Asian freshwater crabs of the family Gecarcinucidae (Brachyura, Potamoidea). In *Crustacean Issues18, Decapod Crustacean Phylogenetics*, ed. J. W. Martin, K. A. Crandall and D. L. Felder. Boca Raton, Florida: Taylor and Francis/CRC Press, pp. 509–31.

Knowles, L. L. (2009). Statistical phylogeography. *Annual Review of Ecology, Evolution and Systematics*, **40**, 593–612.

Knowles, L. L. and Maddison W. P. (2002). Statistical phylogeography. *Molecular Ecology*, **11**, 2623–35.

Köhler, F. and Glaubrecht, M. (2003). Morphology, reproductive biology and molecular genetics of ovoviviparous freshwater gastropods (Cerithioidea, Pachychilidae) from the Philippines, with description of a new genus *Jagora*. *Zoologica Scripta*, **32**, 35–59.

Köhler, F. and Glaubrecht M. (2010). Uncovering an overlooked radiation: molecular phylogeny and biogeography of Madagascar's endemic river snails (Caenogastropoda: Pachychilidae: *Madagasikara* gen. nov.). *Biological Journal of the Linnaean Society*, **99**, 867–94.

Köhler, F., von Rintelen, T., Meyer, A. and Glaubrecht, M. (2004). Multiple origin of viviparity in Southeast Asian gastropods (Cerithioidea, Pachychilidae) and its evolutionary implications. *Evolution*, **58**, 2215–26.

Kottelat, M. (2002). Aquatic systems: neglected biodiversity. In *Terrestrial Ecoregions of the Indo-Pacific*, ed. E. Wikramanayake et al. Washington DC: Island Press, pp. 30–35.

Lemmon, A. R. and Lemmon, E. M. (2008). A likelihood framework for estimating phylogeographic history on a continuous landscape. *Systematic Biology*, **57**(4), 544–61.

Lindenfelser, M. E. (1984). Morphometric and allozymic congruence: evolution in the prawn *Macrobrachium rosenbergii* (Decapoda, Palaemonidae). *Systematic Zoology*, **33**, 195–204.

Lisiecki, L. and Raymo, M. (2005) A Plio-Pleistocene stack of 57 globally distributed benthic delta18 O records. *Paleoceanography*, **20**, 522–33.

Lundberg, J. G. (1993). African–South American freshwater fish clades and continental drift: problems with a paradigm. In *Biotic Relationships Between Africa and South America*, ed. P. Goldblatt. New Haven, CT: Yale University Press, pp. 156–98.

McConnell, S. K. J. (2004). Mapping aquatic faunal exchanges across the Sunda shelf, South-East Asia, using distributional and genetic data sets from the cyprinid fish *Barbodes gonionotus* (Bleeker, 1850). *Journal of Natural History*, **38**, 651–70.

McGuigan, K., Zhu, D., Allen, G. R. and Moritz, C. (2000). Phylogenetic relationships and historical biogeography of melanotaeniid fishes in Australia and New Guinea. *Marine and Freshwater Research*, **51**, 713–23.

Malecha, S. R. (1977). Genetics and selective breeding of *Macrobrachium rosenbergii.* In *Shrimp and Prawn Farming in the Western Hemisphere*, ed. J. A. Hanson and H. L. Goodwin. Stroudsberg, PA: Dowden, Hutchinson and Ross, pp. 328–55.

Malecha, S. R. (1987). Selective breeding and intraspecific hybridization of crustaceans. *Proceedings of the World Symposium on Selection, Hybridization, and Genetic Engineering in Aquaculture*. Vol. 1, Berlin, Germany, pp. 323–36.

Marko, P. B. (2002). Fossil calibration of molecular clocks and the divergence times of geminate species pairs separated by the Isthmus of Panama. *Molecular Biology and Evolution*, **19**, 2005–21.

Mayr, E. (1944). Wallace's line in the light of recent zoogeographic studies. *The Quarterly Review of Biology*, **19**, 1–14.

Mittermeier, R. A., Myers, N., Mittermeier, C. G. and Robles-Gil, P. (1999). *Hotspots: Earth's Biologically Richest and Most Endangered Terrestrial Ecoregions*. Agrupacion, Sierra Madre: Cemex, Conservation International.

Mittermeier, R. A., Gil, P. R., Hoffman, M. et al. (2005). *Hotspots Revisited:*

Earth's Biologically Richest and Most Endangered Terrestrial Ecoregions. Washington DC: Conservation International.

Moss, S. J. and Wilson, M. E. J. (1998). Biogeographic implications of the Tertiary palaeogeographic evolution of Sulawesi and Borneo. In *Biogeography and Geological Evolution of SE Asia*, ed. R. Hall and J. D. Holloway. Leiden, The Netherlands: Backhuys Publishers, pp. 133–63.

Murphy, N. P. and Austin, C. M. (2005). Phylogenetic relationships of the globally distributed freshwater prawn genus *Macrobrachium* (Crustacea, Decapoda, Palaemonidae): biogeography, taxonomy and the convergent evolution of abbreviated larval development. *Zoologica Scripta*, **34**, 187–97.

Myers, N., Mittermeier, R. A., Mittermeier, C. G., da Fonseca, G. A. B. and Kent, J. (2000). Biodiversity hotspots for conservation priorities. *Nature*, **403**, 853–8.

Nelson, G. and Platnick, N. I. (1981). *Systematics and Biogeography; Cladistics and Vicariance.* New York: Columbia University Press.

New, M. B. and Singholka, S. (1985). Freshwater prawn farming: a manual for the culture of *Macrobrachium rosenbergii.* FAO Fisheries Technical Paper, 225.

Nguyen, T. T. T., Na-Nakorn, U., Sukmanomon, S. and Ziming, C. (2008). A study on phylogeny and biogeography of mahseer species (Pisces, Cyprinidae) using sequences of three mitochondrial DNA gene regions. *Molecular Phylogenetics and Evolution*, **48**, 1223–31.

Page, R. D. M. and Charleston, M. A. (1998). Trees within trees: phylogeny and historical associations. *Trends in Ecology and Evolution*, **13**, 356–9.

Page, T. J., von Rintelen, K. and Hughes, J. M. (2007). An island in the stream: Australia's place in the cosmopolitan world of Indo-West Pacific shrimp (Decapoda, Atyidae, *Caridina*). *Molecular Phylogenetics and Evolution*, **43**, 645–59.

Park, J. K. and Kim, W. (2003). Two *Corbicula* (Corbiculidae, Bivalvia) mitochondrial lineages are widely distributed in Asian freshwater environment. *Molecular Phylogenetics and Evolution*, **29**, 529–39.

Ponder, W. and Haase, M. (2005). A new genus of hydrobiid gastropods with Australian affinities from Lake Poso, Sulawesi (Gastropoda: Caenogastropoda: Rissooidea). *Molluscan Research*, **25**, 27–36.

Posada, D. and Crandall, K. A. (2001). Intraspecific gene genealogies: trees grafting into networks. *Trends in Ecology and Evolution*, **16**, 37–45.

Posamentier, H. W. (2001). Lowstand alluvial bypass systems: Incised vs. unincised. *AAPG Bulletin*, **85**(10), 1771–93.

Rainboth, W. J. (1996). The taxonomy, systematics and zoogeography of *Hypsibarbus*, a new genus of large barbs (Pisces, Cyprinidae) from the rivers of Southeastern Asia. *University of California Publications in Zoology*, **129**, xxiii–199.

Rawlings, L. H. and Donnellan, S. C. (2003). Phylogeographic analysis of the green python, *Morelia viridis*, reveals cryptic diversity. *Molecular Phylogenetics and Evolution*, **27**, 36–44.

Ree, R. H. and Sanmartín, I. (2009). Prospects and challenges for parametric models in historical biogeographical inference. *Journal of Biogeography*, **36**, 1211–20.

Reeves, J. M., Chivas, A. R., Garcia, A. and De Deckker, P. (2007). Palaeoenvironmental change in the Gulf of Carpentaria (Australia) since the last interglacial based on Ostracoda. *Palaeogeography, Palaeoclimatology, Palaeoecology*, **246**, 163–87.

Reeves, J. M., Chivas, A. R., Garcia, A. et al. (2008). The sedimentary record of Palaeoenvironments and sea-level change in the Gulf of Carpentaria, Australia, through the last glacial cycle. *Quaternary International*, **183**, 3–22.

Rogers, A. R. and Harpending, H. (1992). Population growth makes waves in the distribution of pairwise genetic differences. *Molecular Biology and Evolution*, **9**, 552–69.

Ruedi, M. and Fumagalli, L. (1996). Genetic structure of Gymnures (genus *Hylomys*; Erinaceidae) on continental islands of Southeast Asia: historical effects of fragmentation. *Journal of Zoological Systematics*, **34**, 153–62.

Sarasin, P. (1896). Die wissenschaftlichen Gesichtspunkte, welche uns bei der Erforschung von Celebes geleitet haben (im Auszug mitgeteilt). *Verhandlungen der Gesellschaft für Erdkunde zu Berlin*, **23**, 337–9.

Sarasin, P. and Sarasin, F. (1898). *Die Süßwassermollusken von Celebes*. Wiesbaden, Germany: Kreidel.

Sarasin, P. and Sarasin, F. (1901). *Entwurf einer geographisch-geologischen Beschreibung der Insel Celebes*. Wiesbaden, Germany: Kreidel.

Sarasin, P. and Sarasin, F. (1905). *Reisen in Celebes ausgeführt in den Jahren 1893–1896 und 1902–1903*. Wiesbaden, Germany: Kreidel.

Sathiamurthy, E. and Voris, H. K. (2006). Maps of Holocene sea level transgression and submerged lakes on the Sunda Shelf. *The Natural History Journal of Chulalongkorn University, Supplement*, **2**, 1–43.

Schliewen, U. K. and Klee, B. (2004). Reticulate sympatric speciation in Cameroonian crater lake cichlids. *Frontiers in Zoology*, **1**, 5.

Schmitt, L. H., Kitchener, D. J. and How, R. A. (1995). A genetic perspective of mammalian variation and evolution in the Indonesian Archipelago: biogeographic correlates in the fruit bat genus *Cynopterus*. *Evolution*, **49**, 399–412.

Shapiro, B., Drummond, A. J., Rambaut, A. et al. (2004). Rise and fall of the Beringian steppe bison. *Science*, **306**, 1561–5.

Shih, H. T., Yeo, D. C. J. and Ng, P. K. L. (2009). The collision of the Indian plate with Asia: molecular evidence for its impact on the phylogeny of freshwater crabs (Brachyura, Potamidae). *Journal of Biogeography*, **36**, 703–19.

Simpson, G. G. (1977). Too many lines; the limits of the Oriental and Australian zoogeographic regions. *Proceedings of the American Philosophical Society*, **121**(2), 107–20.

Slatkin, M. (1989). Detecting small amounts of gene flow from phylogenies of alleles. *Genetics*, **121**, 609–12.

Slatkin, M. and Hudson, R. R. (1991). Pairwise comparisons of mitochondrial DNA sequences in stable and exponentially growing populations. *Genetics*, **129**, 555–62.

Smouse, P. E. (1998). To tree or not to tree. *Molecular Ecology*, **7**, 399–412.

So, N., van Houdt, J. K. J. and Volckaert, F. A. M. (2006a). Genetic diversity and population history of the migratory catfishes *Pangasianodon hypophthalmus* and *Pangasius bocourti* in the Cambodian Mekong River. *Fisheries Science*, **72**, 469–76.

So, N., Maes, G. E. and Volckaert, F. A. M. (2006b). High genetic diversity in cryptic populations of the migratory sutchi catfish *Pangasianodon hypophthalmus* in the Mekong River. *Heredity*, **96**, 166–74.

Sparks, J. S. and Smith, W. L. (2004). Phylogeny and biogeography of the Malagasy and Australasian rainbowfishes (Teleostei: Melanotaenioidei): Gondwanan vicariance and evolution in freshwater. *Molecular Phylogenetics and Evolution*, **33**, 719–34.

Takehana, Y., Naruse, K. and Sakaizumi, M. (2005). Molecular phylogeny of the medeka fishes genus *Oryzias* (Beloniformes, Adrianichthyidae) based on nuclear and mitochondrial DNA sequences. *Molecular Phylogenetics and Evolution*, **36**, 417–28.

Templeton, A. R. (2009). Statistical hypothesis testing in intraspecific phylogeography: nested clade phylogeographical analysis vs. approximate Bayesian computation. *Molecular Ecology*, **18**, 319–31.

Thiele, J. (1929). *Handbuch der Systematischen Weichtierkunde*. Jena, Germany: Gustav Fischer Verlag.

Torgersen, T., Hutchinson, M. F., Searle, D. E. and Nix, H. A. (1983). General bathymetry of the Gulf of Carpentaria and the Quaternary physiography of Lake Carpentaria. *Palaeogeography, Palaeoclimatology, Palaeoecology*, **41**, 207–25.

Torgersen, T., Jones, M. R., Stephens, A. W., Searle, D. E. and Ullman, W. J. (1985). Late Quartenary hydrological changes in the Gulf of Carpentaria. *Nature*, **313**, 785–7.

Usmani, S., Tan, S. G., Siraj, S. S. and Yusoff, K. (2003). Population structure of the Southeast Asian river catfish *Mystus nemurus*. *Animal Genetics*, **34**, 462–4.

van Oosterzee, P. (1997). *Where Worlds Collide. The Wallace Line*. Ithaca, NY: Cornell University Press.

von Martens, E. (1897). *Süß- und Brackwasser-Mollusken des Indischen Archipels*. Leiden, The Netherlands: Brill.

von Rintelen, K., von Rintelen, T. and Glaubrecht, M. (2007). Molecular phylogeny and diversification of freshwater shrimps (Decapoda, Atyidae, Caridina) from ancient Lake Poso (Sulawesi, Indonesia): the importance of being colourful. *Molecular Phylogenetics and Evolution*, **45**, 1033–41.

von Rintelen, K., Karge, A. and Klotz, W. (2008). News from a small island: first record of a freshwater shrimp (Decapoda, Atyidae, *Caridina*) from Peleng, Banggai Islands, Indonesia. *Journal of Natural History*, **42**, 2243–56.

von Rintelen, K., Glaubrecht, M., Schubart, C. D., Wessel, A. and von Rintelen, T. (2010). Adaptive radiation and ecological diversification of Sulawesi's ancient lake shrimps. *Evolution*, **64**, 3287–99.

von Rintelen, T. and Glaubrecht, M. (2005). Anatomy of an adaptive radiation: a unique reproductive strategy in the endemic freshwater gastropod Tylomelania (Cerithioidea, Pachychilidae) on Sulawesi, Indonesia, and its biogeographic implications. *Biological Journal of the Linnaean Society*, **85**, 513–42.

von Rintelen, T. and Glaubrecht, M. (2006). Rapid evolution of sessility in an endemic species flock of the freshwater bivalve *Corbicula* from ancient lakes on Sulawesi, Indonesia. *Biology Letters*, **2**, 73–7.

Voris, H. K. (2000). Maps of Pleistocene sea levels in Southeast Asia: shorelines, river systems and time durations. *Journal of Biogeography*, **27**, 1153–67.

Wallace, A. R. (1859). Letter from Mr Wallace concerning the geographical distribution of birds. *Ibis*, **1**, 449–54.

Wallace, A. R. (1860). On the zoological geography of the Malay Archipelago. *Journal of the Proceedings of the Linnean Society*, **4**, 172–84.

Wallace, A. R. (1862). Narrative of search after birds of paradise. *Proceedings of the Zoological Society of London 1862*, 153–61.

Wallace, A. R. (1869). *The Malay Archipelago*. London: Macmillan.

Wallace, A. R. (1880). *Island Life, or the Phenomena and Causes of Insular Faunas and Floras, Including a Revision and Attempted Solution of the Problem of Geological Climates*. London: Macmillan.

Wallis, G. P. and Trewick, S. A. (2001). Finding fault with vicariance: a critique of Heads (1998). *Systematic Biology*, **50**(4), 602–609.

Weber, M. (1902). *Der Indo-australische Archipel und die Geschichte seiner Tierwelt*. Jena, Germany: Gustav Fischer.

Whitten, A. J., Bishop, K. D., Nash, S. V. and Clayton, L. (1987). One or more extinctions from Sulawesi, Indonesia? *Conservation Biology*, **1**, 42–8.

Wilson, M. E. J. and Moss, S. J. (1999). Cenozoic palaeogeographic evolution of Sulawesi and Borneo. *Palaeogeography, Palaeoclimatology, Palaeoecology*, **145**, 303–37.

Woodruff, D. S. (2003). Neogene marine transgressions, palaeogeography and biogeographic transitions on the Thai-Malay Peninsula. *Journal of Biogeography*, **30**, 551–67.

Woodruff, D. S. (2010). Biogeography and conservation in Southeast Asia: how 2.7 million years of repeated environmental fluctuations affect today's patterns and the future of the remaining refugial-phase biodiversity. *Biodiversity and Conservation*, **19**(4), 919–41.

Woodruff, D. S. and Turner, L. M. (2009). The Indochinese–Sundaic zoogeographic transition: a description of terrestrial mammal species distributions. *Journal of Biogeography*, **36**, 803–21.

Wowor, D., Muthu, V., Meier, R. et al. (2009). Evolution of life history traits in Asian freshwater prawns of the genus *Macrobrachium* (Crustacea, Decapoda, Palaemonidae) based on multilocus molecular phylogenetic analysis. *Molecular Phylogenetics and Evolution*, **52**, 340–50.

Wüster, W., Dumbrell, A. J., Hay, C. et al. (2005). Snakes across the Strait: trans-Torresian phylogeographic relationships in three genera of Australasian snakes (Serpentes, Elapidae: *Acanthophis*, *Oxyuranus*, and *Pseudechis*). *Molecular Phylogenetics and Evolution*, **34**, 1–14.

Yeo, D. C. J., Shih, H. T., Meier, R. and Ng, P. K. L. (2007). Phylogeny and biogeography of the freshwater crab genus *Johora* (Crustacea, Brachyura, Potamidae) from the Malay Peninsula, and the origins of its insular fauna. *Zoologica Scripta*, **36**, 255–69.

Zink, R. M. (2002). Methods in comparative phylogeography, and their application to studying evolution in the North American aridlands. *Integrative and Comparative Biology*, **42**(5), 953–9.

14

Patterns of biodiversity discovery through time: an historical analysis of amphibian species discoveries in the Southeast Asian mainland and adjacent island archipelagos

Rafe M. Brown and Bryan L. Stuart

'Amphibians have not evolved for the convenience of taxonomists'

Robert F. Inger, 2007

14.1 Introduction

Because of its dynamic geological history and partitioned insular geography (Hamilton 1979, Hall 1996), Southeast Asia has one of the most diverse and yet poorly understood amphibian faunas on the planet. Despite serving as the backdrop for the birthplace of the field of biogeography (Wallace 1860, 1876, Lomolino et al. 2006, van Wyhe this volume, Chapter 2), an understanding of the region's biodiversity ironically still suffers from unequal sampling and study among areas (Bain et al. 2008), logistical obstacles to fieldwork (Brown et al. 2008, Brown 2009), a lack of communication between workers in the region's many countries (Stuart and Bain 2008), and a lack of sufficient regional syntheses that transcend

Biotic Evolution and Environmental Change in Southeast Asia, eds D. J. Gower et al. Published by Cambridge University Press.

political borders (but see Inger 1999, Bain et al. 2008). Nonetheless, the region is universally recognised as one of the Earth's great strongholds of biological diversity (Mittermeier et al. 1999) due to the dense concentration of numerous subcentres of species diversity (Whitmore 1984, 1987, 1990, Mittermeier et al. 1997, 1998, Reid 1998). In fact, the region's megadiversity combined with an acute conservation crisis of regional proportion (Bawa 1990, Collins et al. 1991, Brooks et al. 2002, Rowley 2010), now designates four (the Philippines, Wallacea, Sundaland and Indo-Burma) of the world's 34 Global Conservation Hotspots (Mittermeier et al. 1997) within the region. Arguably, terrestrial biodiversity in Southeast Asia constitutes one of Earth's greatest conservation urgencies (Bickford et al. this volume, Chapter 17).

Given the immense global significance of the biodiversity of Southeast Asia, it is reasonable to scrutinise the history of the discovery documentation of that diversity. How has our appreciation of Southeast Asia's biodiversity developed? What historical trends can we infer from historical patterns of the recognition of the region's biodiversity? Given the current conservation crisis and the need to accurately and rapidly catalogue biodiversity before it is lost, are there lessons we can learn from the past several centuries of discovery?

In this chapter we focus on the last 200 years of amphibian discovery and species description in Southeast Asia. Because of the immense monographic contributions of early workers (i.e. Boulenger 1920, Taylor 1920, Van Kampen 1923, Smith 1930, 1935, Bourret 1942), one reasonable prediction might be that the vast majority of amphibian species diversity accumulated in the literature after the turn of the last century and that the subsequent rate of species discovery has declined. However, we know that several mid-century works contributed heavily to our knowledge of the fauna (Inger 1954, 1966, Taylor 1965, 1968), and, of course, the many dozens of smaller taxonomic works of many workers have contributed substantially as well (Duellman and Trueb 1994). How have the last two centuries of Southeast Asian species discovery compared to current activity in amphibian taxonomic studies? Are we now experiencing a period of asymptotic decline in rates of discovery and can we assume that the region's fauna is reasonably well known (Giam et al. 2010, Joppa et al. 2011)?

In this contribution we focus on five biogeographic regions of Southeast Asia, species discoveries and descriptions through time, and the kinds of data used by amphibian systematists to recognise species and define boundaries between taxa. We ask where accumulation of Southeast Asian amphibian species diversity has been greatest, at what times have the greatest numbers of currently recognised species been discovered and what types of data have contributed disproportionately to the recognition of species. Finally, given fundamental expectations derived from predictions concerning the geographic distribution of diversity, we ask whether Southeast Asia's island archipelagos are the source of more species diversity compared to the Asian mainland. It is our expectation that this review will contribute to the eventual recognition of the years between 1990 and 2010 as a modern 'Age of Discovery' in Southeast Asian amphibian biodiversity studies.

14.2 Methods

Because our goal was to explore patterns of Southeast Asian amphibian species diversity and discoveries through time, we first assembled a list of all known (to us) currently valid 622 native species of amphibians recognised from the region as of July 2010, arranged by date of publication of description, and ordered these chronologically, beginning with the first description of the native toad *Duttaphrynus melanostictus* (Schneider, 1799). We used the Amphibian Species of the World (version 5.4, April–May, 2010) online taxonomic database (Frost 2010) as an initial guide, and augmented this resource with data from original descriptions, regional and country summaries provided by AmphibiaWeb (2010), and recent literature (published between June 2010 and March 2011) that has not yet been incorporated into these taxonomic and conservation resources. All data used in the analyses are presented in Appendix C.

We recorded the type of data used for the justification of the recognition of each species and classified these into four categories: (1) morphological only, (2) morphological and molecular, (3) morphological and bioacoustical, or (4) morphological, molecular and bioacoustical. These data were used for plotting trends in the types of evidence used in descriptions of Southeast Asian amphibians through time.

For the purpose of comparing patterns in species discoveries across biogeographic regions, and for enabling mainland versus adjacent island archipelago comparisons, we scored the presence of each species in one or more of the following five biogeographic regions: (1) the northern mainland, extending from Myanmar to Vietnam and from the southern border of China to the Isthmus of Kra, (2) the southern mainland, defined as the area south of the Isthmus of Kra that includes southern Thailand, Peninsular Malaysia and Singapore, (3) the islands of Sundaland (Borneo, Java, Sumatra, Bali, east to Wallace's Line), (4) the Philippines (including Palawan), and (5) Sulawesi and neighbouring small islands. Patterns were also compared on the mainland versus island archipelagos by grouping species presence in each of three land mass types: (1) the mainland (consisting of the northern and southern mainland biogeographic regions), (2) the islands of Sundaland (equivalent to the Sundaland biogeographic region), and (3) oceanic islands (all remaining islands).

Some biogeographers may question our necessarily artificial division of Southeast Asia for the purpose of this exercise. We chose the southern border of China as the northern boundary (as did Inger 1999) as a clear demarcation for excluding Eurasian temperate fauna from this analysis. Although this political boundary is relatively arbitrary, we nonetheless chose it as a clear starting place in order to define the

Southeast Asian limits of this study. The Isthmus of Kra and Wallace's Lines are non-controversial, well-established biogeographic boundaries that require little comment (Huxley 1868, Kloss 1929, Inger 1999, Voris 2000, Inger and Voris 2001, Brown and Guttman 2002, Woodruff 2003, 2010) due to their clear relation to geological history of the region (Hall 1998, 2002). We included Sulawesi in this analysis because the majority of its amphibian fauna is derived from Sundaland (Iskandar and Tjan 1996, Evans et al. 2003) but extended our consideration no further east into the Malukus because much of eastern Indonesia's biota is derived from Papuan elements (Allison 1996, Brown 1997, Whitten et al. 2002). Finally, we included Palawan as part of the Philippines. Although past studies of mammals and birds have suggested Sundaic affinities for Palawan's fauna and treated Palawan's fauna as a nested subset of the Bornean fauna (Huxley 1868, Kennedy et al. 2000, Esselstyn et al. 2004), recent phylogenetic studies (Brown and Guttman 2000, Evans et al. 2003, Brown et al. 2009, Oliveros and Moyle 2010) and faunistic reviews (Brown and Diesmos 2009, Esselstyn et al. 2010, Blackburn et al. 2010, Siler et al. 2012) have called this traditional interpretation into question. These and several other studies suggest that Palawan might best be viewed as a transition or filter zone, with an amalgamation of faunal elements from both sides of Wallace's and Huxley's lines composing its terrestrial vertebrate community. Because virtually every Palawan endemic amphibian that has been included in a phylogenetic analysis has been demonstrated to be more closely related to endemics from the oceanic portions of the Philippines (Brown and Guttman 2002, Evans et al. 2003, Brown et al. 2009, Brown and Diesmos 2009), and because our analysis is limited to amphibians, we classify Palawan as part of the Philippines. To illustrate trends in species discoveries and taxonomic activity per region through time, we plotted descriptions per year as a function of the five narrow and three broad geographic areas, as defined above.

As a measure of the degree to which each species is geographically widespread, we scored each species for its occurrence in one through five of our biogeographic regions. We then plotted the number of biogeographic regions habited by each species against the year of species description.

Finally, as an illustration of patterns of species discoveries and taxonomic activity over the last 210 years, we plotted separate species accumulation curves (cumulative numbers of species versus year of description) for the 11 included amphibian families (Ichthyophiidae, Salamandridae, Bombinatoridae, Megophryidae, Hylidae, Bufonidae, Microhylidae, Ceratobatrachidae, Dicroglossidae, Rhacophoridae and Ranidae) with family data pooled across all regions.

14.3 Results

When the raw data are pooled for a single Southeast Asian species accumulation curve (Fig 14.1), the pattern that emerges is a gradual increase in species

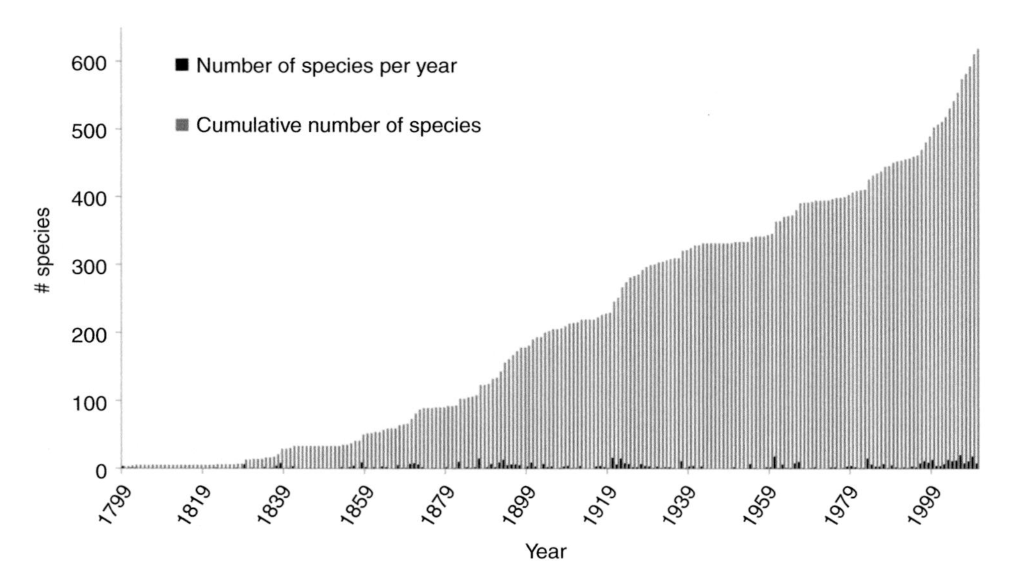

Figure 14.1 Southeast Asian amphibian species descriptions per year (dark bars) and cumulative species accumulation (pale bars; total of 619 taxa) over the past two centuries.

accumulation through the majority of the 1800s, a steady increase in species numbers, with punctuated periods of rapid species accumulation (note rate increases in late 1800s, early 1900s, and mid 1900s), and a dramatically steep increase in species accumulation over the last two decades (Fig 14.1).

The vast majority of species descriptions over the past 210 years have been based solely on morphology (Fig 14.2a). The exceptions are descriptions published over the last 30 years. This recent period (Fig 14.2b) has seen a more frequent occurrence of species descriptions of anurans based on morphology plus bioacoustic data, beginning with Dring (1983, 1987). In later years, especially since the year 2000, many species were described based on combinations of data or including all types of data (morphological, bioacoustical and molecular).

The analysis of species descriptions through time grouped by geographic regions similarly reveals several prevalent trends. Most regions experienced taxonomic work, in fits and starts, but somewhat steadily through time. The Sunda Region, the northern mainland, and the southern mainland, in particular, are represented with species discoveries nearly every decade (Fig 14.3a). In contrast, Sulawesi exhibits a modest number of species descriptions through the 1800s and early 1900s but only one since 1933 (Riyanto et al. 2011). Interestingly, the northern mainland shows a particularly rapid increase in species discovery over the past decade (Fig 14.3a). Species numbers have steadily increased on the mainland, Sundaland, and other oceanic islands (Fig 14.3b), with highest species diversity (and rapid recent species discovery) on the mainland.

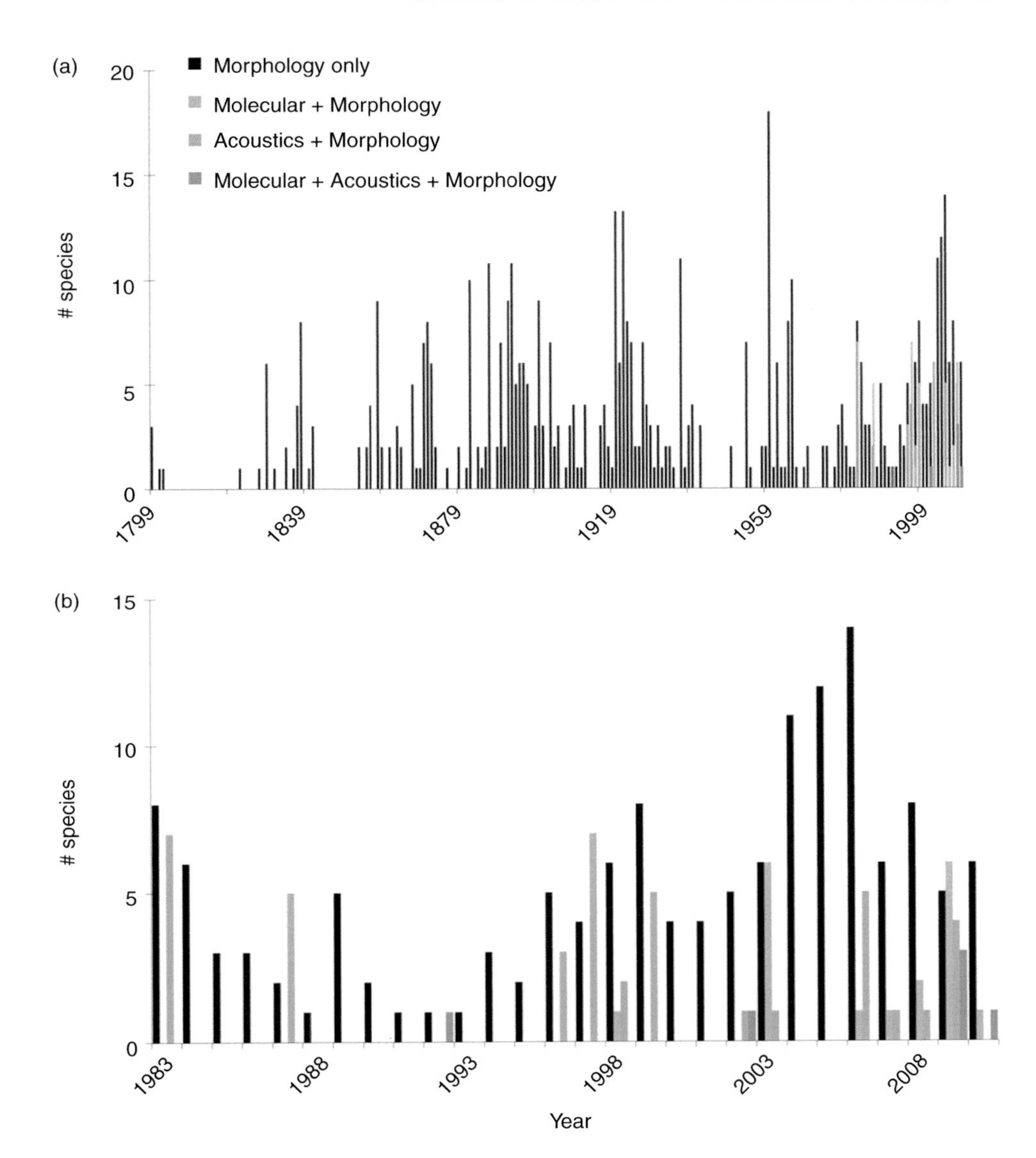

Figure 14.2 (a) Sources of data for recognising Southeast Asian amphibian species over the past two centuries. Note concentration of diversity of data types over the last three decades. (b) Expansion of Fig 15.2a to include descriptions using more than one data type (beginning in 1983 with the publication of Dring's anuran descriptions based on morphology and advertisement calls). See plate section for colour version.

The analysis of species descriptions through time categorised by numbers of biogeographic regions inhabited by each species identifies several trends (Fig 14.4a). First, most widespread species (those inhabiting 3–5 zoogeographic regions) were discovered and described in the early history of work in the region, principally in

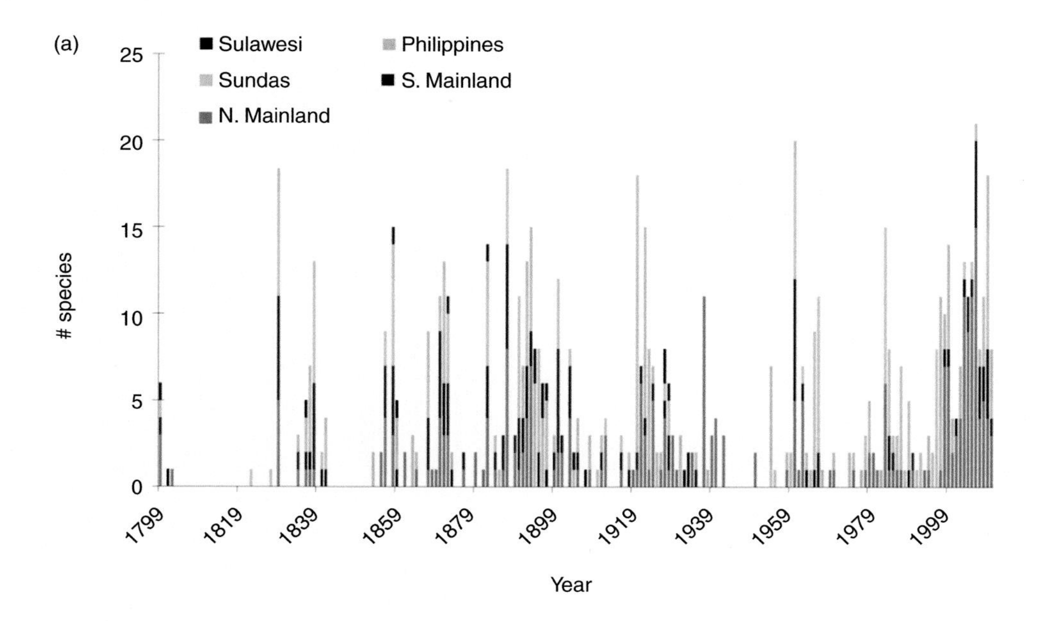

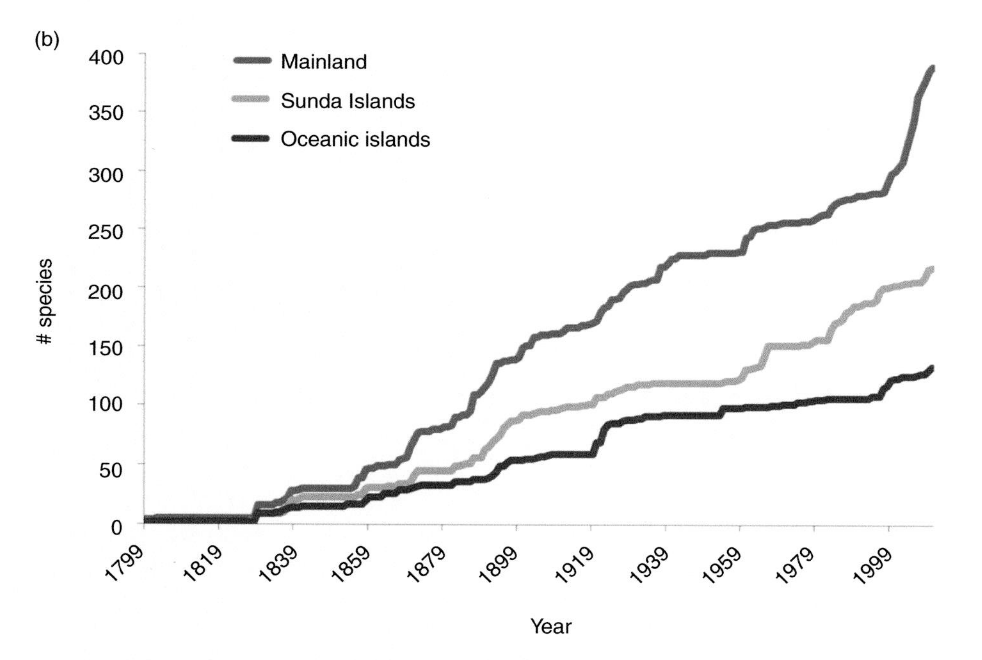

Figure 14.3 (a) Frequency distribution of Southeast Asian amphibian species descriptions over the past two centuries according to geographical region. Note the near absence of activity in Sulawesi after the 1930s and intense activity over the past decade in the northern mainland. (b) Species accumulation curves of Southeast Asian amphibian descriptions in each of three land mass types. See plate section for colour version.

the 1800s. Only a few species described in the 1900s are species found over three geographic regions (Fig 14.4a) but for most of the past century, the vast majority of species described have been those with limited geographic distributions found within one or, at most, two geographic regions (Fig 14.4a). Almost all two-region species described over the past century have been species that span the Isthmus of Kra in southern Thailand.

Finally, the analysis of family accumulation curves (Fig 14.4b) reveals several important trends, the first of which is a fundamental dichotomy in patterns of species discoveries over the past two centuries. In four families (Ichthyophiidae, Hylidae, Salamandridae and Bombinatoridae), initial species accumulation has been followed by a levelling-off in rates of discovery and a subsequent period of taxonomic stasis (Fig 14.4b). In contrast, in all other families, the last century has been witness to a steady increase in species numbers, with particularly rapid rates of species accumulation in the last two decades (Fig 14.4b); this last burst of taxonomic activity has been particularly intense in the families Ranidae, Dicroglossidae and Rhacophoridae.

14.4 Discussion

Despite over two centuries of taxonomic study, new amphibian species continue to be discovered in Southeast Asia at an increasingly rapid rate (Figs 14.1, 14.3b, 14.4b). This discovery is both geographically and taxonomically widespread, with a few notable exceptions. Species continue to be rapidly discovered in all of our pre-defined biogeographic regions except Sulawesi, where only one species has been described since the frogs *Limnonectes heinrichi* Ahl, 1933, and *Oreophryne zimmeri* Ahl, 1933. The lack of recently discovered diversity in Sulawesi is clearly sociological rather than biological (Inger 1999), because this island has a complex geological history (formed by accretions of land masses of different origins) with numerous endemics in other taxonomic groups and intra-island barriers to gene flow (Evans et al. 2003a, b, 2008, McGuire et al. 2007). New discoveries in amphibian diversity are therefore expected on Sulawesi; some additional new species currently await description (Iskandar and Tjan 1996, Evans et al. 2003a, 2008, D.T. Iskandar and J.A. McGuire pers. comm. 2010).

At a coarser scale, amphibians continue to be discovered at a rapid rate on the mainland, Sundaland, and oceanic islands, with an especially dramatic increase over the past decade on the mainland (notably the northern mainland; Fig 14.3b). The Southeast Asian archipelagos have played an extremely important role in the field of biogeography (i.e. the works of Wallace and Huxley), because these islands provide natural opportunities for studies of vicariance and dispersal given the apparent geographical boundaries for amphibians (land-positive areas with

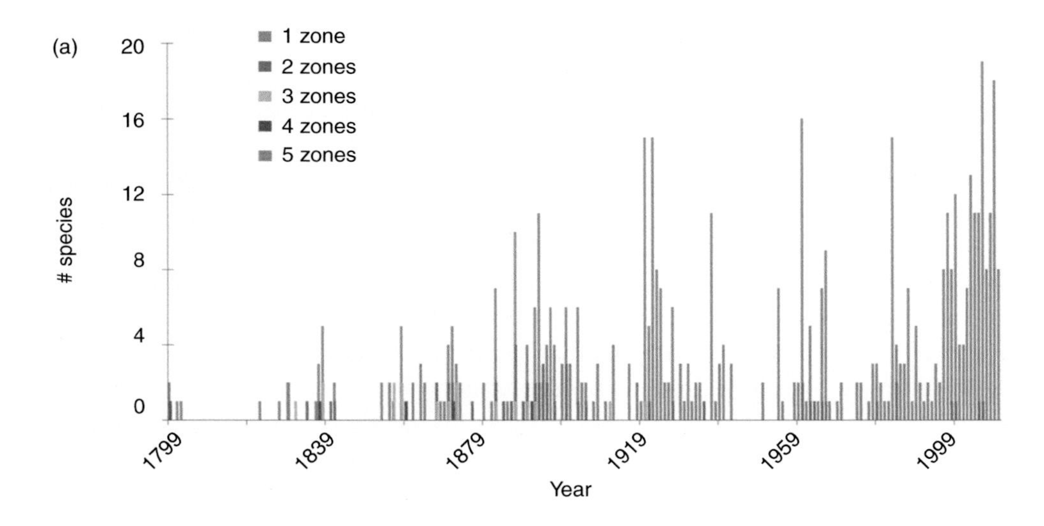

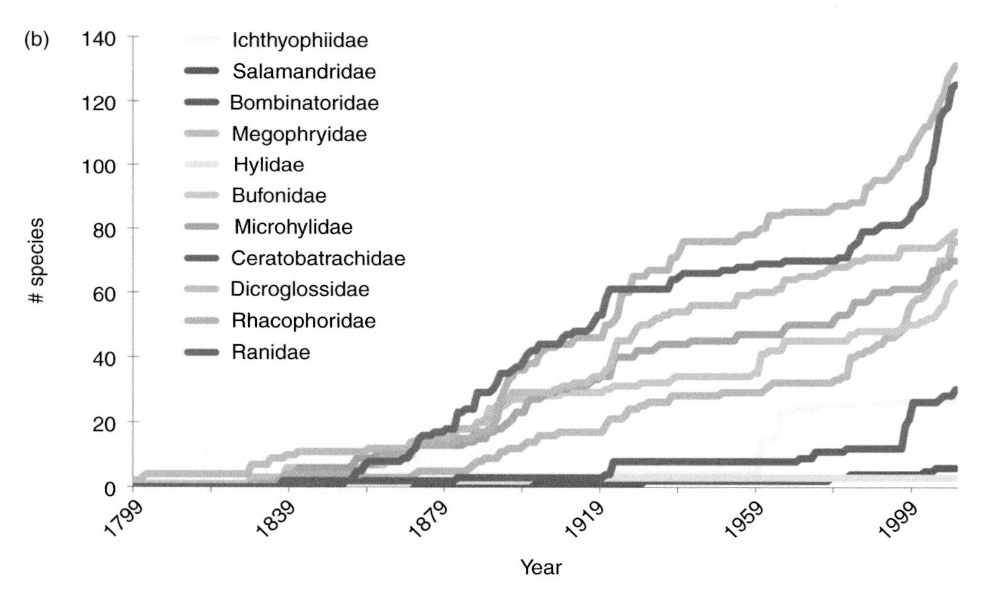

Figure 14.4 (a) Southeast Asian amphibian species descriptions through time categorised by the number of biogeographic zones inhabited (zones defined in main text). Note that widespread species (2–5 zones) were all discovered and described in the 1800s and in most cases, species described in the last century are found in one, or at most two, biogeographic regions. (b) Southeast Asian amphibian species accumulation curves by family. See plate section for colour version.

saltwater margins; Brown and Diesmos 2002, 2009). As a result of this known history of habitat fragmentation, and if sampling efforts were equal across biogeographic units, we might have expected the highest rates of historical species discovery per unit area to be centred on the oceanic portions of the island archipelagos. Similarly, given an equitable sampling effort, we might have expected intermediate levels of species discovery on the land-bridge islands of the Sunda Region that have been periodically connected to the mainland and then isolated during glacial cycles. Finally, given the apparent lack of absolute biogeographic boundaries across which amphibians would be unlikely to disperse, we might have predicted lower levels of species discovery on the Asian mainland. However, the mainland was identified as the most species-rich biogeographic landscape type in our analyses (Fig 14.3b), with rapid, recent discovery of species (Stuart and Bain 2008), particularly in Vietnam, Thailand and Laos.

Caution should be taken with regards to interpretation of trends detected here. First, the Southeast Asian mainland is the largest geographical unit delineated here, and so it may not be surprising that species accumulation appears greater on the mainland. Additionally, our general sense is that this region is now receiving intensive sampling efforts (Stuart and Bain 2008), certainly when compared to the islands of Indonesia where logistical obstacles to research restrict large-scale survey work (Brown 2009). The landscape across the Southeast Asian mainland is continuous without saltwater barriers, but it clearly contains island-like pockets of endemism with elevational and/or climatic barriers to dispersal, all of which may contribute to increased diversification (Stuart et al. 2006a). We also note that over the last decade many recent clarifications of species complexes on the mainland have involved international collaborations working on cross-border taxonomic problems (e.g., Bain et al. 2006, 2007, 2009a, 2009b, Stuart et al. 2006b, 2010, Chan et al. 2009, Chan and Grismer 2010). We expect that both greater numbers of amphibian systematists working today and the increasingly international nature of recent collaborations (Stuart and Bain 2008, Rodrigues et al. 2010) have resulted in elevated numbers of species descriptions published on mainland amphibians (Joppa et al. 2011). Despite the caveats that sampling is unequal and sociological and political factors still influence sampling effort, further analyses of the biogeographical relations of amphibians across the Southeast Asian mainland are warranted (Inger 1999, Inger and Voris 2001); we expect that future analyses will make for particularly interesting comparisons between the mainland, the adjacent continental shelf islands of the Sunda Region, and oceanic islands of the Philippines and eastern Indonesia.

Species continue to be discovered at a notable rate in 7 of the 11 families of Southeast Asian amphibians, with the exceptions being Ichthyophiidae, Salamandridae, Hylidae and Bombinatoridae, which are generally experiencing a period of taxonomic stasis (and/or a balancing of description versus synonymy) in Southeast Asia. Ichthyophiid caecilian species boundaries are difficult to detect, and while new species continue to be discovered in species-rich parts of tropical South Asia, where the family originated (Gower et al. 2002), there has been minimal taxonomic activity in Southeast Asia since the work by E.H. Taylor. The fact that a single worker (Taylor) described such a large portion of the known diversity may also contribute to the apparent stasis in this group. We anticipate that alternative perspectives from other researchers applying differing types of data to the problem of caecilian diversity will result in the recognition of additional taxa. At present several new Southeast Asian species await description (D.J. Gower pers. comm. 2010). Salamandridae is a temperate clade with a few representatives that marginally enter Southeast Asia at higher elevations in the northern mainland. In contrast, Hylidae originated in tropical South America and only recently colonised Asia (Wiens et al. 2006), also with only a few species in the region found at higher elevations in the northern mainland. Bombinatoridae is a species-poor clade with only eight extant species (Frost 2010), an ancient evolutionary origin, and a controversial biogeographic history (Blackburn et al. 2010).

Despite the overall scientific accumulation of Southeast Asian amphibian species, taxonomic activity over the past 210 years has not been gradual. Rather, activity has been concentrated into several bursts followed by years of relative stasis, before activity soared again. These multimodal curves in the descriptions by years frequency distribution help identify the careers of various herpetologists (Figs 14.1, 14.3b, 14.4b). For example, a surge in species descriptions near the end of the 1800s and early 1900s is indicative of the careers of European age-of-discovery taxonomists, in particular G.A. Boulenger (Brown and Diesmos 2002, Brown et al. 2002, 2008). A surge of activity in the 1920s and 1930s appears to be related to the early careers of E.H. Taylor and R. Bourret, while a burst of species descriptions in the 1950s and 1960s is related to the late career descriptions of E.H. Taylor and the early career work of R.F. Inger. The effects of notable herpetologists can also be seen within the taxonomic history of particular biogeographic regions. For example, the Philippines is characterised by a burst of descriptions in the early 1900s (Taylor), another in the mid 1900s (Inger), a less concentrated spread of descriptions across the 1970s and 1980s (W.C. Brown and A.C. Alcala) and then a burst in the last two decades (Brown 2007, Brown et al. 2002, 2008, Brown and Diesmos 2009).

The continued and increasingly rapid discovery of species (as indicated by the sharp terminal increase in species accumulation curves; Figs 14.1, 14.3b, 14.4b) suggests that amphibian diversity in Southeast Asia is underestimated. Prior to Dring (1983, 1987), all new Southeast Asian amphibians were distinguished solely on the basis of morphological differences, and new species continue to be recognised with this single type of data. New fieldwork in previously unexplored parts of the region is yielding unique forms to science; yet the application of new types of data (acoustic characters, molecular data) also appears to be related to the recent increase in species diversity (Giam et al. 2010, Joppa et al. 2011).

Morphological stasis (speciation not accompanied by morphological change) is widespread within amphibians, probably because most amphibians rely on acoustic or chemical signals rather than visual signals for mate recognition (Ryan 2001, Bickford et al. 2006, Gerhardt and Huber 2007, Narins et al. 2007, Wells 2007). The increased use of molecular and acoustic tools for recognising amphibian species in the region since 1983 has made it apparent that many single 'species' are actually complexes of cryptic species (two or more species erroneously classified as a single species, Bickford et al. 2006, Brown et al. 2009, Inger et al. 2009). In many cases, surprising new suites of cryptic species have been discovered in sympatry (Evans et al. 2003a, Stuart et al. 2006a, Inger et al. 2009), providing yet another clue to exceptionally high levels of diversity and species accumulation on the mainland (Stuart et al. 2006a, McLeod 2010).

Although geographically widespread species (those occurring in two or more biogeographic zones in our analyses) were mostly identified early in taxonomic history (as expected), many of these may eventually be found to be cryptic species complexes containing hidden diversity as non-traditional data continue to be evaluated (Stuart et al. 2006a, McLeod 2010, Chan and Grismer 2010). The authors of phylogeographic studies of Southeast Asian amphibians (i.e. those assessing genetic variation across the ranges of species) invariably conclude the presence of cryptic species in their analyses (Stuart et al. 2006a), suggesting that many unrecognised taxa still masquerade in out-of-date taxonomy. In general, we suspect that there may be very few naturally widespread amphibian species in Southeast Asia. Current exceptions most likely include large complexes of species in need of taxonomic review (Stuart et al. 2006a, McLeod 2010) or invasive lineages of single species introduced as transplantations due to the activities of humans (Brown et al. 2010).

As an increasingly urgent conservation crisis unfolds in Southeast Asian amphibian biodiversity (Sodhi et al. 2008, Brown and Diesmos 2009, Rowley et al. 2010), the need for accurate taxonomy and diversity estimates will become increasingly acute. Uninformed conservation priority setting will only be as effective as the underlying knowledge of existing biodiversity and will be limited by underestimation of biodiversity. The results of this study suggest that some regions have experienced a greater rate of species discovery and biodiversity description (northern mainland Southeast Asia), while others have undergone many decades of neglect on the part of taxonomists (Sulawesi). Our review has also demonstrated the importance of new and varied types of data for species delimitation, most notably the application of molecular and bioacoustic tools for ascertaining species boundaries. Our work demonstrates that some families currently exhibit levels of species diversity that represent asymptotic declines in species discoveries (Ichthyophiidae, Bombinatoridae) while others show high probabilities of increased species discoveries (Rhacophoridae, Ranidae, Dicroglossidae). Finally, this work provides lessons for evolutionary biologists and biogeographers applying centuries of expectations to the process of species discovery and analyses of species diversity across partitioned geographical landscapes. Although islands are textbook examples of evolutionary history in isolation and resulting highly endemic faunas, the highest rates of species discovery detected in our study were found on the Southeast Asian mainland where numerous regional faunas come together. The results of this study, therefore, suggest that we still have much to learn about Southeast Asia's geographical patterns of amphibian diversity – and that we have yet to fully comprehend the evolutionary processes that have given rise to these elusive patterns.

14.5 Conclusions

In the last several centuries Southeast Asian amphibian discoveries have not accumulated at a constant, equitable rate. Instead, they have occurred in periods of rapid species accumulation (followed by periods of relative stasis), concentrated in the time periods corresponding to the careers of prominent herpetologists, according to their taxonomic interests, and in the geographic regions in which these herpetological luminaries worked. Recent descriptions have moved away from the first 170 years' reliance on solely morphological data, and the last three decades have seen increasingly diverse types of data brought to bear on recognising species. Most lineages (particularly ranid and rhacophorid frogs) continue to be discovered and described with no evidence of a decline in rates of discovery in recent years. Most geographic centres of endemism have exhibited variable numbers of species discoveries over the past two centuries, but the two extremes in these periodic fluctuations have been Sulawesi, with no formal species descriptions since the 1930s and the northern mainland, where rates of species descriptions have been exceedingly high and shown no signs of abating. Finally, with a few exceptions, most geographically widespread species were discovered and described by the end of the 1800s. Although a small number of descriptions over the past century have identified species distributed over two biogeographic zones, most of the last 100 years of new species have been identified as taxa that are limited to a small geographic area in a single biogeographic zone.

Acknowledgements

We appreciate the invitation to contribute this chapter and thank our numerous friends and colleagues who have contributed to the field surveys, faunistic studies, taxonomic work and synthetic analyses that laid the foundation for this summary. We thank J. Weghorst for help with the figures and J. Rowley, B. Evans, R. Bain and D. Gower for constructively critical reviews of earlier versions of the manuscript. During the preparation of this manuscript we were supported by grants from the MacArthur Foundation (BJS) and the U.S. National Science Foundation (RMB).

References

Allison, A. (1996). Zoogeography of amphibians and reptiles of New Guinea and the Pacific region. In *The Origin and Evolution of Pacific Island Biotas, New Guinea to Eastern Polynesia: Patterns and Processes*, ed. A. Keast and S. E. Miller. The Netherlands: SPB Academic Publishing, pp. 407–36.

AmphibiaWeb. (2010). *AmphibiaWeb: Information on Amphibian Biology and Conservation*. Berkeley, California: Available at: http://amphibiaweb.org/.

Bain, R., Stuart, B. and Orlov, N. (2006). Three new Indochinese species of cascade frogs (Amphibia: Ranidae) allied to *Rana archotaphus*. *Copeia*, **2006**, 43–59.

Bain, R., Nguyen, T. Q. and Doan, K. (2007). New herpetofaunal records from Vietnam. *Herpetological Review*, **38**, 107–17.

Bain, R., Biju, S. D. Brown, R. M. et al. (2008). Amphibians of the Indomalayan Realm. In *Threatened Amphibians of the World*, ed. S. N. Stuart, M. Hoffmann, J. S. Chanson et al. Barcelona, Spain: Lynx Ediciones; Gland, Switzerland: The World Conservation Union; and Arlington, VA: Conservation International, pp. 74–9.

Bain, R., Nguyen, T. Q., and Doan, K. (2009a). A new species of the genus *Theloderma* Tschudi, 1838 (Anura: Rhacophoridae) from Northwestern Vietnam. *Zootaxa*, **2191**, 58–68.

Bain, R., Stuart, B. L., Nguyen, T. Q., Che, J. and Ra, D.-Q. (2009b). A New *Odorrana* (Amphibia: Ranidae) from Vietnam and China. *Copeia*, **2009**, 348–62.

Bawa, K.S., Primack, R. and Woodruff, D. (1990). Conservation of biodiversity: a southeast Asian perspective. *Trends in Ecology and Evolution*, **5**, 394–6.

Bickford, D., Lohman, D. J., Sodhi, N. S. et al. (2006). Cryptic species as a window on diversity and conservation. *Trends in Ecology and Evolution*, **22**, 148–55.

Blackburn, D. C., Bickford, D. P., Diesmos, A. C., Iskandar, D. T. and Brown, R. M. (2010). An ancient origin for the enigmatic flat-headed Frogs (Bombinatoridae: *Barbourula*) from the islands of Southeast Asia. *PLoS ONE* doi:10.1371/journal.pone.0012090.t002.

Boulenger G. A. (1920). A monograph of the South Asian, Papuan, Melanesian and Australian frogs of the genus *Rana*. *Records of the Indian Museum*, **20**, 1–126.

Bourret, R. (1942). Notes herpétologiques sur l'Indochine francaise. XIV. Les batraciens de la collection du Laboratoire des Sciences Naturelles de l'Université. Descriptions de quinze espéces ou variétés nouvelles. *Général de l'Instruction Publique Annexe*, **4**, 5–56.

Brooks, T. M., Mittermeier, R. A., Mittermeier, C. G. et al. (2002). Habitat loss and extinction in the hotspots of biodiversity. *Conservation Biology*, **16**, 909–23.

Brown, R. M. (2007). Introduction to Robert F. Inger's Systematics and Zoogeography of Philippine Amphibia. Foreword to the reprint of Inger's 1954 monograph, *Systematics and Zoogeography of Philippine Amphibia*. Kota Kinabalu: Natural History Publications, pp. 1–17.

Brown, R. M. (2009). Frogs. In *Encyclopedia of Islands*. ed. R. Gillespie and D. Clague. Berkeley, CA: University of California Press, pp. 347–51.

Brown, R. M. and Diesmos, A. C. (2002). Application of lineage-based species concepts to oceanic island frog populations: the effects of differing taxonomic philosophies on the estimation of Philippine biodiversity. *The Silliman Journal*, **42**, 133–62.

Brown, R. M. and Diesmos, A. C. (2009). Philippines, Biology. In *Encyclopedia of Islands*, ed. R. Gillespie, and D. Clague. Berkeley, CA: University of California Press, pp. 723–32.

Brown, R. M. and Guttman S. I. (2002). Phylogenetic systematics of the *Rana signata* complex of Philippine and Bornean stream frogs: reconsideration of Huxley's modification of Wallace's Line at the Oriental-Australian faunal zone interface. *Biological Journal of the Linnaean Society*, **76**, 393–461.

Brown, R. M., McGuire, J. A. and Diesmos A. C. (2000). Status of some Philippine frogs related to *Rana everetti* (Anura: Ranidae), description of a new species, and resurrection of *Rana igorota* Taylor 1922. *Herpetologica*, **56**, 81–104.

Brown, R. M., Diesmos, A. C. and Alcala A. C. (2002). The state of Philippine herpetology and the challenges for the next decade. *The Silliman Journal*, **42**, 18–87.

Brown, R. M. Diesmos, A. C. and Alcala, A. C. (2008). Philippine amphibian biodiversity is increasing in leaps and bounds. In *Threatened Amphibians of the World*, ed. S. N. Stuart, M. Hoffmann, J. S. Chanson et al. Barcelona, Spain: Lynx Ediciones; Gland, Switzerland: The World Conservation Union; Arlington Virginia: Conservation International, pp. 82–83.

Brown, R. M. Siler, C. D., Diesmos, A. C. and Alcala, A. C. (2009). Philippine frogs of the genus *Leptobrachium* (Anura, Megophryidae): taxonomic revision, phylogeny-based species delimitation, and descriptions of three new species. *Herpetological Monographs*, **23**, 1–44.

Brown, R. M., Linkem, C. W., Siler, C. D. et al. (2010). Phylogeography and historical demography of *Polypedates leucomystax* in the islands of Indonesia and the Philippines: evidence for recent human-mediated range expansion? *Molecular Phylogenetics and Evolution*, **57**, 598–619.

Brown, W. C. (1997). Biogeography of amphibians in the islands of the Southwest Pacific. *Proceedings of the California Academy of Sciences*, **50**, 21–38.

Collins, N. M., Sater, J. A. and Whitmore, T. C. (1991). *The Conservation Atlas of Tropical Forests: Asia and the Pacific*. New York: Simon and Schuster.

Chan, K. O. and Grismer, L. L. (2010). Re-assessment of the Reinwardt's Gliding Frog, *Rhacophorus reinwardtii* (Schlegel 1840) (Anura: Rhacophoridae) in Southern Thailand and Peninsular Malaysia and its re-description as a new species. *Zootaxa*, **2505**, 40–50.

Chan, K. O., Grismer, L. L., Ahmed, N. and Belabut, D. (2009). A new species of *Gastrophrynoides* (Anura: Microhylidae): an addition to a previously monotypic genus and a new genus for Peninsular Malaysia. *Zootaxa*, **2124**, 63–8.

Dring, J. (1983). Some new frogs from Sarawak. *Amphibia-Reptilia*, **4**, 103–15.

Dring J. (1987). Bornean treefrogs of the genus *Philautus* (Rhacophoridae). *Amphibia-Reptilia*, **8**, 19–47.

Duellman, W. E. and Trueb, L. (1994). *Biology of Amphibians*. New York: McGraw-Hill.

Esselstyn, J. A., Widmann, P. and Heaney, L. R. (2004). The mammals of Palawan Island, Philippines. *Proceedings of the Biological Society of Washington*, **117**, 271–302.

Esselstyn, J. A. Oliveros, C. H. Moyle, R. G. et al. (2010). Integrating phylogenetic and taxonomic evidence illuminates complex biogegraphic patterns along

Huxley's modification of Wallace's Line. *Journal of Biogeography*, **37**, 2054–66.

Evans, B. J., Brown, R. M., McGuire, J. A. et al. (2003a). Phylogenetics of fanged frogs (Anura, Ranidae, *Limnonectes*): testing biogeographical hypotheses at the Asian–Australian faunal zone interface. *Systematic Biology*, **52**, 794–819.

Evans, B. J. Supriatna, J., Andayani, N. et al. (2003b). Monkeys and toads define areas of genetic endemism on the island of Sulawesi. *Evolution*, **57**, 1436–43.

Evans, B. J., McGuire, J. A., Brown, R. M., Andayani, N. and Supriatna. J. (2008). A coalescent framework for comparing alternative models of population structure with genetic data: evolution of Celebes toads. *Biology Letters*, **4**, 430–3.

Frost, D. R. (2010). Amphibian Species of the World: an Online Reference. Version 5.4 (8 April, 2010). Electronic database available at http://research.amnh.org/vz/herpetology/amphibia/American Museum of Natural History, New York, USA.

Giam, X., Ng, T. H., Yap, V. B. and Tan, H. T. W. (2010). The extent of undiscovered species in Southeast Asia. *Biodiversity and Conservation*, **19**, 943–54.

Gower, D. J., Kupfer, A., Oommen, O. V. et al. (2002). A molecular phylogeny of ichthyophiid caecilians (Amphibia: Gymnophiona: Ichthyophiidae): out of India or out of South East Asia? *Proceedings Royal Society London B*, **269**, 1563–9.

Gerhardt, H. C. and Huber, F. (2007). *Acoustic Communication in Insects and Anurans*. Chicago IL: The University of Chicago Press.

Hall, R. (1996). Reconstructing Cenozoic SE Asia. In *Tectonic Evolution of Southeast Asia*, ed. R. Hall and D. J. Blundel. Bath, UK: Geological Society Publishing House, pp.153–83.

Hall, R. (1998). The plate tectonics of Cenozoic SE Asia and the distribution of land and sea. In *Biogeography and Geological Evolution of SE Asia*, ed. R. Hall and J. D. Holloway. Leiden, The Netherlands: Backhuys Publishers, pp. 99–131.

Hall, R. (2002). Cenozoic geological and plate tectonic evolution of SE Asia and the SW Pacific: computer-based reconstructions and animations. *Journal of Asian Earth Sciences*, **20**, 353–434.

Hamilton, W. (1979). Tectonics of the Indonesian region. *United States Geological Survey Professional Papers*, **1078**, 1–345.

Huxley, T. H. (1868). On the classification and distribution of the Alectoromorphae and Heteromorphae. *Proceedings of the Zoological Society of London*, **1868**, 296–319.

Inger, R. F. (1954). Systematics and zoogeography of Philippine Amphibia. *Fieldiana*, **33**, 181–531.

Inger, R. F. (1966). The systematics and zoogeography of the Amphibia of Borneo. *Fieldiana*, **52**, 1–402.

Inger, R. F. (1999). Distribution of amphibians in southern Asia and adjacent islands. In *Patterns of Distribution of Amphibians: A Global Perspective*, ed. W. E. Duellman Baltimore, MD: Johns Hopkins Press, pp. 445–82.

Inger, R. F. and Voris, H. K. (2001). The biogeographical relations of the frogs and snakes of Sundaland. *Journal of Biogeography*, **28**, 863–91.

Inger, R. F., Stuart, B. L. and Iskandar, D. T. (2009). Systematics of a widespread Southeast Asian frog, *Rana chalconota* (Amphibia: Anura: Ranidae). *Zoological Journal of the Linnean Society*, **155**, 123–47.

Iskandar, D. T. and Tjan, K. N. (1996). The amphibians and reptiles of Sulawesi, with notes of the distribution and chromosomal number of frogs. In *Proceedings of the First International Conference on Eastern Indonesian-Australian Vertebrate Fauna*, ed. D. J. Kitchner and A. Suyanto. Western Australia: Western Australia Museum, pp. 39–46.

Joppa, L. N., Roberts, D. L. and Pimm, S. L. (2011). The population ecology and social behaviour of taxonomists. *Trends in Ecology and Evolution*, **26**, 551–3.

Kennedy, R. S. Gonzales, P. C., Dickinson, E. C., Miranda, H. C. and Fisher, T. H. (2000). *A Guide to the Birds of the Philippines*. Oxford: Oxford University Press.

Kloss, B. (1929). The zoo-geographical boundaries between Asia and Australia and some Oriental sub-regions. *Bulletin of the Raffles Museum*, **2**, 1–10.

Lomolino, M. V., Riddle, B. R. and Brown, J. H. (2006). *Biogeography*, 3rd edn. Sunderland, MA: Sinauer Associates.

McGuire, J. A. Brown, R. M., Mumpuni, Riyanto, A. and Andayani, N. (2007). The flying lizards of the *Draco lineatus* group (Squamata: Iguania: Agamidae): a taxonomic revision with descriptions of two new species. *Herpetological Monographs*, **21**, 180–213.

McLeod, D. S. (2010). Of least concern? Systematics of a cryptic species complex: *Limnonectes kuhlii* (Amphibia: Anura: Dicroglossidae). *Molecular Phylogenetics and Evolution*, **56**, 991–1000.

Mittermeier, R. A., Robles P. Gil, P. R. and Mittermeier, C. G. (1997). *Megadiversity: Earth's Biologically Wealthiest Nations*. Monterrey, Mexico: CEMEX.

Mittermeier, R. A., Myers, N., Thomsen, J. B., da Fonseca G. A. B. and Olivieri, S. (1998). Biodiversity hotspots and major tropical wilderness areas: approaches to setting conservation priorities. *Conservation Biology*, **12**. 516–20.

Mittermeier, R. A., Myers, N., Gil, P. R. and Mittermeier, C. G. (1999). *Hotspots: Earth's Biologically Richest and Most Endangered Terrestrial Ecosystems*. Monterrey, Mexico: CEMEX.

Narins, P. M., Feng, A. S., Fay, R. R. and Popper A. N. (2007). *Hearing and Sound Communication in Amphibians*. New York: Springer.

Oliveros, C. H. and Moyle, R. G. (2010). Origin and diversification of Philippine bulbuls. *Molecular Phylogenetics and Evolution*, **54**, 822–32.

Reid, W. V. (1998). Biodiversity hotspots. *Trends in Ecology and Evolution*, **13**, 275–80.

Riyanto, A., Mumpuni and McGuire, J. A. (2011). Morphometry of striped tree frogs, *Polypedates leucomystax* (Gravenhorst, 1829) from Indonesia with description of a new species. *Russian Journal of Herpetology*, **18**, 29–35.

Rodrigues, A. S. L., Gray, C. L., Crowter, B. J. et al. (2010). A global assessment of amphibian taxonomic effort and expertise. *BioScience*, **60**, 798–806.

Rowley, J., Brown, R., Bain, R. et al. (2010). Impending conservation crisis for Southeast Asian amphibians. *Biology Letters*, **6**, 336–8.

Ryan, M. J. (2001). *Anuran Communication*. Washington DC: Smithsonian Institution Press.

Siler, C. D., Oaks, J. R., Linkem, C. W. et al. (2012). Did geckos ride the Palawan raft to the Philippines? *Journal of Biogeography*, doi: 10.1111/j.1365-2699.2011.02680.X.

Smith M. A. (1930). The Reptilia and Amphibia of the Malay Peninsula. *Bulletin of the Raffles Museum*, **3**, 1–149.

Smith M. A. (1935). *The Fauna of British India, Including Ceylon and Burma. Reptilia and Amphibia, II: Sauria*. London: Taylor and Francis.

Sodhi, N. S., Bickford, D., Diesmos, A. C. et al. (2008). Measuring the meltdown: drivers of global amphibian extinction and decline. *PLoS ONE*, **3**, doi:10.1371/journal.pone.0001636

Stuart, B. L. and Bain, R. H. (2008). Amphibian species discovery in mainland Southeast Asia. In *Threatened Amphibians of the World*, ed. S. N. Stuart, M. Hoffmann, J. S. Chanson et al. Barcelona, Spain: Lynx Edicions; Gland, Switzerland: The World Conservation Union; Arlington, Virginia: Conservation International, pp. 83–4.

Stuart, B. L., Inger, R. F., and Voris, H. K. (2006a). High level of cryptic species diversity revealed by sympatric lineages of Southeast Asian forest frogs. *Biology Letters*, **2**, 470–4.

Stuart, B. L., Sok, K., and Neang, T. (2006b). A collection of amphibians and reptiles from hilly eastern Cambodia. *Raffles Bulletin of Zoology*, **54**, 129–55.

Stuart, B. L., Bain, R. H., Phimmachak, S. and Spence, K. (2010). Phylogenetic systematics of the *Amolops monticolaa* group (Amphibia: Ranidae) with description of a new species from northwestern Laos. *Herpetologica*, **66**, 52–66.

Taylor, E. H. (1920). Philippine Amphibia. *Philippine Journal of Science*, **16**, 213–359.

Taylor, E. H. (1965). New Asiatic and African caecilians with redescriptions of certain other species. *University of Kansas Science Bulletin*, **45**, 253–302.

Taylor, E. H. (1968). *The Caecilians of the World, A Taxonomic Review*. Lawrence, KS: University of Kansas Press.

Van Kampen, P. N. (1923). *The Amphibia of the Indo-Australian Archipelago*. Leiden, The Netherlands: E. J. Brill Ltd.

Voris, H. K. (2000). Maps of Pleistocene sea levels in Southeast Asia: shorelines, river systems and time durations. *Journal of Biogeography*, **27**, 1153–67.

Wallace, A. R. (1860). On the zoological geography of the Malay Archipelago. *Proceedings of the Linnean Society, Zoology*, **4**, 172–84.

Wallace, A. R. (1876). *The Geographical Distribution of Animals, with a Study of the Relations of Living and Extinct Faunas as Elucidating the Past Changes of the Earth's Surface*. New York: Harper and Brothers.

Wells, K. D. (2007). *The Ecology and Behavior of Amphibians*. Chicago IL: The University of Chicago Press.

Whitmore, T. C. (1984). *Tropical Rainforests of the Far East*. Oxford: Clarendon Press.

Whitmore, T. C. (1987). *Biogeographical Evolution of the Malay Archipelago*. Oxford: Clarendon Press.

Whitmore, T. C. (1990). *An Introduction to Tropical Rainforests*. Oxford: Clarendon Press.

Whitten, T. G., Henderson, S. and Mustafa, M. (2002). *The Ecology of Sulawesi*. Hong Kong: Periplus Editions Ltd.

Wiens, J. J., Graham, C. H., Moen, D. S., Smith, S. A. and Reeder, T. W. (2006). Evolutionary and ecological causes of the latitudinal diversity gradient in hylid frogs: treefrog trees unearth the roots of high tropical diversity. *The American Naturalist*, **168**, 579–96.

Woodruff, D. S. (2003). Neogene marine transgressions, palaeogeography and biogeographic transitions on the Thai–Malay Peninsula. *Journal of Biogeography*, **30**, 551–67.

Woodruff, D. S. (2010). Biogeography and conservation in Southeast Asia: how 2.7 million years of repeated environmental fluctuations affect today's patterns and the future of the remaining refugial-phase biodiversity. *Biodiversity Conservation*, **19**, 919–41.

Appendix C

The taxonomy, year of description, type of data used in the original description and distribution (as biogeographic region, defined in the main text) for the 622 species of Southeast Asian amphibians considered in this analysis. Abbreviations: A, acoustics; Ms, molecules; morphology, My; NM, Northern Mainland; P, Philippines; Sd, Sundas; SM, Southern Mainland; Sw, Sulawesi.

Taxon	Year	Data Type	Region
Ichthyophiidae			
Caudacaecilia asplenia	1965	My	SM, Sd
Caudacaecilia larutensis	1960	My	SM
Caudacaecilia nigroflava	1960	My	SM
Caudacaecilia paucidentula	1965	My	Sd
Caudicaecilia weberi	1920	My	P
Ichthyophis acuminatus	1960	My	NM
Ichthyophis atricollaris	1965	My	Sd
Ichthyophis bannanicus	1984	My	NM
Ichthyophis bernisi	1975	My	Sd
Ichthyophis biangularis	1965	My	Sd
Ichthyophis billitonensis	1965	My	Sd
Ichthyophis dulitensis	1965	My	Sd
Ichthyophis elongatus	1965	My	Sd
Ichthyophis glandulosus	1923	My	P
Ichthyophis humphreyi	1965	My	Sd
Ichthyophis hypocyaneus	1827	My	Sd
Ichthyophis javanicus	1960	My	Sd
Ichthyophis kohtaoensis	1960	My	NM, SM
Ichthyophis laosensis	1969	My	NM
Ichthyophis mindanaoensis	1960	My	P
Ichthyophis monochrous	1858	My	Sd
Ichthyophis paucisulcus	1960	My	SM

Taxon	Year	Data Type	Region
Ichthyophis singaporensis	1960	My	SM
Ichthyophis sumatranus	1960	My	Sd
Ichthyophis supachaii	1960	My	SM
Ichthyophis youngorum	1960	My	NM
Salamandridae			
Laotriton laoensis	2002	My	NM
Paramesotriton deloustali	1934	My	NM
Paramesotriton guangxiensis	1983	My	NM
Tylototriton asperrimus	1930	My	NM
Tylototriton verrucosus	1871	My	NM
Tylototriton vietnamensis	2005	My	NM
Bombinatoridae			
Barbourula busuangensis	1924	My	P
Barbourula kalimantanensis	1978	My	Sd
Bombina maxima	1905	My	NM
Megophryidae			
Borneophrys edwardinae	1989	My	Sd
Brachytarsophrys carinense	1889	My	NM, SM
Brachytarsophrys feae	1887	My	NM
Brachytarsophrys intermedia	1921	My	NM
Leptobrachella baluensis	1931	My	Sd
Leptobrachella brevicrus	1983	My, A	Sd
Leptobrachella mjobergi	1925	My	Sd
Leptobrachella natunae	1895	My	Sd
Leptobrachella palmata	1992	My	Sd
Leptobrachella parva	1983	My, A	Sd
Leptobrachella serasanae	1983	My, A	Sd
Leptobrachium abbotti	1926	My	Sd

Taxon	Year	Data Type	Region
Leptobrachium ailaonicum	1983	My	NM
Leptobrachium banae	1998	My	NM
Leptobrachium buchardi	2004	My	NM
Leptobrachium chapaense	1937	My	NM
Leptobrachium gunungense	1996	My, A	Sd
Leptobrachium hasseltii	1838	My	Sd
Leptobrachium hendricksoni	1962	My	SM, Sd
Leptobrachium lumadorum	2009	My, Ms, A	P
Leptobrachium mangyanorum	2009	My, Ms, A	P
Leptobrachium montanum	1885	My	Sd
Leptobrachium mouhoti	2006	My	NM
Leptobrachium ngoclinhense	2005	My	NM
Leptobrachium nigrops	1963	My	SM, Sd
Leptobrachium pullum	1921	My	NM
Leptobrachium smithi	1999	My	NM, SM
Leptobrachium tagbanorum	2009	My, Ms, A	P
Leptobrachium xanthospilum	1998	My	NM
Leptolalax applebyi	2009	My, A	NM
Leptolalax arayai	1997	My, A	Sd
Leptolalax bourreti	1983	My	NM
Leptolalax dringi	1987	My	Sd
Leptolalax fuliginosus	2006	My, A	NM
Leptolalax gracilis	1872	My	NM, SM, Sd
Leptolalax hamidi	1997	My, A	Sd
Leptolalax heteropus	1900	My	SM
Leptolalax kajangensis	2004	My	SM
Leptolalax kecil	2009	My, A	SM
Leptolalax lateralis	1871	My	NM

Taxon	Year	Data Type	Region
Leptolalax maurus	1997	My	Sd
Leptolalax melanoleucus	2006	My, A	NM, SM
Leptolalax melicus	2010	My, Ms, A	NM
Leptolalax nahangensis	1998	My, Ms	NM
Leptolalax oshanensis	1950	My	NM
Leptolalax pelodytoides	1893	My	NM, SM
Leptolalax pictus	1992	My, A	Sd
Leptolalax pluvialis	2000	My	NM
Leptolalax solus	2006	My, A	SM
Leptolalax sungi	1998	My	NM
Leptolalax tuberosus	1999	My	NM
Megophrys kobayashii	2002	My	Sd
Megophrys ligayae	1920	My	P
Megophrys montana	1822	My	Sd
Megophrys nasuta	1858	My	SM, Sd
Megophrys stejnegeri	1920	My	P
Ophryophryne gerti	2003	My	NM
Ophryophryne hansi	1903	My	NM
Ophryophryne microstoma	1985	My	NM
Ophryophryne pachyproctus	2006	My	NM
Ophryophryne synoria	2003	My	NM
Scutiger adungensis	1979	My	NM
Xenophrys aceras	1903	My	SM
Xenophrys auralensis	2002	My	NM
Xenophrys baluensis	1899	My	Sd
Xenophrys brachykolos	1961	My	NM
Xenophrys dringi	1995	My	Sd
Xenophrys glandulosa	1990	My	NM

Taxon	Year	Data Type	Region
Xenophrys kuatunensis	1983	My	NM
Xenophrys lekaguli	1929	My	NM
Xenophrys longipes	2006	My	NM
Xenophrys major	1886	My	SM
Xenophrys minor	1908	My	NM
Xenophrys omeimontis	1926	My	NM
Xenophrys pachyproctus	1981	My	NM
Xenophrys palpebralespinosa	1937	My	NM
Xenophrys parallela	2005	My	Sd
Xenophrys parva	1893	My	NM
Hylidae			
Hyla annectans	1870	My	NM
Hyla chinensis	1858	My	NM
Hyla simplex	1901	My	NM
Bufonidae			
Ansonia albomaculata	1960	My	Sd
Ansonia echinata	2009	My	Sd
Ansonia endauensis	2006	My	SM
Ansonia fuliginea	1890	My	Sd
Ansonia guibei	1966	My	Sd
Ansonia hanitschi	1960	My	Sd
Ansonia inthanon	1998	My	NM
Ansonia jeetsukumarani	2008	My	SM
Ansonia kraensis	2005	My	NM, SM
Ansonia latidisca	1966	My	Sd
Ansonia latiffi	2008	My	SM
Ansonia latirostra	2006	My	SM
Ansonia leptopus	1872	My	Sd, SM

Taxon	Year	Data Type	Region
Ansonia longidigita	1960	My	Sd
Ansonia malayana	1960	My	NM, SM
Ansonia mcgregori	1922	My	P
Ansonia minuta	1960	My	Sd
Ansonia muelleri	1887	My	P
Ansonia penangensis	1870	My	SM
Ansonia platysoma	1960	My	Sd
Ansonia siamensis	1985	My	SM
Ansonia spinulifer	1890	My	Sd
Ansonia tiomanica	1966	My	SM
Ansonia torrentis	1983	My, A	Sd
Bufo cryptotympanicus	1962	My	NM
Bufo pageoti	1937	My	NM
Duttaphrynus crocus	2003	My, A	NM
Duttaphrynus melanostictus	1799	My	NM, SM, Sd, Sw
Duttaphrynus stuarti	1929	My	NM
Duttaphrynus sumatranus	1871	My	Sd
Duttaphrynus totol	2010	My	Sd
Duttaphrynus valhallae	1909	My	Sd
Ingerophrynus biporcatus	1829	My	Sw, Sd, SM
Ingerophrynus celebensis	1858	My	Sw
Ingerophrynus claviger	1871	My	Sd
Ingerophrynus divergens	1871	My	SM, Sd
Ingerophrynus galeatus	1864	My	NM
Ingerophrynus gollum	2007	My	SM
Ingerophrynus kumquat	2001	My	SM
Ingerophrynus macrotis	1887	My	NM, SM
Ingerophrynus parvus	1887	My	NM, SM, Sd

Taxon	Year	Data Type	Region
Ingerophrynus philippinicus	1887	My	P
Ingerophrynus quadriporcatus	1887	My	SM, Sd
Leptophryne borbonica	1838	My	SM, Sd
Leptophryne cruentata	1838	My	Sd
Pedostibes everetti	1896	My	Sd
Pedostibes hosii	1892	My	SM, Sd
Pedostibes rugosus	1958	My	Sd
Pelophryne albotaeniata	1938	My	P
Pelophryne api	1983	My, A	Sd
Pelophryne brevipes	1867	My	P, SM
Pelophryne exigua	1900	My	Sd
Pelophryne guentheri	1882	My	Sd
Pelophryne lighti	1920	My	P
Pelophryne linanitensis	2008	My	Sd
Pelophryne misera	1890	My	Sd
Pelophryne murudensis	2008	My	Sd
Pelophryne rhopophilia	1996	My	Sd
Pelophryne saravacensis	2009	My	Sd
Pelophryne signata	1895	My	SM, Sd
Phrynoidis aspera	1829	My	NM, SM, Sd
Phrynoidis juxtasper	1864	My	Sd
Pseudobufo subasper	1838	My	SM
Sabahphrynus maculatus	1890	My	Sd
Microhylidae			
Calluella brooksii	1904	My	Sd
Calluella flava	1984	My	Sd
Calluella guttulata	1856	My	NM, SM
Calluella minuta	2004	My	SM

Taxon	Year	Data Type	Region
Calluella smithi	1916	My	Sd
Calluella volzi	1905	My	SM, Sd
Calluella yunnanensis	1919	My	NM
Chaperina fusca	1892	My	P, Sd, SM
Gastrophrynoides borneensis	1897	My	Sd
Gastrophrynoides immaculatus	2009	My	SM
Glyphoglossus molossus	1869	My	NM
Kalophrynus baluensis	1984	My	Sd
Kalophrynus bunguranus	1895	My	Sd
Kalophrynus elok	2003	My	Sd
Kalophrynus heterochirus	1900	My	Sd
Kalophrynus interlineatus	1855	My	NM
Kalophrynus intermedius	1966	My	Sd
Kalophrynus menglienicus	1980	My	NM
Kalophrynus minusculus	1988	My	Sd
Kalophrynus nubicola	1983	My, A	Sd
Kalophrynus palmatissimus	1984	My	SM, Sd
Kalophrynus pleurostigma	1838	My	P, Sd, SM
Kalophrynus punctatus	1871	My	Sd
Kalophrynus robinsoni	1922	My	SM
Kalophrynus subterrestris	1966	My	Sd
Kalophrynus yongi	2009	My, A	SM
Kaloula aureata	1989	My	SM
Kaloula baleata	1836	My	NM, SM, P, Sd, Sw
Kaloula conjuncta conjuncta	1863	My	P
Kaloula conjuncta meridionalis	1954	My	P
Kaloula conjuncta negrosensis	1922	My	P

Taxon	Year	Data Type	Region
Kaloula conjuncta stickeli	1954	My	P
Kaloula kalingensis	1922	My	P
Kaloula kokacii	1922	My	P
Kaloula mediolineata	1917	My	NM
Kaloula picta	1841	My	P
Kaloula pulchra	1831	My	NM, SM, Sd
Kaloula rigida	1922	My	P
Kaloula walteri	2002	My, A	P
Metaphrynella pollicaris	1890	My	SM, Sd
Metaphrynella sundana	1867	My	Sd
Microhyla achatina	1838	My	Sd
Microhyla annamensis	1923	My	NM
Microhyla annectens	1900	My	SM
Microhyla berdmorei	1856	My	NM, SM, Sd
Microhyla borneensis	1928	My	Sd, SM
Microhyla butleri	1900	My	NM, SM
Microhyla erythropoda	1994	My	NM
Microhyla fissipes	1884	My	NM, SM
Microhyla fowleri	1934	My	NM
Microhyla fusca	1942	My	NM
Microhyla heymonsi	1911	My	NM, SM, Sd
Microhyla maculifera	1989	My	Sd
Microhyla mantheyi	2007	My	SM
Microhyla marmorata	2004	My	NM
Microhyla nanopollexa	2004	My	NM
Microhyla ornata	1841	My	Sd, SM
Microhyla palmipes	1897	My	SM, Sd
Microhyla perparva	1979	My	Sd

Taxon	Year	Data Type	Region
Microhyla petrigena	1979	My	Sd, P
Microhyla picta	1901	My	NM
Microhyla pulchra	1861	My	NM
Microhyla pulverata	2004	My	NM
Microhyla superciliaris	1928	My	SM, Sd
Micryletta inornata	1890	My	NM, SM, Sd
Oreophryne annulata	1908	My	P
Oreophryne celebensis	1894	My	Sw
Oreophryne nana	1967	My	P
Oreophryne variabilis	1896	My	Sw
Oreophryne zimmeri	1933	My	Sw
Phrynella pulchra	1887	My	SM, Sd
Ceratobatrachidae			
'*Ingerana*' *baluensis*	1912	My	NM
'*Ingerana*' *mariae*	1954	My	P
Platymantis banahao	1997	My, A	P
Platymantis bayani	2010	My	P
Platymantis biak	2010	My	P
Platymantis cagayanensis	1999	My, A	P
Platymantis cornutus	1922	My	P
Platymantis corrugatus	1853	My	P
Platymantis diesmosi	2006	My, A	P
Platymantis dorsalis	1853	My	P
Platymantis guentheri	1882	My	P
Platymantis hazelae	1920	My	P
Platymantis indeprensus	1999	My, A	P
Platymantis insulatus	1970	My	P
Platymantis isarog	1997	My	P

Taxon	Year	Data Type	Region
Platymantis lawtoni	1974	My	P
Platymantis levigatus	1974	My	P
Platymantis luzonensis	1997	My, A	P
Platymantis mimulus	1997	My, A	P
Platymantis montanus	1922	My	P
Platymantis naomiae	1998	My, A	P
Platymantis negrosensis	1997	My, A	P
Platymantis paengi	2007	My, A	P
Platymantis panayensis	1997	My	P
Platymantis polillensis	1922	My	P
Platymantis pseudodorsalis	1999	My, A	P
Platymantis pygmaeus	1998	My, A	P
Platymantis rabori	1997	My, A	P
Platymantis sierramadrensis	1999	My, A	P
Platymantis spelaeus	1982	My	P
Platymantis subterrestris	1922	My	P
Platymantis taylori	1999	My, A	P
Dicroglossidae			
Euphlyctis cyanophlyctis	1799	My	NM
Fejervarya cancrivora	1829	My	NM, SM, Sd, P, Sw
Fejervarya limnocharis	1829	My	NM, SM, Sd, Sw
Fejervarya pulla	1870	My	SM
Fejervarya schlueteri	1893	My	Sd
Fejervarya triora	2006	My	NM
Fejervarya vittigera	1834	My	P
Hoplobatrachus rugulosus	1834	My	NM, SM
Hoplobatrachus tigerinus	1802	My	NM

Taxon	Year	Data Type	Region
Ingerana borealis	1896	My	Sd
Ingerana liui	1983	My	NM
Ingerana tasanae	1921	My	NM, SM
Ingerana tenasserimensis	1892	My	NM, SM
Limnonectes acanthi	1923	My	P
Limnonectes arathooni	1928	My	Sw
Limnonectes asperatus	1996	My	Sd
Limnonectes bannaensis	2007	My, Ms	NM
Limnonectes blythii	1920	My	NM, SM, Sd
Limnonectes dabanus	1922	My	NM
Limnonectes diuatus	1977	My	P
Limnonectes doriae	1887	My	NM, SM
Limnonectes ferneri	2010	My	P
Limnonectes finchi	1966	My	Sd
Limnonectes grunniens	1801	My	Sw
Limnonectes gyldenstolpei	1916	My	NM
Limnonectes hascheanus	1870	My	NM, SM
Limnonectes heinrichi	1933	My	Sw
Limnonectes ibanorum	1964	My	Sd
Limnonectes ingeri	1978	My	Sd
Limnonectes kenepaiensis	1966	My	Sd
Limnonectes khammonensis	1929	My	NM
Limnonectes kohchangae	1922	My	NM
Limnonectes kuhlii	1838	My	NM, SM, Sd
Limnonectes laticeps	1882	My	SM, Sd
Limnonectes leporinus	1923	My	Sd
Limnonectes leytensis	1893	My	P
Limnonectes limborgi	1892	My	NM

Taxon	Year	Data Type	Region
Limnonectes macrocephalus	1954	My	P
Limnonectes macrodon	1841	My	Sd
Limnonectes macrognathus	1917	My	NM, SM
Limnonectes magnus	1909	My	P
Limnonectes malesianus	1984	My	SM, Sd
Limnonectes megastomias	2008	My, Ms	NM
Limnonectes micrixalus	1923	My	P
Limnonectes microdiscus	1892	My	Sd
Limnonectes microtympanum	1907	My	Sw
Limnonectes modestus	1882	My	Sw
Limnonectes nitidus	1932	My	SM
Limnonectes palavanensis	1894	My	P, Sd
Limnonectes paramacrodon	1966	My	SM, Sd
Limnonectes parvus	1920	My	P
Limnonectes plicatellus	1873	My	SM
Limnonectes poilani	1942	My	NM
Limnonectes rhacodus	1996	My	Sd
Limnonectes shompenorum	1996	My	Sd
Limnonectes tweediei	1935	My	SM, Sd
Limnonectes visayanus	1954	My	P
Limnonectes woodworthi	1923	My	P
Nanorana aenea	1922	My	NM
Nanorana arnoldi	1975	My	NM
Nanorana bourreti	1987	My	NM
Nanorana delacouri	1928	My	NM
Nanorana feae	1887	My	NM
Nanorana yunnanensis	1879	My	NM
Occidozyga baluensis	1896	My	Sd

Taxon	Year	Data Type	Region
Occidozyga celebensis	1927	My	Sw
Occidozyga diminutiva	1922	My	P
Occidozyga laevis	1859	My	P, Sd, SM, Sw
Occidozyga lima	1829	My	NM, SM, Sd, Sw
Occidozyga magnapustulosa	1958	My	NM
Occidozyga martensii	1867	My	NM, SM
Occidozyga semipalmata	1927	My	Sw
Occidozyga vittata	1942	My	NM
Quasipaa acathophora	2009	My	NM
Quasipaa boulengeri	1889	My	NM
Quasipaa fasciculispina	1970	My	NM
Quasipaa verrucospinosa	1937	My	NM
Sphaerotheca breviceps	1799	My	NM
Rhacophoridae			
Chiromantis doriae	1893	My	NM
Chiromantis laevis	1924	My	NM
Chiromantis nongkhorensis	1927	My	NM
Chiromantis punctatus	2003	My	NM
Chiromantis samkosensis	2007	My	NM
Chiromantis vittatus	1887	My	NM
Feihyla palpebralis	1924	My	NM
Gracixalus gracilipes	1937	My	NM
Gracixalus jinxiuensis	1978	My	NM
Gracixalus quyeti	2008	My, Ms	NM
Gracixalus supercornutus	2004	My	NM
Kurixalus ananjevae	2004	My	NM
Kurixalus baliogaster	1999	My	NM
Kurixalus banaensis	1939	My	NM

Taxon	Year	Data Type	Region
Kurixalus bisacculus	1962	My	NM
Kurixalus carinensis	1893	My	NM
Kurixalus naso	1912	My	NM
Kurixalus odontotarsus	1993	My	NM
Kurixalus verrucosus	1893	My	NM, SM
Nyctixalus margaritifer	1882	My	Sd
Nyctixalus pictus	1871	My	NM, SM, Sd, P
Nyctixalus spinosus	1920	My	P
Philautus abditus	1999	My	NM
Philautus acutirostris	1867	My	P
Philautus acutus	1987	My, A	Sd
Philautus alticola	1931	My	P
Philautus amoenus	1931	My	Sd
Philautus aurantium	1989	My	Sd
Philautus aurifasciatus	1837	My	Sd
Philautus bunitus	1995	My	Sd
Philautus cardamonus	2002	My	NM
Philautus cinerascens	1870	My	NM
Philautus cornutus	1920	My	Sd
Philautus davidlabangi	2009	My, A	Sd
Philautus disgregus	1989	My	Sd
Philautus erythrophthalmus	2000	My	Sd
Philautus gunungensis	1996	My, A	Sd
Philautus hosii	1895	My	Sd
Philautus ingeri	1987	My, A	Sd
Philautus jacobsoni	1912	My	Sd
Philautus kerangae	1987	My, A	Sd
Philautus leitensis	1897	My	P

Taxon	Year	Data Type	Region
Philautus longicrus	1894	My	P, Sd
Philautus maosonensis	1937	My	NM
Philautus mjobergi	1925	My	Sd
Philautus refugii	1996	My	Sd
Philautus pallidipes	1908	My	Sd
Philautus petersi	1900	My	SM, Sd
Philautus petilus	2004	My	NM
Philautus poecilus	1994	My	P
Philautus saueri	1996	My, A	Sd
Philautus schmackeri	1892	My	P
Philautus similis	1923	My	Sd
Philautus surdus	1863	My	P
Philautus surrufus	1994	My	P
Philautus tectus	1987	My, A	Sd
Philautus truongsonensis	2005	My	NM
Philautus tytthus	1940	My	NM
Philautus umbra	1987	My, A	Sd
Philautus vermiculatus	1900	My	SM
Philautus vittiger	1897	My	Sd
Philautus williamsi	1922	My	P
Philautus worcesteri	1905	My	P
Polypedates colletti	1890	My	SM, Sd
Polypedates chlorophthalmus	2005	My	Sd
Polypedates hecticus	1863	My	P
Polypedates leucomystax	1829	My	NM, SM, Sd, P, Sw
Polypedates macrotis	1891	My	NM, SM, Sd, P

Taxon	Year	Data Type	Region
Polypedates megacephalus	1861	My	NM
Polypedates mutus	1940	My	NM
Polypedates otilophus	1893	My	Sd
Pseudophilautus gryllus	1924	My	NM
Pseudophilautus longchuanensis	1979	My	NM
Pseudophilautus parvulus	1893	My	NM
Rhacophorus angulirostris	1927	My	Sd
Rhacophorus annamensis	1924	My	NM
Rhacophorus appendiculatus	1858	My	SM, Sd, P
Rhacophorus baluensis	1954	My	Sd
Rhacophorus barisani	2002	My	Sd
Rhacophorus belalongensis	2008	My, A	Sd
Rhacophorus bifasciatus	1923	My	Sd
Rhacophorus bimaculatus	1867	My	P, Sd, SM
Rhacophorus bipunctatus	1927	My	NM, SM
Rhacophorus burmanus	1939	My	NM
Rhacophorus calcaneus	1924	My	NM
Rhacophorus chuyangsinensis	2008	My	NM
Rhacophorus cyanopunctatus	1998	My	NM, SM
Rhacophorus dennysi	1881	My	NM
Rhacophorus dorsoviridis	1937	My	NM
Rhacophorus duboisi	2000	My	NM
Rhacophorus dugritei	1872	My	NM
Rhacophorus dulitensis	1892	My	Sd
Rhacophorus edentulus	1894	My	Sw

Taxon	Year	Data Type	Region
Rhacophorus everetti	1894	My	P, Sd
Rhacophorus exechopygus	1999	My	NM
Rhacophorus fasciatus	1895	My	Sd
Rhacophorus feae	1893	My	NM
Rhacophorus gauni	1955	My	Sd
Rhacophorus georgi	1904	My	Sw
Rhacophorus harrissoni	1959	My	Sd
Rhacophorus hoanglienensis	2001	My	NM
Rhacophorus jarujini	2006	My	NM
Rhacophorus kajau	1983	My, A	Sd
Rhacophorus kio	2006	My, Ms	NM
Rhacophorus margariter	1837	My	Sd
Rhacophorus marmoridorsum	2008	My	NM
Rhacophorus maximus	1858	My	NM
Rhacophorus modestus	1920	My	Sd
Rhacophorus monticola	1896	My	Sw
Rhacophorus nigropalmatus	1895	My	SM, Sd
Rhacophorus norhayatii	2010	My	SM
Rhacophorus orlovi	2001	My	NM
Rhacophorus pardalis	1858	My	SM, Sd, P
Rhacophorus penanorum	2008	My	Sd
Rhacophorus poecilonotus	1920	My	Sd
Rhacophorus prominanus	1924	My	SM
Rhacophorus reinwardti	1840	My	SM, Sd
Rhacophorus rhodopus	1960	My	NM
Rhacophorus robinsonii	1903	My	SM
Rhacophorus rufipes	1966	My	Sd

Taxon	Year	Data Type	Region
Rhacophorus spelaeus	2010	My	NM
Rhacophorus turpes	1940	My	NM
Theloderma andersoni	1927	My	NM
Theloderma asperum	1886	My	NM, SM
Theloderma bicolor	1937	My	NM
Theloderma corticale	1903	My	NM
Theloderma gordoni	1962	My	NM
Theloderma horridum	1903	My	SM, Sd
Theloderma lateriticum	2009	My	NM
Theloderma leporosum	1838	My	SM
Theloderma licin	2007	My	SM
Theloderma phrynoderma	1927	My	NM
Theloderma rhododiscus	1962	My	NM
Theloderma ryabovi	2006	My	NM
Theloderma stellatum	1962	My	NM
Ranidae			
Amolops akhaorum	2010	My, Ms	NM
Amolops archotaphus	1997	My	NM
Amolops compotrix	2006	My	NM
Amolops cremnobatus	1998	My	NM
Amolops cucae	2006	My	NM
Amolops daorum	2003	My, Ms	NM
Amolops gerbillus	1912	My	NM
Amolops iriodes	2004	My	NM
Amolops kaulbacki	1940	My	NM
Amolops larutensis	1899	My	SM
Amolops longimanus	1939	My	NM
Amolops marmoratus	1855	My	NM

Taxon	Year	Data Type	Region
Amolops minutus	2007	My	NM
Amolops panhai	2006	My, A	NM
Amolops ricketti	1899	My	NM
Amolops spinapectoralis	1999	My	NM
Amolops splendissimus	2007	My	NM
Amolops viridimaculatus	1983	My	NM
Amolops vitreus	2006	My	NM
Babina adenopleura	1909	My	NM
Babina chapaensis	1937	My	NM
Clinotarsus alticola	1882	My	NM, SM
Huia cavitympanum	1893	My	Sd
Huia masonii	1884	My	Sd
Huia melasma	2005	My	NM
Huia modiglianii	1999	My	Sd
Huia sumatrana	1991	My	Sd
Humerana humeralis	1887	My	NM
Humerana miopus	1918	My	SM
Humerana oatesii	1892	My	NM
Hylarana attigua	1999	My	NM
Hylarana banjarana	2003	My	SM
Hylarana baramica	1900	My	SM, Sd
Hylarana celebensis	1872	My	Sw
Hylarana chalconota	1837	My	Sd
Hylarana crassiovis	1920	My	Sd
Hylarana cubitalis	1917	My	NM
Hylarana debussyi	1910	My	Sd
Hylarana erythraea	1837	My	NM, SM, Sd, P
Hylarana eschatia	2009	My, Ms	SM

Taxon	Year	Data Type	Region
Hylarana faber	2002	My	NM
Hylarana glandulosa	1882	My	NM, SM, Sd
Hylarana grandocula	1920	My	P
Hylarana guentheri	1882	My	NM
Hylarana labialis	1887	My	SM
Hylarana laterimaculata	1916	My	SM
Hylarana leptoglossa	1868	My	NM
Hylarana luctuosa	1871	My	SM, Sd
Hylarana macrodactyla	1858	My	NM, SM
Hylarana macrops	1897	My	Sw
Hylarana macquardi	1901	My	Sw
Hylarana mangyanum	2002	My, Ms, A	P
Hylarana maosonensis	1937	My	NM
Hylarana margariana	1879	My	NM
Hylarana megalonesa	2009	My, Ms	Sd
Hylarana melanomenta	1920	My	P
Hylarana milleti	1921	My	NM
Hylarana moellendorffi	1893	My	P
Hylarana montivaga	1921	My	NM
Hylarana mortenseni	1903	My	NM
Hylarana nicobariensis	1870	My	NM, SM
Hylarana nigrovittata	1856	My	NM, SM, Sd
Hylarana parvaccola	2009	My, Ms	Sd
Hylarana picturata	1920	My	Sd
Hylarana raniceps	2009	My, Ms	Sd
Hylarana rufipes	2009	My, Ms	Sd
Hylarana siberu	1990	My	SM

Taxon	Year	Data Type	Region
Hylarana signata	1872	My	NM, SM, Sd
Hylarana similis	1873	My	P
Hylarana taipehensis	1909	My	NM
Meristogenys amoropalamus	1984	My	Sd
Meristogenys jerboa	1872	My	Sd
Meristogenys kinabaluensis	1966	My	Sd
Meristogenys macrophthalmus	1986	My	Sd
Meristogenys maryatiae	1985	My	Sd
Meristogenys orphnocnemis	1986	My	Sd
Meristogenys phaeomerus	1983	My	Sd
Meristogenys poecilus	1983	My	Sd
Meristogenys whiteheadi	1887	My	Sd
Odorrana absita	2005	My	NM
Odorrana andersonii	1882	My	NM
Odorrana aureola	2006	My	NM
Odorrana bacboensis	2003	My, Ms	NM
Odorrana banaorum	2003	My, Ms	NM
Odorrana bolavensis	2005	My	NM
Odorrana chapaensis	1937	My	NM
Odorrana chloronota	1876	My	NM, SM
Odorrana geminata	2009	My, Ms	NM
Odorrana gigatympana	2006	My	NM
Odorrana grahami	1917	My	NM
Odorrana graminea	1900	My	NM
Odorrana heatwolei	2005	My	NM
Odorrana hmongorum	2003	My, Ms	NM
Odorrana hosii	1891	My	NM, SM, Sd

Taxon	Year	Data Type	Region
Odorrana indeprensa	2005	My	NM
Odorrana junlianensis	2001	My	NM
Odorrana khalam	2005	My	NM
Odorrana livida	1856	My	NM
Odorrana margaretae	1950	My	NM
Odorrana megatympanum	2003	My, Ms	NM
Odorrana monjerai	2006	My	SM
Odorrana morafkai	2003	My, Ms	NM
Odorrana nasica	1903	My	NM
Odorrana orba	2005	My	NM
Odorrana schmackeri	1892	My	NM
Odorrana tabaca	2004	My	NM
Odorrana tiannanensis	1980	My	NM
Odorrana trankieni	2003	My	NM
Odorrana yentuensis	2008	My	NM
Pelophylax lateralis	1887	My	NM
Pterorana khare	1986	My	NM
Rana johnsi	1921	My	NM
Rana nicobariensis	1870	My	P, Sd, SM
Sanguirana albotuberculata	1954	My	P
Sanguirana aurantipunctata	2011	My	P
Sanguirana everetti	1882	My	P
Sanguirana igorota	1920	My	P
Sanguirana luzonensis	1896	My	P
Sanguirana sanguinea	1893	My	P
Sanguirana tipanan	2000	My	P
Staurois guttatus	1858	My	Sd
Staurois latopalmatus	1887	My	Sd

Taxon	Year	Data Type	Region
Staurois natator	1859	My	P
Staurois nubilis	1890	My	P
Staurois parvus	1959	My	Sd
Staurois tuberilinguis	1918	My	Sd

15

Wildlife trade as an impediment to conservation as exemplified by the trade in reptiles in Southeast Asia

VINCENT NIJMAN, MATTHEW TODD AND CHRIS R. SHEPHERD

15.1 Introduction

Wildlife trade is at the heart of biodiversity conservation and sustainable development. It includes all sales or exchanges of wild animal and plant resources by people (Broad et al. 2003, Abensperg-Traun 2009). Wildlife trade involves live animals and plants or a diverse range of products derived from them and needed or prized by humans — including luxury goods, medicinal ingredients, food and pets. It generates considerable revenue and may provide or supplement incomes for some of the least economically affluent people (Ng and Tan 1997, Shunichi 2005, TRAFFIC 2008). The principal motivating factor for wildlife traders is economic, ranging from small-scale local income generation to major profit-oriented business. Wildlife can be traded locally (within a village or region) or nationally (that is within the political borders of a country or state) but a large volume of wildlife is traded internationally (Green and Shirley 1999, Wood 2001, Stoett 2002, Auliya 2003, Blundell and Mascia 2005, Schlaepfer et al. 2005, Nijman and Shepherd 2007). Between collectors/harvesters of wildlife and end users, any number of middlemen may be involved in the wildlife trade, including specialists involved in transporting, storage, handling, manufacturing, industrial production, marketing

Biotic Evolution and Environmental Change in Southeast Asia, eds D. J. Gower et al. Published by Cambridge University Press.

and the export and retail businesses, and these may operate both domestically and internationally (TRAFFIC 2008). Intrinsically linked to economic growth, the demand for wildlife has increased and, exacerbated by ongoing globalisation, the scale and extent of wildlife trade likewise may have enlarged (Nijman 2010). Human population growth, increasing buyer power and globalisation have led to a rise in demand for exotic wildlife (hence international trade) and this has occurred in developed, emerging and developing nations alike. In the absence of strong regulatory mechanisms, and given large financial gains, these demands are often fulfilled, putting a strain on wildlife populations. In the most extreme cases this may lead to the extinction of populations or even species (e.g. Shepherd and Ibarrondo 2005).

One of the best-known regulatory policy instruments for international wildlife trade is the Convention on International Trade in Endangered Species of Wild Fauna and Flora (CITES) called to life in 1973 and to which 175 Parties are signatories. With the aim of preventing species from becoming (economically and ecologically) extinct as a result of unsustainable trade, thousands of species have been put on one of the three Appendices that preclude (Appendix I) or regulate (II and III) international commercial trade. When CITES laws are passed in a contracting party, the police, customs inspectors, wildlife officials and/or other government officers are empowered to enforce CITES regulations (Nijman and Shepherd 2011).

Although levels of wildlife trade are rarely quantified and specified, it is clear that for many species groups and different areas huge quantities are traded annually (Li and Li 1998, van Dijk et al. 2000, Auliya 2003, Zhou and Jiang 2004, Schlaepfer et al. 2005, Engler and Parry-Jones 2007). Reptiles are traded globally in large volumes to supply the demand especially for skins, food, traditional medicines and pets (Nijman 2010). With other factors, such as habitat loss and degradation (Gibbons et al. 2000, Gardner et al. 2007), anthropogenic disturbance (Garbeer and Burger 1995), climate change (Janzen 1994, Araujo et al. 2006) and pollutants (Guillette et al. 1994), the collection of reptiles from the wild for commercial purposes has been invoked as a contributing factor to the declines or even extinction of individual species (Gibbons et al. 2000, Schlaepfer et al. 2005, Shepherd and Ibarrondo 2005, Stuart et al. 2006). Extant 'reptiles' (as used here) are a paraphyletic group comprising all non-avian and non-mammalian amniotes, with four orders being recognised: Squamata (lizards, snakes ~8000 species), Crocodylia (alligators, crocodiles, gavials ~25 species), Testudines (turtles, tortoises ~300 species) and Sphenodontia (tuataras, 2 species).

Here we provide four case studies of domestic and international trade in reptiles for various purposes in four Southeast Asian countries: Thailand, Vietnam, Malaysia and Indonesia. These four countries differ economically, with Thailand and Malaysia being more affluent than Vietnam and Indonesia; geographically, with the island-nation Indonesia being somewhat isolated from the remainder of Southeast Asia, and Vietnam and Thailand offering a gateway into China; and culturally, with Malaysia and Thailand appearing to be more internationally orientated than Indonesia and Vietnam.

Combined, the case studies give a representative picture of wildlife trade in Southeast Asia, showing similarities and highlighting some differences among types of trade and countries. Our data are derived from various sources using different methodologies, illustrating the diversity in data collection for monitoring wildlife trade in this respect. Specifically, we report on the sale of exotic tortoises for the pet trade in Thailand, the trade in stuffed marine turtles and *bekko* ('tortoiseshell', i.e. products made out of the carapace of hawksbill turtles used for decorative purposes) in Vietnam, the export of monitor lizard meat from Malaysia, and the commercial 'captive-breeding' of reptiles for the international pet trade in Indonesia. We show that in the first three case studies there is a consistent, open and substantial illegal trade in these protected animals, and that in the latter case study there are clear indications that wild-caught individuals are exported under the disguise of being bred in captivity. These cases demonstrate how wildlife trade acts as an impediment to the conservation of these reptiles.

15.2 Methods

15.2.1 Data acquisition

Data were collected during surveys by the wildlife trade monitoring network TRAFFIC, and were conducted in the period 2002–2010. Data for the trade in exotic (non-native) tortoises in Thailand were collected during five surveys of Chatuchak market in Bangkok in January and August 2006, April 2007, June 2009 and January 2010. The trade in stuffed marine turtles and *bekko* was assessed during two surveys in May 2002 and April–May 2008; focusing on two major Vietnamese trade hubs, Ho Chi Minh City and Ha Tien. The *bekko* is made exclusively from the carapaces of hawksbill turtles (*Eretmochelys imbricata*) and trade in stuffed marine turtles was restricted to hawksbill turtles and green turtles (*Chelonia mydas*). In both countries surveys typically lasted several days and all shops specialising in reptiles and amphibians (Thailand) or *bekko* (Vietnam) were visited and species composition and numbers were recorded. The recorder in all cases was a westerner with considerable experience with the local wildlife trade and in wildlife trade monitoring, and while showing a clear interest in the wildlife for sale would not make themselves known as a researcher. Where possible, the authenticity of the products for sale or the species identity was checked. Prices were recorded when displayed or otherwise requested from vendors; it should be noted that prices reported here are initial quotes and that prices will normally go down with bargaining or when more items are purchased at once. All are converted to USD using the exchange rate at the time of the survey.

Data on seizures of clouded monitor lizards (*Varanus nebulosus*) in peninsular Malaysia in the period 2005–09 were compiled from a number of sources, including press releases and information obtained directly from government agencies. The data from 2005 and 2006 were based exclusively on summary statistics provided by the Malaysian Department of Wildlife and National Parks, but documents from other ministries provided conflicting data; we take a conservative approach and report the lowest numbers.

In August and November 2006, a TRAFFIC researcher and a member of the Indonesian Ministry of Forestry (MoF) visited all active reptile captive-breeding facilities in Indonesia and their observations were compared with numbers presented in the monthly breeding reports that these facilities had submitted to the MoF. We focus on five species that are commercially captive-bred for the international pet trade in at least three of these facilities, the pig-nosed turtle (*Carettochelys insculpta*), frillneck lizard (*Chlamydosaurus kingii*), green tree-python (*Chondropython viridis*), emerald monitor lizard (*Varanus prasinus*) and Timor monitor lizard (*V. timorensis*). Together we consider commercial captive breeding of these species as illustrative of the pet reptile trade from Indonesia.

Details of the survey methods can be found in reports by TRAFFIC (2004), van Dijk and Shepherd (2004), Nijman and Shepherd (2007, 2010a), Shepherd and Nijman (2008b), Stiles (2008), Shepherd and John (2010) and Todd (2010).

15.2.2 Legality of trade

The trades in our case study species in the different countries vary in the extent to which they are legal. The four countries are all Party to CITES with the Convention entering into force in Malaysia in January 1978, in Indonesia in March 1979, in Thailand in April 1983 and in Vietnam in April 1994. CITES is implemented at the national level through national legislation, and parties must have legislation that allows the implementation and enforcement of the Convention. Although Thailand, Vietnam and Indonesia's national legislation generally meets all the requirements for implementing CITES (but there are some ambiguities, see below), Malaysia's legislation is somewhat deficient (see e.g. CITES 2000).

Thai wildlife laws with respect to non-native species are ambiguous. Species not native to Thailand are covered under Chapter 4 of Wild Animal Reservation and Protection Act B.E. 2535: Importation, Exportation, Transitory Movement of Wild Animals and Wild Animal check point. Importation and exportation of protected and reserved wild animals and carcasses are prohibited (unless these were obtained from captive breeding) but there is no mention of 'possession' or 'domestic trade' of species on the prohibition list, only import and export. Exotic tortoises that are listed in Appendix I of CITES are not allowed to be traded commercially, and here we restrict our discussion to those species.

Hawksbill and green turtles could be legally exploited in Vietnam prior to April 2002, but subsequently commercial use and exploitation in the wild was prohibited (note that the first survey included here took place in May 2002). Circular 02/2006/TT-BTS of March 2006 provides a list of marine resources that are prohibited from commercial trade, including hawksbill turtles and green turtles. Both species are included in Appendix I of CITES.

The clouded monitor lizard is a Totally Protected Animal in peninsular Malaysia under the Protection of Wild Life Act 1972 (Act No. 76). Totally Protected Animals are species that shall not be killed, taken or be held in anyone's possession. The clouded monitor lizard is listed in CITES Appendix I and therefore international commercial trade is prohibited.

All five species of reptiles we focus on in the Indonesian case study are protected under Indonesian law. It is illegal to keep these animals or to trade in them, but the eight captive-breeding facilities we assessed were given permission by the Indonesian MoF to breed them for commercial purposes. The resultant offspring can be sold internationally but trade in wild-caught specimens of protected species by these facilities is not allowed. Four of the five species are included in Appendix II of CITES, regulating all commercial international trade.

15.3 Results

15.3.1 Tortoise trade in Thailand

We observed a total of 475 individuals of six species of Appendix I-listed tortoises for sale at Chatuchack market (Table 15.1). These included some of the most endangered species of tortoise in the world, such as the Critically Endangered radiated (*Astrochelys radiata*), ploughshare (*A. yniphora*) and spider (*Pyxis arachnoides*) tortoises, all endemic to Madagascar. Apart from Madagascar, we also observed Appendix I-listed species from the Indo-Burmese region, viz. the Indian roofed turtle (*Kachuga tecta*) and the Burmese eyed turtle (*Morenia ocellata*). The numbers of individuals we observed for each species differed substantially, with hundreds of radiated tortoises and dozens of spider tortoises and spotted pond turtles (*Geoclemys hamiltonii*) but smaller numbers of the other species. Most individuals were openly displayed in the shops although during later surveys some of the rarer species were kept at nearby houses (and were observed upon request). Prices ranged from ~USD 20 for a small Indian roofed turtle to ~USD 2 000 for a medium-sized, unusual yellow radiated tortoise, but individuals of most species could be bought for ~USD 100–300.

Table 15.1 Appendix-I tortoises observed for sale at Chatuchak market, Bangkok, Thailand in the period 2006–10 showing a consistent supply of some of the world's rarest tortoises. IUCN Red List classifications abbreviated as follows: CR, Critically Endangered; VU, Vulnerable; LR, Lower Risk/Least Concern.

Species	IUCN	Jan-06	Aug-06	Apr-07	Jun-09	Jan-10
Radiated tortoise *Astrochelys radiata*	CR	32	187	50	18	106
Spotted pond turtle *Geoclemys hamiltonii*	VU	0	2	4	28	1
Indian roofed turtle *Kachuga tecta*	LR	2	0	0	0	3
Burmese eyed-turtle *Morenia ocellata*	VU	4	0	0	0	0
Spider tortoise *Pyxis arachnoides*	CR	0	4	0	0	31
Ploughshare tortoise *Astrochelys yniphora*	CR	0	0	0	0	3

15.3.2 Marine turtle and *bekko* trade in Vietnam

Informants in the markets confirmed that hawksbill turtles are rare in Vietnamese waters, more so in the 2008 survey than in the 2002 survey. Raw scutes used in *bekko* manufacture are illegally imported primarily from Malaysia and Indonesia. The number of *bekko* products observed for sale decreased somewhat between the surveys suggesting a decrease in demand and/or availability/supply (Table 15.2), but both in Ha Tien and Ho Chi Minh City *bekko* is still widely and openly available for sale. In total ~12 000 items were recorded, two-thirds of which were in Ho Chi Minh City. The most expensive *bekko* items on sale were handbags with asking prices of USD 300 and USD 450 in Ha Tien and Ho Chi Minh City, respectively. Most items were considerably less expensive, with bangles, bracelets, necklaces and combs typically selling for USD 3–30.

A total of 200 stuffed turtles were observed, ranging from 20–40 cm maximum width and many of them, particularly the hawksbills, were very young individuals. A 56 cm wide hawksbill in Ho Chi Minh City was selling for USD 485, but smaller stuffed green and hawksbill turtles were on offer for USD 20–30 each.

15.3.3 Clouded monitor lizard trade in Malaysia

In the 5 years (2005–09) ~38 000 clouded monitor lizards were confiscated in 33 seizures in 8 of the 13 States and Federal Territories of Malaysia (Table 15.3). The

Table 15.2 *Bekko* ('tortoise shell') and stuffed marine turtles for sale in two Vietnamese cities in 2002 and 2008.

City	Ho Chi Minh City		Ha Tien		
Year (shops selling *bekko*)	2002 (21)	2008 (22)	2002 (3)	2008 (10)	Total
Bracelet / bangle	3092	812	939	622	5465
Hair clips, bands, pins	550	135	663	581	1929
Finger ring	443	556	425	316	1740
Necklace	261	34	8	4	307
Spectacle frames	284	120	20	18	442
Earrings (pair)	225	51	32	0	308
Comb	147	76	143	186	552
Cigarette holder/ filter	116	24	8	38	186
Brooches, pendants	103	4	0	86	193
Name seal	102	13	0	0	115
Ornamental box	74	18	5	11	108
Hand fan	59	6	12	6	83
Lighter holder	39	0	2	0	41
Pipe	33	5	0	27	65
Cigarette box	31	7	3	1	42
Purse	22	6	0	3	31
Miscellaneous	94	35	97	39	265
Total *bekko*	5696	1924	2360	1948	11928
Whole stuffed turtle	24	35	90	58	207

highest numbers came from three of the east-coast States (Terangganu, Johor and Pahang). Individual seizures ranged from single to >7000 individuals, with median seizures ranging from 24 to 2215 individuals in different years. Eight seizures comprised >2000 lizards each. The monitors were seized from dealers in warehouse/ cold rooms, in houses, shop lots and at airports and jetties – being transported or awaiting transport. The shipments were apparently destined for China, or, to a lesser extent, to be used in local 'exotic' meat restaurants.

Table 15.3 Number of clouded monitor lizard individuals in seizures in peninsular Malaysia in the period 2005–09. *Malaysia's Ministry of Natural Resources and Environment (2009) reports the confiscation of 9265 clouded monitor lizards for 2006; the data presented here are from the Department of Wildlife and National Parks as reported by CITES (2007).

Year	Total monitors	Median seizure (N)	Maximum seizure
2005	5400	–	–
2006	4612*	2215 (3)	2390
2007	9604	433 (10)	2910
2008	14568	222 (11)	7093
2009	3803	24 (8)	2330
2005–2009	37987	181 (32)	

15.3.4 Commercial captive breeding of reptiles in Indonesia

Of the eight operations included in this assessment, only two facilities were run at what we consider a professional level, with knowledgeable staff and adequate facilities to breed at least some species of reptiles in captivity. All other facilities did not appear to be suitable for captive breeding and/or appeared to be rarely used for captive breeding.

For all five species there were large discrepancies between stock/offspring data that the breeding facilities reported on a monthly basis to the Indonesian MoF and what was observed to be present (Fig 15.1). A total of 1200 individuals of the five species were observed whereas the breeding reports indicated that ~4000 individuals should have been present. There were five cases (from three facilities) where a facility reported a stock of some 20–40 animals and regular production of ~100 second-generation offspring, where not a single individual of that species was observed to be present during the inspection. Illustrative is the case of Timor monitor lizards: from the monthly breeding reports it appeared that five facilities were breeding the species. Upon inspection, two of these facilities did not have the species present, one facility owner made it clear that he did not breed the species at all (but housed 19 individuals) and two stated that breeding did take place. The monthly breeding reports of one of these facilities stated that 445 Timor monitor lizards were present but upon inspection only 20 individuals were located and facilities for breeding were absent. The other facility housed five individuals (39 in their breeding report) and was judged to be capable of producing offspring. For the years 2006 and 2007, Indonesia reported the export of 606 and 615 captive-bred Timor monitor lizards, respectively, mainly to the USA.

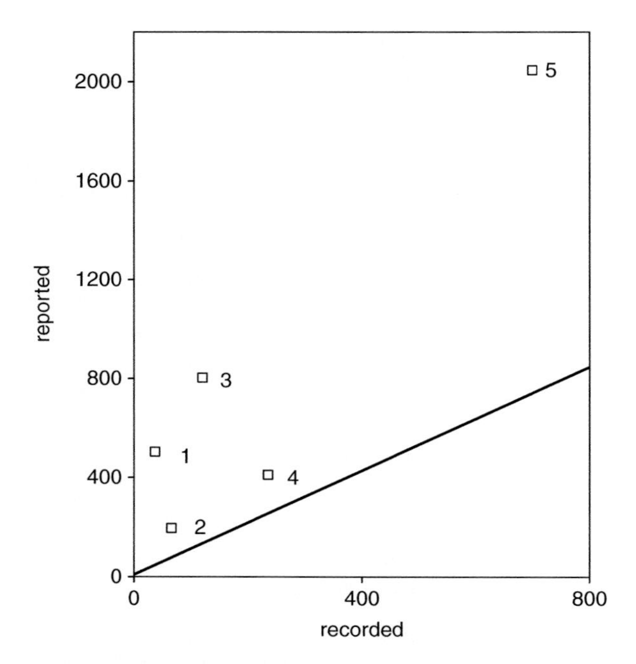

Figure 15.1 Captive-breeding of reptiles in Indonesia, showing numbers of individuals reported in monthly breeding reports and numbers recorded during site visits. The line of unity represents the line where reported numbers equal recorded numbers. Key: 1 = Timor monitor lizard *Varanus timorensis*; 2 = pig-nosed turtle *Carettochelys insculpta*; 3 = frillneck lizard *Chlamydosaurus kingii*; 4 = emerald monitor lizard *V. prasinus*, 5 = green tree-python *Morelia viridis*.

15.3.5 Wildlife trade links and wildlife trade enforcement

Although we here focus on the trade in reptiles, it is important to note that much of the trade in the above-mentioned species occurs concurrently with (legal and illegal) trade in other wildlife. The Thai traders in Appendix I-listed tortoises traded in a large range of pet reptiles, often from similar range countries. For example, during the 2010 survey at Chatuchak market, more than 400 Madagascar chameleons, geckos, snakes and frogs were also observed. Chatuchak is also known to be a major outlet for mammalian products as diverse as elephant ivory (Stiles 2009) and live slow lorises (Nekaris et al. 2010).

Especially during the 2002 survey, but also to a lesser extent during the 2008 survey, we found a strong link between the trade in *bekko* and ivory in Vietnam. In 2002, of the 36 shops with *bekko* for sale in Ho Chi Minh City, 16 also had elephant ivory for sale. Furthermore, a number of business cards collected from dealers depicted both sea turtles and elephants, suggesting the availability of both ivory and *bekko* (or other marine turtle products). In some cases, clouded monitor lizards were seized on their own, but more often they were seized with a variety of other vertebrate species also destined for the exotic meat market, including East Asian porcupines (*Hystrix brachyura*), Malayan pangolins (*Manis javanica*), leopards (*Panthera pardus*), barn owls (*Tyto alba*), estuarine crocodiles (*Crocodylus porosus*) and reticulated pythons (*Python reticulatus*). Hence, while the shipments of clouded monitor lizards were substantial, the traders dealing in them were also dealing illegally in other protected wildlife.

The popular international media often link wildlife trade crimes directly to the illicit drugs or arms trade or to trafficking of women and children, but it is worth noting that in the four case studies addressed here we found no such link. Traders in animals and animal parts were, by and large, specialised in various types of wildlife trade and it is our impression that most are indeed only involved in wildlife trade or other legal commercial trade.

15.4 Discussion

The trade in live reptiles to supply the demand for the international pet trade involves large numbers (e.g. Hoover 1998, Bridges et al. 2001, Auliya 2003). The curio trade (including *bekko*) affects certain species disproportionately (Heppell and Crowder 1996), as does the trade in reptiles for traditional Chinese medicine (Chen et al. 2009) but numbers of individual animals are dwarfed by the quantities traded in reptile skins (Dodd 1986, Zhou et al. 2004). Increasingly the trade in reptiles to supply the demand for the 'exotic' meat markets is becoming more apparent (van Dijk et al. 2000, Zang et al. 2008) with, for instance, tonnes of turtles being traded into China (Shepherd 2000, Shoppe 2009). The case studies presented here, in our experience, give a fair and illustrative representation of the wildlife trade throughout Southeast Asia. Although the often illicit wildlife trade in Indonesia, Vietnam and Thailand occurs very much in the open (e.g. in large open markets), or in specialised shops, it is somewhat more hidden in Malaysia and other Southeast Asian countries. Increasingly the internet is becoming an important vehicle for buying and selling wildlife, though here we see even larger differences between countries (this may be related to the countries' internet penetration rate, i.e. the percentage of the total population of a given country that uses the internet, which ranges from 12% in Indonesia, 26% in Thailand, 27% in Vietnam to 65% in Malaysia: Anonymous 2010).

Our four case studies all clearly indicate deficiencies in wildlife trade regulation and show that the scale of this international trade in threatened species can be a substantial impediment to conservation. The trade in *bekko* and stuffed marine turtles in Vietnam is currently sourced largely from Indonesian and Malaysian vessels because marine turtles are largely depleted from Vietnamese waters

(Hamann et al. 2005). Globally, hawksbill turtles are Critically Endangered, with trade in *bekko* being identified as one of the main threats to the survival of the species (Mortimer and Donnelly 2008). Likewise, harvesting of green turtles has been identified as the main threat to this species, contributing to its Endangered status (Seminoff 2004). The *bekko* industry in Vietnam supplies demand from other Asian countries and, to a lesser extent, Asian nationals living elsewhere (Stiles 2008).

As with the trade in *bekko*, the trade in pet tortoises in Thailand is truly international, with animals sourced from Madagascar and Indo-Burma (India, Bangladesh, Myanmar) only to be re-exported illegally to countries such as Japan, Malaysia and Singapore (Nijman and Shepherd 2007, Shepherd and Nijman 2008b, Todd 2010). This trade involves some of the world's most endangered tortoises. For the most part, tortoises are openly displayed in shops, suggesting ineffective enforcement.

The data on trade in clouded monitor lizards were obtained from the Malaysian authorities. While it is encouraging to see that these seizures are made, the numbers involved (nearly 40 000 lizards in some 30 confiscations) and the intended destination (with large numbers bound for China) are cause for concern. With limited efforts going into wildlife trade enforcement in Malaysia, and indeed large parts of Southeast Asia, the amounts seized are generally considered the 'tip of the iceberg'. While this leaves unknown how much remains undetected, it indicates that there is a substantial illegal trade in wildlife meat. To have large numbers of a CITES I-listed Totally Protected Animal being captured, killed and packed in Malaysia, and then shipped abroad is hard to view as anything other than an indictment against the law enforcement efforts of the relevant agencies, both in Malaysia, countries along the trade chain and importing countries such as China.

Although commercial captive breeding of reptiles may relieve some pressure on wild populations, this is valid only if the animals exported as captive-bred are indeed bred in controlled captive conditions from parent stock that themselves were bred in similar conditions. When wild-caught individuals are exported labelled as 'captive-bred' this undermines both the rules and the intentions of CITES (Nijman and Shepherd 2010b). Commercial captive breeding has a positive conservation impact only if it is coupled with efficient enforcement in situ, ensuring wild animals are no longer entering the trade.

In recent years 'captive-bred' specimens of species that are clearly unsuitable for commercial captive breeding have been exported from Indonesia. Prime examples are the pig-nosed turtles reported here (females not reaching maturity until ~20 years of age, producing small clutches and generally difficult to keep in captivity; Cann 1998) or spiny turtles (*Heosemys spinosa*, both sexes may attain maturity at the age of ~10 years, producing small clutches; Herman 1993) reported by Nijman and Shepherd (2010a). It is unlikely that these species can be bred commercially in viable quantities and we urge importing parties to be vigilant when accepting

Table 15.4 Major source countries and end destinations of trade in reptiles and their derivatives. Taiwan PoC, Taiwan, Province of China; Hong Kong SAR, Hong Kong Special Administrative Region.

Taxon	Type	Origin – source	Trader – exporter	Destination
Hawksbill turtles	*bekko* Stuffed	Indonesia Malaysia	Vietnam	Taiwan, PoC, China, Hong Kong SAR
Tortoises	Pets	Madagascar Indo-Burma	Thailand	Japan, Malaysia, Singapore
Clouded monitor lizard	Meat	Malaysia	Malaysia	China
Reptiles	Pets	Indonesia	Indonesia	USA, EU

difficult-to-breed species. Furthermore, we urge the Indonesian authorities to clamp down on farms wrongly claiming to produce viably captive-bred reptiles.

This research is based on a number of surveys conducted by TRAFFIC in the last decade, and has a bias towards wildlife trade with international aspects (Table 15.4). The illegal flow of legally protected and CITES-listed species from one country to the next often in substantial numbers suggests that enforcement efforts need to be stepped up. Source countries (Madagascar, Myanmar, Indonesia and Malaysia among others) should ensure that protected wildlife is not exported illegally, importing and re-exporting countries in the region (Thailand, Vietnam, China, Singapore) must ensure that imports and exports do not violate existing legislation. End destinations (Japan, USA, EU, among others) must assure that the imports do not violate the rules and intentions of CITES and that imported specimens were acquired in accordance with national legislation in the country of origin. With limited resources being spread out over a vast geographic area and with great monetary gains to be made from the illegal wildlife trade (and generally low risks of detection and prosecution), this is a daunting task. Furthermore, the weight of legal instruments to control the trade is undermined when local harvesters realise that little action is being taken against known traders and observe high-ranking officials trading or consuming wildlife products. Low rates of prosecution, low penalties and imposition of below-maximum fines all act as a limiting factor to enforcement success (Nijman, 2005).

As shown here and elsewhere (Wang et al. 1996, Li and Li 1998, Davies 2005, Karesh et al. 2007, Shepherd and Nijman 2008a), it is important to realise that most wildlife trade streams pass through a limited number of trade hubs (although these vary for different species-groups and may change over time). These trade streams and trade hubs provide ample opportunities to maximise the effects of regulatory efforts, as demonstrated with domestic animal trading systems (e.g. processing plants and wholesale and retail markets). Only through targeted and well-informed actions will authorities be able to substantially reduce the illegal reptile trade in Asia.

Acknowledgements

Daniel Stiles, Mark Auylia and Elizabeth John conducted parts of the TRAFFIC surveys from which we drew our examples: we thank them all. Loretta Ann Shepherd is thanked for information and comments. The TRAFFIC surveys were financially supported by grants from WWF Netherlands, WWF UK and Royal Danish Embassy. We thank the reviewers for constructive comments on our paper and for suggestions for improvement.

References

Abensperg-Traun, M. (2009). CITES, sustainable use of wild species and incentive-driven conservation in developing countries, with an emphasis on southern Africa. *Biological Conservation*, **142**, 948–63.

Anonymous (2010). Internet world stats: usage and population statistics. Available at: http://www.internetworldstats.com/, accessed 24 October 2010.

Auliya, M. (2003). *Hot Trade in Cool Creatures: A Review of the Live Reptile Trade in the European Union in the 1990s*. Brussels: TRAFFIC Europe.

Araujo, M. B., Thuiller, W. and Pearson, R. G. (2006). Climate warming and the decline of amphibians and reptiles in Europe. *Journal of Biogeography*, **33**, 1712–28.

Blundell, A. G. and Mascia, M. B. (2005). Discrepancies in reported levels of international wildlife trade. *Conservation Biology*, **19**, 2020–25

Bridges, V., Kopral, C. and Johnson, R. (2001). *The Reptile and Amphibian Communities in the United States*. Fort Collins, CO: USDA; APHIS; VS Centers for Epidemiology and Animal Health

Broad, S., Mulliken, T. and Roe, D. (2003). The nature and extent of legal and illegal trade in wildlife. In *The Trade in Wildlife. Regulation for Conservation*, ed. S. Oldfield. London: Flora and Fauna International, Resource Africa and TRAFFIC International, pp. 3–22

Cann, J. (1998). *Australian Freshwater Turtles*. Singapore: Beauworth Publishing Ltd.

Chen, T. H., Chang, H. C. and Lue, K. Y. (2009). Unregulated trade in turtle shells for Chinese traditional medicine in East and Southeast Asia: the case of Taiwan. *Chelonian Conservation Biology*, **8**, 11–18.

CITES (2000). Doc. 11.21.1 Interpretation and implementation of the Convention, National laws for implementation of the Convention, National Legislation

Project. Available at: http://www.cites.org/eng/cop/11/doc/21_01.pdf, accessed 27 October 2010.

CITES (2007). Notification No. 2005/035. Available at: http://www.cites.org/common/resources/reports/pab/05-06Malaysia.pdf, accessed 17 March 2009.

Davies, B. (2005). *Black Market: Inside the Endangered Species Trade in Asia*. San Rafael, CA: Earth Aware Editions.

van Dijk, P.-P. and Shepherd, C. R. (2004). *Shelled Out? A Snapshot of Bekko Trade in Selected Locations in South-east Asia*. Petaling Jaya, Malaysia: TRAFFIC Southeast Asia.

van Dijk, P.-P., Stuart, B. L. and Rhodin, A. G. J. (2000). *Asian Turtle Trade: Proceedings of a Workshop on Conservation and Trade of Freshwater Turtles and Tortoises in Asia*. Lunenberg, MA: Chelonian Research Foundation.

Dodd, C. K. Jr. (1986). Importation of live snakes and snake products into the United States, 1977–1983. *Herpetological Review*, **17**, 76–9.

Engler, M. and Parry-Jones, R. (2007). *Opportunity or Threat: The Role of the European Union in Global Wildlife Trade*. Brussels: TRAFFIC Europe.

Garber, S. D. and Burger, J. (1995). A 20-year study documenting the relationship between turtle decline and human recreation. *Ecological Applications*, **5**, 1151–62.

Gardner, T. A., Barlowa, J. and Peres, C. A. (2007). Paradox, presumption and pitfalls in conservation biology: the importance of habitat change for amphibians and reptiles. *Biological Conservation*, **138**, 166–79.

Gibbons, J. W., Scott, D. E., Ryan, T. J. et al. (2000). The global decline of reptiles, déjà vu amphibians. *BioScience*, **50**, 653–66.

Green, E. P. and Shirley, F. (1999). *The Global Trade in Corals*. Cambridge: World Conservation Monitoring Centre.

Guillette, L. J., Gross, T. S., Masson, G. R. et al. (1994). Developmental abnormalities of the gonad and abnormal sex hormone concentrations in juvenile alligators from contaminated and control lakes in Florida. *Environmental Health Perspectives*, **102**, 680–8.

Hamann, M., Hien, B. T. T., Cox, N. et al. (2005). Marine turtle conservation in Viet Nam: towards 2010. *Marine Turtle Newsletter*, **107**, 5–6.

Heppell, S. S. and Crowder, L. B. (1996). Analysis of a fisheries model for harvest of hawksbill sea turtles (*Eretmochelys imbricata*). *Conservation Biology*, **10**, 874–80.

Herman, D. W. (1993). Reproduction and management of the Southeast Asian spiny turtle (*Heosemys spinosa*) in captivity. *Herpetological Natural History*, **1**, 97–100.

Hoover, C. (1998). *The US Role in the International Live Reptile Trade: Amazon Tree Boas to Zululand Dwarf Chameleons*. Washington DC: TRAFFIC North America.

Janzen, F. J. (1994). Climate change and temperature dependent sex determination in reptiles. *Proceedings of the National Academy of Sciences*, **91**, 7487–90.

Karesh, W. B., Cook, R. A., Gilbert, M. and Newcomb, J. (2007). Implications of wildlife trade on the movement of avian influenza and other infectious diseases. *Journal of Wildlife Diseases*, **43**, 55–9.

Li, Y. and Li, D. (1998). The dynamics of trade in live wildlife across the

Guangxi border between China and Vietnam during 1993–1996 and its control strategies. *Biodiversity and Conservation*, **7**, 895–914.

Ministry of Natural Resources and Environment (2009) *Response to the Article. Kuala Lumpur: Corporate Communication Unit of the Ministry of Natural Resources and Environment.* Available at: http://www.nre.gov.my/BM/Maklumbalas/RTTA%20Wildlife%2010%2012%2013%20August.pdf, accessed 17 March 2009.

Mortimer, J. A. and Donnelly, M. (2008). *Eretmochelys imbricata*. In *IUCN Red List of Threatened Species. Version 2010.2*. Available at: www.iucnredlist.org, accessed 27 August 2010.

Nekaris, K. A. I., Starr, C. R., Shepherd, C. R. and Nijman, V. (2010). Revealing culturally-specific patterns in wildlife trade via an ethnoprimatological approach: a case study of slender and slow lorises (*Loris* and *Nycticebus*) in South and Southeast Asia. *American Journal of Primatology*, **72**, 877–86.

Ng, P. K. L. and Tan, H. H. (1997). Freshwater fishes of Southeast Asia: potential for the aquarium fish trade and conservation issues. *Journal of Aquarium Science and Conservation*, **1**, 79–90.

Nijman, V. (2005). *In full swing. An assessment of the Trade in Gibbons and Orangutans on Java and Bali, Indonesia*. Petaling Jaya, Malaysia: TRAFFIC Southeast Asia.

Nijman, V. (2010). An overview of the international wildlife trade from Southeast Asia. *Biodiversity and Conservation*, **19**, 1101–14.

Nijman, V. and Shepherd, C. R. (2007). Trade in non-native, CITES-listed, wildlife in Asia, as exemplified by the trade in freshwater turtles and tortoises (Chelonidae) in Thailand. *Contributions to Zoology*, **76**, 207–11.

Nijman, V. and Shepherd, C. R. (2010a). *Wildlife trade from ASEAN to the EU: Issues with the Trade in Captive-bred Reptiles from Indonesia*. Brussels: TRAFFIC Europe.

Nijman, V. and Shepherd, C. R. (2010b). The role of Asia in the global trade in CITES II-listed poison arrow frogs: hopping from Kazakhstan to Lebanon to Thailand and beyond. *Biodiversity and Conservation*, **19**, 1963–70.

Nijman, V. and Shepherd, C. R. (2011). The role of Thailand in the international trade in live CITES-listed reptiles and amphibians. *PLoS One*, **6**(3), e17825.

Schlaepfer, M. A., Hoover, C. and Dodd, C. K. (2005). Challenges in evaluating the impact of the trade in amphibians and reptiles on wild populations. *BioScience*, **55**, 256–64.

Schoppe, S. (2009). *Status, Trade Dynamics and Management of the Southeast Asian Box Turtle in Indonesia*. Petaling Jaya, Malaysia: TRAFFIC Southeast Asia.

Seminoff, J. A. 2004. *Chelonia mydas*. In *IUCN Red List of Threatened Species. Version 2010.2*. Available at: www.iucnredlist.org, accessed 27 August 2010.

Shepherd, C. R. (2000). Export of live freshwater turtles and tortoises from North Sumatra and Riau, Indonesia: a case study. In *Asian Turtle Trade: Proceedings of a Workshop on Conservation and Trade of Freshwater Turtles and Tortoises in Asia*, ed. P.-P. van Dijk, B. L. Stuart and A. G. J. Rhodin. Lunenberg, MA: Chelonian Research Foundation, pp. 112–19.

Shepherd, C. R. and Ibarrondo, B. (2005). *The Trade of the Roti Island Snake-necked Turtle Chelodina mccordi,*

Indonesia. Petaling Jaya, Malaysia: TRAFFIC Southeast Asia.

Shepherd, C. R. and John, E. (2010). Clouded monitor lizard seizures in Peninsula Malaysia. *TRAFFIC Bulletin*, **22**, 102–103.

Shepherd, C. R. and Nijman, V. (2008a). Trade in bear parts from Myanmar: an illustration of the in-effectiveness of enforcement of international wildlife trade regulations. *Biodiversity and Conservation*, **17**, 35–42.

Shepherd, C. R. and Nijman, V. (2008b). *An Assessment of Trade in Freshwater Turtles and Tortoises at Chatucak Animal Market, Bangkok, Thailand*. Petaling Jaya, Malaysia: TRAFFIC Southeast Asia.

Shunichi, T. (2005). *The State of the Environment in Asia 2005–2006*. Tokyo: Springer-Verlag, Japan Environmental Council.

Stiles, D. (2008). *An Assessment of the Marine Turtle Products Trade in Viet Nam*. Petaling Jaya, Malaysia: TRAFFIC Southeast Asia.

Stiles, D. (2009). The status of the ivory trade in Thailand and Viet Nam. *TRAFFIC Bulletin*, **22**, 83–91.

Stoett, P. (2002). The international regulation of trade in wildlife: institutional and normative considerations. *International Environmental Agreements: Politics, Law and Economics*, **2**, 195–210.

Stuart, B. L., Rhodin, A. G., Grismer, L. L. and Hansel, T. (2006). Scientific description can imperil species. *Science*, **312**, 1137.

Todd, M. (2010). *Trade in Malagasy Reptiles and Amphibians in Thailand*. Petaling Jaya, Malaysia: TRAFFIC Southeast Asia.

TRAFFIC (2004). *The Trade in Marine Turtle Products in Viet Nam*. Ha Noi, Vietnam: TRAFFIC Southeast Asia-Indochina.

TRAFFIC (2008). *What's Driving the Wildlife Trade?* Washington DC: The World Bank.

Wang, Z., Chen, H. and Wu, D. (1996). The status on live wildlife trade near the port areas in Yunnan. In *Conserving China's Biodiversity*, ed. P. J. Schei, W. Sung and X. Yan. Beijing: China Environmental Science Press, pp. 197–210.

Wood, E. M. (2001). *Collection of Coral Reef Fish for Aquaria: Global Trade, Conservation Issues and Management Strategies*. Ross-on-Wye, UK: Marine Conservation Society.

Zhang, L., Ning, H. and Sun, S. (2008). Wildlife trade, consumption and conservation awareness in southwest China. *Biodiversity and Conservation*, **17**, 1493–516.

Zhou, Z. and Jiang, Z. (2004). International trade status and crisis for snake species in China. *Conservation Biology*, **18**, 1386–94.

16

The tropical peat swamps of Southeast Asia: human impacts on biodiversity, hydrology and carbon dynamics

SUSAN PAGE, ALJOSJA HOOIJER, JACK RIELEY,
CHRIS BANKS AND AGATA HOSCILO

16.1 Introduction

Peatlands are important terrestrial carbon stores and vital components of global carbon soil–atmosphere exchange processes; they also play a significant role in supporting biodiversity, with a unique combination of habitats and endemic and endangered species. In both these regards, tropical peatlands, most of which are located in the Southeast Asian coastal lowlands, are particularly important. These ecosystems have developed over millennial time scales and, in their natural state, support a vegetation cover of peat swamp forest. The substrate on which this forest grows consists of partially decomposed organic matter, i.e. peat, derived mostly from the remains of the forest vegetation (tree roots, leaves, dead branches and even trunks). Under the prevailing wet, acidic and nutrient-poor conditions, there is limited decay of this organic material and it accumulates to form vast domes of woody peat, often extending over many tens of kilometres. Owing to their considerable extent and thickness, up to 15 to 20 m thick, these peatlands are significant terrestrial carbon stores. Besides biodiversity maintenance, hydrological regulation and carbon storage, this unique dual system of forest and peatland also

Biotic Evolution and Environmental Change in Southeast Asia, eds D. J. Gower et al.
Published by Cambridge University Press.

provides a range of other valuable ecosystem goods and services, both directly and indirectly, including forestry and fishery products, flood mitigation and climate regulation. The range of ecosystem services that tropical peatlands provide to society at local to global scales is, however, decreasing at a rapid rate. Over the last two decades, the connected processes of land use change, drainage and fire have contributed to loss and degradation of this habitat, resulting in a severely reduced biodiversity, increased atmospheric greenhouse gas (GHG) emissions, local and regional pollution, and increased flooding.

By synthesising recent information on tropical peatland biodiversity, hydrology and carbon dynamics, this chapter provides the context for future studies of these globally important but little known ecosystems. It also addresses likely responses of tropical peatlands to a changing climate, especially for an ecosystem that is now fragmented and heavily disturbed, and concludes by considering briefly the scope for mitigative action, for example through ecosystem rehabilitation and new initiatives to reduce deforestation and decrease the associated GHG emissions.

16.2 The tropical peatlands of Southeast Asia

16.2.1 Peatland location, age and development

Tropical peatlands have, at a best estimate, a global area of 441 025 km^2 (Page et al. 2011; Fig 16.1). They reach their greatest extent in Southeast Asia, where there are 247 778 km^2, representing 56% of the total global tropical resource; South America and Africa rank second and third, with 107 486 km^2 (24%) and 55 860 km^2 (13%), respectively. Within the Southeast Asian region, Indonesia has 206 950 km^2 of peatland (i.e. 84% of the region's and 47% of the world's tropical peatland resource by area), followed by Malaysia (25 889 km^2) and Papua New Guinea (10 986 km^2).

Most peatlands in Southeast Asia are found along the east coast of Sumatra, on the island of Borneo (principally in Kalimantan and Sarawak), along the south coast of West Papua and on the coastal lowlands of peninsular Malaysia (Page et al. 2006). The peat is mostly ombrotrophic (i.e. rain-fed, and hence nutrient-poor and acidic), although there are some minerotrophic (groundwater-fed) peatlands, for example the Tasek Bera peatland in central peninsular Malaysia (Wüst and Bustin 2004). Most contemporary peat deposits began formation during the Late Pleistocene or Holocene (Page et al. 2004, Wüst et al. 2007) and although there are older peat or organic-rich deposits within the sedimentary rock record (Anderson and Muller 1975, van der Kaars 2001), there is no continuous stratigraphic record owing to major sea-level and climatic shifts. Palaeoenvironmental studies indicate that coastal deposits, for example, the extensive peatlands along the east coast of Sumatra, are the youngest peatlands, with peat accumulation in these locations

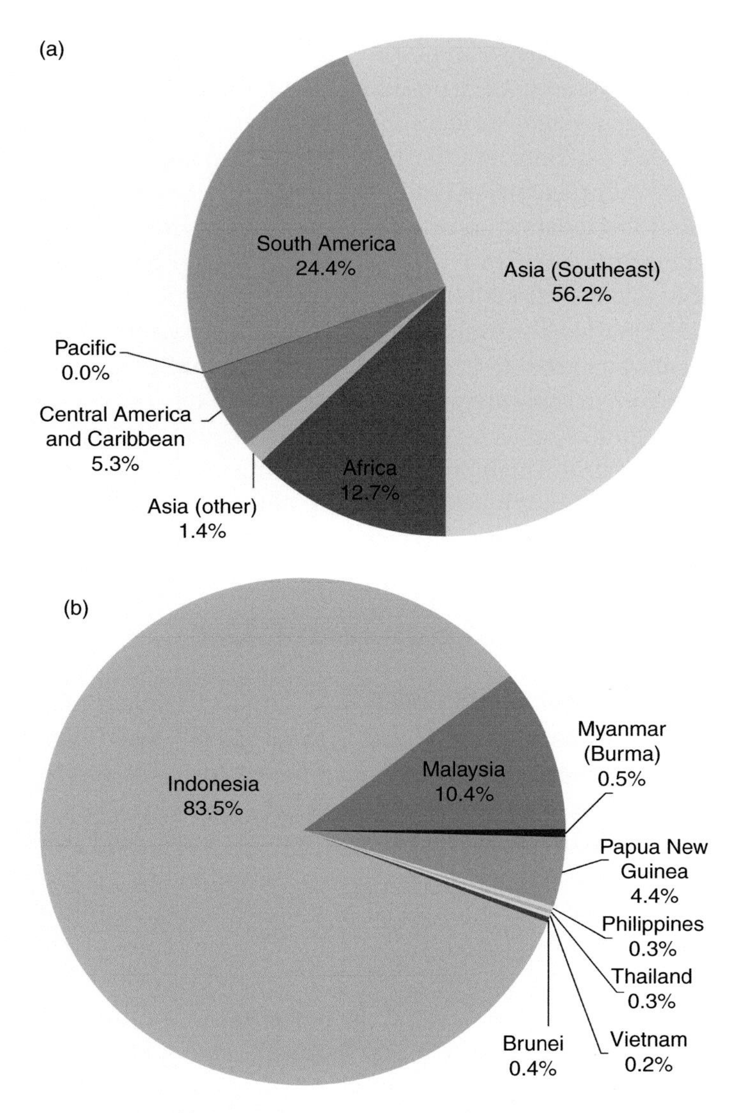

Figure 16.1 (a) Global distribution of tropical peatlands based on area, indicating the contribution made by Southeast Asian peatlands to the total resource. (b) Peatland distribution in Southeast Asia based on area, indicating the contribution made by Indonesian peatlands to the regional resource (data derived from Page et al. 2011).

commencing around 3.5–6.0 cal ka. On the other hand, most sub-coastal and inland peatlands started forming earlier, in the Late Pleistocene (~40 ^{14}C ka to ~23 ^{14}C ka), for example, the Danau Sentarum peatland in West Kalimantan (Anshari et al. 2001) and the Sebangau peatland in Central Kalimantan (Page et al. 2004), and through into the early Holocene (10–8 cal ka), for example, the peatlands of the Baram Delta in Sarawak (Staub and Esterle 1994).

Changes in climate and related fluctuations in sea level have both had important influences on the rate and nature of peatland development and peat preservation (Page et al. 2004). Palynological studies have shown that at least two different pathways of development occurred, commencing in either brackish (marine) or freshwater environments. Anderson and Muller (1975) showed that the Marudi peat swamp in Sarawak originated in a mangrove environment; their palynological profile indicated a gradual change from a mixed mangrove community through a transitional community to a true peat swamp association. A similar developmental sequence was recorded by Yulianto et al. (2005) for a peat swamp in South Kalimantan. By contrast, in Central Kalimantan, Morley (1981) described peat formation occurring abruptly over freshwater swamp and Hope et al. (2005) and Brady (1997) demonstrated similar freshwater, as opposed to mangrove, origins for peatlands in East Kalimantan and Riau, respectively.

The palaeo-record indicates a complex peat accumulation and degradation pattern across the lowland peatlands of this region. Most of the inland swamps, for example, in West and Central Kalimantan, began their development earlier than the coastal swamps and, in several cases, well before the Last Glacial Maximum (LGM; ~18 ka). Drier conditions during the LGM inhibited peat accumulation and may have led to deflation and loss of some deposits as a result of water-table drawdown and increased peat decomposition. Renewed peat accumulation followed a rise in sea level at the end of the LGM (~13–14 ka), in combination with increased precipitation and reduced drainage from the land (Page et al. 2004, 2010). Towards the end of this period of rapid accumulation for the inland peats (~6.0 cal ka) large, relatively flat areas of land were exposed around the coast as the rising sea levels stabilised and fell back slightly; in combination with favourable climatic conditions, this led to the initiation of today's extensive coastal and sub-coastal peat deposits in Southeast Asia (Page et al. 2004, 2010). While peat accumulation rates at coastal locations were increasing, however, the palaeoecological record for some inland sites in Central Kalimantan indicates that around 2 ka these deposits ceased accumulation and began to degrade (Siefferman et al. 1988). These changes were probably linked to an alteration in the rainfall pattern, in particular more seasonality perhaps associated with increasingly strong El Niño droughts (Gagan et al. 2004) and an increased drainage gradient as a result of sea-level fall from the post-glacial maximum to the present level about 5 m lower.

16.2.2 Biodiversity

Several early accounts of the peat swamp forests of Southeast Asia referred to low species diversity, attributed to the acidic, nutrient-poor environment and low ecosystem productivity (Janzen 1974, Johnson 1967). More recent studies, however, have demonstrated relatively high species richness of many plant and animal taxa, including a number of species with narrow niche requirements and

Figure 16.2 View inside undisturbed peat swamp forest in Riau province, Sumatra. See plate section for colour version.

restricted distributional range. The global centre of tropical peat swamp forest distribution lies in the Indo-Malayan region, in Malaysia, Indonesia, Vietnam, Thailand, Papua New Guinea and the Philippines. Thus the extensive peatlands of Malaysia and western Indonesia lie within the Sundaland biodiversity hotspot (Myers et al. 2000).

Owing to their waterlogged nature, tropical peat swamps provide a mixture of terrestrial, semi-aquatic and aquatic habitats (Fig 16.2). There is a continuum of at least two or three, and sometimes up to seven or more forest vegetation types from the edge to the centre of the peatland dome that reflect progressive increase in peat thickness and degree of waterlogging (Anderson 1983, Page et al. 1999). Tree cover ranges from a tall, closed-canopy swamp forest on shallow, marginal peat through to stunted, open-canopy pole or 'padang' forest on the thickest, most waterlogged peat, where *Pandanus* spp. (screw palms) are often a feature of the sub-canopy. Many of the trees show adaptations to the waterlogged, anoxic environment: buttress and stilt roots provide stability in the soft ground; knee roots and pneumatophores allow tree roots to 'breathe' in the anaerobic, waterlogged peat; and thick mats of fine roots close to the peat surface enhance nutrient-uptake in the zone where nutrient concentrations are highest. Studies of peat geochemistry and nutrient cycling have shown that the uppermost peat layer, to a depth of 2 m, is enriched with essential plant nutrient elements (calcium, magnesium, phosphorus, sulfur and potassium), reflecting efficient recycling and concentration by rooted plants (Weiss et al. 2002). With increasing distance onto the peat dome, the trees show additional evidence of adaptation to the ombrotrophic environment: average tree height and diameter decrease; leaf area decreases, while leaf cuticle thickness and the levels of toxic secondary compounds within the leaves (for example, tannins, phenols and resins) increase, possibly indicating nutrient-conservation defence strategies against herbivory in the nutrient-poor environment (Yule and Gomez 2008) and/or a response to water stress (alternating drought and flooding). Humidity is high throughout all peat swamp forest types; light levels are low in tall, closed canopy forest, but are higher under the more open canopy of low pole forest.

There is an undulating 'hummock–hollow' surface topography on the forest floor, which provides another suite of habitats. Hummocks are formed from tree roots and bases and can be 0.5–0.75 m in height, while the intervening hollows are usually devoid of vegetation other than protruding breathing roots from trees growing on adjacent hummocks. Trees growing on top of tall hummocks include flood-intolerant species, such as *Shorea teysmanniana*, while lower situations are occupied by flood-tolerant species, including *Eugenia rhizophora*, *Stemonurus secundiflorus* and *Knema intermedia* (Shimamura and Momose 2005).

In common with other rainforest types within the region, the tree species composition of peat swamp forest shows considerable variation according to geographical location at both a regional and a local scale. Unfortunately, there have been insufficient detailed studies to determine clear biogeographical patterns, but it is evident that the peat swamps of Borneo support a greater diversity of both peat swamp forest vegetation types and tree species than those elsewhere in the region. Even at a local scale, there is a surprising level of variation in species assemblages between different peatland domes. Several habitat endemic tree species have been identified; these include *Archidendron clypearia*, *Dactylocladus stenostachys*, *Durio carinatus*, *Gonystylus bancanus*, *Horsfieldia crassifolia*, *Shorea albida* (specific to the peat swamps of northern Borneo) and *Shorea teysmanniana* (Page et al. 2006).

Despite the more complex structure, greater plant species diversity and ecosystem productivity of lowland rainforests on mineral soils, mammal diversities and densities in peat swamp forest are relatively high. This ecosystem provides an important habitat for several primates, including the charismatic orang-utan (*Pongo pygmaeus*); some of the largest remaining populations of this primate are found in the peat swamps of Central Kalimantan (Morrogh-Bernard et al. 2003). Gibbons (*Hylobates* spp.) are also widespread (Buckley et al. 2006, Cheyne et al. 2007, Wich et al. 2008), together with more localised occurrences of red langur (*Presbytis rubicunda*) and proboscis monkey (*Nasalis larvatus*), which is endemic to peatland riverine and coastal fringes in some parts of Borneo (Meijaard and Nijman 2000), while the isolated peat swamps on the Mentawai Islands, off the west coast of Sumatra, support Kloss's gibbon (*Hylobates klossii*), Mentawai langur (*Presbytis potenziani*) and pig-tailed langur (*Simias concolor*), all three of which are narrow range species (Quinten et al. 2008). Recent studies have also emphasised the importance of peat swamp forests for felids: the Bornean clouded leopard (*Neofelis diardi*) has been recorded in the Sabangau peatland in Central Kalimantan; in the peat swamps of Riau province, Sumatra there are records of the endangered Sumatran tiger (*Panthera tigris sumatrae*); and marbled cat (*Pardofelis marmorata*), leopard cat (*Felis bengalensis)* and flat-headed cat (*Prionailurus planiceps*) are known from this and other peat swamp forests across the region (Cheyne 2008). Among smaller mammals there is a high diversity of bats including several regionally rare and threatened species (Struebig et al. 2006). While none of the mammals recorded from peat swamp forest can be considered true habitat endemics, the loss of forested habitat elsewhere in the tropical lowlands of Southeast Asia has given the remaining peat swamp forests an increased conservation importance for these species.

Studies of peat swamp forest avifauna have concluded that this habitat harbours lower bird species diversity than the region's dipterocarp forests (e.g. Gaither 1994, Janzen 1974). There are fewer primary forest species such as trogon (Trogonidae), broadbill (Eurylaimidae), pheasant (Phasianinae), pitta (Pittidae) and jungle babbler (*Turdoides striata*) but, conversely, an abundance of secondary growth species, including fantails (*Rhipidura* spp.) and tit-babblers (*Macronius* spp.). Both peat and freshwater swamps do, however, provide a vital refuge for several specialist wetland species; Storm's, milky and lesser adjutant storks (*Cicona stormi, Mycteria cinerea* and *Leptoptilos javanicus)* are all listed as highly endangered and are now restricted to wetland forests where there is little human disturbance.

The amphibians and reptiles of Southeast Asian peat swamps have been little studied and remain poorly known (Das 2006). The number of endemics is thus unclear, though, for example, the crocodile *Crocodylus raninus* (Borneo), false gharial (*Tomostoma schlegelli*) (peninsular Malaysia, Sumatra and Borneo), toad *Ingerophrynus kumquat* (peninsular Malaysia) and caecilian *Ichthyophis monochrous* (Borneo) are among those species possibly restricted to this habitat (Bezuijen et al. 2001, Das and Lim 2001, Das 2006, D.J. Gower pers. comm. 2011). It seems that herpetofaunal communities of peat swamp forests have a greater proportion of arboreal taxa than in rainforests (Das 2006).

The peat swamp aquatic environment provides important habitats for some 200 to 300 species of freshwater fish, around 100 of which are stenotopic (narrow range endemic) species adapted to the highly acidic and low nutrient conditions (Dennis and Aldhous 2004, L. Rüber pers. comm. 2010). These include rare species of *Encheloclarias, Bihunichthys, Betta* and *Parosphromenus*, as well as an unusually high number of miniature fish, which are mostly species of cyprinid genera such as *Paedocypris* and *Sundadanio* (Ng and Kottelat 1992, Ng and Lim 1993, Ng et al. 2004). The species *Paedocypris progenetica*, which was discovered in a Sumatran peat swamp in 2006, is the world's smallest fish and one of the world's smallest vertebrates (Kottelat et al. 2006).

This section has highlighted the critical role that peat swamp forests play in maintaining regional biodiversity and providing a habitat for a wide range of species, some with very specific ecological requirements. Many taxa, particularly invertebrate groups, remain understudied and no doubt further habitat endemics, for example, species of blackwater fish, remain to be discovered. Given current rates of loss and degradation of the peat swamp forest ecosystem, it seems inevitable, however, that some endemic species with restricted range have become extinct already, while others are on the verge of disappearing (Ng et al. 1994). Studies in other ecosystems indicate that ecosystem degradation not only impacts on the regional biota but also affects the delivery of other key ecosystem services (e.g. Balvanera et al. 2006). In peatland environments these include carbon sequestration and storage functions, as well as hydrological regulation and the supply of forest products to local communities. The link between biodiversity and carbon storage and the crucial need for urgent habitat conservation efforts in order to safeguard these vital functions is returned to in a later section.

16.2.3 Hydrology

Peat contains far more water than organic matter – in many cases in excess of 90–95% by volume. In their natural state, all tropical, ombrotrophic peatlands in Southeast Asia have high water tables, just above or below the peat surface for most of the time. It should be noted, however, that some peatlands in the south of Borneo and Sumatra experience much more severe and regular dry seasons, and therefore have lower minimum and average water levels, than others nearer the equator (Fig 16.3). The peatland water table will rise with rainfall and fall as a result of evapotranspiration (evaporation of water from peat and vegetation). Once the peat is saturated, excess water flows radially across the surface of the peat dome into local waterways and the convex shape of the peat surface prevents flooding from adjacent riverine or marine systems. Except at the peatland margins, inflow of nutrient-enriched river water is therefore excluded and the only source of water, and nutrients, to the greater part of the peat surface is derived from precipitation, which is relatively dilute in solutes compared to river water.

During the wet season, rainfall exceeds the combination of evaporation and groundwater run-off and the water table may come above the peat surface, at least in the hollows, for several months of the year (Ong and Yogeswaran 1992, Takahashi et al. 2002). During the drier months, when rain-free periods may last for weeks or even several months during El Niño events, the water level can drop well below the peat surface. Large peatlands tend to have wide, almost flat bog plains on the top of the dome that can store rainwater above the peat surface for

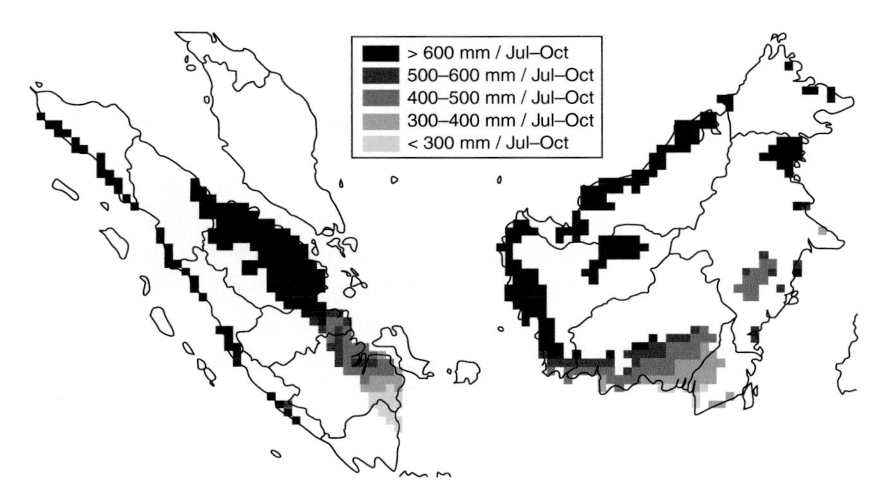

Figure 16.3 Average July–October (dry season) rainfall on peatland areas in Sumatra and Borneo, 2002–08, as calculated from 3-hourly TRMM satellite rainfall data (Hooijer et al. unpublished data), indicating that the peatlands of northern Sumatra and northern Borneo experience a much wetter dry season than the peatlands of southern Sumatra and southern Kalimantan.

long periods of time. Water flow over the peat surface is slow as a result of the very low surface gradient, but is also retarded by the high surface roughness caused by the typical hummock-hollow microtopography of an intact peat swamp forest. Surface flow can continue for several weeks after rainfall, possibly months in the case of very large undrained peat domes (few if any of which remain intact), slowly feeding into the 'blackwater' streams and rivers that drain the peatland catchment. During prolonged periods of low rainfall, outflow from the peat will be limited as water will be retained within the peat matrix (Hooijer 2005); clearly the peat system has evolved to retain water during such dry periods. Some larger peat domes contain blackwater lakes on or close to the highest part of the peat dome. Formation of such lakes may be triggered when surface wetness exceeds a certain threshold, whereby the vegetation becomes less productive and organic matter accumulation is slowed down to below the rate on the surrounding more productive peatland. Similar features are also found on temperate and boreal peatlands, though there they tend to remain the size of pools rather than large lakes.

The movement of water, and hence nutrients, within a tropical peatland has a strong influence on the vegetation. At the centre of the peat dome, where the gradient is low, the total amount of plant nutrient elements is limiting to tree growth and the peat swamp forest is dominated by small diameter trees of low stature (low biomass). The openness of the canopy allows dense thickets of pandan to develop underneath. This nutrient-stressed forest is similar in appearance to 'savannah' woodland on nutrient-poor podzol (Anderson 1964, 1983). Away from the central dome, both the gradient and the rate of water flow increase. Transitional areas receive a larger amount of total nutrients because the rainwater falling on them is enhanced with surface water flowing from the dome, although the concentrations of elements in the surface water may not indicate this. Towards the margins of the peat dome, which may be 20 km or more from the centre, the amount of water flowing off the peatland is greatest, the rate is fastest and the total amount of nutrients supplied in any period of time is greatest enabling peat swamp forest trees to achieve a higher biomass.

16.2.4 Carbon storage

In tropical peatlands, both the forest vegetation and underlying peat constitute a large and highly concentrated carbon pool that, upon degradation, releases GHGs to the atmosphere (mainly as carbon dioxide, CO_2). In undrained peat swamp forest, the water table fluctuates according to the balance between rainfall, evapotranspiration and run-off. In periods of low rainfall the uppermost peat layer becomes aerated, promoting aerobic decomposition of the organic material and the release of CO_2 through microbial respiration. During the wet season, however, the water table oscillates close to or above the peat surface and anaerobic conditions prevail, certainly in the hollows, reducing the decomposition rate and resulting in net accumulation of peat and carbon over time. Peat formation continues as long as organic matter input exceeds loss. Impacts on peatland ecology or hydrology (i.e. disturbance to the forest vegetation or lowering of the water table because of either changes in climate, in particular reduction or increased seasonality of precipitation, or anthropogenic drainage) will result, however, in changes to the rate of peat accumulation and lead to increased CO_2 emissions.

A recent study has concluded that the peatlands of Southeast Asia store 68.5 Gt (billion tonnes) of carbon, which is 77% of the total tropical and between 11 and 19% of the global peatland carbon store (Page et al. 2011). Indonesia holds the largest share of the tropical peatland carbon pool with an estimated 57.4 Gt, followed by Malaysia with 9.1 Gt; much smaller amounts are stored in the peatlands of other countries in the region (1.4 Gt in Papua New Guinea; 0.3 Gt in Brunei; 0.2 Gt in the Philippines and less than 0.1 Gt in both Myanmar and Thailand). In both Indonesia and Malaysia, the amount of peat carbon is more than twice that stored in the entire forest vegetation of those countries; 20 Gt and 4 Gt carbon are stored in the forests of Indonesia and Malaysia, respectively (Brown et al. 1993). These data emphasise the important role played by peat swamp forests in forest carbon storage and the consequences that anthropogenic destabilisation of these pools can have for the overall regional carbon balance.

16.3 Human impacts on tropical peatlands: cause and consequence

16.3.1 Drivers of forest loss and degradation

Over the last two decades, the hydrological and carbon sequestration and storage functions of most Southeast Asian peatlands have been either severely impaired or completely destroyed. While some inland peatlands may have ceased peat accumulation some 2–3 ka (Sieffermann et al. 1988), the degradation of most peatlands has occurred only in the last two to three decades. A combination of forest loss and land drainage has destroyed the carbon sequestration mechanism (peat swamp forest) and led to lowered water tables. Drainage leads to aerobic conditions that favour microbial activity in the peat above the water table, resulting in enhanced CO_2 loss by peat oxidation (Jauhiainen et al. 2005). Forest disturbance and peat drainage also increase the risk of fire (Page et al. 2002, 2009b, Siegert et al. 2001) promoting additional, rapid loss of carbon from the peatland store.

The Southeast Asian region is experiencing a relative rate of forest loss that is greater than that for other tropical regions (Achard et al. 2002). Hansen et al. (2009) estimated that the Sumatran and Kalimantan lowlands experienced an increase in an annual rate of deforestation from 1.27% per year in 1990–2000 to 1.4% per year in 2000–05. Even forests in protected areas are being lost – those in Kalimantan, for example, by more than 56% (2.9 million ha) between 1985 and 2001 (Curran et al. 2004). Langner et al. (2007) showed that Borneo, in particular, experienced very rapid deforestation between 2002 and 2005 at an average rate of 1.7% per year; while the peat swamp forests were being cleared or degraded at the even higher rate of 2.2% per year. In Sumatra, Uryu et al. (2008) described a similar situation for the Province of Riau where more than 4 million ha of forest (65% of the original area) were lost between 1982 and 2007, including 57% (1 831 193 ha) of the province's peat swamp forests. Between 2005 and 2006, an 11% deforestation rate was reported in Riau, which is one of the highest deforestation rates in the world. The studies of Langner et al. (2007) and Uryu et al. (2008) confirm that while the loss of tropical peat swamp forest is accelerating, the deforestation of other rainforest types on mineral soils is reducing.

The principle drivers of peat swamp forest loss and degradation are timber extraction and development of plantation estates for export crops (palm oil and paper pulp). In contrast, direct population pressure plays only a minor role because these peatlands have traditionally not been inhabited and are still considered unsuitable for human habitation. Peat swamp forests contain a number of valuable timber trees and they have been used for legal (controlled) concession-based logging for at least four to five decades, both in Malaysia and Indonesia. This type of selective timber extraction, using low impact logging methods, has the potential to cause limited ecological damage. A lack of enforcement of basic logging regulations, especially in Indonesia, has, however, led to over-cutting, extraction outside concession boundaries (Kartawinata et al. 2001) and application of unsuitable log transport techniques using canals, which have lowered peatland water tables. Endemic peat swamp tree species with a high timber value, including *Gonystylus bancanus* (ramin) and *Shorea albida* (alan), have been heavily logged and are now severely threatened with extinction (IUCN 2009). Logging has also greatly increased the risk of fire (Siegert et al. 2001, Page et al. 2002), discussed in more detail below.

In Southeast Asia, rates of forest conversion to agriculture (plantations and food-crops) are among the highest in the world and these trends are expected to continue for a number of decades (IPCC 2000). Plantation development has led to the widespread loss of peat swamp forest in Sumatra (Uryu et al. 2008), and a similar trend is becoming established in Kalimantan and Sarawak. Up until 20 years ago, most cultivation took place on shallow peat soils and on a relatively limited scale but, despite the inherent problems associated with intensive cultivation of nutrient-poor, acidic peat soils, millions of hectares of deep peats in both Indonesia and Malaysia have been deforested and drained to make way for plantations of oil palm and pulpwood (fast-growing *Acacia*) which produce raw materials for the vegetable oil, bio-fuel, pulp and paper industries. Indonesia and Malaysia currently account for 86% of global palm oil production (Basiron 2007). More than a quarter of oil palm concessions in Indonesia are on peat (Hooijer et al. 2006, 2010), accounting for a land area of ~28 000 km^2, while in Malaysia at least 4200 km^2 of peatland have been developed for this purpose. Plantation methods dictate that the peat is drained to considerable depth; the optimal water table for oil palm and pulp wood production is around 0.6–0.8 m, but depths of 1.0–1.5 m are commonly observed (Hooijer et al. 2006). Thus a deep surface layer of peat is exposed to oxidative degradation, resulting in enhanced CO_2 emissions to the atmosphere and rapid land surface subsidence (Fig 16.4).

In recent years, forest fires have become a regular feature of the Southeast Asian region, with some of the most extensive occurring in peatland areas (Langner et al. 2007, Langner and Siegert 2009, Page et al. 2002). Contemporary land-use and land-cover changes have played a significant role in the enhanced frequency and intensity of fires, and the increasing incidence of fires in peatland areas has been linked to logging, drainage and increased human access (Langner et al. 2007, Langner and Siegert 2009, Page et al. 2009a, 2009b).

In the last 10 years, degraded peatlands, particularly in Kalimantan, have become the focus of widespread and uncontrolled fires owing to increased forest degradation and peatland drainage; the former Mega Rice Project (MRP) area in Central Kalimantan, where fires are now of regular occurrence, is a case in point. Conditions in the MRP area have changed considerably over the past

Figure 16.4 One of the authors (CB) standing next to a peat subsidence measurement pole in drained peatland in Johor, southern peninsular Malaysia. Around 2.3 m of peat have been lost over a period of 28 years, a consequence of drainage, compaction and peat oxidation. See plate section for colour version.

15–20 years. This degradation process began with widespread logging, followed by construction of massive drainage systems (with canals that cut across peat domes) and widespread, frequent fires. For example, within a 4500 km^2 study area in the western section of the MRP (Hoscilo et al. 2008a), fire affected 24% of the area over the period 1973–96. Most burn scars were located along forest edges (i.e. in disturbed forest), usually in close proximity to human settlements, which provided a source of ignition. More remote, intact forests were unaffected by fire, even during extended ENSO (El Niño Southern Oscillation) droughts. This situation changed markedly following construction of the extensive MRP canal network in 1995–96. For example, the fires of 1997 affected about 33% of the study area, 10% more than had burned during the previous 23 year period while, in 2002, fires affected 22% of the area. Two years later, in 2004 and then in 2005 (non-El Niño years) fires affected again 14% and 12% of the area, respectively. In total, 45% of the study area was subject to multiple fires, with 37% burnt twice and 8% burnt three or more times during 1997–2005. Further extensive fires occurred subsequently in 2006 (Field and Shen 2008, van der Werf et al. 2008) and 2009, indicating the susceptibility of drained, deforested peatlands to fire. As well as high emissions of GHGs and particulates, peatland fires also result in land subsidence and an enhanced risk of prolonged flooding.

16.3.2 Effects of peat swamp forest degradation and loss on biodiversity

Outright conversion of peat swamp forest to other uses has led to a considerable reduction in the area of this habitat. A recent study involving analysis of remote sensing data for the peatland areas of Kalimantan and Sumatra has shown that only around 4% of the peat swamp forests in these two regions of Indonesia remain in good condition (Miettinen and Liew 2010), with a further 46% comprising slightly to heavily degraded and secondary forests and 48% deforested land (the remaining 2% is made up of open water). In a similar smaller-scale study in Sarawak, only 1.5% of that State's peat swamps were found to be relatively untouched, with the remainder logged or converted to other land uses (van der Meer et al. 2005). This scale of habitat loss, fragmentation and degradation threatens the survival of many characteristic and rare peat swamp forest species. In addition, the longer-term sustainability of the remaining blocks of peat swamp forest is now in doubt since off-site impacts, in particular deep drainage for plantations, can have an effect several kilometres beyond the edge of the agricultural land, threatening the hydrological integrity and sustainability of the remnant forest blocks as a result of increased tree fall, alterations to species assemblages and enhanced risk of wildfire.

The immediate effects of logging on peat swamp forest are changes in the forest structure, notably reduction in canopy height and an increase in canopy openness; this in turn affects the forest microclimate, increasing the temperature, reducing humidity and altering the light levels reaching the ground. Logging can damage the hummock–hollow microtopography, as logs are dragged out across the peat surface, with consequences for site hydrology through increased surface run off. Timber extraction also removes vital nutrients from the forest, which may impede subsequent forest recovery. Studies of peat swamp forest regeneration following logging activities have shown that fast-growing tree species, such as *Xylopia coriifolia*, *Litsea* spp., *Dactylocladus stenostachys*, *Cratoxylum arborescens* and *C. glaucum*, are favoured over slower-growing species, including *Combretocarpus rotundatus*, *Gluta* spp., *Palaquium* spp. and *Gonystylus bancanus* (Lee 1979, Whitmore 1984, Dwiyono and Rachman 1996, van der Meer et al. 2005). In heavily logged areas, the regenerating forest may become dominated by stands of fast-growing species, notably *Macaranga* spp. (Ibrahim 1996). The effects of logging on the fauna of peat swamp forest are less well understood. In a review of the effects of timber concession management on animal species across a range of forest types in Borneo, Meijaard et al. (2005) concluded that well-managed production forests could provide adequate habitat for most animal species. Mammals showing highest tolerance to logging included species with a wide ecological niche, i.e. those that could easily adapt to the disturbed conditions and feed on a range of different food items. In contrast, the fragmentation of large contiguous forests into smaller forest blocks affected a much broader range of species, with populations of virtually all species reduced by the loss of habitat. They noted that an increase in the ratio of forest edge to area had significant negative effects on population and community dynamics in the remnant forests and that the effects of logging extended for considerable distances into surrounding undisturbed forests.

More severe and long-lasting changes to the structure and species composition of peat swamp forest occur following fire. Peat swamp forest trees are not fire-adapted; most are thin barked, so post-fire tree mortality is high. In addition, many trees fall once the supporting peat is burnt away. In studies carried out on post-fire vegetation succession on peatland in Central Kalimantan, Page et al. (2009a, 2009b) and Hoscilo et al. (2008b) have shown that sites subject to a single, low intensity fire, subsequently undergo succession to secondary peat swamp forest, achieving a biomass equivalent to about 10% of undisturbed forest within 9 years. Compared to forest fires on mineral soils, however, the rate of recovery is slow (Nykvist 1996) and there is a large loss of tree species diversity, with only ~16 species recorded in 9 year old secondary regrowth, compared to ~250 species in the undisturbed forest. In addition, the secondary regrowth is mainly dominated by just two pioneer species: *Combretocarpus rotundatus* and *Cratoxylon glaucum*. Following multiple fires, the numbers of tree species and individual trees, saplings and seedlings within the secondary vegetation are greatly reduced and, following three or more fires over a short time period (<10 years), secondary succession back to forest is diverted to plant communities dominated by ferns with few or no trees (Fig 16.5). At this point, woody biomass and hence carbon sequestration potential are greatly reduced. Spessa et al. (2009) demonstrated that the occurrence of uncontrolled fires is favoured by the increased fire susceptibility of both over-drained peatland and previously disturbed forests. The greater the loss of tree cover, for example, the greater the risk of fire, a consequence of both the amount of dead and damaged vegetation left behind by logging and changes in the forest microclimate (Siegert et al. 2001). The early stages of the transformation of previously tree-dominated ecosystems to mostly fern-dominated ecosystems can promote more fires as part of a so-called 'vicious positive feedback loop' (Cochrane et al. 1999, Cochrane 2003) since fern-dominated communities dry out quickly during periods of low rainfall and thus burn more easily, enhancing flammability (Page et al. 2009a, 2009b). Once fern communities are established, however, their fuel load is very low and thus it is less likely that surface fires will generate sufficient heat to establish ground fires within the peat.

Figure 16.5 Fire-impacted peatland in Central Kalimantan. This part of the former Mega Rice Project area (see text for details) has been burnt at least three times since 1997. The vegetation, which was once species-diverse peat swamp forest, is now dominated by ferns and just two or three tree species capable of withstanding the greatly altered environmental conditions. See plate section for colour version.

A further consequence of repeated fires on tropical peatland is land subsidence, a result of peat dewatering, biological oxidation and combustion, which creates depressions on the surface and an increased risk of permanent or semi-permanent flooding. Van Eijk and Leenman (2004) and Wösten et al. (2006) describe post-fire vegetation trends in the Berbak peatland, Sumatra, where sites subject to multiple burns experienced prolonged flooding and had a poorly developed and low diversity vegetation dominated by flood-tolerant, non-woody species.

There have been very few studies of fire impacts on peat swamp forest fauna. Larger, more mobile animals are able to move away from the immediate location of a fire, but populations of smaller, less mobile species will probably decline. Personal observations indicate that some taxa increase in abundance in the immediate aftermath of a fire (tree shrews and certain invertebrate groups such as arachnids, for example); similar increases in abundance and diversity have been noted in studies of other types of lowland forests (e.g. Barlow and Peres 2004, Slik and van Balen 2006). Fire may also have longer-term and less direct impacts on animal populations and behaviour through reductions in the size of populations and mortality rates (O'Brien et al. 2003). Cheyne et al. (2007), for example, noted that gibbons (*Hylobates albibarbis*) sing less frequently in smoky conditions, which could interfere with territorial spacing and, ultimately, reproduction. A reduction in mammal and bird species involved in seed dispersal will also have long-term implications for forest regeneration.

16.3.3 Effects of forest degradation and loss on peatland hydrology, carbon storage and emissions

A major part of the carbon stock in peat swamp forest is located in the peat carbon pool rather than in the aboveground vegetation, but the preservation of the forest and its diversity are essential in order to maintain the carbon sequestration function of the peatland, i.e. the assimilation of CO_2 from the atmosphere and its ultimate addition, as leaf litter, woody material and dead roots, to the accruing peat deposit. Undisturbed peat swamps are rich in both biodiversity and carbon, whilst degraded peat swamps support a reduced biodiversity and can no longer function as a net carbon sink (Jauhiainen et al. 2005), thus there is a link between these two attributes. Individual tree species characteristics will determine how much carbon is assimilated from the atmosphere and how much is made available for long-term storage within the peat. It is unclear, however, how a change in the tree species composition of a peat swamp forest, which might follow disturbance by logging or drainage, affects the ecosystem carbon sequestration function. Tree species that promote a high C:N ratio within plant litter may retard decomposition and increase carbon storage within the peat, but alteration of forest structure and diversity will undoubtedly lead to changes in the quality and quantity of the organic matter being produced. There have been insufficient studies of these aspects of peat swamp ecosystem carbon dynamics to establish precise relationships and there is a vital need to understand better the link between species (particularly tree species) diversity and ecosystem functioning. In addition, the effects of anthropogenic disturbance on microbiological diversity and the fate of below-ground carbon remains largely unexplored.

When an intact peatland is drained, for example by the construction of canals, the most immediate effects are that the residence time of water on the peat surface is reduced and surface water depths in wet periods are reduced. This lowering of surface water levels causes decomposition of hummocks and reduces the hummock–hollow surface roughness leading to a further increase in the speed of water run-off. When forest is removed before or during drainage, these processes of increased surface flow and reduced water levels are accelerated and where fires have occurred the surface microtopography is lost even faster. Permanent lowering of the peatland water table leads to enhanced aerobic decomposition and hence oxidation of stored carbon; additional effects include peat surface compaction and subsidence.

At the present time, around 130 000 km^2 (i.e. ~50%) of peatland in Southeast Asia have been deforested and drained to provide land for plantation or smallholder agriculture (Hooijer et al. 2006). The CO_2 emissions arising from peat oxidation in drained peatlands are in the range 15–86 t ha^{-1} yr^{-1}, depending on the average drainage depth, and are estimated to give rise to an annual carbon loss of 173 Mt a^{-1} (range 97–239 Mt yr^{-1}, i.e. 355–874 Mt CO_2 yr^{-1}) (Hooijer et al. 2006, 2010). This level of carbon emission is equivalent to 2.1% of current global fossil fuel emissions and is predicted to increase in coming decades unless current peatland management practices and development plans are changed.

Peatland fires, which ignite both the surface vegetation and, if the peat water table reaches a critical depth, also the underlying peat, are a further source of carbon emissions. As a result of peat combustion, these fires release substantially larger amounts of CO_2 and particulates into the atmosphere than fires on mineral soils that contain lower amounts of soil organic carbon. These emissions contribute to climate change processes and reduce air quality, cause human health problems and destabilise regional economies (Schweithelm 1999, Heil and Goldammer 2001, Lohman et al. 2007, Field et al. 2009). The devastating 1997–98 Indonesian fires were one of the largest peak emissions events in the recorded history of fires in equatorial Southeast Asia, if not globally (van der Werf et al. 2006, Schultz et al. 2008). Page et al. (2002), for example, conservatively estimated that the Indonesian fires in 1997 released more than 870 Mt of carbon to the atmosphere, which was equivalent to 14% of the average global annual fossil fuel emissions released during the 1990s. The most severe fires of recent years have been linked to ENSO events, which cause extended drought periods, particularly across the peatlands of southern Sumatra and southern Kalimantan. Undisturbed peat swamp forest is at low risk of fire, even during dry periods, because humidity remains high underneath the canopy and the water table rarely falls below 0.8 m from the surface (Page et al. 1999), i.e. the depth that needs to be exceeded before peat above the water table dries out sufficiently in order to ignite. Following deforestation and drainage the water table in tropical peatland can fall to much lower levels, often exceeding 1.5 m; under these conditions the surface peat to a depth of 0.5 m or more becomes so dry that it can ignite if a surface vegetation fire breaks out (Page et al. 2009b).

Data on the carbon losses associated with tropical peatland development have been used in several studies investigating the link between the growing demand for biofuels and climate change. Fargione et al. (2008) demonstrated that it would take 420 years for the carbon emissions saved by the use of palm oil as a biofuel to compensate for the carbon lost through deforestation and peat decomposition following conversion of peat swamp forest to plantation. In a second study, Danielsen et al. (2009) estimated this time period to be longer, at 600 years, and also stressed the consequences of plantation development for biodiversity, with significant reductions in both floral and faunal communities. Both studies highlight, yet again, the global consequences of tropical peatland development.

16.3.4 Future prospects: climate change impacts

Deforestation, drainage and fire are currently the dominant drivers of biodiversity loss and carbon emissions from tropical peatlands but future climate change will add further pressure. The peatlands of Southeast Asia lie within the inter-tropical convergence zone that experiences a wet tropical climate with annual rainfall generally in excess of 2000 mm, up to 4000 mm. The seasonality of rainfall varies, however, from moderate to strong (Fig 16.3), with the peatlands of northern Sumatra and Borneo experiencing a shorter, wetter July–October 'dry' season than the peatland areas further south. Significant inter-annual climatic variation is also associated with the 'warm' and 'cold' phases of the ENSO cycle and the positive phase of the Indian Ocean Dipole (Fuller and Murphy 2006, van der Werf et al. 2008). These events have an important and well-documented role in controlling rainfall variation across Southeast Asia but, particularly, across Indonesia (Field et al. 2009), and can periodically result in an extended dry season with a rainfall deficit during more than 6 months in some years. The peatland areas of southern Sumatra and southern Kalimantan are particularly sensitive to these climate extremes and can experience severe drought conditions with water tables dropping to 1 m below the peat surface even in undrained areas, with implications for their future sustainability, particularly where they are degraded by logging and drainage.

Climate change predictions for the Southeast Asian region are for a median warming of 2.5°C by the end of the twenty-first century accompanied by a predicted mean precipitation increase of about 7% and a rise in sea level (IPCC 2007). Wet season rainfall will likely increase and dry season rainfall decrease, with the peatlands in the southern parts of Borneo and Sumatra, which are already the driest, experiencing a further increase in the frequency and severity of drought conditions (Li et al. 2007). The future behaviour of ENSO is uncertain, but a recent study by Abram et al. (2007) indicates that Indonesia as a whole could expect more frequent and longer droughts in the future, again threatening the longer-term survival of fragmented and degraded peat swamp forests.

These climate predictions foretell that there is a considerable risk of continuing or increasing carbon emissions from Southeast Asian peatlands in the coming decades, both from peat oxidation but particularly from peat fires, unless large-scale improvements in water and fire management are possible. The biodiversity, carbon stocks and hydrological regulatory capacity of Southeast Asian peatlands will become increasingly vulnerable. The main effects of climate change on tropical peatland will most likely be through changes in temperature and precipitation, which will influence peat hydrology and thus forest tree species composition. Decreased rainfall during the dry season will result first in lower peatland water tables, thus exposing larger carbon stocks to aerobic conditions and increased decomposition through enhanced microbial activity, and, second, an enhanced risk of fire. The palaeo-record indicates that peat swamp forests can certainly adapt to wetter conditions, which could stimulate increased rates of peat and carbon accumulation but, if higher wet season rainfall is accompanied by less during a subsequent prolonged dry season, there are likely to be net peat and carbon losses. In addition, enhanced wet season rainfall will place degraded peatlands that are subsiding due to peat decomposition and fire at a greater risk of flooding. The effects of a rising sea level on peatland sustainability are less easy to predict; coastal peatlands will experience reduced drainage which could lead to increased peat accumulation rates at the least degraded sites, but where there has been land conversion for agriculture the result will be decreased drainability and an increased risk of both coastal and riverine flooding.

In addition to longer-term impacts of climate change, deforestation itself, i.e. large-scale alterations in land cover, may lead to local climate changes. Henderson-Sellers and Gomitz (1984) ran a simulation of the effect of transforming forested regions of Southeast Asia to scrub grassland. They predicted a reduction in rainfall, although the simulated effect was relatively weak. In a later modelling study, Werth and Avissar (2005) simulated a similar effect as a result of deforestation. They predicted local precipitation would be significantly reduced for each month of the year, with Borneo experiencing one of the largest changes in February, at the peak of the rainy season. This would have implications for peatland forest functioning, peat decomposition rates and fire risk, as well as the feasibility of current initiatives to rehabilitate peatlands degraded by deforestation and drainage and to conserve the remaining peat carbon store.

16.4 Conclusions

Tropical peat swamp forests are distinctive and irreplaceable habitats that support a diverse assemblage of habitats and wildlife. Owing to the unique environmental conditions found in these forests, there is a high degree of habitat endemism among certain taxa but they also provide an increasingly important refuge for a large number of more widely distributed lowland forest species. These forested peatlands have a very high carbon density and contain a globally significant terrestrial carbon store. The important goods and services that these peatlands provide are, however, under threat. Only a small proportion of the original peat swamp forest cover in the region remains in anything like a near-natural condition and even in forests with protected area status, ineffective management has resulted in widespread timber extraction, incursions for land conversion, fire and on- and off-site damage caused by drainage. The current prospects for intact areas of this ecosystem surviving into the second half of the twenty-first century are not promising, and GHG emissions from degraded peatlands are likely to continue at a substantial level for some decades to come.

New opportunities for protection of the remaining tropical peat swamp forests seem likely to arise as a result of policy initiatives to reduce greenhouse gas emissions through avoided deforestation in developing countries (Reduced Emissions from Deforestation and Degradation – REDD – and the voluntary carbon market) (Murdiyarso et al. 2008). In this respect, the greatest savings in terms of carbon emissions will be obtained by protection of the remaining forested peatlands (van der Werf et al. 2009), but rehabilitation of degraded peatland landscapes also offers opportunities for rewetting, fire control and reforestation under either formal or informal financial agreements to reduce GHG emissions. Initial restoration trials in the former MRP in Central Kalimantan have been undertaken to investigate the effectiveness of various forms of intervention to protect the remaining carbon store (Page et al. 2009a) and a REDD demonstration project, funded by the Australian Government, is now under way in this area. Technical constraints on these types of activities in a highly degraded landscape are, however, numerous, ranging from the problems associated with constructing water-retention structures across drainage canals in peat which has low load-bearing strength and sometimes a high hydraulic conductivity, through to establishing effective fire control and developing appropriate arboricultural techniques for large-scale reforestation. When combined with natural constraints, such as severe and prolonged, and possibly increasing, rainfall deficits and/or an increased risk of flooding, the challenges for peatland rehabilitation are significant and the long-term prospects for rehabilitation schemes are difficult to predict.

Nevertheless, these new initiatives when combined with an increasing political willingness to limit further development of forested peatlands, could ensure that at least some peat swamp forests are maintained for future generations, whilst the ability of degraded peatlands to deliver a range of ecosystem services could be enhanced. Whether these initiatives are too little and too late for an ecosystem that is highly vulnerable to environmental change remains, however, uncertain at the present time.

Acknowledgements

Contributions by Page, Banks, Hoscilo and Rieley were, in part, carried out under the EU-funded Restorpeat and Carbopeat projects (FP6 INCO-DEV 510931 and FP6 INCO-CT 043743). Contributions by Hooijer were provided under the Deltares PEAT-CO_2 RandD project and the Singapore Delft Water Alliance (SDWA) Peatland Programme.

References

Abram, N. J., Gagan, M. K., Liu, Z. et al. (2007). Seasonal characteristics of the Indian Ocean Dipole during the Holocene epoch. *Nature*, **445**, 299–302.

Achard, F., Eva, H. D., Stibig, H.-J. et al. (2002). Determination of deforestation rates of the world's humid tropical forests. *Science*, **297**, 999–1002.

Anderson, J. A. R. (1964). The structure and development of peat swamps of Sarawak and Brunei. *Journal of Tropical Geography*, **18**, 7–16.

Anderson, J. A. R. (1983). The tropical peat swamps of western Malesia. In *Ecosystems of the World: Mires: Swamp, Bog, Fen and Moor. 4B, Regional Studies*, ed. A. J. P. Gore. New York: Elsevier, pp. 181–99.

Anderson, J. A. R. and Muller, J. (1975). Palynological study of a Holocene peat and a Miocene coal deposit from NW Borneo. *Review of Paleobotany and Palynology*, **19**, 291–351.

Anshari, G., Kershaw, A. P. and van der Kaars, S. (2001). A Late Pleistocene and Holocene pollen and charcoal record from peat swamp forest, Lake Sentarum Wildlife Reserve, West Kalimantan, Indonesia. *Paleogeography, Paleoclimatology, Paleoecology*, **171**, 213–28.

Balvanera, P., Pfisterer, V. N., He, B. et al. (2006). Quantifying the evidence for biodiversity effects on ecosystem functioning and services. *Ecology Letters*, **9**, 1146–56.

Barlow, J. and Peres, C. A. (2004). Avifaunal responses to single and recurrent wildfires in Amazonian forests. *Ecological Applications*, **14**, 1358–73.

Basiron, Y. (2007). Palm oil production through sustainable plantations. *European Journal of Lipid Science Technology*, **109**, 289–95.

Bezuijen, M. R., Webb, G. J. W., Hartoyo, P. and Samedi. (2001). Peat swamp forest and the false gharial *Tomistoma schlegelii* (Crocodilia, Reptilia) in the Merang River, eastern Sumatra, Indonesia. *Oryx*, **35**, 301–7.

Brady, M. A. (1997). Effects of vegetation changes on organic matter dynamics in three coastal peat deposits in Sumatra, Indonesia. In *Biodiversity and Sustainability of Tropical Peatlands*, ed. J. O. Rieley and S. E. Page. Cardigan, UK: Samara Publishing, pp. 113–34.

Brown, S., Iverson, L. R., Prasad, A. and Liu, D. (1993). Geographical distributions of carbon in biomass and soils of tropical Asian forests. *Geocarto International*, **4**, 45–59.

Buckley, C., Nekaris, K. A. I. and Husson, S. J. (2006). Survey of *Hylobates agilis albibarbis* in a logged peat-swamp forest: Sabangau catchment, Central Kalimantan. *Primates*, **47**, 327–35.

Cheyne, S. M. (2008). http://outrop.blogspot.com/2008/08/clouded-leopard-discovery-is-top-story.html.

Cheyne, S. M., Thompson, C. J. H., Phillips, A. C., Hill, R. M. C. and Limin, S. H. (2007). Density and population estimate of gibbons (*Hylobates albibarbis*) in the Sabangau catchment, Central Kalimantan, Indonesia. *Primates*, **24**, 60–80.

Cochrane, M. A. (2003). Fire science for rainforests. *Nature*, **421**, 913–19.

Cochrane, M. A., Alencar, A., Schulze, M. D. et al. (1999). Positive feedbacks in the fire dynamic of closed canopy tropical forests. *Science*, **284**, 1832–5.

Curran, L. M., Trigg, S. N., McDonald, A. K. et al. (2004). Lowland forest loss in protected areas of Indonesian Borneo. *Science*, **303**, 1000–1003.

Danielsen, F., Beukema, H., Burgess, N. D. et al. (2009). Biofuel plantations on forested lands: double jeopardy for biodiversity and climate. *Conservation Biology*, **23**, 348–58.

Das, I. (2006). The reptile fauna. In *The Biodiversity of a Peat Swamp Forest in Sarawak*, ed. F. Abang and I. Das. Kota Samarahan, Malaysia: Universiti Malaysia Sarawak, pp. 115–28.

Das, I. and Lim, K. K. P. (2001). A new *Bufo* (Anura: Bufonidae) from the peat swamps of Selangor, west Malaysia. *The Raffles Bulletin of Zoology*, **49**, 1–6.

Dennis, C. and Aldhous, P. (2004). A tragedy with many players. *Nature*, **430**, 396–8.

Dwiyono, A. and Rachman, S. (1996) Management and Conservation of the tropical peat forest of Indonesia. In *Tropical Lowland Peatlands of Indonesia*, ed. E. Maltby, C. P. Immirzi and R. J. Safford. Gland, Switzerland: IUCN, pp. 103–17.

Fargione, J., Hill, J., Tilman, D., Polasky, S. and Hawthorne, P. (2008). Land clearing and the biofuel carbon debt. *Science*, **319**, 1235–8.

Field, R. D. and Shen, S. S. P. (2008). Predictability of carbon emissions from biomass burning in Indonesia from 1997 to 2006. *Journal of Geophysical Research*, **113**, doi:10.1029/2008jg000694

Field, R. D., van der Werf, G. R. and Shen, S. S. P. (2009). Human amplification of drought-induced biomass burning in Indonesia since 1960. *Nature Geoscience*, **2**, 185–8.

Fuller, D. O. and Murphy, K. (2006). The ENSO-fire dynamic in insular Southeast Asia. *Climatic Change*, **74**, 435–55.

Gagan, M. K., Hendy, E. J., Haberle, S. G. and Hantoro, W. S. (2004). Post-glacial evolution of the Indo-Pacific Warm Pool and El Niño-Southern oscillation. *Quaternary International*, **118–19**, 127–43.

Gaither, Jr., J. C. (1994). Understory avifauna of a Bornean peat swamp forest: is it depauperate? *Wilson Bulletin*, **106**, 381–90.

Hansen, M. C., Stehman, S. V., Potapov, P. V. et al. (2009). Quantifying changes in the rates of forest clearing in Indonesia from 1990 to 2005 using remotely sensed data sets. *Environmental Research Letters*, **4**, doi:10.1088/1748-9326/4/3/034001.

Heil, A. and Goldammer, J. G. (2001). Smoke-haze pollution: a review of the 1997 episode in South-east Asia. *Regional Environmental Change*, **2**, 24–37.

Henderson Sellars, A. and Gomitz, V. (1984). Possible climatic impacts of land cover transformations with an emphasis on tropical deforestation. *Climatic Change*, **6**, 231–58.

Hooijer, A. (2005). Hydrology of tropical wetland forests: recent research results from Sarawak peatswamps. In *Forests–Water–People in the Humid Tropics*, ed. M. Bonell and L. A. Bruijnzeel. Cambridge: Cambridge University Press, pp. 447–61.

Hooijer, A., Silvius, M., Wösten, H. and Page, S. (2006). *PEAT-CO_2, Assessment of CO_2 Emissions from Drained Peatlands in SE Asia*. Delft, The Netherlands: Delft Hydraulics report Q3943, 36 pp.

Hooijer, A., Page, S. E., Canadell, J. et al. (2010) Current and future CO_2 emissions from drained peatlands in SE Asia. *Biogeosciences*, **7**, 1505–14.

Hope, G., Chokkalingam, U. and Anwar, S. (2005). The stratigraphy and fire history of the Kutai Peatlands, Kalimantan, Indonesia. *Quaternary Research*, **64**, 407–17.

Hoscilo, A., Page, S. E. and Tansey, K. J. (2008a). Spatial and temporal alterations of tropical peatlands (Indonesia) due to widespread and repeated fires over the last 30 years. In *Proceedings of the Remote Sensing and Photogrammetry Society Conference*, Exeter, UK, September 2008.

Hoscilo, A., Page, S. E. and Tansey, K. J. (2008b) Development of post-fire vegetation in the tropical ecosystem of Central Kalimantan, Indonesia. In *Proceedings of the 13th International Peat Congress, Tullamore*, Ireland, June 2008, pp. 202–205.

Ibrahim, S. (1996). Forest management systems in peat swamp forest: a Malaysian perspective. In *Tropical Lowland Peatlands of Southeast Asia*, ed. E. Maltby, C. P. Immirzi and R. J. Safford. Gland, Switzerland: Proceedings of a Workshop on Integrated Planning and Management of Tropical Lowland Peatlands, IUCN Wetlands Programme, pp. 175–180.

IPCC (2000). *IPCC Special Report on Emissions Scenarios*. Available at: http://www.grida.no/publications/other/ipcc_sr/?src=/climate/ipcc/emission/.

IPCC (2007). *Climate Change 2007 Synthesis Report, Impacts, Adaptation and Vulnerability*. Contribution of Working Groups I, II and III to the Fourth Assessment Report of the Intergovernmental Panel on Climate Change. Cambridge: Cambridge University Press.

IUCN (2009). *IUCN Red List of Threatened Species. Version 2009.2.* Available at: www.iucnredlist.org, accessed 16 February 2010.

Janzen, D. H. (1974). Tropical blackwater rivers, animals, and mast fruiting by the Dipterocarpaceae. *Biotropica*, **6**, 69–103.

Jauhiainen, J., Takahashi, H., Heikkinen, J. E. P., Martikainen, P. J. and Vasander, H. (2005). Carbon fluxes from a tropical peat swamp forest floor. *Global Change Biology*, **11**, 1788–97.

Johnson, D. S. (1967). Distributional patterns in Malayan freshwater fish. *Ecology*, **48**, 722–30.

Kartawinata, K., Riswan, S., Gintings, A. N. and Puspitojati, T. (2001). An overview of post-extraction secondary forests in Indonesia. *Journal of Tropical Forest Science*, **13**, 621–38.

Kottelat, M., Britz, R., Tan, H. H. and Witte, K. E. (2006). *Paedocypris*, a new genus of Southeast Asian cyprinid fish with a remarkable sexual dimorphism, comprises the world's smallest vertebrate. *Proceedings of the Royal Society B-Biological Sciences*, **273**, 895–99.

Langner, A. and Siegert, F. (2009). Spatiotemporal fire occurrence in Borneo over a period of 10 years. *Global Change Biology*, **15**, 48–62.

Langner, A., Miettinen, J. and Siegert, F. (2007). Land cover change 2002–2005 in Borneo and the role of fire derived from MODIS imagery. *Global Change Biology*, **13**, 2329–40.

Lee, H. S. (1979). Natural regeneration and reforestation in the peat swamp forests of Sarawak. *Tropical Agricultural Research Center*, **12**, 51–60.

Li, W., Dickinson, R. E., Fu, R. et al. (2007). Future precipitation changes and their implications for tropical peatlands. *Geophysical Research Letters*, **34**, L01403, doi:10.1029/2006GL028364.

Lohman, D. J., Bickford, D. and Sodhi, N. (2007). The burning issue. *Science*, **316**, 376.

Meijaard, E. and Nijman, V. (2000). Distribution and conservation of the proboscis monkey (*Nasalis larvatus*) in Kalimantan, Indonesia. *Biological Conservation*, **92**, 15–24.

Meijaard, E., Sayer, J. A. and Inger, R. F. (2005) *Life after Logging: Reconciling Wildlife Conservation and Production Forestry*. Jakarta, Indonesia: CIFOR and UNESCO.

Miettinen, J. and Liew, S. C. (2010): Status of peatland degradation and development in Sumatra and Kalimantan. *Ambio*, 10.1007/s13280-010-0051-2

Morley, R. J. (1981). Development and vegetation dynamics of a lowland ombrogenous peat swamp in Kalimantan Tengah, Indonesia. *Journal of Biogeography*, **8**, 383–404.

Morrogh-Bernard, H., Husson, S., Page, S. E. and Rieley, J. O. (2003). Population status of the Bornean orang-utan *(Pongo pygmaeus)* in the Sebangau peat swamp forest, Central Kalimantan, Indonesia. *Biological Conservation*, **110**, 141–52.

Murdiyarso, D., Suryadiputra, N., Dewi, S. and Agus, F. (2008). How can REDD scheme support the management of vulnerable carbon pools of Indonesian peatlands. In *Proceedings of the 13th International Peat Congress, Tullamore, Ireland, 10th June 2008*, eds C. Farrell and J. Feehan. Jyväskylä, Finland: International Peat Society, pp. 230–32.

Myers, N., Mittermeier, R. A., Mittermeier, C. G., da Fonseca, G. A. B. and Kent, J. (2000). Biodiversity hotspots for conservation priorities. *Nature*, **403**, 853–8.

Ng, P. K. L. and Kottelat, M. (1992). *Betta livida*, a new fighting fish (Teleostei: Belontiidae) from blackwater swamps in Peninsular Malaysia. *Ichthyological. Exploration of Freshwaters*, **3**, 177–82.

Ng, P. K. L. and Lim, K. K. P. (1993). The Southeast Asian catfish genus *Encheloclarias* (Teleostei: Clariidae), with descriptions of four new species. *Ichthyological. Exploration of Freshwaters*, **4**, 21–37.

Ng, P. K. L., Tay, J. B. and Lim, K. K. P. (1994). Diversity and conservation of blackwater fishes in Peninsular Malaysia, particularly in the North Selangor peat swamp forest. *Hydrobiologia*, **285**, 203–18.

Nykvist, N. (1996). Regrowth of secondary vegetation after the 'Borneo fire' of 1982–1983. *Journal of Tropical Ecology*, **12**, 307–12.

O'Brien, T. G, Kinnaird, M. F., Nurcahyo, A., Prasetyaningrum, M. and Iqbal, M. (2003). Fire, demography and the persistence of siamang (*Symphalangus syndactylus*: Hylobatidae) in a Sumatran rainforest. *Animal Conservation*, **6**, 115–21.

Ong, B. Y. and Yogeswaran, M. (1992). Peatland as a resource for water supply in Sarawak. In *Proceedings of the International Symposium on Tropical Peatland. Kuching, Sarawak*. Malaysia: MARDI, pp. 225–68.

Page, S. E., Rieley, J. O., Shotyk, W. and Weiss, D. (1999). The interdependence of peat and vegetation in tropical peat swamp forest. *Philosophical Transactions of the Royal Society, Series B*, **354**, 1–13.

Page, S. E., Siegert, F., Rieley, J. O. et al. (2002). The amount of carbon released from peat and forest fires in Indonesia during 1997. *Nature*, 4**20**, 61–5.

Page, S. E., Wüst, R. A., Weiss, D. et al. (2004). A record of Late Pleistocene and Holocene carbon accumulation and climate change from an equatorial peat bog (Kalimantan, Indonesia): implications for past, present and future carbon dynamics. *Journal of Quaternary Science*, **19**, 625–35.

Page, S. E., Rieley, J. O. and Wüst, R. (2006). Lowland tropical peatlands of Southeast Asia. In *Peatlands: Basin Evolution and Depository of Records on Global Environmental and Climatic Changes*, ed. P. Martini, A. Martinez-Cortizas and W. Chesworth. Amsterdam: Elsevier (Developments in Earth Surface Processes series), pp. 145–72.

Page, S. E., Hoscilo, A., Wösten, H. et al. (2009a). Ecological restoration of tropical peatlands in Southeast Asia. *Ecosystems*, **12**, 888–905.

Page, S. E., Hoscilo, A., Langner, A. et al. (2009b). Chapter 9: Tropical peatland fires in Southeast Asia. In *Tropical Fire Ecology: Climate Change, Land Use and Ecosystem Dynamics*, ed. M. A. Cochrane. Heidelberg, Germany: Springer-Praxis, pp. 263–87.

Page, S. E., Wüst, R. and Banks, C. (2010) Past and present carbon accumulation and loss in Southeast Asian peatlands. *PAGES News*, **18**, 25–6.

Page, S. E., Rieley, J. O. and Banks, C. J. (2011) Extent and global significance of tropical peat carbon pools. *Global Change Biology*, **17**, 789–818.

Quinten, M. C., Waltert, M., Syamsuri, F. and Hodges, J. K. (2008). Peat swamp forest supports high primate densities on Siberut Island, Sumatra, Indonesia. *Oryx*, **44**, 147–51.

Schultz, M. G., Heil, A., Hoelzemann, J. J. et al. (2008). Global wildland fire emissions from 1960 to 2000. *Global Biogeochemical Cycles*, **22**, B2002.

Schweithelm, J. (1999). The fire this time. An overview of Indonesia's forest fires in 1997/98. Jakarta, Indonesia: WWF Discussion Paper, 51 pp.

Shimamura, T. and Momose, K. (2005). Organic matter dynamics control plant species coexistence in tropical peat swamp forest. *Proceedings of the Royal Society, Series B*, **272**, 1503–10.

Sieffermann, G., Fournier, M., Triutomo, S., Sadelman, M. T. and Semah, A. M. (1988). Velocity of tropical forest peat accumulation in Central Kalimantan Province, Indonesia (Borneo). *In Proceedings VIII International Peat Congress, Leningrad, 1988*, Leningrad, USSR: International Peat Society. Section I, pp. 90–8.

Siegert, F., Ruecker, G., Hinrichs, A. and Hoffmann, A. A. (2001). Increased damage from fires in logged forests during droughts caused by El Niño. *Nature*, **414**, 437–40.

Slik, F. and van Balen, S. (2006). Bird community changes in response to single and repeated fires in a lowland tropical rainforest of eastern Borneo. *Biodiversity Conservation*, **15**, 4425–51.

Spessa, A., Weber, U., Langner, A., Siegert, F. and Heil, A. (2009). Fire in the vegetation and peatlands of Borneo, 1997–2007: Patterns, drivers and emissions. Poster presented at the European Geophysical Union conference, April 2009, Vienna, Austria.

Staub, J. R. and Esterle, J. S. (1994). Peat-accumulating depositional systems of Sarawak, East Malaysia. *Sedimentary Geology*, **89**, 91–106.

Struebig, M. J., Galdikas, B. M. F. and Suatma (2006). Bat diversity in oligotrophic forests of southern Borneo. *Oryx*, **40**, 447–55.

Takahashi, T., Shimada, S., Ibie, B. F., Usup, A., Yudha and Limin, S. H. (2002). Annual changes of water balance and drought index in a tropical peat swamp forest of Central Kalimantan, Indonesia. In *Peatlands for People: Natural Resource Functions and Sustainable Management. Proceedings of the International Symposium on Tropical Peatland, 22–23 August 2001, Jakarta*, ed. J. O. Rieley, S. E. Page and B. Setiadi. Jakarta, Indonesia: BPPT and Indonesian Peat Association, pp. 63–7.

Uryu, Y. et al. (2008). *Deforestation, Forest Degradation, Biodiversity Loss and CO_2 Emissions in Riau, Sumatra, Indonesia*. Jakarta, Indonesia: WWF Indonesia Technical Report.

van Eijk, P. and Leenman, P. H. (2004). *Regeneration of Fire Degraded Peat Swamp Forest in Berbak National Park and Implementing Replanting Programmes*. Water for food and ecosystems programme on: 'Promoting the river basin and ecosystem approach for sustainable management of SE Asian lowland peat swamp forests'. Available at: http://www.waterfoodecosystems.nl.

van der Kaars, W. A., Penny, D., Tibby, J., Fluin J., Dam, R. A. C. and Suparan, P. (2001). Late Quaternary palaeoecology, palynology and palaeolimnology of a tropical lowland swamp: Rawa Danau, West Java, Indonesia. *Palaeogeography, Palaeoclimatology, Palaeoecology*, **171**, 129–45.

van der Meer, P. J., Chiew, F. C. Y., Hillegers, P. J. M. and Manggil, P. (2005). *Sustainable Management of Peat Swamp Forests of Sarawak: Competition Report*. Wageningen, The Netherlands: Alterra.

van der Werf, G. R., Randerson, J. T., Giglio, L., Collatz, G. J. and Kasibhatla, P. S. (2006). Interannual variability in global biomass burning emission from 1997 to 2004. *Atmospheric Chemistry and Physics*, **6**, 3423–41.

van der Werf, G. R., Dempewolf, J., Trigg, S. N. et al. (2008). Climate regulation of fire emissions and deforestation in equatorial Asia. *Proceedings of the National Academy of Sciences of the United States of America*, **105**, 20350–5.

van der Werf, G. R., Morton D. C., DeFries, R. S. et al.(2009). CO2 emissions from forest loss. *Nature Geoscience*, **2**, 737–8.

Weiss, D., Shotyk, W., Rieley, J. O. et al. (2002). The geochemistry of major and selected trace elements in a forested peat bog, Kalimantan, SE-Asia, and its implications for past atmospheric dust deposition. *Geochimica et Cosmochimica Acta*, **66**, 2307–23.

Werth, D. and Avissar, R. (2005). The local and global effects of Southeast Asian deforestation. *Geophysical Research Letters*, **12**, L20702, doi: 10.1029/2005GL022970.

Whitmore, T. C. (1984). *Tropical Rainforests of the Far East*, 2nd edn. Oxford: Clarendon Press.

Wich, S. A., Meijaard, E., Marshall, A. J. et al. (2008). Distribution and conservation status of the orang-utan (*Pongo* spp.) on Borneo and Sumatra: how many remain? *Oryx*, **42**, 329–39.

Wösten, J. H. M., van Denberg, J., van Eijk, P. et al. (2006). Interrelationships between hydrology and ecology in fire degraded tropical peat swamp forests. *Water Resources Development*, **22**, 157–74.

Wüst, R. A. J. and Bustin, R. M. (2004). Late Pleistocene and Holocene development of the interior peat-accumulating basin of tropical Tasek Bera, Peninsular Malaysia. *Paleogeography, Paleoclimatology, Paleoecology*, **211**, 241–70.

Wüst, R., Rieley, J., Page, S. E. and van der Kaars, S. (2007). Peatland evolution in SE Asia over the last 35 000 yrs: implications for evaluating their carbon storage potential. In *Carbon Climate-Human Interaction on Tropical Peatland, Proceedings of the International Symposium and Workshop on Tropical Peatland, Yogyakarta, August 2007, EU CARBOPEAT and RESTORPEAT Partnership*, ed. J. O. Rieley, C. J. Banks and B. Radjagukguk. Yogyakarta, Indonesia and Leicester, UK: Gadjah Mada University and University of Leicester.

Yule, C. M. and Gomez, L. N. (2008). Leaf litter decomposition in a tropical peat swamp forest in Peninsular Malaysia. *Wetlands Ecology and Management*, **17**, 231–41.

Yulianto, E., Rahardjo, A. T., Noeradi, D., Siregar, D. A. and Hitakawa, K. (2005). A Holocene pollen record of vegetation and coastal environmental changes in the coastal swamp forest at Batulicin, South Kalimantan, Indonesia. *Journal of Asian Earth Sciences*, **25**, 1–8.

17

Southeast Asian biodiversity crisis

David Bickford, Sinlan Poo and Mary Rose C. Posa

17.1 Introduction

The International Year of Biodiversity, 2010, finds us in a state of severe ecological dysfunction and in the midst of a global biodiversity crisis. With our climate changing (IPCC 2007), our natural habitats decreasing in size and quality (Vitousek et al. 1997), our economically important species being over-harvested (Pauly et al. 1998) and our air and water becoming increasingly polluted (Hungspreugs 1988, Nriagu 1996, Galloway et al. 2008), we face some of the biggest challenges ever encountered by humanity. Globally we are currently losing species up to 1000 times faster than 'normal background' rates (Baillie et al. 2004). Among nine key 'planetary boundaries' that define a safe operating space for humanity, we have already far exceeded the threshold for biodiversity loss, threatening the functioning and resilience of the ecosystems on which we depend (Rockstrom et al. 2009). Despite scientists' best efforts to highlight the negative consequences of a 'business as usual' scenario and suggest ways to mitigate extinctions (e.g. Lubchenco 1998), global biodiversity continues to be lost at alarming rates (Baillie et al. 2004), leaving us now at a crossroads. While, on one hand, we are aware of what needs to be done to conserve and protect our future (Dietz et al. 2009), on the other hand, efforts thus far aimed at changing the destructive and wasteful lifestyles of our societies are inadequate to reverse the negative effects of humanity's actions on the planet.

Biotic Evolution and Environmental Change in Southeast Asia, eds D. J. Gower et al. Published by Cambridge University Press.

Within the larger context of a global biodiversity crisis, Southeast Asia stands apart – unfortunately outperforming all other regions in terms of deforestation, human population growth and biodiversity loss (Hannah et al. 1995, Laurance 1999, Achard et al. 2002, Sodhi and Brook 2006, Bradshaw et al. 2010). Although there have been several notable and recent reviews of this subject (Sodhi et al. 2004, Sodhi and Brook 2006, Sodhi et al. 2010), we aim to highlight some new insights into the main drivers, synergies and effects of the biodiversity crisis, the endemic problems of Southeast Asia, and suggest directions towards what we feel are some of the best and most realistic solutions. Specifically, we hypothesise that education and public awareness; restoration and reconnection of protected areas; and rethinking priorities and enhancement via incentives, economic policy and enforcement are the methods most likely to ensure a viable future for biodiversity. Only with a renewed sense of urgency, the right tools and analytical approaches, and coordinated efforts across local, regional and global scales, can we deal with a largely uncertain future.

17.2 Southeast Asian biodiversity and threat

Due to its complex geological history and island biogeography, Southeast Asia supports high levels of species richness and endemism, both in terrestrial and marine biomes (Sodhi and Brook 2006, Hoeksema 2007, Hall 2009). It has a diversity of tropical ecosystems, some of which, such as mangrove and peat swamp forests, reach their greatest extent and diversity within this region (FAO 2006, Page et al. this volume, Chapter 16). It also contains the 'coral triangle', an area regarded as the global centre of tropical marine biodiversity containing over 500 species of reef building coral and high fish species diversity (Hoeksema 2007, Tun et al. 2008, Bellwood et al. this volume, Chapter 9). However, concomitant with these high levels of biodiversity is a high degree of endangerment. The whole of Southeast Asia is comprised of four biodiversity 'hotspots', namely, Indo-Burma, Sundaland, Wallacea, and the Philippines (Myers et al. 2000). Moreover, the Philippines, part of the coral triangle, has also been identified as a top marine biodiversity hotspot with the highest level of threat (Roberts et al. 2002). Coral reef cover across the region continues to show an overall decline, with bleak prospects for recovery (Tun et al. 2008). An examination of species across various habitats within the region shows that nearly 30% of all species are threatened, with many species threatened in all major habitats (Fig 17.1). The high levels of endemism coupled with the small geographical ranges, and other physiological and life history characteristics of species have also made the Southeast Asian biota more susceptible to a number of extinction drivers (Fordham and Brook 2010). If current rates of habitat loss and exploitation continue unabated, widespread species declines and

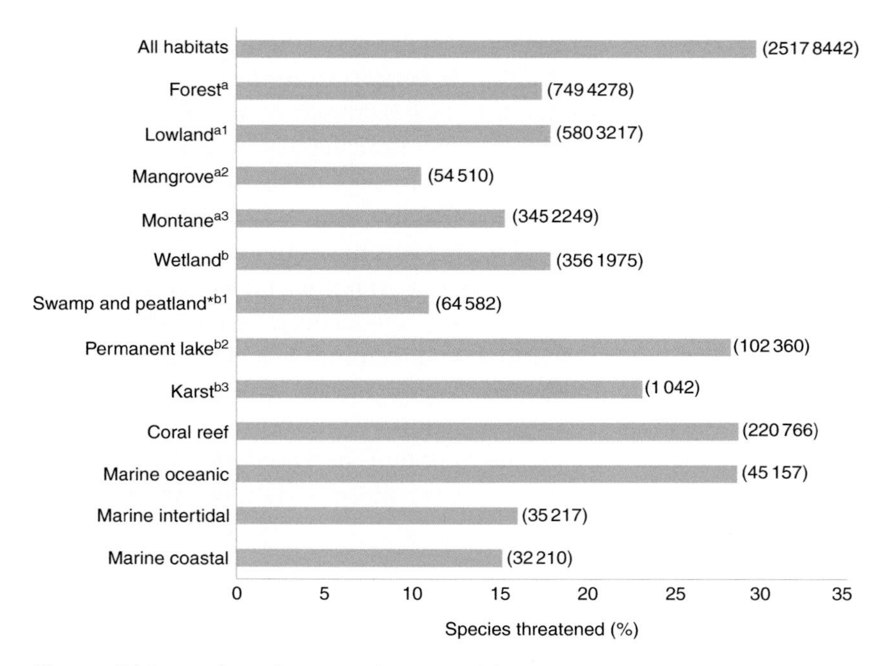

Figure 17.1 Number of species threatened (IUCN categories = Critically Endangered, Endangered, or Vulnerable) in major habitats within Southeast Asia. Bars represent the percentage of species threatened, while numbers in parentheses represent threatened and total number of species on the IUCN Red List (IUCN 2010).

[a] include species from [a1], [a2], and [a3]. [b] include species from [b1], [b2], and [b3]. * include species listed under tropical forest swamps.

losses are predicted to occur (Brook et al. 2006, Bradshaw et al. 2009, Sodhi et al. 2010). To examine the causes behind biodiversity loss in region, we will discuss the major drivers of extinction threat in the next section.

17.3 Drivers of biodiversity loss

17.3.1 Human population growth

The growth of human populations and increased demand for resources are the ultimate driving forces behind the loss of tropical biodiversity (Vitousek et al. 1997). In Southeast Asia, population density has been increasing steadily over the past decade at a rate of 1.4% per annum (UN Statistics Division 2008), bringing about land-use change and over-exploitation of natural resources (Giri et al. 2003, Sodhi et al. 2010). Human population density was found to be negatively correlated with percentage of remaining natural forest and positively correlated with percentage of threatened bird species within Southeast Asian countries (Sodhi et al. 2010). Similar trends are seen in increasing urban populations, and found to be positively correlated with percentage of plant and vertebrate species listed as Vulnerable, Endangered and Critically Endangered on the IUCN Red List (Sodhi et al. 2010). Even though these correlations are indirect measures, the link between human population growth and biodiversity loss is evident.

Some have suggested that the migration of rural dwellers to urban centres in tropical regions may alleviate the pressure that human population growth exerts on natural habitats (Wright and Muller-Landau 2006). Others, however, have emphasised uncertainties in this prediction because the relationship between rural populations and their surrounding environments is changing in response to the rapid industrialisation in developing nations (Laurance 2007). Despite the fact that rural populations are expected to decrease in certain regions, if overall human population continues to rise, the demand for natural resources is unlikely to decrease over time. Moreover, with economic globalisation, the bulk of demand for products has now shifted from local communities to international corporations (Hughes et al. 2003, Butler and Laurance 2008). A recent analysis showed that between 2000 and 2005, forest loss in the tropics was positively correlated with urban population growth and agricultural exports, demonstrating the trend towards industrial-scale, export-oriented production (DeFries et al. 2010). Thus, unless consumption patterns change, tropical forests will continue to face huge pressures to satisfy demands from a growing global urban population.

One endemic aspect of human population growth and improper governance in Southeast Asia is the *transmigrasi* program in Indonesia (Fearnside 1997), where people in overpopulated Java are relocated by means of a government programme, to other areas in Indonesia that are less populated. Unfortunately, this seemingly good idea has many unforeseen consequences that marginalise many local people that were originally living in these areas. The direct impacts of development and the accidental secondary effects of invasive species, fires and over-harvesting are all vastly increased in *transmigrasi* areas.

17.3.2 Loss, fragmentation and degradation of habitats

Habitat loss has had a more detrimental impact on tropical ecosystems than any other extinction driver (Sala et al. 2000, Sodhi and Brook 2006, Sodhi et al. 2008). Deforestation is currently one of the largest threats to terrestrial biodiversity in Southeast Asia (Sodhi et al. 2004). Forest areas are lost to commercial logging, conversion to agriculture, expansion of settlements, road building, mining and urbanisation. Analyses of satellite images show that relative annual deforestation rates in Southeast Asia were higher compared to all other tropical areas, both for the period of 1990–2000 (Mayaux et al. 2005) and 2000–2005 (Hansen et al. 2008), with forest loss being particularly severe in the Indonesian provinces of Sumatra and Kalimantan. The same trend was observed for mangrove forests (Mayaux et al. 2005) and approximately 50% of their original area has been lost (Tun et al. 2008).

Timber has represented one of the major exports from Southeast Asian countries since the 1950s, and Southeast Asia is still among the top producers of wood products for the global market (ITTO 2008). Due to poor forestry policies, logging was conducted at unsustainable levels for many decades (Ross 2001), using methods such as clear-cutting whereby most or all of the trees in a harvest area are cut down (Barbier 1993). Selective logging, where only particular tree species and sizes are extracted and a proportion of the forest is left intact, can still result in substantial forest degradation, affecting canopy height, surface area and species richness of trees in the harvest area (Foody and Cutler 2003, Okuda et al. 2003). A number of studies have examined the effects of selective logging on forest fauna, and even though there is evidence that logged forests can still support some species (e.g. Wells et al. 2007, Meijaard and Sheil 2008), the amount of species richness retained varies greatly across different taxa. More importantly, the community structure and composition of almost all secondary forests are markedly different from those of primary forests (Barlow et al. 2007).

Freshwater and wetland habitats are also being diminished and fragmented. Many rivers are being diverted to provide water supplies for agriculture and human settlements. The construction of dams, such as the Pak Mun Dam in Thailand, fragment fish populations by blocking movement, despite the addition of fish ladders (Roberts 2001). This is an especially critical issue in the Mekong River and its tributaries, because many fishes here make upstream migrations for breeding (Dudgeon 2000). Deforestation alters natural flow patterns of rivers and streams, causing large-scale changes in flow patterns, increased run-off, sedimentation and flash floods (Dudgeon 2000). These hydrological alterations are among the activities that threaten rivers, gallery forests and animals associated with freshwater habitats, such as endangered crocodilians, turtles and mammals, including orang-utans (*Pongo pygmaeus*) and proboscis monkeys (*Nasalis larvatus*).

Limestone karsts, a neglected biome that covers around 10% of the land area in Southeast Asia and recently recognised as important for harbouring high species diversity with many site endemics, are threatened by quarrying for use in cement (Clements et al. 2006). In Bornean karsts disturbed by quarrying, timber extraction and burning with secondary vegetation, the endemic Prosobranchia snail fauna was found to decrease in abundance (Schilthuizen et al. 2005). Extinctions of at least 18 plant species and perhaps two land snail species have been observed, with many more species possibly lost before they could be discovered due to the complete quarrying of karsts (Clements et al. 2006).

Currently, less than 10% of Southeast Asia's forests are under any form of protection (Sodhi et al. 2010). In addition, protected areas are not sufficiently conserved, and many appear to be 'paper parks' where weak governmental control and corruption leave room for illegal exploitation of natural resources within park boundaries (Laurance 2004, Smith et al. 2003). For example, in Indonesia, roughly 64% of reported protected areas are actually 'unclassified' and have no biodiversity protection (Bickford et al. 2008b). Ground-truthing in national parks and nature reserves showed signs of severe habitat degradation, overexploitation and hunting pressure in nearly all of them (Bickford et al. 2008b). More than 56% of protected lowland forests in Kalimantan was lost between 1985 and 2001, primarily due to illegal logging, including areas in national parks such as Gunung Palung where 38% of lowlands were deforested between 1999 and 2002 (Curran et al. 2004). On Sumatra, it was found that while protected areas had lower deforestation rates than unprotected areas, logging still continues within their boundaries (Gaveau et al. 2009). Adding to the direct pressure of habitat loss, protected areas are also becoming more ecologically isolated as deforestation and degradation continue in buffer zones and throughout the wider landscape (DeFries et al. 2005), raising many questions as to how effective terrestrial protected areas are in conserving Southeast Asia's forest biodiversity.

17.3.3 Wild fires

Natural forest fires have occurred in Southeast Asia since at least the Pleistocene, but the combination of human altered ecosystems and droughts caused by El Niño Southern Oscillation events has led to an increase in the number and intensity of catastrophic fires (Goldammer 2007, Field et al. 2009). Fire is also the cheapest, fastest and most efficient method of land clearance and is used in both small-scale shifting agriculture and large-scale plantations. As a result, anthropogenic fires have become a major driver of forest loss and degradation in the region.

Fires in Borneo during the 1982–83 drought consumed around 3.5 million ha of forest in East Kalimantan alone (Goldammer 2007). Similarly, over the 10-year period of 1997–2006, forest fires across Borneo affected 0.3 million ha in normal years and 1 million ha during El Niño years (Langner and Siegert 2009). Worryingly, during this period, fires also affected an average of 0.8% of the protected areas in normal years, increasing by up to five fold in Kalimantan during El Niño years (Langner and Siegert 2009). Impacts were found to be particularly severe in previously logged areas and timber concessions, caused by the addition of suitable fuel material during timber extraction (Goldammer 2007, Langner and Siegert 2009). In recent years there has also been a shift towards more fire events occurring in swamp forests compared to lowland dipterocarp forest.

Repeated burning increases fire susceptibility and can initiate the formation of fire climax grasslands, leading to cascading effects on animal populations and higher trophic levels. Furthermore, smoke from forest fires often has widespread effects across the Southeast Asian region (Lohman et al. 2007). For example, biomass burning in Indonesia has subsequently resulted in elevated levels of formate and acetate in rainwater collected in Singapore (Zhong et al. 2001). Because forest fires release gases that contribute to global warming and rainwater acidity, as well as produce smoke particles (polynuclear aromatic hydrocarbons) that could affect human health (Radojevic 2003), their wide impact on both the source and surrounding countries is profound.

17.3.4 Monoculture plantations

During the period 1990–97, agriculture was the main driver for land conversion in Southeast Asian tropical forests (Achard et al. 2002). Conversion to agriculture has detrimental impacts on overall ecosystem functioning, because intense agricultural activity depletes soil nutrients and induces erosion (Grubb et al. 1994). Furthermore, the shift from traditional farming practices to large-scale monoculture plantations has contributed to the loss of biodiversity that would otherwise be sustained in swidden ecosystems (Rerkasem et al. 2009).

High profitability and low suitability as a habitat for forest-dwelling species have made oil palm agriculture among the greatest immediate threats to biodiversity in Southeast Asia (Wilcove and Koh 2010). With the rising demand for vegetable oil and biofuel, oil palm has become one of the world's most rapidly expanding crops, with increases in production often coming at the expense of forestland (Fitzherbert et al. 2008). This trend is particularly apparent in Malaysia and Indonesia, which jointly produce more than 80% of the world's palm oil (Koh and Wilcove 2008, Wilcove and Koh 2010). For instance, in Malaysia, conversion of forests to oil palm accounted for 94% of the country's deforestation from 1990 to 2005 (Wilcove and Koh 2010). Oil palm plantations support low levels of faunal and floral diversity. For example, they now cover more than 15% of Sabah's land area but only sustain about 5% of the ground-dwelling ant species of the forest interior (Brühl and Eltz 2010). Reviewing results of various studies assessing the biodiversity value of oil palm plantations, Danielsen et al. (2009) found that plantations contained 23% and 31% of forest vertebrate and invertebrate species, respectively, compared with natural forests. Community composition of both flora and fauna change drastically, and plantations are usually dominated by a few abundant generalists and non-forest species, including invasives and pests (Fitzherbert et al. 2008, Danielsen et al. 2009). One compromise that has been suggested to mitigate such biodiversity losses is to restrict future expansion of oil palm agriculture to degraded habitats and pre-existing croplands (Koh and Wilcove 2008).

17.3.5 Overexploitation

Hunting pressure has increased immensely with increased human population density (Sodhi et al. 2004), resulting in extinctions caused by overhunting of wildlife for meat (Bennett 2002). In communities living close to tropical forests, bushmeat is still a major source of protein (Bennett and Robinson 2000). In Sarawak, for example, bushmeat is found in 67% of all meals of highland people (Bennett et al. 2000). With recent industrial developments, exploitation of local fauna has been further facilitated by increased access to forested areas through road building, deforestation and other land-use changes. Currently, wildlife is being extracted from tropical forests at approximately six times the sustainable rate (Bennett 2002). In Vietnam, this has caused the extirpation of up to 12 large vertebrate species (Milner-Gulland and Bennett 2003).

Overfishing and unsustainable fishing practices in freshwater rivers and lakes as well as in marine environments are causing population declines (Dudgeon 2000, Tun et al. 2008). The limited data available for freshwater fisheries indicate that fish stocks are declining, with decrease in fish size at maturity and reductions in total catches (Dudgeon 2000).

Fish biomass in coastal fisheries was found lowered to 5–30% of their unexploited level, with declines seen in all areas studied (Silvestre et al. 2003). With 70% of Southeast Asia's population living near the coastline (Lao PDR is the only landlocked country), marine fisheries are an important source of food. Many fishermen, especially in Indonesia, the Philippines, Thailand, Vietnam and Sabah, are resorting to destructive methods such as dynamite fishing and cyanide to maintain short-term income and obtain food (Tun et al. 2008).

In addition to local consumption, hunting is also driven by the international wildlife trade in commodities, pets, skins, ornaments and medicinal ingredients. Within the region, wildlife trade generates considerable revenue, and Southeast Asia remains one of the major centres for global wildlife traffic (e.g. Nijman et al. this volume, Chapter 15). An examination of international trade in CITES-listed animals showed that more than 30 million animals (representing around 300 species) and 18 million pieces of coral were exported from Southeast Asia between 1998 and 2007 (Nijman 2010). More alarmingly, undocumented illegal trade can be significantly greater than official exports (Nijman 2010). Therefore, the impact of wildlife trade on targeted species is likely to be much greater than our current estimates. For instance, at present, Indonesia is the largest exporter of frogs' legs in the world (Teixeira et al. 2001, Kusrini and Alford 2006), but there are already indications that harvesting patterns are unsustainable, warning that they may follow a similar trend of population crashes as previously observed in this industry in the US, Europe and South Asia (Warkentin et al. 2009). Moreover, the domestic Indonesian consumption of frogs is largely undocumented and has been estimated to be up to seven times the reported export volume (Kusrini and Alford 2006).

The export of many plants and animal parts from Southeast Asia for use in traditional Chinese medicine elsewhere includes everything from roots, bark, stems, leaves and flowers of plants to marine invertebrates and large terrestrial predators.

Turtle shells, for example, are made into a medicinal dessert and consumed for general health in Sinocultures. In Taiwan alone, an estimated 1989 metric tons of shells were imported from China, Indonesia and Vietnam between 1999 and 2008, with no evidence of reduction in trade volume in response to CITES listing of main target species (Chen et al. 2009b). The harvesting of sea cucumbers, another item considered as a medicinal and culinary delicacy, has also increased in the past few decades, leading to population collapses in various areas within the region (Manez and Ferse 2010). Threats from overexploitation are especially severe for charismatic animals and in Southeast Asia have resulted in the endangerment of the Sumatran rhino (*Dicerorhinus sumatrensis*), Asian elephant (*Elephas maximus*) and Malayan tiger (*Panthera tigris jacksoni*) (Clements et al. 2010). In addition to direct impacts on targeted species, loss of predators such as the Malayan tiger exert top-down effects that could potentially result in changes to predator–prey dynamics and trophic cascades.

17.3.6 Invasive species

With globalisation, an increasing number of exotic species are being introduced into new habitats, some of which become invasive and have negative effects on local ecosystems, possibly resulting in the extinction of native species. Therefore, invasive species are now recognised as an important threat to global biodiversity (Sala et al. 2000, Chapin et al. 2000, Molnar et al. 2008) and one of the leading causes of extinctions (Clavero and García-Berthou 2005). Tropical ecosystems have traditionally been regarded as less vulnerable to invasive species because of the more complete use of available resources and space by species-rich communities (Rejmanek 1996, Stachowicz et al. 1999). At present, the overall threat of invasive species on Southeast Asian biodiversity loss is still perceived to be low and there have been fewer reported harmful invasions relative to other regions. However, recent reviews hint that they may be a substantial but overlooked problem, even within relatively intact terrestrial ecosystems (Peh 2010). The number of marine invasive species in the region is likely to be higher than reported, given the correlation of invasions with high seaport traffic (Molnar et al. 2008). Where invasions have occurred, they can have serious consequences. In wetland communities in Thailand, for example, the invasive golden apple snail (*Pomacea canaliculata*) has resulted in the almost complete collapse of the aquatic plant community; altering the state and function of the ecosystems they occupy (Carlsson et al. 2004). Compared to other tropical regions, Southeast Asia may be more susceptible to invasion of exotic species, because it is composed of a large proportion of oceanic islands (Blackburn et al. 2004, Laurance and Useche 2009, Fordham and Brook 2010). However, the lack of research on invasive species in the region prevents a comprehensive assessment of their actual impact.

17.3.7 Climate change

Climate change will have widespread effects on the region and synergise with habitat loss, degradation and fragmentation as well as overexploitation in both terrestrial and aquatic ecosystems. This is because temperatures will increase and precipitation patterns will change. Predictions are that 20–30% of plant and animal species are likely to be at increased risk of extinction with a temperature rise of 1.5 to 2.5°C (IPCC 2007). Corals expel their zooanthellae symbionts in response to stress and overheating, and when thermal stress is severe and prolonged, most corals on a reef can bleach and die. Increases in carbon dioxide and temperature are expected to change the composition of coral reefs, because some species are more tolerant to climate change and coral bleaching than others (Hughes et al. 2003). Furthermore, coastal and island ecosystems will be exposed to increasing threats from rising sea levels and erosion (IPCC 2007, Fordham and Brook 2010). Besides marine habitats, freshwater systems will also be severely impacted. Models also suggest that changes in precipitation may result in both more severe floods and drier dry seasons, altering annual patterns that river systems have adapted to for centuries (Dudgeon 2000).

With global temperature predicted to rise by up to 6°C by the end of the century, the resilience of many ecosystems is likely to be exceeded (IPCC 2007). Though temperature increase in the tropics is expected to be less than in extra-tropical regions, the magnitude of impact on biodiversity will depend on species' physiological sensitivity to warming and their options for behavioural and physiological compensation (Huey et al. 2009). Given that tropical species normally experience less temperature variability, they may be less adaptive to temperature changes (Deutsch et al. 2008, Wright et al. 2009). Furthermore, some tropical species are already living in climate conditions close to or slightly above their thermal optima (Deutsch et al. 2008, Hughes et al. 2003, Huey et al. 2009) so that even slight increases in temperature may exceed their thermal tolerances (Colwell et al. 2008). Temperatures are projected to increase faster than species can evolve new adaptations, so many will need to shift their ranges to cooler habitats in order to survive. A study on the distribution and habitat association of mammals indicates that the distance to the nearest cool refuge is greater for tropical species, exceeding 1000 km for 20% of species examined by the end of the century (Wright et al. 2009). For the extensive coral reefs in Southeast Asia, we can expect large-scale disruptions, because they are immobile and have slow growth rates. Their temperature tolerance has developed over longer time scales than current rates of change, and thus they may be unable to evolve quickly (Hughes et al. 2003). Thus, the impact of temperature change on tropical biota may be much higher compared to that of extra-tropical regions.

Apparent upward shifts in terrestrial fauna of the region have already been noted in the elevational distribution of 94 common resident birds over a 28-year period

(Peh 2007), and in 102 species of moths over a 42-year period (Chen et al. 2009a). Elevation shift and poleward movement within the next century is likely to result in an increase in biodiversity in temperate regions and a decrease in Southeast Asia for the following reasons. First, in contrast to temperate ecosystems, tropical lowlands lack a source pool of species adapted to higher temperatures to migrate and occupy the empty niches once the original species are extinct or extirpated (Colwell et al. 2008). Second, the topography of Southeast Asia – island archipelagos with few mountains – prevents both altitudinal and latitudinal range shifts for most terrestrial species. The impact of changes in temperature and precipitation is particularly significant for amphibians and reptiles, because of their ectothermic metabolic rates and often temperature-dependent sex ratios and aquatic life stages. Based on current temperature change predictions, most of the amphibians and reptiles in Southeast Asia may reach the limit in their ability to adapt to climate change effects in less than 50 years (Bickford et al. 2010).

17.3.8 Pollution

Previously thought to be of less importance as an extinction driver (Sodhi et al. 2004), pollution has also become an emerging threat to biodiversity in Southeast Asia. Perhaps the greatest impact of pollution can be seen in marine and freshwater ecosystems (e.g. Cumberlidge et al. 2009). As countries develop economically and the scale of urbanisation and industrialisation expands, the expansion of agriculture, farming, deforestation, and near shore development has led to increased discharge of various pollutants into rivers and oceans (Hungspreugs 1988, Todd et al. 2010). Pollution is already ubiquitous in freshwater habitats in tropical Asia (Dudgeon 2000) and marine pollution is a widespread problem in the region (Todd et al. 2010). Increased sediment loads, eutrophication, toxic industrial compounds and litter can disrupt processes at the organismal level and cause direct mortality, as well as disrupt the structure and function of communities at all trophic levels (Todd et al. 2010). For example, sediment pollution in streams has resulted in downstream decreases in macroinvertebrates in Borneo (Yule et al. 2010), and agricultural practices and gold mining activities reduce viable habitat for already threatened species such as the flat-headed cat (*Prionailurus planiceps*) (Wilting et al. 2010) and the lungless frog (*Barbourula kalimantanensis*) (Bickford et al. 2008a). Coral reef degradation from marine pollutants could lead to the extinction of associated taxa, because although most coral species have a wide distribution in the region, lobsters and half of the fish and snail species have relatively restricted ranges (Tun et al. 2008).

There is still a lack of empirical studies on the impacts of acid rain, nitrogen deposition, and sulfur deposition on biodiversity in Southeast Asia (Todd et al. 2010). A global synthesis suggested that roughly 30% of the vegetation in Southeast Asia is affected by nitrogen and sulfur deposition (Dentener et al. 2006).

In particular, the mountainous regions in Southeast Asia, where soil sensitivity is relatively high, are thought to be more vulnerable to acidification (Bhatti et al. 1992) and nitrogen deposition (Bobbink et al. 2010). Recent studies indicate that, contrary to nitrogen-limited temperate ecosystems, the increase in nitrogen in tropical ecosystems may have negative effects on forest productivity (Barron et al. 2009), leading to further soil acidification and nutrient loss (Matson et al. 1999). Moreover, the reduction of nitrogen heterogeneity by atmospheric nitrogen deposition remains a factor contributing to vulnerability in plants, especially at the regeneration phase (Jones et al. 2002). Unlike Europe, North America or East Asia, the combination of low-speed near-surface winds, abundance of rain events and inefficient vertical mixing in convective clouds results in a relatively short transport range of areal pollutants in Southeast Asia (Engardt et al. 2005). In the case of sulfur, 60–70% of a country's emissions are deposited within its boundaries (Engardt et al. 2005). Therefore, as Southeast Asian countries continue to develop, acid rain may also become an increasing threat to the ecosystems in the region.

17.4 Future projections

17.4.1 Estimates of future biodiversity loss

Overall biodiversity loss in Southeast Asia is projected to be between 13 and 85% of known species by 2100 (Sodhi et al. 2010), with the majority of species considered already doomed to extinction because of past land conversion that has removed roughly 60% of the original forest cover in the region (Brook et al. 2006). Using country-specific deforestation rates, Sodhi and Brook (2006) estimated that 24–63% of Southeast Asian endemic taxa (859–4815 species of vertebrates and 8343–48 043 species of vascular plants) are at risk of extinction by the end of this century. Combining the current human footprint with future projections of temperature and precipitation change, biodiversity loss across Southeast Asia appears to be particularly elevated in centers of high human population density and along coastal regions (Fig 17.2).

17.4.2 Synergies among extinction drivers

Despite possible biases and limitations of future predictions, Sodhi et al. (2010) suggest that current projections, which are based solely on species-area relationships and current deforestation rates, may still be underestimates because they do not take into account the complexity of interactions between multiple extinction drivers. Extinction drivers are often examined separately, but recent evidence indicates that synergistic effects of multiple drivers are common in tropical forests (Stork et al. 2009) and likely to have a grave effect on species that are already threatened (Bradshaw et al. 2009). When synergistic effects are taken into account,

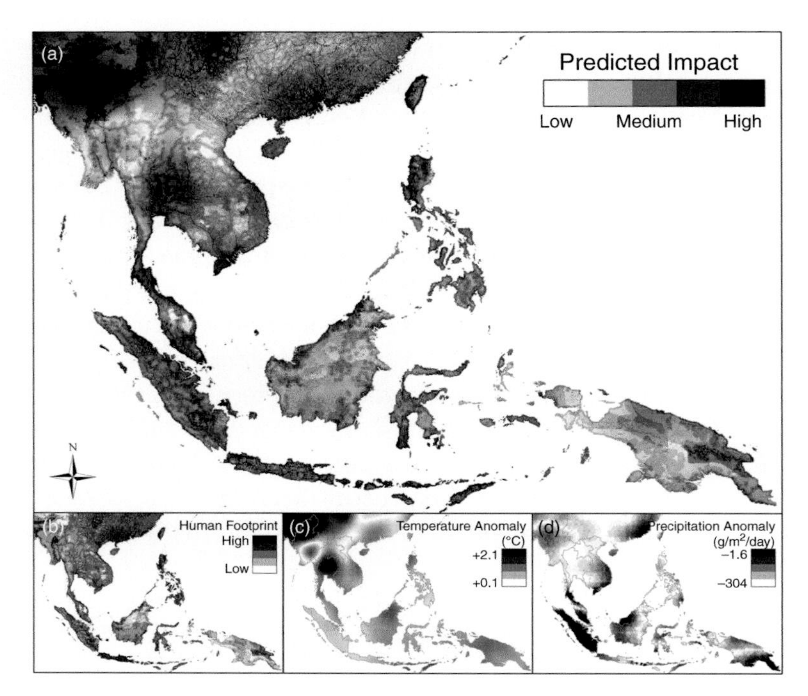

Figure 17.2 Predictions for biodiversity loss across Southeast Asia; (a) simplified model of predicted areas of impact on Southeast Asian biodiversity; (b) current human footprint; (c) predicted maximum monthly surface air temperature anomalies by 2050; (d) predicted maximum monthly precipitation flux anomalies by 2050.

In order to better visualise the areas that might be at higher possible threat due to climate change, we gathered the best available data on factors most responsible for changes to biotic systems. Temperature and precipitation data were downloaded from www.gisclimatechange.org (accessed on 25 August 2009) from the global CCSM projection dataset bounded at 22° to -12° latitude and 94° to 154° longitude. To develop the 50-year anomaly values of precipitation and temperature, ensemble average Climate Change Commitment scenario data for Surface Air Temperature (tas) and Precipitation flux (pr) were downloaded for years 2000 and 2050. Data were imported into Microsoft Excel (Microsoft Ltd.) and the 50-year monthly anomalies were calculated. For each latitude and longitude point (approximately 1.4° grid) the maximum monthly tas and maximum negative pr were identified and exported into Arcmap 9.3 (ESRI Ltd.).

Human impact data were downloaded from http://www.ciesin.columbia.edu/wild_areas/ (accessed 22 April 2010) and imported to Arcmap. World borders (wb) were downloaded from http://mappinghacks.com/data/world_borders.zip (accessed 25 August 2009). The tas and pr data were interpolated to raster format using inverse distance weighting, second power and six local points (Arcmap, spatial analyst extension) and cropped, along with elevation data, to land area using the wb mask. The tas, pr and elevation were then mapped as shown.

Calculation of impact was as follows: pr was squared to make the direction of reduced precipitation positive to match tas and human footprint layers. Tas, pr and human population declines and biodiversity losses will be greater than projections based on individual drivers (Brook et al. 2008). A pairwise combination of ten different extinction drivers on IUCN Red List threatened mammals, birds, and amphibians in tropical forests revealed significantly higher frequency of synergistic effects, suggesting that they may be the norm rather than the exception in causing biodiversity losses (Laurance and Useche 2009). Results from this study further indicated that the most important synergisms involve interactions between habitat loss or alteration, and other anthropogenic disturbances such as hunting, fire, exotic-species invasions or pollution (Laurance and Useche 2009). A closer look at current extinction rates also points to the rising threat of synergies with climate change, because it can interact with other extinction drivers on both local and global scales (Stork et al. 2009). Human impacts from overfishing, coastal development and pollution have been drivers of the massive decrease of coral reef species and have undermined reef resilience, already seen in the increase in bleaching events in the past 30 years, making them even more at risk from future climate change (Hughes et al. 2003).

The combination of extinction drivers and the resulting synergistic effects may deal a final blow to a species by pushing it below its minimum viable population, thus increasing its susceptibility to extinction via stochastic events (Brook et al. 2008). Moreover, with the growing populations and economic demands, rates of anthropogenic disturbances are likely to accelerate in the future. As such, the actual rate and number of extinctions in Southeast Asia could be much higher (Sodhi et al. 2010).

17.4.3 Adaptive responses

While the overall picture looks bleak, the resilience of natural systems has not been fully appreciated and we know little of the buffering capacity of ecosystems, their long-term resilience to stress and their response to restoration or reconnection. Moreover, there are many suites of species that have taken advantage of a human-dominated landscape, adapting to become human commensals and, in turn, increasing their ranges and niche breadth. These plants and animals are mostly considered 'weeds', but have become the most common organisms in the region (e.g. the common myna, *Acridotheres tristis*; catclaw mimosa, *Mimosa pigra*; and the fire ant, *Solenopsis invicta*). In all organisms, adaptive responses range from

Caption for Figure 17.2 (*cont.*) footprint were reclassified (slice, spatial analyst) to 255 equal interval units. Impact was then predicted as [0.5 × human footprint + 0.3 × tas + 0.2 × pr] based on our a priori perception that impacts on biodiversity will be based mostly on anthropomorphic disturbance, followed by temperature and precipitation changes. Maps are displayed overlaid with wb edges for clarity.

ecological to evolutionary in time, and from molecular to population size in space, so generalising is difficult. We do know that some species seem pre-adapted to generalised life histories and these species are most of the benefactors of a changing and human-dominated landscape. What we do not fully understand are the potential changes that more specialised species can make in response to variation in abiotic and biotic characteristics. Although most habitat specialists will probably not be able to adapt in a meaningful timeframe (relative to anthropogenic habitat changes), some might. This innate adaptive ability is under-appreciated and impossible to quantify across the range of species that co-exist in the habitats and ecosystems of Southeast Asia. Learning what species are able to modify their behaviours, diet preferences, phenologies, elevational and geographic ranges, for example, and how they are able to adapt, will be critical in our progress towards understanding how biodiversity will change in the future and what our best strategies will be for conserving biodiversity.

17.5 Country case studies

We feel that there are two poignant examples on either end of a continuum of responses to environmental problems in Southeast Asia. On one hand, Singapore embodies much of the biodiversity crisis in a modern city-state that sacrificed its biodiversity for economic development, while the Philippines underscores many fundamental problems and possible solutions. Although our tone remains realistic, we have chosen to embrace a more optimistic outlook and so balance these case studies to reflect both the harsh reality and the possible positive responses from both ends of the spectrum.

17.5.1 Singapore

Singapore presents a case in this region whereby maximised economic development has taken place at the expense of biodiversity (Sodhi and Brook 2006). The city-state has made tremendous gains in economic growth, but in the process has devastated its natural resource base and forested lands (now <5% of the land area). Researchers have found high percentage of local extinctions across a wide range of taxa; overall observed biodiversity loss was 28%, but simulations reveal that Singapore has lost as much as 50% of its species, if inferred extinctions are taken into account (Brook et al. 2003). Protected forests, comprising only 0.25% of the country's land area, harbour over 50% of its residual biodiversity (Brook et al. 2003). Despite the economic transformation, and the usual trend for highly developed countries to reverse some of their environmental legacy, Singapore unfortunately has recently been shown to have the highest environmental impact per capita (Bradshaw et al. 2010). This rank of environmental degradation is independent of economic factors, making Singapore a focal target for conservation efforts aimed at the societal level. As the epitome of the over-consumption epidemic, Singapore has some particularly easily targeted sectors for better performance in reducing environmental impact. Notable among these are the education and awareness sectors. Singapore can take a lead in becoming a sustainable tropical city and has all the expertise and technical abilities to do so within a reasonable timeframe.

17.5.2 The Philippines

The Philippines exemplifies the biodiversity crisis, with its remarkably high levels of endemism (Mallari et al. 2001), marine fish and invertebrate diversity (Roberts et al. 2002, Carpenter and Springer 2005), and its teetering position on the brink of ecological ruin due to the long history of natural resource extraction and exploitation that has already devastated its rainforest, mangrove and reef ecosystems (Bankoff 2007, Gomez et al. 1994, Primavera 2000). Of the 1007 Philippine vertebrate species assessed for the 2006 IUCN Red List, nearly 21% are classified as Threatened, as are 215 of the 323 plants that have been evaluated (Posa et al. 2008) – making the Philippines one of the hottest conservation hotspots (Myers et al. 2000, Roberts et al. 2002). Despite the dire situation and continuing socio-economic problems (e.g. widespread poverty, increasing population, large national debt) and political dysfunction (e.g. corruption, opposition by commercial interest groups), there are signs of progress and hope for biodiversity in the country (Posa et al. 2008). One of the driving positive factors has been the emergence of an environmental consciousness leading to the formation of civil society groups, which have advocated for better environmental policies and the decentralization of natural resource management (Utting 2000). Local communities are becoming more involved in managing their resources sustainably through a range of schemes including community based forestry and marine protected areas, as well as active representation in protected area management boards (Custodio and Molinyawe 2001, White et al. 2002, Lasco and Puhlin 2006). Notably, there is also a renewed interest by academics in biodiversity research and exploration, shown by the increase in the number of publications, and the many new species still being found and described (Posa et al. 2008).

17.6 Challenges and future directions

Given the amount of habitat that has already been lost and degraded in the region combined with the impacts of other extinction drivers mentioned above, the biodiversity crisis in Southeast Asia is accelerating (Sodhi et al. 2004, Bradshaw et al. 2009). There is an urgent need for conservation to err on the side of caution (Laurance 2007). Though there remains a level of uncertainty in the predictions of future extinctions (Sodhi et al. 2010), misplaced optimism may result in dire consequences for the preservation of Southeast Asian biodiversity (Laurance and Useche 2009). Urgent informed action is needed if the biodiversity crisis is to be averted. In this section we discuss the main hindrances and make suggestions regarding conservation actions by various sectors of society.

17.6.1 Poverty and economic priorities

Economic hardship in many Southeast Asian countries can contribute to biodiversity loss (Adams et al. 2004). Livelihoods of poor communities are often dependent on exploiting natural resources, so areas that are poverty stricken are found to overlap greatly with areas experiencing high levels of biodiversity loss (Fisher and Christopher 2007). Though the percentage of population in poverty has decreased in recent years, the percentage of people living on less than USD 1.25 a day is still considerably high in countries such as Cambodia (40.2%), Laos (44.0%), Philippines (22.6%) and Vietnam (21.5%) (Asian Development Bank 2009). Conflicts between maintaining the livelihood of communities in poverty and efforts to preserve biodiversity can arise when protected areas are set aside for conservation (Christie 2004, West et al. 2006). In light of the current economic status of Southeast Asian countries, their goal of having higher living standards through economic growth and development is only reasonable. It is perhaps no surprise, therefore, that short-term economic gains overshadow measures of sustainable management or protection of natural resources. The relationship between biodiversity conservation and poverty alleviation is complex and case-specific, encompasses both moral and pragmatic considerations, and requires cooperation and coordination between various government, conservation and humanitarian agencies (Adams et al. 2004). However, it is imperative that protection of the environment becomes a priority for Southeast Asian nations. Strong government commitment to reducing poverty and dependence of rural communities on resource exploitation through sustainable practices is needed. Relevant social issues must be addressed concurrently with biodiversity protection, such as land tenure and providing alternative livelihoods. Maintaining biodiversity and, in turn, the structures and functions of natural ecosystems, will ultimately be beneficial for human populations.

17.6.2 Education and environmental awareness

One of the most effective ways to produce far-reaching and long-term progress towards the conservation of biodiversity in Southeast Asia is to strengthen public education and ecological awareness. Better appreciation of the direct and long-term benefits of intact ecosystems, including the functional roles they play in flood protection, sustainable food production and clean water supply (Bradshaw et al. 2009, Kareiva and Marvier 2007) will build towards higher levels of conservation incentives and increased success for conservation in all related domains (i.e. governmental, local, NGO, etc.). Elsewhere, for example in Puerto Rico and Hispaniola, progress in the training of local students and wildlife professionals has promoted conservation awareness at local and national levels, subsequently influencing a growing ecotourism industry and driving national park management and conservation efforts (Latta and Faaborg 2009). Similarly in the Philippines, capacity building has led to significant developments in biodiversity conservation (Posa et al. 2008). To facilitate regional environmental education and awareness, high income countries, such as Brunei and Singapore, should invest more in the improvement of conservation initiatives in poorer countries in the region (e.g. Indonesia, Philippines), with much needed funds, training, and expertise in environmental management (Sodhi et al. 2010).

The use of charismatic endangered species can be extremely effective for educational and public awareness campaigns. Traditionally, medium- and large-sized mammals or birds have been used as flagship species, but concentrating efforts on these species results in a limited focus, neglecting other species groups or habitats that do not directly support these species. An example of other important potential areas for positive education and awareness action is in the recently rediscovered lungless frog in Kalimantan, Indonesia (Bickford et al. 2008a). As a highly unusual habitat specialist, the lungless frog captures interest because of its strange appearance and highly specialised ecology. There are excellent opportunities for using the lungless frog and other unfamiliar species as focal taxa for awareness campaigns and educational programmes. Moreover, the Indonesian government could take an active role in the conservation of this biodiversity flagship species, making a giant step forward in a socially responsible project in a remote area with very little infrastructure.

17.6.3 Certification schemes and public-pressure campaigns

In areas where the driving force behind land-use change has shifted from rural communities to large international corporations, public-pressure campaigns by conservation organisations (Butler and Laurance 2008) and certification programmes to reinforce international regulations (Warkentin et al. 2009) have been suggested as a means of coercing large corporations to become ecologically more responsible. Certification schemes, such as the Forest Steward Council (FSC) certification for timber extraction and the Roundtable of Sustainable Palm Oil (RSPO) for oil palm plantations, are currently present in a number of Southeast Asian countries, though overall implementation is still much in need (Dennis et al. 2008, Yaap et al. 2010). In these situations, establishing a reporting system to closely monitor each part of the exploration process will be necessary to maintain integrity of certification schemes (Warkentin et al. 2009, Wilcove and Koh 2010). The possibility of gaining enough public support in Southeast Asia to demand a change in the market remains debatable. Without increased environmental awareness, public-pressure for certifications is expected to have less effect in Southeast Asia compared to regions comprised of high-income countries such as Europe and North America (Butler and Laurance 2008). Nonetheless, successful examples can be seen in the Philippines where public advocacy has resulted in considerable progress in environmental protection legislations (Posa et al. 2008).

17.6.4 Local government law reinforcement

It is well known that political instability and lack of infrastructure in Southeast Asia have resulted in anaemic national institutions, lack of forest protection laws and poor enforcement of legislation (Bickford et al. 2008b, Laurance 1999, Sodhi et al. 2006). Furthermore, corruption in natural resource management (Smith et al. 2003) has led to overexploitation of forests, wildlife, fisheries and other natural resources, and consequently to biodiversity loss in this region (Laurance 2004). With the exception of climate change, major threats to tropical biodiversity are largely the responsibility of local governments and local populations (Wright et al. 2009). Conservation efforts can be made more effective through the establishment of integrated local management, local environmental legislation, as well as on-the-spot enforcement of measures against illegal activities. For example, improvements in regulating tropical logging practices is an immediate way to conserve functional tropical forest (Bradshaw et al. 2009), while overall improvements of park management is crucial to the success of protected areas in conserving native habitats and biodiversity in Southeast Asia (Sodhi et al. 2004, Bickford et al. 2008b). This can be further strengthened by regional cooperation that tackles threats that transcend national borders, such as the illegal wildlife trade and emissions from forest fires.

17.6.5 International monetary aid

As funding continues to be a limiting factor for conservation efforts in all places and at all levels (Posa et al. 2008), financial and technological support from international organisations and high-income nations will be critical to establish good governance of tropical biological resources through strong multilateral policy and concomitant socio-economic and administrative aid (Bradshaw et al. 2009). In particular, there are high hopes that Reduced Emissions from Deforestation and Degradation (REDD), relying on market and financial incentives to reduce deforestation and greenhouse gas emissions in the tropics, may provide a straightforward way to mitigate the effects of global climate change. REDD payments can be used to increase viable habitat for forest species by funding the protection of forests slated for development or expanding forest over marginal agricultural lands that have been abandoned (Venter et al. 2009, Wright et al. 2009). However, reduction of deforestation by REDD needs to be implemented with international and local policies that take biodiversity into account so as to maximise the quantity and, more importantly, the quality of conserved habitats (Grainger et al. 2009). Care must also be taken for REDD to be integrated into existing forestry management schemes that have already proven to be effective, because there is already concern that it will interrupt the trend of decentralisation, giving governments an incentive to take control of monetised forest carbon (Phelps et al. 2010). In addition to carbon credits for maintaining forested areas, biodiversity credits have also been proposed as a useful means of monetary incentive for conservation by treating biodiversity as a global market commodity or societal investment (Ferraro and Kiss 2002). The competitiveness of carbon-offset schemes, REDD, and biodiversity credits will ultimately depend on their present and future prices in relation to the profits from other land-use types (Wilcove and Koh 2010).

17.7 Overall suggestions

Besides conservation and rehabilitation of forested and degraded areas, future conservation efforts should invest in reforestation, species' reintroduction, and the restoration of habitat connectivity (Sodhi et al. 2010). Even though secondary forests appear to sustain fewer species, they are nevertheless able to buffer some of the effects of deforestation (Wright and Muller-Landau 2006). Just as the cessation of destructive activities, such as mining, has been shown to lead to partial recovery of biodiversity in previously affected areas (Yule et al. 2010), many degraded areas can be regenerated and their capacity to provide ecosystem services restored. Given the limited economic, logistical and technological capacity and low current commitment of Southeast Asian countries to conserve environmental assets in the face of developmental advancements, there can be no simple cure-for-all solution to decrease biodiversity loss (Bradshaw et al. 2009). To reach conservation aims, social issues will need to be integrated into conservation planning and tailored to each country and region (Bickford et al. 2008b, Sodhi et al. 2010).

17.8 Conclusions

Although progress towards the conservation of Southeast Asian biodiversity is slow and many obstacles still remain, it is clear that large steps have been made in both societal awareness, capacity building and science-based conservation decision making, and that a number of mechanisms for resource management and biodiversity protection are now in place. Increased involvement, organisation and networking of stakeholders from many sectors have resulted in encouraging trends for conservation in the region, although most are nascent. With a biodiversity crisis looming globally, and concentrated in the tropics, it is crucial to learn from the experience of countries such as the Philippines and Singapore, both regarding the perils of pursuing economic progress at the expense of biodiversity, as well as the strategies that may be effective in counteracting ecological degradation in the face of immense obstacles. Above all, it will be of vital importance to change the working philosophies of corporations and large institutions that have become more powerful than governments and can make lasting impacts simply through the vision of their leaders. The 'business-as-usual' schemes cannot continue by any definition. Science has the ability to get the answers and to find the solutions to our environmental problems; however, it is the will of the people that needs to change in order for the science to make a difference. Our next step is to find out how to determine the best possible ways to change peoples' minds, behaviours and interactions with their natural environment. Once we learn what these methods are, we then need to implement them across a broad range of human communities, targeting urban centres in particular.

Acknowledgements

MRCP was supported by the Singapore–Delft Water Alliance research programme (R 264-001-648 004-272) during the preparation of this manuscript. SP was supported by the Singapore International Graduate Award. We thank S. Howard for preparing the map for predicted impact on biodiversity (Fig 17.2). This research uses data provided by the Community Climate System Model project (http://www.ccsm.ucar.edu), supported by the Directorate for Geosciences of the National Science Foundation and the Office of Biological and Environmental Research of the U.S. Department of Energy. NCAR GIS Initiative provided CCSM data in a GIS format through GIS Climate Change Scenarios portal (http://www.gisclimatechange.org), Last of the Wild Data Version 1 2002 (LWP-1): Global Human Footprint Dataset (Geographic), and Wildlife Conservation (WCS) and Center for International Earth Science Information Network (CIESIN). We thank Menno Schilthuizen and three anonymous reviewers for their comments and edits.

References

Achard, F., Eva, H. D., Stibig, H. J. et al. (2002). Determination of deforestation rates of the world's humid tropical forests. *Science*, **297**, 999–1002.

Adams, W. M., Aveling, R., Brockington, D. et al. (2004). Biodiversity conservation and the eradication of poverty. *Science*, **306**, 1146–9.

Asian Development Bank (2009). *Key Indicators for Asia and the Pacific 2009*. Mandaluyong City, Philippines: Asian Developmental Bank.

Baillie, J. E. M., Hilton-Taylor, C. and Stuart, S. (2004). *IUCN Red List of Threatened Species: A Global Species Assessment*. Cambridge: IUCN.

Bankoff, G. (2007). One island too many: reappraising the extent of deforestation in the Philippines prior to 1946. *Journal of Historical Geography*, **33**, 314–34.

Barbier, E. B. (1993). Economic aspects of tropical deforestation in Southeast Asia. *Global Ecology and Biogeography Letters*, **3**, 215–34.

Barlow, J., Gardner, T. A., Araujo, I. S. et al. (2007). Quantifying the biodiversity value of tropical primary, secondary, and plantation forests. *Proceedings of the National Academy of Sciences of the United States of America*, **104**, 18555–60.

Barron, A. R., Wurzburger, N., Bellenger, J. P. et al. (2009). Molybdenum limitation of asymbiotic nitrogen fixation in tropical forest soils. *Nature Geoscience*, **2**, 42–5.

Bennett, E. L. (2002). Is there a link between wild meat and food security? *Conservation Biology*, **16**, 590–2.

Bennett, E. L. and Robinson, J. G. (2000). *Hunting of Wildlife in Tropical Forests: Implications for Biodiversity and Forest Peoples*. Washington DC: The World Bank.

Bennett, E. L., Nyaoi, A. J. and Sompud, J. (2000). Saving Borneo's bacon: the sustainability of hunting in Sarawak and Sabah. In *Hunting for Sustainability in Tropical Forests*, ed. E. L. Bennett and J. G. Robinson. New York: Columbia University Press, pp. 305–24.

Bhatti, N., Streets, D. G. and Foell, W. K. (1992). Acid rain in Asia. *Environmental Management*, **16**, 541–62.

Bickford, D., Iskandar, D. and Barlian, A. (2008a). A lungless frog discovered on Borneo. *Current Biology*, **18**, 374–5.

Bickford, D., Supriatna, J., Andayani, N. et al. (2008b). Indonesia's protected areas need more protection: suggestions from island examples. In *Biodiversity and Human Livelihoods in Protected Areas: Case Studies from the Malay Archipelago*, ed. N. S. Sodhi, G. Acciaioli, M. Erb and A.-J. Tan. Cambridge: Cambridge University Press, pp. 53–77.

Bickford, D., Howard, S. D., Ng, D. J. J. and Sheridan, J. A. (2010). Impacts of climate change on the amphibians and reptiles of Southeast Asia. *Biodiversity and Conservation*, **19**, 1043–62.

Blackburn, T. M., Cassey, P., Duncan, R. P., Evans, K. L. and Gaston, K. J. (2004). Avian extinction and mammalian introductions on oceanic islands. *Science*, **305**, 1955–8.

Bobbink, R., Hicks, K., Galloway, J. et al. (2010). Global assessment of nitrogen deposition effects on terrestrial plant diversity: a synthesis. *Ecological Applications*, **20**, 30–59.

Bradshaw, C. J. A., Sodhi, N. S. and Brook, B. W. (2009). Tropical turmoil: a biodiversity tragedy in progress. *Frontiers in Ecology and the Environment*, **7**, 79–87.

Bradshaw, C. J. A., Giam, X. and Sodhi, N. S. (2010). Evaluating the relative environmental impact of countries. *PLos One*, **5**, 10440.

Brook, B. W., Sodhi, N. S. and Ng, P. K. L. (2003). Catastrophic extinctions follow deforestation in Singapore. *Nature*, **424**, 420–3.

Brook, B. W., Bradshaw, C. J. A., Koh, L. P. and Sodhi, N. S. (2006). Momentum drives the crash: mass extinction in the tropics. *Biotropica*, **38**, 302–5.

Brook, B. W., Sodhi, N. S. and Bradshaw, C. J. A. (2008). Synergies among extinction drivers under global change. *Trends in Ecology and Evolution*, **23**, 453–60.

Brühl, C. and Eltz, T. (2010). Fuelling the biodiversity crisis: species loss of ground-dwelling forest ants in oil palm plantations in Sabah, Malaysia (Borneo). *Biodiversity and Conservation*, **19**, 519–29.

Butler, R. A. and Laurance, W. F. (2008). New strategies for conserving tropical forests. *Trends in Ecology and Evolution*, **23**, 469–72.

Carlsson, N. O. L., Bronmark, C. and Hansson, L. A. (2004). Invading herbivory: the golden apple snail alters ecosystem functioning in Asian wetlands. *Ecology*, **85**, 1575–80.

Carpenter, K. E. and Springer, V. G. (2005). The center of the center of marine shore fish biodiversity: the Philippine Islands. *Environmental Biology of Fishes*, **72**, 467–80.

Chapin, F. S., Zavaleta, E. S., Eviner, V. T. et al. (2000). Consequences of changing biodiversity. *Nature*, **405**, 234–42.

Chen, I.-C., Shiu, H.-J., Benedick, S. et al. (2009a). Elevation increases in moth assemblages over 42 years on a tropical mountain. *Proceedings of the National Academy of Sciences*, **106**, 1479–83.

Chen, T. H., Chang, H. C. and Lue, K. Y. (2009b). Unregulated trade in turtle shells for Chinese traditional medicine in East and Southeast Asia: the case of Taiwan. *Chelonian Conservation and Biology*, **8**, 11–18.

Christie, P. (2004). Marine protected areas as biological successes and social failures in southeast Asia. In *Aquatic Protected Areas as Fisheries Management Tools*, ed. J. B. Shipley. Bethesda, MD: American Fisheries Society, pp. 155–64.

Clavero, M. and García-Berthou, E. (2005). Invasive species are a leading cause of animal extinctions. *Trends in Ecology and Evolution*, **20**, 110.

Clements, R., Sodhi, N. S., Schilthuizen, M. and Ng, P. K. L. (2006). Limestone karsts of Southeast Asia: important arks of biodiversity. *BioScience*, **56**, 733–42.

Clements, R., Rayan, D. M., Zafir, A. W. A. et al. (2010). Trio under threat: can we secure the future of rhinos, elephants and tigers in Malaysia? *Biodiversity and Conservation*, **19**, 1115–36.

Colwell, R. K., Brehm, G., Cardelus, C. L., Gilman, A. C. and Longino, J. T. (2008). Global warming, elevational range shifts, and lowland biotic attrition in the wet tropics. *Science*, **322**, 258–61.

Cumberlidge, N., Ng,P. K. L., Yeo, D. C. J. et al. (2009). Freshwater crabs and the biodiversity crisis: importance, threats, status, and conservation challenges. *Biological Conservation*, **142**, 1665–73.

Curran, L. M., Trigg, S. N., McDonald, A. K. et al. (2004). Lowland forest loss in protected areas of Indonesian Borneo. *Science*, **303**, 1000–1003.

Custodio, C. C. and Molinyawe, N. M. (2001). The NIPAS Law and the management of protected areas in the Philippines: Some observations and critique. *Silliman Journal*, **42**, 202–28.

Danielsen, F., Beukema, H., Burgess, N. D. et al. (2009). Biofuel plantations on forested lands: double jeopardy for biodiversity and climate. *Conservation Biology*, **23**, 348–58.

DeFries, R., Hansen, A., Newton, A. C. and Hansen, M. C. (2005). Increasing isolation of protected areas in tropical forests over the past twenty years. *Ecological Applications*, **15**, 19–26.

DeFries, R. S., Rudel, T., Uriarte, M. and Hansen, M. (2010). Deforestation driven by urban population growth and agricultural trade in the twenty-first century. *Nature Geoscience*, **3**, 178–81.

Dennis, R. A., Meijaard, E., Nasi, R. and Gustafsson, L. (2008). Biodiversity conservation in Southeast Asia timber concessions: a critical evaluation of policy mechanisms and guidelines. *Ecology and Society*, **12**, 25.

Dentener, F., Drevet, J., Lamarque, J. F. et al. (2006). Nitrogen and sulfur deposition on regional and global scales: a multimodel evaluation. *Global Biogeochemical Cycles*, **20**(4), doi:10.1029/2005GB002672.

Deutsch, C. A., Tewksbury, J. J., Huey, R. B. et al. (2008). Impacts of climate warming on terrestrial ectotherms across latitude. *Proceedings of the National Academy of Sciences of the United States of America*, **105**, 6668–72.

Dietz, T., Gardner, G. T., Gilligan, J., Stern, P. C. and Vandenbergh, M. P. (2009). Household actions can provide a behavioral wedge to rapidly reduce US carbon emissions. *Proceedings of the National Academy of Sciences of the United States of America*, **106**, 18452–6.

Dudgeon, D. (2000). Large-scale hydrological changes in tropical Asia: prospects for riverine biodiversity. *BioScience*, 5**0**, 793–806.

Engardt, M., Siniarovina, U., Khairul, N. I. and Leong, C. P. (2005). Country to country transport of anthropogenic sulphur in Southeast Asia. *Atmospheric Environment*, **39**, 5137–48.

FAO (United Nations Food and Agriculture Organization) (2006). Global forest resources assessment 2005: progress towards sustainable forest management. Forestry paper 147. Rome: FAO.

Fearnside, P. M. (1997). Transmigration in Indonesia: lessons from its environmental and social impacts. *Environmental Management*, **21**, 553–70.

Ferraro, P. J. and Kiss, A. (2002). Direct payments to conserve biodiversity. *Science*, **298**, 1718–19.

Field, R. D., van der Werf, G. R. and Shen, S. S. P. (2009). Human amplification of drought-induced biomass burning in Indonesia since 1960. *Nature Geoscience*, **2**, 185–8.

Fisher, B. and Christopher, T. (2007). Poverty and biodiversity: measuring the overlap of human poverty and the biodiversity hotspots. *Ecological Economics*, **62**, 93–101.

Fitzherbert, E. B., Struebig, M. J., Morel, A. et al. (2008). How will oil palm expansion affect biodiversity? *Trends in Ecology and Evolution*, **23**, 538–45.

Foody, G. M. and Cutler, M. E. J. (2003). Tree biodiversity in protected and logged Bornean tropical rainforests and its measurement by satellite remote sensing. *Journal of Biogeography*, **30**, 1053–66.

Fordham, D. A. and Brook, B. W. (2010). Why tropical island endemics are acutely susceptible to global change. *Biodiversity and Conservation*, **19**, 329–42.

Galloway, J. N., Townsend, A. R., Erisman, J. W. et al. (2008). Transformation of the nitrogen cycle: recent trends, questions, and potential solutions. *Science*, **320**, 889–92.

Gaveau, D. L. A., Epting, J., Lyne, O. et al. (2009). Evaluating whether protected areas reduce tropical deforestation in Sumatra. *Journal of Biogeography*, **36**, 2165–75.

Giri, C., Defourny, P. and Shrestha, S. (2003). Land cover characterization and mapping of continental Southeast Asia using multi-resolution satellite sensor data. *International Journal of Remote Sensing*, **24**, 4181–96.

Goldammer, J. (2007). History of equatorial vegetation fires and fire research in Southeast Asia before the 1997–98 episode: a reconstruction of creeping environmental changes. *Mitigation and Adaptation Strategies for Global Change*, **12**, 13–32.

Gomez, E. D., Alino, P. M., Yap, H. T. and Licuanan, W. Y. (1994). A review of the status of Philippine reefs. *Marine Pollution Bulletin*, **29**, 62–8.

Grainger, A., Boucher, D. H., Frumhoff, P. C. et al. (2009). Biodiversity and REDD at Copenhagen. *Current Biology*, **19**, 974–6.

Grubb, P. J., Turner, I. M. and Burslem, D. (1994). Mineral nutrient status of coastal hill dipterocarp forest and adinandra belukar in Singapore: analysis of soil, leaves and litter. *Journal of Tropical Ecology*, **10**, 559–77.

Hall, R. (2009). Southeast Asia's changing palaeogeography. *Blumea*, **54**, 148–61.

Hannah, L., Carr, J. L. and Landerani, A. (1995). Human disturbance and natural habitat: a biome level analysis of a global data set. *Biodiversity and Conservation*, **4**, 128–55.

Hansen, M. C., Stehman, S. V., Potapov, P. V. et al. (2008). Humid tropical forest clearing from 2000 to 2005 quantified by using multitemporal and multiresolution remotely sensed data. *Proceedings of the National Academy of Sciences*, **105**, 9439–44.

Hoeksema, B. W. (2007). Delineation of the Indo-Malayan centre of maximum marine biodiversity: the coral triangle. In *Biogeography, Time, and Place: Distributions, Barriers, and Islands*, ed. W. Renema. Dordrecht, The Netherlands: Springer, pp. 117–78.

Huey, R. B., Deutsch, C. A., Tewksbury, J. J. et al. (2009). Why tropical forest lizards are vulnerable to climate warming. *Proceedings of the Royal Society B: Biological Sciences*, **276**, 1939–48.

Hughes, T. P., Baird, A. H., Bellwood, D. R. et al. (2003). Climate change, human impacts and the resilience of coral reefs. *Science*, **301**, 929–33.

Hungspreugs, M. (1988). Heavy metals and other non-oil pollutants in Southeast Asia. *Ambio*, **17**, 178–82.

Intergovernmental Panel on Climate Change (IPCC) (2007). *Climate Change 2007: Synthesis Report*. Geneva: IPCC.

International Tropical Timber Organization (ITTO) (2008). *Annual Review and Assessment of the World Timber Situation. Document GI-7/08*, Yokohama, Japan: ITTO.

Jones, M. L. M., Oxley, E. R. B. and Ashenden, T. W. (2002). The influence of nitrogen deposition, competition and desiccation on growth and regeneration of *Racomitrium lanuginosum* (Hedw.) Brid. *Environmental Pollution*, **120**, 371–8.

Kareiva, P. and Marvier, M. (2007). Conservation for the people: pitting nature and biodiversity against people makes little sense. Many conservationists now argue that human health and well-being should be central to conservation efforts. *Scientific American*, **297**, 50–7.

Koh, L. P. and Wilcove, D. S. (2008). Is oil palm agriculture really destroying tropical biodiversity? *Conservation Letters*, **1**, 60–64.

Kusrini, M. D. and Alford, R. A. (2006). Indonesia's exports of frogs' legs. *Traffic Bulletin*, **21**, 13–24.

Langner, A. and Siegert, F. (2009). Spatiotemporal fire occurrence in Borneo over a period of 10 years. *Global Change Biology*, **15**, 48–62.

Lasco, R. D. and Puhlin, J. M. (2006). Environmental impacts of community-based forest management in the Philippines. *International Journal of Environment and Sustainable Development*, **5**, 46–56.

Latta, S. C. and Faaborg, J. (2009). Benefits of studies of overwintering birds for understanding resident bird ecology and promoting development of conservation capacity. *Conservation Biology*, **23**, 286–93.

Laurance, W. F. (1999). Reflections on the tropical deforestation crisis. *Biological Conservation*, **91**, 109–117.

Laurance, W. F. (2004). The perils of payoff: corruption as a threat to global biodiversity. *Trends in Ecology and Evolution*, **19**, 399–401.

Laurance, W. F. (2007). Have we overstated the tropical biodiversity crisis? *Trends in Ecology and Evolution*, **22**, 65–70.

Laurance, W. F. and Useche, D. C. (2009). Environmental synergisms and extinctions of tropical species. *Conservation Biology*, **23**, 1427–37.

Lohman D. J., Bickford, D. and Sodhi N. S. (2007). The burning issue. *Science*, **316**, 376.

Lubchenco, J. (1998). Entering the century of the environment: a new social contract for science. *Science*, **279**, 491–7.

Mallari, N. A. D., Tabaranza, B. R. and Crosby, M. C. (2001). *Key Conservation Sites in the Philippines*. Manila, Philippines: Bookmark.

Manez, K. S. and Ferse, S. C. A. (2010). The history of Makassan trepang fishing and trade. *PLoS One*, **5**.

Matson, P. A., McDowell, W. H., Townsend, A. R. and Vitousek, P. M. (1999). The globalization of N deposition: ecosystem consequences in tropical environments. *Biogeochemistry*, **46**, 67–83.

Mayaux, P., Holmgren, P., Achard, F. et al. (2005). Tropical forest cover change in the 1990s and options for future monitoring. *Philosophical Transactions of the Royal Society B: Biological Sciences*, **360**, 373–84.

Meijaard, E. and Sheil, D. (2008). The persistence and conservation of Borneo's mammals in lowland rainforests managed for timber: observations, overviews and opportunities. *Ecological Research*, **23**, 21–34.

Milner-Gulland, E. J. and Bennett, E. L. (2003). Wild meat: the bigger picture. *Trends in Ecology and Evolution*, **18**, 351–7.

Molnar, J. L., Gamboa, R. L., Revenga, C. and Spalding, M. D. (2008). Assessing the global threat of invasive species to marine biodiversity. *Frontiers in Ecology and the Environment*, **6**, 485–92.

Myers, N., Mittermeier, R. A., Mittermeier, C. G., da Fonseca, G. A. B. and Kent, J. (2000). Biodiversity hotspots for conservation priorities. *Nature*, **403**, 853–8.

Nijman, V. (2010). An overview of international wildlife trade from Southeast Asia, *Biodiversity and Conservation*, **19**, 1101–14.

Nriagu, J. O. (1996). A history of global metal pollution. *Science*, **272**, 223–4.

Okuda, T., Suzuki, M., Adachi, N. et al. (2003). Effect of selective logging on canopy and stand structure and tree species composition in a lowland dipterocarp forest in peninsular Malaysia. *Forest Ecology and Management*, **175**, 297–320.

Pauly, D., Christensen, V., Dalsgaard, J., Froese, R. and Torres, F. C., Jr. (1998). Fishing down marine food webs. *Science*, **279**, 860–3.

Peh, K. S. H. (2007). Potential effects of climate change on elevational distributions of tropical birds in Southeast Asia. *Condor*, **109**, 437–41.

Peh, K. S. H. (2010). Invasive species in Southeast Asia: the knowledge so far. *Biodiversity and Conservation*, **19**, 1083–99.

Phelps, J., Webb, E. L. and Agrawal, A. (2010). Does REDD+ threaten to recentralize forest governance? *Science*, **328**, 312–13.

Posa, M. R. C., Diesmos, A. C., Sodhi, N. S. and Brooks, T. M. (2008). Hope for threatened tropical biodiversity: lessons from the Philippines. *BioScience*, **58**, 231–40.

Primavera, J. H. (2000). Development and conservation of Philippine mangroves: institutional issues. *Ecological Economics*, **35**, 91–106.

Radojevic, M. (2003). Chemistry of forest fires and regional haze with emphasis on Southeast Asia. *Pure and Applied Geophysics*, **160**, 157–87.

Rejmanek, M. (1996). Species richness and resistance to invasions. *Biodiversity and Ecosystem Processes in Tropical Forests*, **122**, 153.

Rerkasem, K., Lawrence, D., Padoch, C. et al. (2009). Consequences of swidden transitions for crop and fallow biodiversity in Southeast Asia. *Human Ecology*, **37**, 347–60.

Roberts, C. M., McClean, C. J., Veron, J. E. N. et al. (2002). Marine biodiversity hotspots and conservation priorities for tropical reefs. *Science*, **295**, 1280–4.

Roberts, T. R. (2001). On the river of no returns: Thailand's Pak Mun dam and its fish ladder. *Natural History Bulletin of the Siam Society*, **49**, 189–230.

Rockstrom, J., Steffen, W., Noone, K. et al. (2009). A safe operating space for humanity. *Nature*, **461**, 472–5.

Ross, M. L. (2001). *Timber Booms and Institutional Breakdown in Southeast Asia*. Cambridge: Cambridge University Press.

Sala, O. E., Chapin, F. S., Armesto, J. J. et al. (2000). Global biodiversity scenarios for the year 2100. *Science*, **287**, 1770–4.

Schilthuizen, M, Liew, T.-S., Berjaya, E. and Lackman-Ancrenaz, I. (2005). Effects of karst forest degradation on pulmonate and prosobranch land snail communities in Sabah, Malaysian Borneo. *Conservation Biology*, **19**, 949–54.

Silvestre, G. T., Graces, L. R., Stobutzki, I. et al. (2003). South and South-East Asian coastal fisheries: their status and directions for improved management: conference synopsis and recommendations. In *Assessment, Management and Future Directions for Coastal Fisheries in Asian Countries*, ed. G. Silvestre, L. Garces, I. Stobutski et al.. World Fish Center Conference Proceedings 67, pp. 1–40.

Smith, R. J., Muir, R. D. J., Walpole, M. J., Balmford, A. and Leader-Williams, N. (2003). Governance and the loss of biodiversity. *Nature*, **426**, 67–70.

Sodhi, N. S. and Brook, B. W. (2006). *Southeast Asian Biodiversity in Crisis*. Cambridge: Cambridge University Press.

Sodhi, N. S., Koh, L. P., Brook, B. W. and Ng, P. K. L. (2004). Southeast Asian biodiversity: an impending disaster. *Trends in Ecology and Evolution*, **19**, 654–60.

Sodhi, N. S., Brooks, T. M., Koh, L. P. et al. (2006). Biodiversity and human livelihood crises in the Malay archipelago. *Conservation Biology*, **20**, 1811–13.

Sodhi, N. S., Bickford, D., Diesmos, A. C. et al. (2008). Measuring the meltdown:

drivers of global amphibian extinction and decline. *PLoS One*, **3**.

Sodhi, N. S., Posa, M. R. C., Lee, T. M. et al. (2010). The state and conservation of Southeast Asian biodiversity. *Biodiversity and Conservation*, **19**, 317–28.

Stachowicz, J. J., Whitlatch, R. B. and Osman, R. W. (1999). Species diversity and invasion resistance in a marine ecosystem. *Science*, **286**, 1577–9.

Stork, N. E., Coddington, J. A., Colewell, R. K. et al. (2009). Vulnerability and resilience of tropical forest species to land-use change. *Conservation Biology*, **23**, 1438–47.

Teixeira, R. D., Silva, C. R., Mello, P. and Lima dos Santos, C. A. M. (2001). *The World Market for Frog Legs*. Rome: Food and Agriculture Organization, Globefish Research Programme.

Todd, P. A., Ong, X. Y. and Chou, L. M. (2010). Impacts of pollution on marine life in Southeast Asia. *Biodiversity and Conservation*, **19**, 1063–82.

Tun, K., Chou, L. M., Yeemin, T. et al. (2008). Status of coral reefs in Southeast Asia. In *Status of Coral Reefs of the World: 2008*, ed. C. Wilkinson. Townsville, Australia: Global Coral Reef Monitoring Network and Reef and Rainforest Research Center, pp. 131–44.

United Nations Department of Economic and Social Affairs Statistics Division (UN Statistics Division) (2008). *Statistical Year Book* 52nd Issue. New York: United Nations.

Utting, P. (2000). An overview of the potential pitfalls of participatory conservation. In *Forest Policy and Politics in the Philippines: The Dynamics of Participatory Conservation*, ed. P. Utting. Quezon City, Philippines: Ateneo de Manila and United Research Institute for Social Development, pp. 171–215.

Venter, O., Meijaard, E., Possingham, H. et al. (2009). Carbon payments as a safeguard for threatened tropical mammals. *Conservation Letters*, **2**, 123–9.

Vitousek, P. M., Mooney, H. A., Lubchenco, J. and Melillo, J. M. (1997). Human domination of Earth's ecosystems. *Science*, **277**, 494–9.

Warkentin, I. G., Bickford, D., Sodhi, N. S. and Bradshaw, C. J. A. (2009). Eating frogs to extinction. *Conservation Biology*, **23**, 1056–9.

Wells, K., Kalko, E. K. V., Lakim, M. B. and Pfeiffer, M. (2007). Effects of rainforest logging on species richness and assemblage composition of small mammals in Southeast Asia. *Journal of Biogeography*, **34**, 1087–99.

West, P., Igoe, J. and Brockington, D. (2006). Parks and peoples: the social impact of protected areas. *Annual Review of Anthropology*, **35**, 251–77.

White, A. T., Courtney, C. A. and Salamanca, A. (2002). Experience with Marine Protected Area planning and management in the Philippines. *Coastal Management*, **30**, 1–26.

Wilcove, D. S. and Koh, L. P. (2010). Addressing the threats to biodiversity from oil-palm agriculture. *Biodiversity and Conservation*, **19**, 999–1007.

Wilting, A., Cord, A., Hearn, A. J. et al. (2010). Modelling the species distribution of flat-headed cats (*Prionailurus planiceps*), an endangered South-East Asian small felid. *PLoS ONE*, **5**.

Wright, S. J. and Muller-Landau, H. C. (2006). The future of tropical forest species. *Biotropica*, **38**, 287–301.

Wright, S. J., Muller-Landau, H. C. and Schipper, J. (2009). The future of tropical species on a warmer planet. *Conservation Biology*, **23**, 1418–26.

Yaap, B., Struebig, M. J., Paoli, G., and Koh, L. P. (2009). Mitigating the biodiversity impacts of oil palm. *CAB Reviews: Perspectives in Agriculture, Veterinary Science, Nutrition and Natural Resources*, **5**, 19.

Yule, C. M., Boyero, L. and Marchant, R. (2010). Effects of sediment pollution on food webs in a tropical river (Borneo, Indonesia). *Marine and Freshwater Research*, **61**, 204–13.

Zhong, Z. C., Victor, T. and Balasubramanian, R. (2001). Measurement of major organic acids in rainwater in Southeast Asia during burning and non-burning periods. *Water Air and Soil Pollution*, **130**, 457–62.

Index

Systematics Association Publications

1. Bibliography of Key Works for the Identification of the British Fauna and Flora, 3rd edition (1967)†
   Edited by G. J. Kerrich, R. D. Meikie and N. Tebble
2. The Species Concept in Palaeontology (1956)†
   Edited by P. C. Sylvester-Bradle
3. Function and Taxonomic Importance (1959)†
   Edited by A. J. Cain
4. Taxonomy and Geography (1962)†
   Edited by D. Nichols
5. Speciation in the Sea (1963)†
   Edited by J. P. Harding and N. Tebble
6. Phenetic and Phylogenetic Classification (1964)†
   Edited by V. H. Heywood and J. McNeill
7. Aspects of Tethyan Biogeography (1967)†
   Edited by C. G. Adams and D. V. Ager
8. The Soil Ecosystem (1969)†
   Edited by H. Sheals
9. Organisms and Continents through Time (1973)*
   Edited by N. F. Hughes
10. Cladistics: A Practical Course in Systematics (1992)‡
    P. L. Forey, C. J. Humphries, I. J. Kitching, R. W. Scotland, D. J. Siebert and D. M. Williams
11. Cladistics: The Theory and Practice of Parsimony Analysis, 2nd edition (1998)‡
    I. J. Kitching, P. L. Forey, C. J. Humphries and D. M. Williams

---

† Published by the Systematics Association (out of print)
* Published by the Palaeontological Association in conjunction with the Systematics Association
‡ Published by Oxford University Press for the Systematics Association

Systematics Association Special Volumes

1. The New Systematics (1940)[a]
   Edited by J. S. Huxley (reprinted 1971)
2. Chemotaxonomy and Serotaxonomy (1968)[*]
   Edited by J. C. Hawkes
3. Data Processing in Biology and Geology (1971)[*]
   Edited by J. L. Cutbill
4. Scanning Electron Microscopy (1971)[*]
   Edited by V. H. Heywood
5. Taxonomy and Ecology (1973)[*]
   Edited by V. H. Heywood
6. The Changing Flora and Fauna of Britain (1974)[*]
   Edited by D. L. Hawksworth
7. Biological Identification with Computers (1975)[*]
   Edited by R. J. Pankhurst
8. Lichenology: Progress and Problems (1976)[*]
   Edited by D. H. Brown, D. L. Hawksworth and R. H. Bailey
9. Key Works to the Fauna and Flora of the British Isles and Northwestern Europe, 4th edition (1978)[*]
   Edited by G. J. Kerrich, D. L. Hawksworth and R. W. Sims
10. Modern Approaches to the Taxonomy of Red and Brown Algae (1978)[*]
    Edited by D. E. G. Irvine and J. H. Price
11. Biology and Systematics of Colonial Organisms (1979)[*]
    Edited by C. Larwood and B. R. Rosen
12. The Origin of Major Invertebrate Groups (1979)[*]
    Edited by M. R. House
13. Advances in Bryozoology (1979)[*]
    Edited by G. P. Larwood and M. B. Abbott
14. Bryophyte Systematics (1979)[*]
    Edited by G. C. S. Clarke and J. G. Duckett

15. The Terrestrial Environment and the Origin of Land Vertebrates (1980)[*]
    Edited by A. L. Panchen
16. Chemosystematics: Principles and Practice (1980)[*]
    Edited by F. A. Bisby, J. G. Vaughan and C. A. Wright
17. The Shore Environment: Methods and Ecosystems (2 volumes) (1980)[*]
    Edited by J. H. Price, D. E. C. Irvine and W. F. Farnham
18. The Ammonoidea (1981)[*]
    Edited by M. R. House and J. R. Senior
19. Biosystematics of Social Insects (1981)[*]
    Edited by P. E. House and J.-L. Clement
20. Genome Evolution (1982)[*]
    Edited by G. A. Dover and R. B. Flavell
21. Problems of Phylogenetic Reconstruction (1982)[*]
    Edited by K. A. Joysey and A. E. Friday
22. Concepts in Nematode Systematics (1983)[*]
    Edited by A. R. Stone, H. M. Platt and L. F. Khalil
23. Evolution, Time and Space: The Emergence of the Biosphere (1983)[*]
    Edited by R. W. Sims, J. H. Price and P. E. S. Whalley
24. Protein Polymorphism: Adaptive and Taxonomic Significance (1983)[*]
    Edited by G. S. Oxford and D. Rollinson
25. Current Concepts in Plant Taxonomy (1983)[*]
    Edited by V. H. Heywood and D. M. Moore
26. Databases in Systematics (1984)[*]
    Edited by R. Allkin and F. A. Bisby
27. Systematics of the Green Algae (1984)[*]
    Edited by D. E. G. Irvine and D. M. John
28. The Origins and Relationships of Lower Invertebrates (1985)[‡]
    Edited by S. Conway Morris, J. D. George, R. Gibson and H. M. Platt
29. Intraspecific Classification of Wild and Cultivated Plants (1986)[‡]
    Edited by B. T. Styles
30. Biomineralization in Lower Plants and Animals (1986)[‡]
    Edited by B. S. C. Leadbeater and R. Riding
31. Systematic and Taxonomic Approaches in Palaeobotany (1986)[‡]
    Edited by R. A. Spicer and B. A. Thomas
32. Coevolution and Systematics (1986)[‡]
    Edited by A. R. Stone and D. L. Hawksworth
33. Key Works to the Fauna and Flora of the British Isles and Northwestern Europe, 5th edition (1988)[‡]
    Edited by R. W. Sims, P. Freeman and D. L. Hawksworth
34. Extinction and Survival in the Fossil Record (1988)[‡]
    Edited by G. P. Larwood

35. The Phylogeny and Classification of the Tetrapods (2 volumes) (1988)[‡]
Edited by M. J. Benton

36. Prospects in Systematics (1988)[‡]
Edited by J. L. Hawksworth

37. Biosystematics of Haematophagous Insects (1988)[‡]
Edited by M. W. Service

38. The Chromophyte Algae: Problems and Perspective (1989)[‡]
Edited by J. C. Green, B. S. C. Leadbeater and W. L. Diver

39. Electrophoretic Studies on Agricultural Pests (1989)[‡]
Edited by H. D. Loxdale and J. den Hollander

40. Evolution, Systematics and Fossil History of the Hamamelidae (2 volumes) (1989)[‡]
Edited by P. R. Crane and S. Blackmore

41. Scanning Electron Microscopy in Taxonomy and Functional Morphology (1990)[‡]
Edited by D. Claugher

42. Major Evolutionary Radiations (1990)[‡]
Edited by P. D. Taylor and G. P. Larwood

43. Tropical Lichens: Their Systematics, Conservation and Ecology (1991)[‡]
Edited by G. J. Galloway

44. Pollen and Spores: Patterns and Diversification (1991)[‡]
Edited by S. Blackmore and S. H. Barnes

45. The Biology of Free-Living Heterotrophic Flagellates (1991)[‡]
Edited by D. J. Patterson and J. Larsen

46. Plant–Animal Interactions in the Marine Benthos (1992)[‡]
Edited by D. M. John, S. J. Hawkins and J. H. Price

47. The Ammonoidea: Environment, Ecology and Evolutionary Change (1993)[‡]
Edited by M. R. House

48. Designs for a Global Plant Species Information System (1993)[‡]
Edited by F. A. Bisby, G. F. Russell and R. J. Pankhurst

49. Plant Galls: Organisms, Interactions, Populations (1994)[‡]
Edited by M. A. J. Williams

50. Systematics and Conservation Evaluation (1994)[‡]
Edited by P. L. Forey, C. J. Humphries and R. I. Vane-Wright

51. The Haptophyte Algae (1994)[‡]
Edited by J. C. Green and B. S. C. Leadbeater

52. Models in Phylogeny Reconstruction (1994)[‡]
Edited by R. Scotland, D. I. Siebert and D. M. Williams

53. The Ecology of Agricultural Pests: Biochemical Approaches (1996)[**]
Edited by W. O. C. Symondson and J. E. Liddell

54. Species: The Units of Diversity (1997)[**]
Edited by M. F. Claridge, H. A. Dawah and M. R. Wilson

55. Arthropod Relationships (1998)[**]
Edited by R. A. Fortey and R. H. Thomas

56. Evolutionary Relationships among Protozoa (1998)[**]
Edited by G. H. Coombs, K. Vickerman, M. A. Sleigh and A. Warren

57. Molecular Systematics and Plant Evolution (1999)[‡‡]
Edited by P. M. Hollingsworth, R. M. Bateman and R. J. Gornall

58. Homology and Systematics (2000)[‡‡]
Edited by R. Scotland and R. T. Pennington

59. The Flagellates: Unity, Diversity and Evolution (2000)[‡‡]
Edited by B. S. C. Leadbeater and J. C. Green

60. Interrelationships of the Platyhelminthes (2001)[‡‡]
Edited by D. T. J. Littlewood and R. A. Bray

61. Major Events in Early Vertebrate Evolution (2001)[‡‡]
Edited by P. E. Ahlberg

62. The Changing Wildlife of Great Britain and Ireland (2001)[‡‡]
Edited by D. L. Hawksworth

63. Brachiopods Past and Present (2001)[‡‡]
Edited by H. Brunton, L. R. M. Cocks and S. L. Long

64. Morphology, Shape and Phylogeny (2002)[‡‡]
Edited by N. MacLeod and P. L. Forey

65. Developmental Genetics and Plant Evolution (2002)[‡‡]
Edited by Q. C. B. Cronk, R. M. Bateman and J. A. Hawkins

66. Telling the Evolutionary Time: Molecular Clocks and the Fossil Record (2003)[‡‡]
Edited by P. C. J. Donoghue and M. P. Smith

67. Milestones in Systematics (2004)[‡‡]
Edited by D. M. Williams and P. L. Forey

68. Organelles, Genomes and Eukaryote Phylogeny (2004)[‡‡]
Edited by R. P. Hirt and D. S. Horner

69. Neotropical Savannas and Seasonally Dry Forests: Plant Diversity, Biogeography and Conservation (2006)[‡‡]
Edited by R. T. Pennington, G. P. Lewis and J. A. Rattan

70. Biogeography in a Changing World (2006)[‡‡]
Edited by M. C. Ebach and R. S. Tangney

71. Pleurocarpous Mosses: Systematics and Evolution (2006)[‡‡]
Edited by A. E. Newton and R. S. Tangney

72. Reconstructing the Tree of Life: Taxonomy and Systematics of Species Rich Taxa (2006)[‡‡]
Edited by T. R. Hodkinson and J. A. N. Parnell

73. Biodiversity Databases: Techniques, Politics, and Applications (2007)[‡‡]
Edited by G. B. Curry and C. J. Humphries

74. Automated Taxon Identification in Systematics: Theory, Approaches and Applications (2007)[††]
    Edited by N. MacLeod
75. Unravelling the Algae: The Past, Present, and Future of Algal Systematics (2008)[††]
    Edited by J. Brodie and J. Lewis
76. The New Taxonomy (2008)[††]
    Edited by Q. D. Wheeler
77. Palaeogeography and Palaeobiogeography: Biodiversity in Space and Time (in press)[††]
    Edited by P. Upchurch, A. McGowan and C. Slater

---

[a] Published by Clarendon Press for the Systematics Association
[*] Published by Academic Press for the Systematics Association
[†] Published by Oxford University Press for the Systematics Association
[**] Published by Chapman and Hall for the Systematics Association
[††] Published by CRC Press for the Systematics Association

Printed in the United States
by Baker & Taylor Publisher Services